TECHNOSCIENTIFIC IMAGINARIES

RAND McNALLY
International Series
World
ATLANTIC OCEAN
GREENLAND SEA
EUROPE
INDIAN OCEAN
GREENLAND
MAURITANIA
MALI
AFRICA
WEST INDIES
UNITED STATES
PACIFIC OCEAN
NORWEGIAN SEA

World Map 2, © 1989 by Nina Katchadourian

Late Editions 2

Cultural Studies for the End of the Century

TECHNOSCIENTIFIC IMAGINARIES

CONVERSATIONS, PROFILES, AND MEMOIRS

George E. Marcus, EDITOR

University of Chicago Press
Chicago and London

George E. Marcus is professor of anthropology at Rice University. He is coauthor of *Anthropology as Cultural Critique* (University of Chicago Press, 1986) and was the inaugural editor of the journal *Cultural Anthropology*.

The University of Chicago Press, Chicago 60637
The University of Chicago Press, Ltd., London

Printed in the United States of America
04 03 02 01 00 99 98 97 96 95 1 2 3 4 5

ISBN: 0-226-50443-3 (cloth)
0-226-50444-1 (paper)

ISSN: 1070-8987 (for Late Editions)

∞The paper used in this publication meets the minimum requirements of the American National Standard for Information Sciences—Permanence of Paper for Printed Library Materials, ANSI Z39.48-1984.

Contents

George E. Marcus

Introduction

This is the second volume in a series of annuals, lasting until the year 2000, that is absorbed by end-of-the-century changes in society and culture worldwide. It seeks a mode of exposing aspects of these changes to a general, diverse readership that is both alternative and supplementary to the often preemptory assimilation of events and experiences of contemporary social actors through familiar categories employed by scholars, journalists, and others whose task it is to explain and narrate the present world as it unfolds. Indeed, we are responding to what is certainly a widespread sense among even scholars and experts that, more so than in the past, conceptual languages that are available for describing social conditions and cultural states in various locations are inadequate. And we believe that this sense of inadequacy of authoritative conceptual schemes has reached such a level that a simple denial or else a mere hedged acknowledgment that imperfection is a standard condition of describing or translating *any* situation will no longer suffice. Our wager is that staged and edited exposures of the words of persons in various strategic sites of change will inform readers, who are willing to work through and across the pieces in these volumes, in a way that not only supplements but also therapeutically challenges the dominance of the authoritative concept or theoretical construct of scholarly arguments.

We are not so much interested in producing conventional scholarly collections in these annuals that intend to be comprehensive, authoritative, or even representative in covering a field. Rather, each annual is a somewhat opportunistic assemblage around a broad theme that is intended to work primarily through the comparative resonance for readers of the materials that are juxtaposed. Thus our emphasis is upon the suggestion of unexpected or revealed connections among clearly related subjects rather than the definitive display of systematic relationships that define a field, which is the much more usual purpose of the scholarly collection.

The form that these volumes take is inspired partly by the wide-ranging ethnographic conversations of the anthropologist in the field with "key informants," and partly by the effectiveness for broad communication of the journalistic interview or profile. We ask scholars of varied disciplinary backgrounds (but with a heavy representation from anthropology) to provide such relatively unmediated exposures of the embedded testimonies of subjects at various sites and locations through the presentation of interviews, dialogues, conversations, biographical profiles, and memoirs. Typically, interlocutors and subjects heard from in this series will be the rough counterparts of the participating, university-trained, middle-class scholars/correspondents who produce pieces for these annuals, but otherwise in very different circumstances and from diverse backgrounds, embedded existentially in institutional, geographical, and communal sites of distinctive transition and change.

Of course, in this enterprise, how much is to be framed discursively by authors, how much is to be exposed by dialogue, how much tolerance there should be for "loose ends"—diffuse but interesting connections—are signature questions in producing pieces for this series that generate a constant tension for both its writers and readers. This "genre" aspect of the series is nicely expressed in a comment by Michael Fischer, made at our collaborative editorial meeting, on Kim Laughlin's piece included in this year's annual:

> The whole point of the—what was the original word? *entretien?*—was the notion that one could strategically position a conversation with a person that would illuminate the larger context in which that person lives and operates. The idea of presenting dialogue is not to reinforce an ideology of authenticity to the spoken word or to the interaction. Rather, the issue of context itself is indeed the central question. Conversations are fascinating for the linkages they make or suggest. Kim's text is so strong because of all the linkages it makes—the way it brings us in, and its expansiveness is absolutely crucial. Yet, at the same time, this speculative desire to see where material takes us works against the notion of paring a piece down through the focus on the dialogue itself, or on one particular aspect of the text, or even a couple of aspects of the text. This is the visceral tension of the problematic between careful framing and cutting and letting the flow of conversation define its own often unanticipated context that this whole series is trying to instantiate.

In Late Editions 1, *Perilous States,* the social actors focused upon were caught up in traumatic ruptures of their societies. They were directly exposed to "events," unmediated by stable institutions, and their predicament was the challenge of even imagining a future in the midst of the reorganization of civil societies. The fascination of these cases was perhaps in the details of the uncertainty, the suspension into which these subjects were cast.

In this volume, the subjects of interviews and profiles—scientists variously situated—are not as directly exposed to events of change. Rather, they experience the fin de siècle within the routines of institutions and the frame of professional practices, but that themselves are rapidly changing, and are in fact the engines of long-term processes of change and innovation. This is thus a very different set of predicaments for subjects, and overall, another angle on this fin de siècle than that produced in Late Editions 1.

The various kinds of ambiguous and new locations for scientific work and the issues these raise for subjects who no longer are doing science as usual (or at least, as scientific careers and practices might have occurred to them in training or earlier in their professional lives) are what come through most strongly in the papers of this volume, and constitute the broad frame in which we invite readers to create their own comparative, in-depth play of associations among the following profiles and interviews—for example, weapons scientists at the end of the cold war; Lithuanian scientists set free of the Soviet system and facing a strange, new world of multinational technoscience; the creation of brave new science cities; U.S. scientists shaped less by the lab or the university, but more directly than ever by the heady, creative, competitive business of technology-driven capitalism, taking new forms where the scientist/businessman is now at the corporate center; scientist-become-activists at the site of an environmental disaster; the lure of computer images for PET scientists; the binds created for oncologists and patients by constantly emerging regimes of truth in medical science, and the bittersweet retrospective view of all of this—this world of largely bubble-boy science—by certain old-world, older generation scientists.

In initial discussions about doing a Late Editions treatment of fin-de-siècle conditions of work in science and technology, the term *imaginary* emerged effortlessly and just seemed to fit the topic very well. I think this might have been because of immediate associations of scientific practice with the "visual," or "imaging," on one hand, and with visionary, innovative, imagination, on the other—an orientation to imagining futures and the fantastic. Indeed, visualization has always been a defining aspect of Western scientific cognition, and especially now with the increasing prominence of computer graphics and simulation around which there has arisen a varied and intricate microaesthetics of the visual.

As this volume has turned out, however, the technoscientific imaginaries we have in mind here do not quite conform to either of the above senses (although there are strong and representative papers on imaging in this volume, by Joseph Dumit, for example, and several of the papers do deal with the imaginary in the large visionary sense). Indeed, there were questions at our collective editorial meeting about what the predicaments of scientists caught in social changes—of forms of government, or of industry—had to do with the imaginary. Such questions entailed a view of the imaginary which would place contemporary technoscientific thought on a "higher" plane of pure ideas, for example, juxtaposing the reflective, visionary thoughts of scientists to, say, contemporary or classical philosophy, on one side, and the imagination of science-fiction writers on the other. While this would have been an interesting treatment of the theme of the imaginary in contemporary science, we instead were much more interested in the imaginaries of scientists tied more closely to their current positionings, practices, and ambiguous locations in which the varied kinds of science they do are possible at all. This is a socially and culturally embedded sense of the imaginary that indeed looks to the future and future possibility through technoscientific innovation but is equally constrained by the very present conditions of scientific work. The imaginary fills in the cognitive gap and tension that the widespread perceived inadequacy of working practices and concepts create within so many institutions and professions today.

This is indeed the notion of the imaginary that fits with the pervasive fin-de-siècle condition to which this series sees itself responding—often sloganized as a crisis of representation, the fact that available discursive formations fail experiences of the present. Whereas the social actors in Late Editions 1 are currently blocked by traumatic political events in even developing imaginaries in which they can invest belief, those of Late Editions 2 are constantly trying to understand the present by borrowing from a cautiously imagined emergent future, filled with volatility, and uncertainty, but in which faith in practices of technoscience become even more complexly and interestingly constructed in new locations of doing science.

The idea of the imaginary as structures of contingency, then, is what we mostly have in mind in the following chapters. Our interest in the imaginary is in how it is constituted at distinctively ambiguous fin-de-siècle locations of various kinds of scientific practice in their fully embedded social and cultural contexts. This volume probes such locations in its diverse interviews, reflections, and profiles.

A Note on a Retrodesign for the Reading of This Volume

I see no way of myself providing a distanced metaintegration of the following materials, and thus I have avoided the conventional sort of introductory effort to summarize themes or pick out unifying concepts for the reader. Rather, more in the speculative spirit of the series itself, we offer readers another kind of engagement that will substitute for the authoritative, unifying introduction, common to collections: transcribed and edited excerpts from the collective editorial meeting in which earlier drafts of the following pieces were discussed. These substitute for the more conventional thematic integration which the reader, after going through the papers, will be prepared to engage with—

agreeing, disagreeing, noting gaps, offering different takes, personally summing up what connects these papers, in the same way that the participants at this meeting did in their discussions.

Thus, we intend the set of edited discussions following the papers to serve as a deferred introduction, or, so to speak, a *retro*duction to the volume. For example, Hugh Gusterson's critical discussion of Diana Hill's conversation with her father, a physicist in the community of nuclear weapons scientists, and then Kim Laughlin's discussion of Gusterson's own exploration of his ethnographic work in this community poses wide-ranging questions about the ambiguities of such scientific careers that any reader will want to think about in light of what he or she might have encountered in a first, inevitably selective reading through of the diverse contributions to this volume. And then, in turn, certain more synthetic parts of the discussions from the collective editorial meeting, like, for instance, the extended commentaries of Sharon Traweek on the qualities of relationships in technoscientific communities and networks, and of Paul Rabinow on the idea of biosociality, serve to further contextualize or thematically integrate other excerpted discussion fragments like those of Gusterson, Hill, and Laughlin. This edited discussion section as a whole is thus intended to serve as a thematic guide to the volume which requires active responses from readers who have already had some degree of familiarity with the materials and their own impressions of associations among them. The discussions of the contributors about their chapters as a kind of *retro*duction are included in the volume to serve as the major integrating device that is normally expected of collections, even as diverse as this one.

An Additional Agenda of This Particular Volume

While the participants in this volume are at various stages of their careers, have a variety of disciplinary backgrounds, and vary greatly in their research interests, they share a kind of common identity in relation to the STS (science, technology, and society) field, the quite vigorous, multidisciplinary enterprise, institutionalized in academic departments in such U.S. universities as UCLA, the University of California at San Diego, MIT, RPI, and the University of Pennsylvania, as well as various centers in Europe, that is devoted to the social studies of science. Powered by a critical position in relation to perceptions of science as neutral, objective, and hermetic—as the sacralization of rationality and modernity—this field has made substantial contributions in demonstrating the social constructedness of scientific cognition, and the social embeddedness of scientists' careers and institutions. However, almost in seeking the rigor of scientific method itself, these social studies of science have been rather formal and narrow in perspective, favoring, for example, the use of ethnomethodology. Now there is an effort in the name of a *cultural* studies of science to considerably broaden and loosen the kind of critical analyses that seek to place scientific activity in the realm of the social. The biographies and emotional lives of scientists have become a matter of intense interest, as have the relation of various situations of doing science to politics, ethics, capitalism, and systematic sources of inequality in society; the comparative cross-cultural contexts of scientific communities, and the reception and translation of science among various communities of nonscientists—in other words, the sorts of interests represented in the pieces of this volume.

So, there was a particular and positive ethos present among the participants at the 8 May collaborative editorial meeting that arose from a mutual recognition of doing work of various sorts that, while certainly dependent on the kind of questions asked by mainstream social studies of science, was considerably more open and speculative in examining scientific activity within a fuller range of its contemporary social locations and cultural embeddedness in the last decade of the twentieth century. And of course this cultural studies agenda, as a sort of subtext of this year's Late Editions project, fit the somewhat different agenda of the latter perfectly—to use various situations of contemporary scientific activity and imagination as a means, or distinctive access, rather than as an end in itself, to expose from yet another angle fin-de-siècle situations that are deeply embedded in culture and society with non-obvious connections of material in this volume in relation to past and future ones.

The special predicament of the social studies of science is perhaps also the source of a special opportunity for those, like the participants in this volume, who would develop a cultural studies of science—namely, the critical edge of the field has been lost ironically through its key achievement, an effective demonstration that indeed science is socially embedded and constructed. To show that science is not neutral, is fundamentally political, and interested rather than *dis*interested becomes in its repetition and meticulous demonstrations dulling, a cliché. Thus, now what? One new direction in which the construction of science as an object of study becomes less bounded, more a cultural "thing," is through the study of science as popular culture, among nonscientists of various sorts, through the complex articulations and meanings of science across and outside the boundaries of the production of scientific activity (for example, as occurs among new-age groups, environmentalist activists, and computer hackers). Much of what is now called the cultural studies of science deals with this kind of less-bounded social or cultural embedding of science.

It is precisely this kind of study that is not as well represented in this volume (but see the chapters by Mario Biagioli and Kathleen Stewart). Rather, we focus here preponderantly on the broad cultural continua embedded as part of the production of scientific activity itself in various contexts, but within what unambiguously counts as scientific activity and institutions. Here, the critical edge of a cultural studies of science is restored as is evident in a number of the pieces in this volume, through the premise that the insight concerning the social constructedness and embeddedness of scientific activity, provided as the key achievement of the social studies of science field, is also shared to varying degrees by scientists and scientific institutions themselves. This idea of reflexive critical awareness among scientists of their own activities that has various degrees and forms of expression complicates and deepens considerably what it means to pursue a critique of science. In particular, one might suppose that the issue of the social constructedness of their own practices and truth claims is a particularly salient defining predicament of various late-twentieth-century practices of scientists operating in unprecedentedly hybrid locations and collaborations, as virtually every piece in this volume explores.

Thus, the fact that science is political and deeply embedded in events is not simply the now-clichéd, albeit important conclusion of social scientists and historians studying scientists, but it is part of the condition of doing science and forming scientific imaginaries themselves. Some scientists and locations of scientific practice acknowledge their social embeddedness not at all or only in the most indirect and subtle ways; for others it is diversely and strongly expressed. While by no means necessarily leading to virtue, good results, or liberation, the strong recognition of social connection as a feature of fin-de-siècle imaginaries and communities of scientists is pervasive, if only because the possibilities of doing new kinds of science require completely different connections and collaborations that jolt scientists into different awarenesses of their contemporary locations and positionings. These latter perhaps strike them as oddly as they might some readers who possess stereotypical images of scientists as cut off from society and culture, concerned only with the microworlds of labs and their professional networks.

Certainly, such situations, where the reflexive recognition of cultural connections in scientists' careers and organizations are keenest, were of most interest to the pieces of this volume and in the drift of our collective discussions about them. Thus, beginning to ask how scientists have faith in their own activity, or in what ways their perceptions of what they are doing are changing, given some form of distinctive consciousness about the social and cultural construction of their activity, generates a completely transformed and vast field of inquiry on which a distinctly *cultural* studies of science might establish itself. The reflexivity brings a range of new factors *explicitly* into the production of science, and in this sense, makes it more directly cultural, or blended with concerns that were thought to be external to scientific activity.

And it is certainly no accident that the urge to expand the range of the contexts of scientific activity probed by critical scholars of science is generated by the parallel kinds of changes afoot in the conditions of scientific work themselves. These are not unrelated to the nature of changes among those in the social sciences and humanities motivated to pursue research on the physical and biological sciences. This series is indeed founded on such identifications between interviewers and interlocutors, and further, on hoped-for identifications between the resulting conversations and the readers of these volumes.

A Look Backward—Perilous States Revisited

One of the features of this series will be the revisiting of the themes of earlier volumes in later ones. We wager that there will always be a ''context'' for so doing, since our speculation is that the comparative resonances within single volumes will extend among all the volumes and will be most specifically apparent in the ''revisiting'' sections such as this one. But as noted in Late Editions 1, the particular concern of *Perilous States*—the remaking of societies in the midst of great changes in long-standing political orders—will in various guises be especially thematic for the entire series. The quite variable relevance of nation-state organization in constituting contemporary social and institutional orders at different locations, defined by geography, history, and specific activities, is the ''large'' fin-de-siècle issue that will take many forms and replay many times through the quite diverse sites and concerns of the social subjects expressed in these annuals. It is the revelatory ways and in the unexpected idioms that this ''large'' issue keeps reappearing in the materials of this series that give the latter its special cogency.

Both Jim Holston's interview with Craig Hodgetts and Gudrun Klein's autobiographical reflection on her experience of postwar Germany in light of witnessing aspects of reunification have direct and obvious back-linkages with the concerns of Late Editions 1, but they should equally have resonances with this volume. Holston's interview in particular extends technoscience imaginaries to the prospects for urban planning in Los Angeles, a site famous for inspiring apocalyptic visions in fiction (for example, Nathaniel West's *Day of the Locusts*) but also in fact, given the events of 1992 which are the background for the probing of the contemporary predicament of the scientific imaginary of planning in this unimaginable—or else multiply imaginable—city. Klein's piece, while far from the concerns of technoscience, resonates with the memoir genre through which the lives of scientists are being probed (for example, in Fischer's and Polányi's pieces) as a primary means of new access to how scientific activity is deeply embedded in culture.

A Look Forward—a Preview of Late Editions 3

Volume 3 of this series will concern fin-de-siècle transformations in media, especially visual media (the broad and varied uses of video technology, changes in television, and the appearance of new film industries), computer communication (the spread of the "Net," bulletin boards, and so on, and the kinds of personal involvements and consequences for societies that this technology has engendered), and the interfaces between the two in the current imaginings and actual productions of virtual phenomena. A focus on media is of course another way of getting at "perilous states" while touching again upon "technoscientific imaginaries." Just as Late Editions 1 presented social subjects in unmediated existential relationships to traumatic, rapid events of change in the organization of civil societies, and as Late Editions 2, in contrast, presented social subjects securely situated professionally within "foundational" technoscientific institutions of modernity as the very engines of innovation and change that are being reconfigured in uncertain and unexpected ways, so Late Editions 3 will present yet another set of fin-de-siècle locations, but which deal with the *media*tions of events and structural changes to social subjects themselves. That is, increasingly routinized channels of communication through which (mostly middle-class) people daily participate in great structural changes. Such global changes get expressed in and become increasingly an integral part of the everyday lives of persons through new technologies of communication and visualization that have become for users extensions of the body and self-identification.

As a preview of Late Editions 3, we include here a piece by an avid Networker, at once skeptical, soberly realistic, playful, and committed, about the mechanistic extension of himself in which he seems inexorably and reflectively involved. This might be the most perilous state of all.

SCIENTISTS, FAMILIES, AND FRIENDS

These contributions present the lives of scientists in the contexts of family relations and of historic shifts filtered through the poignancy of generational transitions. Livia Polányi's family memoir connects U.S. science of this fin-de-siècle with European science of the previous one through the mature perspective of the remarkable scientists in her family. Michael Fischer's contribution, which argues for the value of autobiography in the cultural studies of science, does much the same as Polányi's in his assessment of his own mother's autobiography, among others.

Cornucopions of History: A Memoir of Science and the Politics of Private Lives

> Cornucopions provide a way to duck the paradox engendered by losing the information forever. "It's a very fine line," said Dr. Strominger, "between saying that the information is lost, and saying it's gone into a region of space-time where you'll never get it again."[1]

Reading and Writing the Double Biography

I do not read very much for pleasure these days. Sometimes I feel as though I read so much in my childhood and adolescence that I need my adult years to absorb all of those words, those images, metaphors, and narratives of others. In particular, I never read about the war. I never read about the bomb. About the cold war. I never read about the Holocaust but I have read almost every book about the Holocaust. I never admit that I am reading them. I stand for hours in book stores denying that I am reading, but, when I leave the shop, the book is no longer unread. One reason I can not read about the war is that there is no new information about the war. It was all there. In the newsreels, in my American-born mother's late evening conversations, in the stories of borders, visas, exit permits, entrance visas, transit visas, temporary visas, denied visas. But I did read one book this past summer. My father was visiting me and we both read it: *Lawrence and Oppenheimer,* by Nuel Pharr Davis, a double biography of Ernest O. Lawrence, father of the cyclotron, and J. Robert Oppenheimer, leader of the Manhattan Project. We fought over this book, actually, very politely appropriating it when the other turned away for a moment and affecting great ignorance when asked about the whereabouts of The Book. It had a hypnotic effect on us.[2] My life story, seemingly in place, solid, became fluid and ill defined once again. I began to interrogate my father's life, my uncle's life, and ultimately my own life—as lives of scientists, lives shaped by the passions and the politics of the scientific microcosm and the external context of national and transnational politics in which science functions.

Geminates

Big Science came into being in the United States during the Second World War. The Manhattan Project devoured money, men, materials, and ideas and produced the atomic bomb, a golem which first poisoned the sands of the Nevada desert and later transformed two Japanese cities to evocations of horror. A double biography of two lives in Big Physics is the topic of *Lawrence and Oppenheimer*, the book which hovers in the background of this double biography of two men in physics, my father and his brother, two trained nuclear physicians who did not work on the Project.

My father and his brother. I am not an impartial reporter. Perhaps I can not be trusted to evaluate the relevance of their stories to deepening our understanding of the nature of practicing science, the nature of the construction of history, the nature of the production of biography, the nature of our century. The stories of these two men of physics will not be told as stories of physics but as a story of this century, our dying century, this century of borders and emigrés, the Holocaust, the Bomb, the Communist Experiment, and in the arts, in physics, the social sciences and the humanities, the century of exploring the inescapable determinacy of the position of the observer.

Two Brothers

My father and his brother were born in Hungary and grew up in fascist Italy. They came to the United States in 1942. For nearly half a century, my father, trained as a theoretical physicist in Rome under Enrico Fermi, worked in rural Massachusetts in the research laboratory of a prominent optical company developing medical instruments, primarily equipment for use in cardiology. His brother, trained also by Fermi in the Institute of Physics at Rome, also worked in commercial research laboratories in New York and Massachusetts, first in mass spectroscopy and later in laser physics. He too made contributions to medicine and is credited as the "father of laser surgery" for his work in adapting laser technology for use in otolaryngology and gynecology. The two men are now retired. They live together with my uncle's wife in the woods in a large wooden house at the top of a hill, with walls of glass facing the view. There is a wood stove and a large open fireplace in the living room, and everywhere there are rugs: kazakhs, bokharas, ivory Chinese rugs with Chinese blue in their ivory fields; runners, prayer rugs, wall rugs, and large rectangular room rugs on top of which the furniture sits. There is a large garden. My uncle gardens. He does the errands. He takes care of the others. He spends a great deal of time in his study where he is writing a history of laser surgery. My father reads and reads and sleeps. He is working on a new invention.

The two brothers look very much alike. People have found the resemblance uncanny at times. Once they were confused about who was who themselves. Walking down a long corridor in a hotel in Chicago each saw himself in the mirror walking towards him. The reflection grew larger the closer the mirror came. Halfway down the corridor, the image assumed its full size. It was the other, of course. Neither had realized his brother would also be at the meeting. They resemble each other closely, but they are actually very different in presentation. One is immaculate: neatly dressed, perfectly coiffed, dapper, and somewhat elegant. The other has never taken care with his appearance. His clothes are casually assembled, often stained from working on the car, in the wood shop, in the lab. Not unlike Lawrence and Oppenheimer, they are forever bound together as much by their differences from one another as by their common loves and concerns. But these twins have not entered the annals of science or the pages of history. The Hungarian brothers pursuing their lives, out of the tempest of Big Science made small but not insignificant contributions. They understood the applicability of theoretical physics to problems of medical diagnosis and treatment—as did Lawrence—and they were never important targets for antileftist attacks, unlike Oppenheimer. Had they been part of the Manhattan Project, would their lives have been very different? In time, we will speculate about those other lives, those subjunctival lives of these two men, in time, after we have told the story of the lives that we will tell as if, indeed, these were the lives that these two men did live.

On Voice and Language

A friend asked if my father, my uncle, the other characters in this account would speak in their own voices? The answer was no, I knew, not really. The conceit of direct speech is all but unavailable to me. Much of my work as a scientist involves artifacts constructed from attempts to capture natural speech. Every um, every gulp, every detail of the talk's prosody, each overlap of more than one speaker talking at once, every gesture with the hand, each facial expression, or shift in body position or eye gaze direction is to find its place, its meaning in the record, to reveal its significance in analysis. They do not, of course. The transcripts do not reenact, let alone recreate, the situation of the fragile moment. But the direct quotation complete with verbs of *he said* and *she said* and off-line commentary about speech volume or attitude is painful to me. It is a tool which has been all but eliminated from texts which I write. While reading, I can barely read dialogue. The artifice humiliates me. I can not write it.

That is not the entire story, however. This story is largely the story of one man, one scientist—my father. His voice, my father's relationship with language, has never been transparent, has never passed without comment. My mother told me once when I was little that my father once had had lockjaw, but that no one at the lab knew anything was wrong. He never opened his mouth; he always mumbled and no one could ever understand what he was saying in any case. My father and I have always spoken together in English, it is our only common language. I do not speak Hungarian, German, Italian, or Spanish and he does not speak Dutch, a language I learned as an adult. From the age of seven or so I would translate what he had said from his English to my English for the waiter or waitress who could not understand him. "He'd like a salad, vinegar and oil, he'll mix it himself," I would chant after him, adjusting the phonology, transforming the syntax, supplying a few key function words, and making eye contact with the server. "Does he want coffee?"—"Coffee later."—"Yes, he'd like coffee, please. Later, after the meal." You see, there will be some dialogue here. My father's voice crept into the text. Texts leak.

My father's first language was Hungarian. He speaks the perfectly grammatical, phonologically impeccable Hungarian of a well-brought-up, privileged child of a century ago. He can neither read nor write Hungarian. German, also a mother tongue, he has refused for almost fifty years to speak or to admit he understands except in emergencies. Nonetheless, while visiting me in Holland, where I lived for some years, he read the Dutch newspapers with little effort and laughed with great amusement at the accents and grammatical displacements which Dutch displays for the German speaker. His accent in German, say those who have heard him speak, is perfect. His father, "the linguist" who spoke all languages perfectly, refused to talk with my father, his oldest son, in any language but Italian. "It's the only language he knows how to speak," he would say. But Italians, when he meets them, citizens of the country he lived in for a quarter of a century since arriving there at six, Italians all say, "You speak very, very good Italian. Perfect Italian, but you are not from here." Because of the Italian which he speaks perfectly, it was easy for him to learn Spanish as well, and he picked up his Spanish-Italian with a Cuban twist when he was in Havana waiting with his brother for an entrance permit to the United States. The other day he stunned a close Colombian friend of mine who has known him well for almost ten years by suddenly speaking to him in Spanish as if they had always conversed in that language. "I didn't know he spoke Spanish," said Nico. "I knew he knew a few words, but actually, I understand his Spanish much more easily than when he talks in English." My father and his brother even today sometimes speak Spanish together, echoing for a moment their Cuban junket.

My father's brother's first language was not Hungarian but German. "At home we all spoke German to him," their sister told me. "We all spoke

Hungarian with Mother but we always spoke German to him.'' My uncle did not learn to speak the language of the country of his birth until his adult years, when he married a young Hungarian refugee whom he met in Chicago, at the home of some cousins who had come to the United States well ahead of the war. Today my uncle speaks reasonably good Hungarian, as well as perfect Italian, native German, some Cuban Spanish, and excellent English. His English does have a Central European flavor. Despite a mild dyslexia, he spells English without a problem and his handwriting can be clear and legible if he makes an effort to write slowly. My father's written communications are themselves artifacts of the diaspora. A simple note from him can take several days to decipher. There are codes encoding codes, codes encoded in codes. Phonetically, the code is undefined Central European; orthographically, it is Italian—he transcribes his own speech sounds phonetically using misspelled Italian sound/symbol correspondences ornamented with final silent *e*s and transposed *gh*s and *th*s from half-remembered spellings of English words. Syntactically his conventions are totally idiosyncratic, but even when one has penetrated the transcription the message can not be read off easily because it will emerge from the chains of orthography encrypting a private joke in family language. To understand the communication, the recipient must recover the full semantics and contexts of the phrase and then reconstruct possible lines of connection between the decades-old punchline of a secret joke to which the text invariably alludes and some situation of possible relevance at the moment. Often one fails, puts away the postcard or shred of paper, and waits for Freudian process to make *éclatant* what my father so laboriously has set on paper.

His scientific papers and patent applications are all composed in our common language, the English in which he has lived for over half of the present century. They emerge elegantly written, perfectly lucid, after torturous composition. As a reader, an editor, a critic of English style, he is perfectly tuned. And for those of us who are competent in his idiolect, or for those who care to try to converse with him, the conversation will be rich and variegated in sound, in texture, and in content. The conversation will not be in English, but one can stretch. While I always knew my father's relationship to language was difficult, one day when I was a child I understood what it was to be an emigré, an orphan without a mother tongue. We were walking up the flight of steep cement steps leading from the lake up to the house where he lived for over forty years. He stubbed his toe viciously on the concrete. I watched as his eyes filled with tears from the pain and I listened as no words came out of his mouth. He smiled a sad, embarrassed smile. "Damn it?'' he finally said softly. There had been a slot to fill with a cry of pain but he had no words, no language with which to fill it. He laughed. I laughed. I helped him up the stairs. He had broken his toe.

My father's voice is not easy to bring to these pages. Without his lexicon, his prosody, his syntax, he would have no voice. With his lexicon, his prosody, his syntax, he would walk these pages into history as a caricature, an absurd ventriloquist's dummy outfitted in ill-fitting paludaments of a language he has never owned. As he could not cry out in pain over a broken toe, I think in these pages he will rather remain silent, mumbling something occasionally, which I will translate to the reader. The voices of others will whisper as well.

Language and Science

I first experienced my father as a scientist some years ago when I announced to him my intention of becoming a linguist. Knowing that I had no great interest in ancient languages or, in fact, in modern languages per se, he was concerned. You become what you do every day. He wanted to know what his daughter was going to do every day, what she was going to become. He asked me exactly what a linguist did. I gave him the standard reply. "A linguist is a scientist who studies the structure of language in general, and, quite often, specializes in the problems of one or another aspect of one particular language." He looked puzzled. "What sort of a science is linguistics?" he asked. *Science* was a term reserved for the systematic investigation of *natural phenomena* and, he argued, linguistics can not be a natural science since language does not exist in nature apart from the human beings who use it. Thus, he reasoned, language was an artifact of human construction and not a preexisting natural object. Language was thus like art and not like physics, and linguistics is thus a humanity concerned with exploring a human representational scheme and not a science concerned with forces and structures existing independently of human participation.

My father spoke of science with passion and conviction as a category of activity not rooted in the social, the conventional. I was taken aback. In my father's family, the talk was of politics not science. The conversation was radical, amused, leftist, aloof, the analytical, cynical, pessimistic, absolute moral conversation of the European fellow traveler. At breakfast, at lunch, at dinner, lying in the sun by the Lake, in the car, discussion of the events of the day, events in the past, events in the future was our talk: a decades-long moral disquisition on history and there is no history, my father said, but the present. Science was not discussed; applied physics was practiced. At breakfast, at lunch, at dinner, lying in the sun by the Lake, in the car my father and his brother sketched designs for apparatus and constructed sequences of equations with long divisors, capital and small roman letters, deltas and sigmas.

My father has always told me, there is no history but present history; there can be no story of these men of their science, their lives or their politics that is not the story of the teller. My work as a linguist has been concerned with stories, with exploring constraints on the story as told which arise from the linguistic, social, and cultural forces present at the moment of telling. This story is thus necessarily and without apology my own story, the story of a family member, a woman, a scientist, a daughter, a wife, a political and social actor, a mother, a linguist, a speaker and an act of speaking, a point in space-time defined by the physical, the social, the cultural, and the linguistic. This essay, thus, about science, about politics, about my father, his brother, about this century and its events—the Communist trajectory, the Bomb, and the Holocaust, which together resulted in the invention of genocide and the development of a discourse of positioning—this present history is both an exploration and a chronicle of two men, one family, several others and the politics and science of the twentieth century. It is my history, her story as she tells it now as a flat-weave rug, a kilim of many colored threads, many patterns, many conjunctions. Without a clear center, a clear point which her science tells her is a central requirement of appropriate storytelling, this flat-weave is a gift of the story of one generation set in one century to be passed on to a next generation, which will learn to read it sometime in the future when the century has turned and this time of present storytelling has itself become history.

Two Lives in Science

My father, I know, is a scientist. My mother uses the word with pride. A physicist, she says carefully. He works in a laboratory. He calls it the "lab." I am curious about what a scientist is, what a physicist is.

The physicist. She may have learned the word from the newsreels, made the word her own from newspapers, *Life* magazine, the movies. For my father, physics was a second choice. He had wanted to be an engineer. When he went to register for the university in Rome, he first tried to enroll in engineering. There was a *numerus clausus* in effect and, because he was a Jew, he would not be allowed to study engineering. He tried physics. The line was shorter. He enrolled. Immediately, he knew he was a physicist. He would never have liked being an engineer.

He studied physics. Fermi, the great professor of the Institute of Physics, taught the first-year course. Hundreds sat in the big hall to hear his opening lectures. Few understood a word he said. He lectured to the few who could understand him. As the year wore on, fewer and fewer students attended. My father got his degree in physics working with Fermi on cosmic rays. He had his work. He loved physics like everyone who was in awe of Fermi. Fermi. He knew.[3] In addition to his work in the lab, he spent a great deal of his time, and much of his passion, with his friends, a small but very active group of anti-Fascist Communist intellectuals. His brother, a few years younger, joined him at the Institute and in the laboratory.

The two Hungarian brothers did what young atomic physicists did in those days at the Institute of Physics until they signed on as deckhands on a Yugoslav freighter and left Italy in 1941. My father's farewell to the Institute: the porter tells him not to worry about the war. "Fermi," the man said, jerking his thumb back up at the building. "The old man's got a pill. Don't worry about the war. Fermi's got a pill."

My father enjoyed the trip. He didn't worry about the war. He read and slept. My uncle worked hard. I never heard if he enjoyed the trip. After a week-long stop at Ellis Island, forbidden to enter the United States, they sailed on to Cuba and jumped ship in Havana. My uncle found a job in a high school teaching English. My father spent his time in the cafes talking about politics, meeting the leftists, the radicals, the opposition of the capital. Soon after arriving in Cuba, he met Verdi's grandson, an anarchist. My father confessed his regret at not having gone to Spain to fight against Franco. Verdi laughed. "The Stalinists would have killed you as soon as you got there," he said. "They kill everyone who's not a Stalinist." "I had not thought so before but when he told me I knew immediately. What he was saying was right." The next day my father went to Havana's Communist Party headquarters and began to read, to hold in his hands and see the newspapers and reports forbidden in Mussolini's Italy. He read decades of *Pravda,* Communist communiqués, Comintern propaganda. A few days in the archives were enough. He emerged a convinced anti-Stalinist. He did not worry about the war. Stalinism was the threat to a Communist Italy. He talked, he read, he argued: he needed to get back to Italy, to warn his friends.

It would be a long time before the Hungarian brothers returned to visit the country where they had been brought into exile in 1919. By the time my father found his friends again they were building the Italian Communist Party, a new Communist world. Today those friends who are still living remain powerful in the de-Stalinized, refurbished, Euro-Communist Italian Party. The party they founded and guided for half a century.

Origins

My father, my uncle, and I are among less than ten living individuals born with the name Polányi. No one named Polányi has lived in Hungary for almost seventy years. We are all in the English-speaking world: Canada, the United States, and England. Yet the name remains an important one in Hungary even today. Recently, Hungarian TV aired a one-hour documentary about the family with well-known Hungarian actors reading from family letters against backdrops of youthful great-grandparents and dead aunts. Beyond history, the program retells a contemporary story.

Hungarian Jewish intellectuals have been at the center of the cultural elite in Budapest since the turn of the century. Always described as "wealthy, influential, and fiercely nationalistic,'' many Hungarian Jews changed their Polish or Russian patronymics to more Hungarian sounding names in the years before the First World War. Thus did Polachek emerge from the Polish and Russian shtetls and enter Budapest society as Polányi. And though no one bearing the name currently lives in Hungary, in Budapest, intellectuals today debate the thoughts of the two well-known Polányis, Karl and Michael, the brothers of my grandfather, Adolf.

Karl, the eldest brother, was acknowledged to be the brilliant one. He is sometimes known as the sociologist and sometimes as the economist. My father has always marvelled that he wrote a whole book about socialism without mentioning socialism or Karl Marx by name. I don't know if that is true. I have never read *The Great Transformation.* Michael was the youngest son. In the family, he was not considered particularly bright and was left to his own devices. He became a medical doctor, a physical chemist, a well-known philosopher, and the father of a Nobel Prize winner. He wrote several books. His best known book, *Personal Knowledge,* is on my list. For some reason, I find it unbearable to read the works of these men.

My grandfather, who had a knack for languages, studied law, was deeply interested in economics, became a journalist, and was in Yokohama at the time of the great earthquake in 1906. While in Japan he wrote a brilliant short monograph on the Japanese economy, and many, many years later, when my father surprised me by eating with chopsticks with great proficiency, I learned that my grandfather had taught his children to eat in the Japanese fashion at an early age. I can not imagine the scene as they ate with their eating sticks. My grandmother had little sense of humor and none whatsoever where table manners are concerned. Even today my uncle prefers to eat lobster with a knife and fork as much as he can. Their mother was a superb cook in the Austro-Hungarian tradition. Her delicious potatoes and perfectly fried Wiener schnitzel were hardly suitable for eating in the oriental manner.

There was one Polányi sister, Laura, called Mousie by the family. She was the first woman to graduate from university in Budapest and became a historian. In the early 1950s, on a trip to visit one of her sons who had returned to Hungary from the United States after the war, she discovered the diary John Smith had kept while a prisoner of the Turks. She thus became renowned among American historians, and, citing the need to continue her fruitful researches, was able to travel back and forth between New York and Budapest during the years when travel was all but forbidden—by both sides—bringing to her fiercely Communist son and daughter-in-law the penicillin, oranges, powdered baby formula, and English detective novels necessary to make life in that ravaged city reasonably comfortable even for staunch members of the Party.

Mousie had an exquisite speaking voice and an enchanting manner which made anyone with whom she spoke feel specially treasured, fascinating, the center of the universe. She used all of her charm and her many connections in 1937 to extricate her daughter from a Stalinist prison. Her daughter returned the favor soon after, using her great beauty and powerful presence to persuade the British Red Cross to move heaven and earth to get her mother released from the Nazis. Mousie was a great beauty even in her seventies when I first knew her. Among the socially desirable beautiful Jewish women of Budapest, her only rival for physical loveliness was my grandmother, the Socialist daughter of the owner of the first French laundry in the Hungarian capital. My grandfather pursued her vigorously. For years she refused his advances. My grandfather wrote to her often from Japan and they married after his return. During his absence she led her own employees out on strike against her parents. My own mother, an Italian-American born in New York City, who owned a successful factory during the war had similar values—she insisted that her workers join the union, she wouldn't employ nonunion labor in her shop. My grandmother was also a public activist. She served as the Hungarian delegate to the First Women's International Socialist Convention in Vienna, where she caught the eye of Angelika Balibanov, one of the most important figures of the Russian Revolution. Balibanov came to Italy soon after being allowed to leave Russia in the 1920s and came to visit the family settled then in Rome, where they had gone after leaving Hungary in 1919. My grandfather, a minister for Béla Kun, had had to leave immediately after the fall of the revolutionary government. Like all of the functionaries of that Eastern European commune, he fled first to Vienna by the Orient Express. In Vienna, half of the train went East to Russia, half to the West. Those who went East flourished for some years and then were swallowed by the purges; those who went West spread out across the world, a thin red diaspora whose influence has far exceeded their numbers or the relative insignificance of the Central European nation from which they came.

My grandfather and his wife and four children went to Italy. The family was moved to a small white house in Castel Gandolfo, the Pope's summer residence on Lago Albano. Castello was a paradise for the two boys, the younger children. They swam and hiked and ran with the village children. When my grandmother first went swimming in the lake, a rumor spread that a circus was in town. Swimming ladies in rural Italy were not a common sight just after the Great War, while in Hungary and Austria, which were in the grip of a massive physical culture fad along with much of northern Europe, a woman who did not swim was not considered truly modern, and my grandmother, a Socialist and advocate of women's rights, was quite the modern woman. Her political convictions did not stop her from waiting on her children and husband and even on my grandfather and his second wife when they visited us many years later in New York.

My grandfather worked as an engineer, a civil engineer of some sort. He took the train everyday into Rome from Castello and returned home in the evening. Eventually he did not return home but moved to Genoa to be with the heavy-featured young Viennese woman who became his second wife. After living in Genoa for many years, but quite before any unpleasantness in Europe, he moved to Brazil with his bride where he designed boilers for steam-powered electrical generation plants. His first wife and four children he abandoned to their fate. All eventually survived the war. Soon after the end of hostilities, my grandmother and my father's eldest sister arrived in Chicago after several years in South America. My grandmother remembered the flocks of flying wild flamingos, pink clouds in the jungle sky; my aunt remembered being lonely and poor and frightened. A trained doctor, she had supported her mother and herself working in a hospital in Paraguay performing full medical duties for the wage of an aide.

There was a second daughter as well. A beautiful young woman but heavy, indolent, she had returned to Hungary soon after the family evacuated to Italy and later married a young man who died in the street, shot to death by the Germans in front of her eyes. She survived the war in Budapest as many Hungarian Jews did—somehow—and eventually moved to Brazil herself after the war ended. She never needed to work. She was supported for the rest of her life by pensions from the German and Soviet governments which had commandeered her comfortable apartment in Pest's fashionable thirteenth district at different times in the 1940s. She and my grandfather both lived in Brazil, but they were not very close, I think. They lived in different cities and did not see one another often. My aunt lived on her pensions, my grandfather supported himself through designing high-pressure steam generation equipment.

I know a bit about boilers and a bit about engineering. My first husband, now the curator of power machinery at a large museum, specialized in steam technology. I learned from him that generating steam is not easy. Boilers must be designed to specific tolerances and turbines must be crafted to fit perfectly on their site in power generation facilities, and then must actually do the job—heating up large quantities of water to very hot temperatures, which is then piped as steam to spin the blades of the turbine. I know how I came to know this but I cannot imagine how my grandfather managed to acquire even this much arcane knowledge, let alone the real expertise needed to carry out the task of boiler design. For me, even today, this remains one of the mysteries of living by your wits. How this lawyer and minister of Béla Kun learned to design boilers for electrical generation plants is still one of the deep secrets of my family's history. No one else has ever found it at all odd. I sometimes wonder if any of these stories ever really happened at all.

What Did You Do in the War, Daddy?

I found I could not put down *Lawrence and Oppenheimer* as I was reading it. The two giants, Ernest Lawrence and Robert Oppenheimer, each had his younger brother, his doppelgänger. John Lawrence, a physician and the first great nuclear radiologist, was the shadow of his older brother, Ernest, while Frank Oppenheimer, a communist and a nuclear physicist who worked under Lawrence in the Berkeley lab, made his brother extremely vulnerable to the charges of being sympathetic to left-wingers. As I read the book, I grew increasingly disoriented. The story of my life became malleable. During the war, every physicist available was being cajoled, dragooned, charmed to work on the Bomb—bright physics undergraduates,[4] established corporation scientists,[5] terrified refugees,[6] and Fermi himself—yet my father and his brother, who had studied high-energy physics with Fermi at the Institute in Rome, had not worked on the bomb. Neither had known the bomb was under development. What had happened? What would have happened if they had worked on the Bomb?

"So, what did you do in the war, Daddy?" I asked.

"I didn't evade the draft," he said.

The Brothers' War

My father did not serve in the armed forces. This, in itself, was not unusual. "Any physicist or chemist at that time had a choice of jobs to keep him out of the Army" (Davis 1986: 161). But my father did not want a deferment fearing, in part, that failing to serve in the military might make him, a refugee, more vulnerable to expulsion, imprisonment, or other harassment. My father has never believed that policemen are your friends. Along with Prohibition, the American's trust of the police and of the government has never failed to reduce him to shy, smiling, mumbling embarrassment.[7] Therefore, when he received a notice from his draft board saying that his request for exemption from the draft because of his work at the laboratories of General Dynamics had been approved, he went to the draft board and demanded that they retract the deferment. "I never requested a deferment and I want it removed from my file that I requested it." When the lady at the draft board refused to remove the deferment, he became not very pleasant. He was sent upstairs to see the captain, an air force officer with half a face. The rest had been burned beyond reconstruction. My father explained. He had never asked for the deferment, he wanted it removed from his file. The captain grew furious listening to the story. He picked up the phone and yelled at the lady downstairs. "Why are you always doing this? If I told you once, I told you a million times. Don't do that with these cases." He glanced up at my father. "Go home," he said. My father went home.

He spent the war variously. There were stories. Before the captain, he had been called for his draft physical. The young would-be soldiers were given all sorts of tests. Of all those tested, he was the best in English. He barely spoke a word of the language. He was sent to see an officer. "The best in English?" "Yes. We need people like you. If you volunteer we cannot promise but you will be an officer, in Intelligence." "The best in English?" "Yes." "I don't know much about the Military. But I know one thing. Never volunteer." He was sent home. He moved to Ithaca. At Cornell he taught Italian to troops destined for the European front. "It was a little stupid." He called on Bethe and soon after found work at the jet propulsion laboratory. Soon after, he was fired for political reasons. The details are unclear. My father and I look like spies and he was a refugee from Fascist Italy carrying a passport from another belligerent—a Communist Hungary over twenty years dead. He was hired at the American Optical Company, in Buffalo. "I had made a little gizmo at the jet propulsion labs to measure the speed of the exhaust from the jet engines. It used a lens and a little flame. I wrote a little paper about it. Somebody saw it and they hired me. I built a machine for them, a refractometer, to measure eyesight. They thought I knew about optics. I always know who to ask."

"Did you know about the Manhattan Project?"

"No. I never suspected it was that. There was something. I thought it was electronics. Everybody was gone."

"Were you ever interviewed?"

"No. Well, Yes. There was a man. I didn't know what the fellow was getting at. They told me to come to New York. I went to a little building in the Village. I talked to a guy there. I never knew what the fellow wanted. He asked me some questions, about Fermi, about the Institute, about the University, about the family. I didn't say much. I never heard from them. I went home."

"What about when you heard about Hiroshima?"

"Then yes of course. I knew how it was done of course. That wasn't a secret. How. The secret was that they could do it. When they did it I knew what had happened. The pill. The porter at the Institute, he knew. Fermi had the pill."

After the War

Two young refugee physicists. One spent the war drifting from research position to research position to university position from here to there. His brother was drafted and served in the army. He had a good war. Sent to the Pacific theater, he was a reporter for an army newspaper. They reached the Philippines just after the Japanese withdrawal. My uncle loved the tropics, the exotic birds and plants. He brought back strange heavy wooden Polynesian statues which I remember staring down on me from the tall bookcase in the study when I would sleep over on visits to my uncle's home a few years later. By then he lived in Bayside, in Queens, and was working in physics on mass spectroscopy at Sylvania Research Labs. He was approached about getting a security clearance. Nothing came of it. My father was not approached. The situation was serious. The House Un-American Activities Committee and McCarthy were abroad in the land. The promised land. America. There was no place else to go.

The two brothers worked in industrial laboratories. My father put away his books, hiding them in cardboard boxes to be stored with my mother's relatives in the Bronx, on Long Island, the volumes of histories of the Communist Party of the Soviet Union, the writings of Bukharin, Trotsky, and Lenin, the novels by James T. Farrell and other progressive writers. The FBI visited. They interviewed my mother. They interviewed the neighbors. They appeared at the lab in Norwalk, Connecticut, where my father was working for American Optical. They came and they went. My father and mother separated. I learned from my mother deep in the night about the camps, about the Holocaust. Stalin died. My grandfather came to visit from South America. He brought my mother a geranium. There were arguments about Stalin. There were discussions of the hearings. My father continued to work at American Optical. He moved with the lab to Southbridge, Massachusetts. He lived on a lake. Like my father who grew up on Lago Albano so many years ago, I passed my summers, too, at the Lake. Reading and reading and reading the books which appeared rehabilitated on the bookshelves released from the boxes hidden in Long Island and the Bronx.

I remember especially the transcripts of the Doctors' Trial in Moscow and the books and short stories by American black writers. I read most American proletarian literature and began to explore the volumes of history and politics and sociology which grew and grew almost as I read. Dreyfus, gulag prison guards, the Scottsboro Boys, Father Divine, Studs Lonigan, Angelika Balibanov, the Little Grandmother of the Russian Revolution, the Knights of Malta, the Medicis, Cotton Mather along with Nancy Drew and Freud borrowed from the local library accompanied me eventually onto the dock and competed for my attention with the fish which could be seen through the cracks between the boards swimming in the clear water. We both read. My father and I. We never discussed what I was reading. We discussed politics, history, this strange place, America. Sometimes we talked about fiber optics and how you could use these strange new devices to understand the processes of the human body. Then my father would begin to draw on the edges of his newspaper with a leaky cartridge pen and I would understand something about the wavelength and differential absorption, but not much.

When his brother came to visit as he did every summer for a few days, sitting down by the Lake between swims there would be more talk and more and more talk. At one point in the late fifties the conversation grew especially excited. My father and his brother talked and talked in English and Italian, sometimes in Spanish, they talked through the night, through breakfast and the dinners we ate each evening outside by the Lake bringing down the meal carefully cooked in the kitchen, the dishes, glasses, cutlery, napkins and tablecloth and then carrying all back upstairs again after an espresso, Chico?[8] One day when my uncle was there Eli came over. On a beautiful Sunday afternoon, he came to the Lake alone, without bringing Shirley, his wife, and the five children over to swim. Eli, a brilliant physicist—a really brilliant physicist. My father had worked to get him hired after hearing him on the radio testify at the McCarthy Hearings, time and again taking the Fifth Amendment refusing to implicate himself, his friends, his brother, any of the others whom he had known in the Party at the University of Chicago.

My father was a fellow traveler, a leftist, a progressive. He was always opposed to the Party. In the United States. In the Soviet Union. In Italy. He did like Eli, however, and he liked his courage. A young physicist with the potential for a brilliant career with five small children under the age of seven, Eli had been fired from several university positions and was all but black-balled from physics when my father contacted him, argued with the managers at the Lab, and arranged for Eli to come to work at American Optical.

By that Sunday at the Lake, Eli was settled in at the Lab and was working on neodymium lasers. Getting the machine to lase was a major effort. Eli and his people worked day and night. Around the clock, weekends and holidays. Competition with the ruby laser and the new gas CO_2 lasers was intense. My father, my uncle, and Eli talked and talked and talked. A few weeks later, my aunt and uncle returned to the Lake. Eli came again. The men talked and talked. Paper napkins and pads of paper were filled with diagrams, equations, doodles, and numbers, stained from endless coffees and the ashes of my uncle's cigarettes. Soon afterwards, my uncle switched from mass spectro-scopy and began to work in lasers. A few years later, he talked late into the night again and again with his sister, my aunt, my father's sister, the doctor. She taught him about physiology, anatomy. He began to work on surgical lasers and to grow very successful professionally. Soon he left General Tele-phone—the company which had bought the Sylvania Company and changed the working atmosphere substantially—and moved on to become the chief scientist of Lasers Incorporated, a new division of American Optical. For a time, the two brothers, the two brothers and Eli, worked for the same orga-nization, and when Lasers Incorporated was closed and moved to Framing-ham to become a division of the main American Optical Laboratories, the three men worked in the same building. My uncle and Eli were Bubble Boys, American style scientists. My father never hustled like that.

Lives in the Bubble

> "Once Lawrence ordered Dennis Gardner and me—Dennis later discovered artificial mesons and died of beryllium poisoning—to build a calutron power source," recalls Roger Hildebrand. "He said we were to get it done in a day and a half. I consider it now a task that should have taken two people several weeks. But we didn't sleep and we improvised and in thirty-six hours we got the power supply going" (Davis 1986: 133).

My first husband, John Bowditch, is a Bowditch of Boston. A direct descendent of the Great Navigator, John's family has been prominent in Boston for centuries. John himself is an eccentric. He has always been an eccentric. I met him in college. I was not yet eighteen when we met, he a few months shy of twenty-one. We began to go together immediately, driving down from Vermont in the dead of winter to go to the opera. John had season tickets to the Boston Opera Company, as did his parents. They described themselves as "sturdy New England types." They had recognized very early that John, their third child, was different from other children. Their present to him on his fourth birthday, among the usual toys and clothes, was permission to wire electric plugs without supervision. Free at last! He was free to expand his explorations in electricity, to wind his own armatures and build his own radios, make his own electric fuses for explosives and kludge together his own electric tools and motors.

When I met him, John was consumed with two principal desires: to build a steam driven motorcycle and to generate a million-volt spark. When not pursuing leads on old boilers or tracking down rumors about large capacitors available from radio amateurs who could get them air force surplus through the MARS Program, he spent his college years building his own state-of-the-art 1923 telephone company, complete with an automatic, stronger switch, central office under our bed and telephone polls connecting the various subscribers with scrounged telephone wires and reclaimed heavy glass insulators. Collecting acoustic operatic and orchestral recordings made before the invention of amplification and discovering early light bulbs and electrical generating equipment also claimed some attention. Vast amounts of time were spent hauling heavy objects, storing the objects, reclaiming the objects, cleaning the objects, and waiting, waiting, waiting for one or another detail to fall into place. John has sometimes waited decades to obtain just the right part to make an early piece of equipment operational. For he is not just a collector, he describes himself as a historical engineer. As he told me, back then when we were so very young, he was not an antiquarian or someone interested in the past. No. He was interested in a specific past. "I am a very up-to-date young man of the 1880's," he would say and then begin another endless adventure involving objects made often of cast iron. Incredibly heavy. Enormously fragile.

I willingly entered into these adventures with him. I spent hours and hours in cold, drafty, musty, dark places, handing wrenches about and slowly winching tons of cast iron covered with one hundred years of grease away from its cozy but threatened home and into a new life among John's stuff. I enjoyed it, of course. I would not have done it had I not enjoyed it. I was fascinated to be visiting a world which was fantastic, and totally absorbing. The world had its rules—how and when outside reality might be accommodated; its villains—those people who were being mean to one or another motor-generator set rusting away happily in a New England field; its heroes—that guy who had one of two existing autogyros and twenty-three Stutz automobiles including three Bearcats; its rhythms—move some portion of the stuff to Vermont in the fall, take it back to John's parents' house in Foxboro in the spring. Move the stuff into the crawl spaces, attics and garage when the parents are in church. Hurry to wash up and look presentable for the wonderful Sunday dinner.

Connected to the concerns of everyday life only tangentially, the world we lived in was totally compelling. Fragile like a soap bubble blown by a child, this world demanded complete loyalty. If we doubted, it would disappear. When it was necessary to earn money, one did so but without involvement. The power of the project, the power of the bubble in which that project arbitrated the meaning of all action was total. John was emperor, citizen, and slave; I with my body, my interest, and willingness to respect the rules was a tolerated consort as long as I did not disturb, did not require attention or introduce my own desires, ambitions, or private realities.

How young men do science in the United States is not particularly different from how they do art or engineering or religion or hobbies. Whether they are in the lab, in the studio, behind the keyboard, in the meditation cell or in the backyard or garage, they are spending long periods of time in a highly constrained, highly restricted world in which almost every action, every relationship, every utterance is interpreted relative to one model and one model alone—the abstract semantic world of the project, be it a complex titration experiment, a metal sculpture, a database subroutine package, a contemplative retreat or calibrating the flyball governor on an old mill engine. The absorption, although cathected onto external objects and processes, is primarily with the self. The model in which action is assigned meaning is highly abstract, highly structured, and very jealous. The prized teamwork of such efforts is brought about by collective embracing of a reduced reality. Each participant is constantly evaluated, his loyalty to the group and to the group's effort determined in part by the others' evaluation of the totality of his immersion in the project bubble world.

When in the project, the young scientist, like his counterparts in other pursuits, will often work sixteen or eighteen hours a day, forgoing sleep, eating badly, and neither copulating nor participating in the arrangement of a social constellation with others of appropriate gender which would lead to copulation. Working at such a feverish pace, the guys are capable of making relatively rapid progress on their projects. ("The boys really hustle," my father says.) The rapidity of progress defined as success permits the successful to set the evaluation criteria for others in their fields.

The successful become key men. Men like Ernest Lawrence; men like Robert Oppenheimer. Key people, of course, do not live in the lab. They live on peer review councils, on the editorial boards of journals, on planes going to and from scientific meetings; they work with other keys—maneuvering for funding and prestige. They live in the key bubble, of power, prestige and influence: like Lawrence, like Robert Oppenheimer. If a woman wants into the game, she must play by the rules or get out.

Tales of Two Sisters

Of course, some women survive apprenticeship in the bubble of bubble boy science. We will take one example, a scientist of my acquaintance who rejects fervently the idea of feminism and is horrified by talk of "women's science." She insists that her science, her life as a scientist, is exactly the same as that of her male colleagues. The details of her life tell another story.

This woman has three children; the youngest, three years old, is still nursing. The scientist runs her lab, does her experiments, oversees her graduate students—lives within the bubble and then she runs the household, makes the arrangements for people to stay with the children when she and her husband are away, figures out the children's transportation to and from school, organizes the registration for school, for extended care, for vacation camp, makes sure that her children have their cough medicine and aspirin when they have a cold, arranges that they see the doctor for their regular checkups, get to the dentist twice a year, to their friends' birthday parties equipped with a wrapped birthday present, and every day is responsible for seeing that they are fed, clothed, and provided with the lunch boxes, T-shirts, and toys necessary for a child to be an ordinary child among others. There are male persons who may also provide for home and children outside the bubble as this woman does. But overwhelmingly those persons who live in this way are women and not men. Those women who do not live this way—who "choose" to live inside the bubble—are censured by their families and often distrusted by their colleagues; while the men who "choose" to live outside the bubble in this way are known to be sabotaging their careers and ruining their lives. They will never get to be one of the key people in the field.

I do know one woman, let's call her Pani, who made it through to become a key person in the field—a very small and abstruse corner of theoretical mathematics. Her story, her life in science, her encounter with the bubble is not a simple tale.

I first learned the details of Pani's story from her when she was my roommate. Pani was born in south India someplace. She was always fairly good in maths, as she put it. First in her year out of 200,000 university math students. Having finished her first degree, she had married a young man whom her parents had picked out for her. At about this time she was invited to apply for a place at India's famed Tata Institute for Advanced Studies, where the most eminent mathematicians in the country work. Each year three or four very promising students are accepted. She applied and heard nothing. One day, while tucking her husband's lunch into his briefcase, she found a letter from the Tata Institute addressed to her. The letter, postmarked weeks earlier, had been opened. In it was the information that she had been accepted for an interview to take place that very next week. Her husband told her when she confronted him with the letter that he and his mother had decided that since they had decided that they would not allow her to accept a position if one were to be offered, they had decided not to tell her of the interview. She made a *tomasha*. She went to the interview and was accepted. It was not what you answered—they were simple problems—it was how you went about answering them. Her mother-in-law decided she could go on when she learned that Pani would be paid a handsome stipend for attending the institute. There were stipulations: she had to live at home, she had to turn over her whole stipend to her husband's family, had to do her share of the housework before she left for the institute. This involved getting up at 3:30 every morning, cooking three complete meals for a household of fourteen people, taking a bus for two hours and arriving at the institute at nine o'clock ready to start work with the other students—all male and all unmarried—who had risen much later, enjoyed a leisurely breakfast prepared by someone else, and then walked over to the institute from their nearby rooms. Pani loved the institute. She loved the opportunity to disappear into the bubble for hours on end—quiet, abstract, total. But outside those hours she was entirely of the world. In the course of her years of study, Pani bore two children. Both were taken away from her by her mother-in-law and given to her husband's childless brother and his wife to raise. Despite her attempts, she was never able to get them back. There was always the threat that she would have to give up mathematics. She could not do that.

Pani has since left India, come to the United States, become a professor of mathematics and a key person in the field. She sits quietly from seven in the morning until six at night in her bubble doing mathematics. She travels widely and serves on the panels and editorial boards. Her children do not talk to her.

She abandoned them, they say. Her mathematics was more important to her than they were. They do not forgive her. She works in her bubble from seven to six. She has a broken heart.

Physics beyond the Bubble

The Institute of Physics in Rome under Fermi was not a world of bubble boys. Thirty years later, the Institute of Physics in Rome was not filled with bubble boys either. Tom Toffoli, a highly innovative researcher on the physics of computing now at M.I.T.'s Laboratory for Computer Science was a student in high energy physics in Rome in the late 1960s and early 1970s. I asked him about the work environment there. Did the guys serious about physics work all night? Did they focus their energies and conversation around the project? Did they define themselves in terms of their commitment to science or willingness to abandon other interests for the good of the work going on in the lab? He knew what I was talking about.

> Well, sometimes we spent the night at Frascati when we were doing our experiments. We were parasites of the beam and we had to be there so we could get on. It was other people's money, so we didn't have much choice. But otherwise, work was important to me, but not existentially important. There was the rest of my life, friends and family. You didn't see people spending day and night on their science though there were some who did. Maybe one out of twenty. The other people resented them. It's unfair. You have family, your friends. Why should somebody set standards that are inhuman so that you become inhuman?

The Tom I know is definitely a bubble boy. He agrees he has that tendency. "I really learned to work like that at graduate school at the University of Michigan. Then I worked all night and all day." He does that at his lab at M.I.T., too, but always finds the time for his kids, his brothers and sisters and friends like me who come into town. When you break into the bubble, you find Tom happy but apologetic. He works like a bubble boy, but he does not think it is right.

Lives in Corporate Research

My father has always had a compiex reaction to the bubble boys. "They really hustle. The boys really hustle," he would say with a mixture of wonder and incredulity. In Rome he had spent more and more time while at the university involved with the anti-Fascists. His work in the lab was never the center of his experience. His brother was different. He sympathized with those working against the government, but he concentrated his energies more fully on his studies, on the experiments, on life in the lab. In later life, he hustled: working late into the night, every weekend, getting up early to drive over to the lab and check things out. Life was for years a series of emergencies when the laser wouldn't lase, when the power supply did not deliver the right power, when the manipulators would come back from the shop with the wrong tolerances and there would be a crucial demonstration the next day to a key man in the field. Often at those times, Gus, the glassblower, would appear and, as in a fairy tale, create a piece of equipment and save the day.

Gus, who is a Hungarian, like many of the best blowers of scientific glass, followed my uncle from one lab to another, from one company to another, from mass spectroscopy into laser research and from laser research into medical laser design and production. Gus and Bob have been there with the family for my entire life. Bob, my uncle's engineer and right-hand man, together with Kaye, his wife, have always been almost members of the family. The *almost* is important, of course. "A good engineer," "a fine detail man," "the best there is when it comes to putting something together," Bob is praised in many ways. But not for his ideas. Never for his ideas. That was the province of the brothers who always consulted each other, always talked through every idea with one another before making an important change, an important design move, an important career decision. Bob always came along, implemented the decisions and was buffeted about by the winds of research management which might one day decide to end a project they had pushed heavily for many years. Bob, instead of moving on to another project, another manager, always followed along my uncle's path. Gus would accept all the changes, all the shifts in policy, but somehow he would always show up at the next company as if nothing had ever happened.

Corporate research is a special world. My father and uncle spent their whole productive careers in such environments. I learned firsthand about the pleasures and terrors of the corporate research environment when I worked for several years in Boston as a scientist for an engineering consulting firm. At its best it is a comfortable life with good resources and great flexibility in regulating your personal working environment. As a researcher, no one notices when you come in, when you go home. A few days extra time away tacked onto a conference, trade show, or business trip will always be overlooked. The question is always one of your perceived usefulness. It always surprised me that my father remained working without conflict for one company for over thirty-five years despite an erratic working style; my uncle, much more dependable, predictable and observably hard working, had a much rougher time.

"Sorry. No idea. Haven't seen him around. Never know where he is. Never tells us his plans. He mumbled something when he left but I didn't catch it. He'll show up at some point."

This is what his coworkers and bosses would say of my father. And eventually, he always did reappear after an hour, an afternoon, a day, a week. Since he worked in the research department of an optical firm, his work involved the application of optical technologies and properties to whatever he found most interesting and was able to persuade research management held promise of reaching the marketplace as a commercial device. Although he began his work designing equipment for measuring eyesight, over the years he worked on many medical instruments, largely for monitoring blood gases. He worked closely with doctors and surgeons in Boston and New York and became knowledgeable about many aspects of physiology and medical practice. He enjoyed his work immensely and took particular pleasure in finding a new twist. Something which would just do things differently and, hopefully, better. Yankee ingenuity. When he needed inspiration he went to the five-and-ten. Eventually he would buy some glue, a few pieces of rubber and some screws, and return to the lab.

My father is often delighted. Snowfall in April, getting stuck in sand on an obscure back road and waiting to see what would happen (something always turns up), getting lost and ending up at an auction of machine tools, thinking up exactly the right Christmas present and wrapping it up in newspaper to put under the tree. Science held the promise of delight for him. Understanding a new problem, coming up with the right set of equations to capture a previously ill-understood phenomenon, building a piece of equipment which really could do the trick. For my uncle, science was always more serious. Problems were set which kept one on the knife edge. Anxiety and the possibility of failure were always there. Always as possible as a Rome winter day without fuel for the stove to ward off the chill.

The family had been poor in Italy. My grandmother repaired oriental rugs to keep the family fed and the boys in school. My father remembers walking all over Rome carrying the rugs from customers to his home for repair and back to their owners' villas when the job was done. He enjoyed the walks and enjoyed the rugs. Throughout his life, oriental rugs have remained an active passion. The family was poor in Italy. My aunts were sent back to Budapest to live as poor relations so that the boys could complete their educations. My father showed little academic promise and spent his days cutting school and exploring the backstreets of Rome with the kids in the neighborhood. Eventually he was sent to a trade school and became a machinist. He entered university late by special examination. My aunt was recalled from Budapest to tutor him in Latin for the test and after several miserable months he passed and was admitted. His brother was a good boy, a good student. He went through school and entered the university at the right time. The two brothers studied together and many, many years later they worked in the same research laboratory for a while. My uncle had a big office, a private secretary, a team of engineers and technicians working for him in several labs. My father had a small office, a small laboratory, and one technician, Dave, a man just my age who began to work in my father's lab when he was still in high school. When the laboratory was moved to upstate New York, Dave and my father were tapped to go. My father had had enough. He was over sixty. He persuaded the bosses to fire him. The severance pay, unemployment, and pension were enough to begin a new adventure. He and Dave went into business on a shoestring, setting up shop to manufacture fiber optic catheters for cardiac diagnosis and angioplasty. Their turnkey operation was housed in the abandoned cellar of Dave's childhood home, where his father had once operated a tavern for the Polish workers in town. Although the business was always wildly undercapitalized, they never owed money to anyone. Much of their equipment was scrounged, bought at auction or at flea markets, or somehow assembled out of nothing. They had few employees and each batch of catheters was made by hand. My father became a small business man, a capitalist in the sixth decade of his life. They may still be in business. It is hard to tell with my father. He never tells anyone where he's going or what his plans are.

About the time my father went into business, my uncle was named president of a start-up company making surgical lasers. Bob, his engineer, became chief engineer. They set up in one room at first, building the machines by hand and then flying to Atlanta to work with a prominent ear, nose, and throat specialist. The talk at the dinner table shifted to discussions of contracts and equity. Cancer, never a subject I had heard mentioned, became a regular topic over coffee and toasted thick-crusted bread prepared at the table by my father who happily burned the bread almost every time. Eventually the laser proved very effective against throat cancers and the names of new doctors and new hospitals in different cities dominated the talk. Lasers, it seemed, held promise to be effective against vaginal and ovarian cancers as well. I found it odd to hear these men who had had the greatest difficulty discussing the need for bathrooms with the architects while the house in the woods was being built talking about these matters over drinks near the fireplace. Martinis for my aunt and uncle, and for my father? "I've fixed you a vermouth, Chico."

Memories of Unlived Lives

One can only speculate, of course, about a future that never became the past. If my father had entered the bubble, what would have changed? What would have happened if the brothers, if even one brother had worked on the Manhattan Project? Could there have been a career in Big Science? Would the FBI have been so disinterested in our lives and activities? Would I have grown up reading and swimming and swimming and reading by the side of the lake with clear, clear water?

There are many possible futures which did not occur. Radiation sickness, involuntary sterilization, birth defects in children conceived after exposure to radiation, very high cancer rates, and accidental death from radioactive poisoning were all attested to in statistically significant numbers among the physicists and other workers at Los Alamos, Oak Ridge, Trinity, Bikini, Berkeley labs, and the other sites of the Manhattan Project. In one future, my father or his brother do not survive the war but die of science, or my father's child is born afflicted, born dead, or not born at all.

In a more likely postwar future, my father and his brother, my father or his brother prosper in Big Physics. The important connections, the important organizations put in place during the war became available resources for them as physicists who had done well at project labs while others outside the bubble would not get the opportunity to do the big projects in high-energy physics which would require major expenditures for equipment and research assistants. The conduits from the project bubbles of the atomic bomb to the big science power bubbles of the postwar world were direct and they can tap into them. The funding agencies were staffed with their colleagues, researchers who had been through the project and who would share their understanding of the technologies and problem areas deemed to be at the cutting edge. The latter were those which had been developed during highly classified bomb development—technologies and problems being pursued after the war, often at the very highly classified labs where the project was carried out. And security after the war was, if anything, even tighter than security during the war.

Security Files Live Long Lives

Throughout the years of my childhood and adolescence, I attended the same private school in Stamford, Connecticut, where I had begun in kindergarten when my father was still working at American Optical in Norwalk. My father never lost his job, the FBI never intruded. My father and his brother built their careers, their lives both inside and outside of science with relatively little interference. While my uncle settled into his life, his work, his world, anxious about his career, his project, all those things which men must find disturbing, my father, who has lived in the same small apartment overlooking the lake for forty years, never shed the air of someone about to pack a small valise and never return.

Security agencies have long memories. In the fifties and sixties, the laboratory at American Optical did some classified work; parts of the Sidewinder missile system were designed and produced there. Security officers were around. One day my father was stopped in the hall. "How is Alex?" he was asked. "And Aunt Mousie, how is she doing?" They were checking up on the first husband of my father's cousin, Eva Zeisel, daughter of his father's sister and very briefly the wife of Alex W., a Communist hero of the Warsaw ghetto and architect of the "trucks for Jews" schemes which exchanged Hungarian Jews for Russian Army trucks. Eichmann was the negotiator for the Germans. My father had met Alex a few times twenty-five years before. He had heard that he'd retired and owned a number of apartment houses in Romania. As for his aunt, he had not spoken with her for many years. He mumbled something in reply. The agent did not quite understand what he said. He did not repeat what he had said. American Optical soon moved their secure operation to another building.

Brothers in Big Physics: Fratricide, Betrayal, Disgrace

Security files live long lives. Security agencies have long memories. Consider the matter of J. Robert and Frank Oppenheimer, brothers by blood, and Frank Oppenheimer and Ernest Lawrence, "both passionate experimenters, disdainful of theory" (Davis 1986: 115), brothers in science:

> In June of 1954, Robert Oppenheimer, the father of the atomic bomb, the man most responsible for the development of the American nuclear capability, was stripped of his security clearance as a result of hearings before the Atomic Energy Commission investigating his connections with "Communists, Communist functionaries and Communists who did engage in espionage." One voice against him was that of his Berkeley and Manhattan Project colleague, Ernest Lawrence, who "left Berkeley to appear at the hearing": "He was keyed up to testify against Oppenheimer and he meant to do it, but he was really ill. He was bleeding badly. Even without Lawrence's personal testimony the judgment went against Oppenheimer: The record contains no direct evidence that Dr. Oppenheimer gave secrets to a foreign nation or that he is disloyal to the United States. However, the record does contain substantial evidence of Dr. Oppenheimer's association with Communists, Communist functionaries, and Communists who did engage in espionage. (Davis 1986: 348–49)

One of the Communists J. Robert Oppenheimer was guilty of associating with was his brother, Frank. Robert Oppenheimer knew that Frank was a Communist. When he had discussed with Lawrence the possibility of the experimentalist hiring his brother to work on the cyclotron, Oppenheimer had admitted that "Frank's been in a lot of left-wing activity." So Lawrence, too, certainly knew of Frank's political leanings when he agreed to take him on in the radiation laboratory in 1940 after he had been fired from his job at Stanford "for radical talk." And though "Lawrence knew Frank well and was more like him than either was like Oppenheimer, when the issue of security clearance arose modestly for [Robert] Oppenheimer in the spring of 1947,

Lawrence fired [Frank] out of hand with a quick brutality that still alters the voices of the physicists when they talk of it. As soon as Frank had any difficulty, Lawrence forbade him to come to the lab and visit. Do you have any idea how quite unusual a thing that is among scientists?'' asks Bethe, the noted atomic physicist (David 1986: 115, 273).

Lawrence betrayed Frank Oppenheimer and was determined to destroy his brother as well. While Lawrence continued to be honored and respected during his lifetime, the men he vilified lived long enough to see him almost forgotten. Ernest O. Lawrence, whose name for many is little more than the first name of the Lawrence-Livermore National Laboratory, died in August 1958, his body destroyed by constant hemorrhaging from ulcerative colitis.

As for Robert Oppenheimer, "on February 18, 1957, he died an official security risk'' in a state of official disgrace (Davis 1986: 354). Frank, Robert Oppenheimer's brother, went on to become the director of the Exploratorium in San Francisco after years of internal exile in Colorado. He never again returned to physics.

The Refugee Brothers: The Emigré and the Immigrant

My uncle immigrated to the United States in 1941. He adjusted, settled in, made a home for himself. My father has arrived only very recently.

My father first went back to Europe for a visit around 1960. There was no big emergency. A scientific conference. He would give a paper at an international conference. Visa, passports, tickets, borders emptied of terror. A betrayal almost of the others, all the others. He returned from his trip elated at the journey, disappointed with the results. His friends from the university, the young anti-Fascists by now grown older, remained faithful Party liners. "*Nil dia sine linea,*'' he would say, quoting the motto of the founder of the Boy Scouts—"never a day without a line.'' "*Sine linea,*'' he would repeat, always amused at the pun, always disappointed in his friends.

My father, an emigré since earliest childhood, has finally immigrated. His earliest memory is leaving Hungary, taking the Orient Express west to Vienna. He has lived all of his life in exile, an exile with no dream of return. Familiar as every apartment interior, every conversation on every street corner was to him when he finally returned to the city of his birth, Budapest was not home, nor was Italy. Much as he loves Italy, he was never Italian. "You speak very good Italian,'' people still say. "Very very good Italian. But you are not from here.'' Not from anywhere until very recently. I believe now, after over fifty years in this country, my father has come home, home to America. It took the end of the Soviet Union, the death of the dream of socialism as embodied throughout most of this century by Russia and her dependents to bring him home. Seven years before the end of the century, seventy-four years in the diaspora, my father has come home. It's a quiet homecoming, I think, with some melancholy and loss, but his brother is here, his daughter is here, his work, his friends, a first grandchild. "It's over," he says from time to time. "It's gone. It's really over." He's come home. The war has ended. He is at work on a new invention.

Notes

1. Cornucopions, explained Dr. Banks, would look to the outside world as though they were a simple point in space, just one Planck constant in diameter. Inside, however, they would contain infinitely long throats that one could think of as an endless horn or cornucopia.

That gives you a place to store lots of information, said Dr. Banks. "Everything that fell into the black hole might leave behind traces. Those traces might run away down that horn at the speed of light and you'd never be able to find out about them from outside, but they wouldn't disappear." Quoted Taube, 1993.

2. And perhaps on others as well. For example, there is no explanation of the decision to reissue in 1986, no preface or introductory material beyond the author's extensive list of acknowledgments from 1968. I can only speculate that someone must have read the book and fallen as much under its spell as I did, as did my father, my husband, my best friend, and decided to reissue it. The book has an odd effect on people. I borrowed a copy originally from my friend Ellen Zweig, for whom one sentence became the basis of a major intermedia work in text, video, still photography, and performance. I have based this memoir of my father and my uncle on this book. I do not think I would have ever been motivated to descend into the horn of space-time of memory otherwise. I also ordered the book from a bookseller in multiple copies, kept one and gave a copy each to my father and to two of my closest friends. This is the sort of thing I often think of doing, but never in fact do.

3. "'I suppose nobody in the world ever completely mastered both the theoretical and the experimental fields except the Italian Fermi'" (Davis 1986: 79).

4. During the week after Pearl Harbor, midyear graduating seniors took their examinations. One of the blithest was a young physics major, Roger Hildebrand: "'When I went into the hall [after finishing the exam] Lawrence grabbed me by the shoulder,'" he says. "'It was a most aggressive physical form of recruiting.'" (Davis 1986: 121).

5. In 1943 at my corporation laboratories I was working on a successful research project unrelated to the bomb when . . . the vice-president in charge of research came and asked me how I'd like to spend some time at Berkeley with Lawrence.

"Fine," I said, "in two or three months the project here will be wound up so I'll be ready to leave."

He looked disturbed. "You'll need to leave in two or three days," he said. "Lawrence says he's fed up with the contingent of physicists we have out there now. He wants new blood."

My plane landed me in California air-sick on a Sunday morning. Some of our disapproved contingent stood waiting at the runway with a worried subdued look on their faces. . . . Within two hours I had a mop in my hand and a bucket at my feet and was swabbing out the decks of the calutrons. So far as I could tell, this work was to be permanent. A day later I located some rubber boots. A fortnight later the vice-president flew out to see me.

"This is a great job you are doing for us here,'' he said. "Lawrence likes it. The president wants you to know we appreciate it. . . . Look, I've got good news for you. Remember that electronics patent you turned over to us last year? The corporation is sending you a check for it. A great big check,'' he said. "I know how you feel.''

"Oh, you don't,'' I said. "If you did, I know what you'd do. Lawrence wants Ph.Ds. The corporation has plenty of good janitors. Why don't you give half a dozen of them honorary Ph.Ds and send them out here? Lawrence would never know the difference'' (Davis 1986: 198–99).

6. At the beginning of 1942, Lawrence and, after him, Compton were the prime movers. . . . The question was where to locate the plutonium reactor program. . . . Lawrence wanted it at Berkeley; Compton wanted it at Chicago. . . . Compton steeled his nerves and told Lawrence he was moving the project to Chicago. . . . [After] Lawrence went back to his calutron, Compton . . . began organizing the . . . Metallurgical Laboratory and called Oppenheimer to a conference defining its two programs, plutonium and bomb design. The Metallurgical Laboratory took shape with quivering anxiety about the German nuclear threat. . . . The laboratory senior staff was made up largely of refugees with nerves on edge from watching the collapse of nations. Compton did less to calm them than did one of their own number, Fermi, who declared that the Germans could not fight a war and build the bomb at the same time. Even so, Fermi had given thought to what country he should escape to next after this one fell (Davis 1986: 125).

7. "It's impossible. They can't really be that naïve. But in a country where everybody drinks whiskey, they voted in you know what. You've heard me tell this story a million times.''

8. "Maybe we should use paper plates and cook out on the fire?''

"It's a little lowbrow.''

"Maybe. But it's a lot easier than carrying all this stuff up and down the steps three times a day.''

"Perhaps, but a little lowbrow.''

References

Davis, Nuel Pharr. 1968. *Lawrence and Oppenheimer.* New York: Simon and Schuster. Repr. 1986. New York: Da Capo Press. Page refs. are to 1986 edition.

Polányi, Karl. 1944. *The Great Transformation.* New York, Toronto: Farrar and Rinehart, Inc.

Polányi, Michael. 1958. *Personal Knowledge.* Chicago: University of Chicago Press.

Taube, G. A. 1993. "The Case of the Disappearing Black Holes.'' *New York Times,* 30 Mar., p. B8.

Fig. 1. Cartoon by Levon Abrahamian depicting Giordorno Bruno. (The cartoons scattered throughout volume 1 of Late Editions are by Abrahamian as well.)

MICHAEL M. J. FISCHER

2

EYE(I)ING THE SCIENCES AND THEIR SIGNIFIERS (LANGUAGE, TROPES, AUTOBIOGRAPHERS): INTERVIEWING FOR A CULTURAL STUDIES OF SCIENCE AND TECHNOLOGY

> How does any information-processing system answer questions about what it is doing? It sets up within itself a model of itself and then examines that model.
>
> —Martin Eger

> Like every intellectual pursuit, physics has both a written and an oral tradition.
>
> —Victor Weisskopf

The cartoon by Armenian anthropologist Levon Abrahamian (fig. 1) is of Giordorno Bruno as a chess bishop being burned at the stake (religious wars of early modern Europe; ethnic strife after the collapse of the Soviet Union). The balloon above his head shows his theoretical model of the world: it is round, not flat (ethnic strife is archaic, irrational, based on nonscientific ideas). But the bishop lives on the chess board: the world may be round, but not in this case.

Anthropology often works its insights through double plays: putting comparative cultural and social empirical details (places, institutions, generations, cultures) in play against universalizing philosophical, theoretical, or moral schemas as critiques of the latter's applicabilities and limitations; putting the rich experiential and emotional complexities of individual lives into play within and against social patterns so as to illuminate both the power and weakness of the latter; putting the forms of narration and explanation that individuals and groups invoke for their actions and motivations in play against alternative narratives and explanations. The fin de siècle of the twentieth century—still reeling from the aftermaths of the second European "World War" and its "cold war" sequel; and simultaneously engaged in the euphoria of the third industrial revolution, that is, of the silicon chip and molecular biology—seems an appropriate moment to reevaluate the promises and challenges of that sort of anthropology. A key arena for such reevaluation is the emergent anthropology of science and technology.

Science and technological innovation have been central to the construction of modernity over the past three centuries. It is odd that the quintessentially modernist discipline of anthropology should have found so relatively little to say beyond the philosophy of science contributions in the 1930s about the ways in which systems of investigation protect themselves from falsification, the contributions of the 1950s and 1960s about the ways in which classificatory and analogical logics operate, a general interest in the sociology of knowledge, and a serious ethnographic interest in traditional healing systems and their associated botanical, psychiatric, and medical knowledges. But this may be changing, as a new generation of anthropologists begins studies of scientific laboratories, biotechnology corporations, science cities, engineering schools, public debates over reproductive and genetic technologies, ecological issues, public epidemiology, nuclear and chemical waste, and the increasing need for public counterauthorities to government funding and control of information about all these issues.

Among the variety of sources for anthropologists to draw upon in engaging science and technology issues are first person accounts by scientists about their own fields, as ways of linking narratives of science to lives of scientists, as ways of locating the scientific imaginary in social communities, cultural anxiety structures, and moral tradition-reworking speculations.[1]

PreTexts: PostModern Times and MetaPhysical Lives

> Who will reassemble these pieces, continued Plato, and restore to us the robe of Socrates?
>
> —Denis Diderot, *Les Bijous indiscrets*

At the beginning of Modern Times—in, as it were, a premonition for a Charlie Chaplin script—Denis Diderot satirized the new empirical sciences in an allegorical tale about an inquisitive Sultan Mangogul whose dreams and diverse forms of experiment seemed only to deconstruct the foundations of deductive, rationalist philosophy. This empirical science, it seemed, could not provide any new eternal synthesis to replace systematic philosophy, only tentative, provisional constructions requiring constant further testing and securing in the physical properties of nature. The constant further testing was allegorized as a continuation of *The Thousand and One Nights,* Shahrazad's multiple substitutions and diversions that keep her tale and life viable and even redemptive to the society around her. The structure of Sultan Mangogul's scopic and sonic perception is voyeuristic, panoptic, third-person listening. It is an obsessive quest of desire, which Diderot literalizes through a tale of a magic ring that allows the Sultan to stage and hear the desires of his harem women, ambiguously playing upon whether his presence otherwise would or would not be any more distorting than this magical instrument of inquiry: sexual desire and *libido sciendi* [the desire to know], as Suzanne Pucci puts it (1990: 157), are allegorizations of one another. The Sultan (the subject, the knower, the wielder of the instruments of power and surveillance, the self) in these tales (both Diderot's and *The Thousand and One Nights*), Pucci points out, is "a perceptual subject [who] is reacted upon by the properties of the physical world and in conjunction with his own physiological processes" (his dreams; 159); he is, in relation to his experiments, a third party (Shahrazad's tales are told to her sister Dinarzad while the sultan Shahriyar listens) who explores the worlds, unknowable directly, through sounding and projecting instruments:

> To the aged philosophers described as garbed in the tattered pieces of Socrates' sacred mantle Mangogul's dream juxtaposes an allegorical figure, an unintimidating infant who nevertheless as he approaches the temple of philosophy grows to giant proportions, becoming the colossus of 'Experiment/Experience.' 'In the progress of his subsequent growth, he appeared to me in a hundred different forms; I saw him direct a long telescope toward the sky, estimate with the aid of a pendulum the fall of bodies, ascertain by a tube filled with mercury the weight of air; and with a prism in his hand, decompose light.' (Diderot, *Bijoux indiscrets,* cited in Pucci 1990: 158.)[2]

If there is a parallel between the eighteenth-century sensibilities—a feeling of transition from stable knowledge no longer viable as a worldly guide into an insecure world knowable only through multiple probes and perspectives—and those of the twentieth century, it does not help to collapse the differences. The late twentieth century is different from the eighteenth century, the romantic period, and the nineteenth century, in part, through the increasing cultural diversity and pluralized perspectives that are no longer one-sided (orientalizing) fantasies or projections (about a Sultan Mangogul) but internal realities on increasingly large demographic scales (migrations of German and Italian scientists to the U.S. and Russia in the 1930s and 1940s; higher education of Indian scientists in the U.S. or Britain), with effects that play themselves out not only in refined cosmopolitanisms of liberal desire, but also in devastating conflicts generated by political reorganizations, disruption through development schemes and wars, economic competition, and "accidents" and epidemiological crises such as Chernobyl, Bhopal, Rocky Flats (Colorado), Woburn (Massachusetts), or Love Canal (New York) have come to represent. The gendering of metaphors about nature and science also no longer function as reflexes of a nigh equal struggle against natural magic and occult forces figured as female.[3] Feminist critiques of science writing and feminist science fiction writing have exposed the contradictions in the tradition of reducing science to pure reason as a metaphysical desire for escaping the body refigured today as "downloading" information in the brain or in the "memory banks" of culture into silicon-based or other hardware. Feminist writing has rather explored how alternative ways of encoding information entails alternative implications for embodiment. Or as Don Ihde puts the contradiction of escaping the body: users want "what the technology gives, but . . . not . . . the limits, the transformations that a technologically extended body implies."[4]

One way to probe these changes of the contemporary world is to use autobiographical and life-history frames to examine the ways the social processes of identity, moral understanding, and scientific worldview are being reproduced and transformed. Autobiographical frames themselves are not given genres: if we wish to use them to extend social theory, to understand the changes of the contemporary world (rather than reaffirming the pieties of the past), we need to be attentive to the variant, culturally diverse formats they may inhabit. Among these are not only autobiographically figured texts (memoirs, the "third genre" of scientific literature or "semipopular" accounts by scientists of conceptual breakthroughs meant to heal the two-cultures gap, biographical texts done by scientists projecting their own dilemmas onto those of an exemplary, demonized, or conflicted other),[5] but also more discursively new languages of science that claim to move beyond modernist sciences—proposed by both scientists, in efforts to seriously bridge the "two cultures" gap, and by humanists and social scientists who would engage in exchange relations with "postmodernist sciences."

ConTexts: Bomb, ProGram, Cybernetics, Chip, Gene, and Other Zones of Cultural Implosion

> Gene, seed, fetus, chip, brain, bomb . . . I'm not interested in boundary crossings, but rather in zones of implosion of nature into culture, of material and trope, sites of transformation that generate narratives of possible worlds and new agencies. Storytelling is not opinion, not all stories are equally good. Gene, seed, fetus, chip, brain, bomb are densities at zones of implosion.
>
> —Donna Haraway

> Wasp and orchid, as heterogeneous elements form a rhizome . . . Something else entirely is going on . . . a capture of code, surplus value of code . . . a virus can . . . move into the cells of an entirely different species but not without bringing with it "genetic information" from the first host . . . our viruses cause us to form a rhizome with other animals. As Francois Jacob says . . .
>
> —Gilles Deleuze and Felix Guattari

> The contemporary biologist speaks of writing and *pro-gram* in relation to the most elemental processes of information within the living cell . . . the entire field covered by the cybernetic *program* will be the field of writing. If the theory of cybernetics is by itself to oust all metaphysical content—which until recently served to separate the machine from man, it must conserve the notion of writing, trace, gramme [written mark], or grapheme, until its own historico-metaphysical character is also exposed.
>
> —Jacques Derrida

As anthropologists begin seriously to undertake studies of the role of the sciences and contemporary technologies, there is at the same time an interesting cross-fertilization between new ideas in the lead sciences of the day (molecular biology, computer sciences), and new formulations in the humanities and social sciences.[6] The day-long, large auditorium filled sessions on "Cyborg Anthropology" at the American Anthropological Association meetings in 1992 and 1993 may be symptomatic.[7] Elicited by fast-paced developments in the biological and information technosciences which promise to radically and rhizomatically reshape the moral, political, and psychological spaces in which we live, and also playfully invoking the popular-culture speculations of science fiction, this new anthropology links biological, social, linguistic, and cultural anthropologies in new ways, rewiring the humanities and social sciences. The range of papers at these sessions included efforts to characterize the changing social organization of modernity or of the late twentieth century; ethnographically situated and cross-culturally juxtaposed accounts of technological sites of mediation (between law and medicine, psychiatry and computer technology; ultrasound in Greece and in New York; law and medicine in Bhopal and in Los Angeles); studies of cohorts (generational, national, gender) of scientists and engineers; and probings of the relations between popular culture and various spheres of authoritative science, including the roles of science in a democratic society.

Several features seem to characterize this new cultural anthropology of science: an interest in the diversity and variants of scientific practices; a mapping of the ways this diversity and sub-versions can rechannel the scientific imaginary to make new kinds of connections across professional boundaries and into lay worlds; a focus on the materiality of scientific knowledge in terms of instruments, skills, and human networks; an acknowledgment of debates over the social implications of science and technology led by scientists and engineers rather than an othering of these actors; and an attention to the modes of information production and diffusion of the new sciences and technologies, especially in the visual and electronic media (see Fischer 1991; Rouse 1993; Haraway 1992, 1993).

Less often noted are the parallels with traditions of scientifically informed philosophical speculations such as those of the history of science tradition to which Michel Foucault belongs (Gaston Bachelard, Georges Canguilhem, Fernand Hallyn) or the post-structuralist writings of Serres, Derrida, and Deleuze and Guattari. Deleuze and Guattari, for instance, foreground the work of Francois Jacob in the opening pages of *A Thousand Plateaus,* and draw a series of their philosophical concepts from microbiology. It is important to their project (and, I think, to the new anthropology) that this new conceptual vocabulary has at least dual (if not multiple) genealogies, only one of which is in the new biology, the other coming as often from medieval and baroque philosophy. Thus *fold* and *expression* both refer to genetics—the folding together of the double helix structure of DNA—and also to Spinoza. The point of this doubling of genealogical references initially is to test whether reconfigurations of conceptual boundaries can be productive and useful. Donna Haraway's use of the term *lively languages* for the metaphorical structures that science draws upon in both its workaday and its promotional faces is similar, and she, too, foregrounds this "liveliness" and promiscuity in order to intervene, to draw attention to alternative routings of thought to which the concepts and the social implications of science might be put. A parallel strategy is pursued by Michel Serres in his play upon *parasite* as simultaneously a term of biology, human relations, and information theory. This particular triplet has been put to good use in Bruno Latour's study of how Pasteur's success was a function of his synergistically becoming allied with popular hygenic movements, and inserting the bacteriological laboratory as a site for making visible a level of interactions that redefine human social bonds below normal human awareness in the nineteenth century:

> The tribunal punishes a criminal with one year's punishment, but he pays for his brief spell in the cell with his life. When a man follows a woman to her hotel, he thinks he is settling the transaction with a coin and ends his days in an asylum. . . . When you bring a woman to birth, you think you are in the presence of three agents—the midwife, the baby, and the mother—but a fourth takes advantage of the situation to pass from your hands to the woman's wounds. . . . You organize a demonstration of Eskimos in the museum. They go out to meet the public, but they also meet cholera and die. (1988:32–33)

In more abstract terms formulated as a challenge to traditional social theory, "the exact sciences elude social analysis not because they are distant or separated from society, but because they revolutionize the very conception of society and what it comprises" (38). He concludes, quoting Serres, that one eliminates a parasite only by introducing another more powerful one (39).

The notions of parasite, leverage, power ratios, are not so different from Foucault's notions of micropolitics, Derrida's notions of displacement, deconstruction by iteration with a difference, Deleuze's molecular rhizomic or viral antigenealogies, and perhaps, in anthropology, Michael Taussig's nervous system's mimetic (also a repetition with a difference, a slippage, a shamanic change). These notions are of interest anthropologically insofar as they are not mere philosophical or literary tactics, but seek their elucidation in levels of ordinary scientific knowledge on a microlevel that mold social relations, yet operate below the purchase of traditional social and cultural theory concepts. Deleuze and Guattari differentiate between molar and molecular approaches, the latter indicating a stress on multiplicities, the former on holism and unities; they advocate the molecular in a provocative way, but by doing so problematize the faith of molecular biology in severe reductionism as well as the faith of social theory in either simple individualism (utilitarian or voluntarist theories) or simple social constructivism (Durkheimian or Wittgensteinian theories). It has been noted by a number of critics of molecular biology that its reductionism effaces the level of the body, of the organism, and leads to both moral dilemmas and ecological ones. This is a point made by anthropologist Paul Rabinow's studies of recombinant DNA technologies. And it begins to raise serious issues about the locus for moral agency.

Sherry Turkle's studies of multiple selves focuses these issues on a series of levels. On the one side, "second selves" created by students through computers can be therapeutic role-playing as well as cognitive play with new ways of working out a sense of self and personhood. On the other side, post-traumatic stress dissociative disorders (including multiple personality disorders) are painful dysfunctions, and seem to be increasing as the "postmodern" disorder of the nineties. Is this a function of changing diagnostic questions, or with virtual worlds' role-playing, and with middlebrow books and media publicity about multiple personality disorders, are these dysfunctions more easily expressible and articulatable as cultural forms? Further, repression of child abuse as a cause in these dysfunctions, as it comes into public view, generates disseminating anxieties and probings with complicated epistemological conundra and ethics of determining what is real, imagined, therapeutic, or iatrogenic.

Raised to yet another level: is there a parallel between the assymetry between disembodied and embodied selves in computer play versus multiple personality disorder, and, on the other hand, the use of smart weapons in technological warfare such as the Gulf War against Iraq and the kind of post-Holocaust morality-theology called for by such thinkers as Emmanuel Levinas? What is the status of embodied conscious human beings: need the choice be merely a dualistic one of the inadequacy of agency of egos in large technological systems (the transgressive play of those who toy with the idea of "actants" as "posthuman subjectivities"[8]) versus the responsibility that cannot be lodged in any but such egos? Or are there multiple goals of responsibility and agency that are both differently embodied and potentially conflict with one another? The technological information society is not just one of simulacra, and not all simulacra are morally equivalent. While there may be a connection between computer games and computer-assisted missile guidance technologies, their users and usages are still subject to moral evaluation. While it is true that humans are now enmeshed in transgenetic, genetically engineered, and chimerically constructed biological modifications to prosthetically engineered external capacities, the questions of humanism remain and directly affect the rules of the polities, ecologies, cultural categories, and public cultures we wish to create for ourselves.

As Byron Good reminds us in his critique of biomedical models of illness which easily incorporate various kinds of cyborgian technologies, "all medicine joins rational and deeply irrational elements, combining an attention to the material body with a concern for the moral dimensions of sickness and suffering. . . . It is thus both the privilege and the obligation of medical anthropology to bring renewed attention to human experience, to suffering, to meaning and interpretation, to the role of narratives and historicity, as well as to the role of social formations and institutions" (1993:52–53). These moral and narrative dimensions are all the more important under conditions in which medicine and molecular biology are tending to recast the body as "an effect of a molecule, an extension or supplement to the deathless bit of immanence that is DNA," in Deleuze's terms, a "virtual" body that is both constituted in the interfaces of wetwares, softwares, and hardwares of a post-vital body, and whose "secret of life" increasingly "implodes" into new dimensions of coding, a "virtual abyss" or "vanishing point within the organism" (Doyle 1993:9).

Deleuze and Guattari work, in part, with a vocabulary drawn from the boundaries between the human and the substratum of the molecular structures that compose humans and other biological species; Haraway's cyborg vocabulary works, in part, by focusing on hybridization between humans and mechanical or computational prostheses. Haraway is concerned with the three boundaries affecting human beings: external cyborg (machine-human) relations, internal biological ones, and those between us and other animals. She poses political and moral questions by committing herself to a series of mutually interrupting and revaluing (Serres's "parasitic") agendas (socialist, feminist, alternative narrational, and commitment to "no-nonsense accounts of the real world"). It is this that generates her search for new metaphors, both that they delight and capture attention (publicity), also that they articulate new political relations and possibilities. As she put it at the end of the day in San Francisco, "I'm not interested in boundary crossings, but rather in zones of implosion, of nature into culture, of the material world into tropes, as sites of transformations that generate narrative possible worlds and new agencies.'' This sort of storytelling, she went on to remind, is not the same as mere opinion: not all stories are equally good. "Gene, seed, fetus, chip, brain, and bomb'' are key terms at the moment. They are key nodes invested with "densities'' of cultural imagination and work.

The linguistic virtuosity of Haraway, Deleuze, Serres, or Derrida is not mere flashy attention seeking; it is also a serious attempt to find conceptualizations for a world in which the traditional boundaries and categories are upset and reworked by the technoscientific infrastructure of organic life, social relations, and communications. Derrida, like Deleuze, draws upon the revolution in molecular biology, which happened roughly at the same time as the revolution of structuralism and post-structuralism in the humanities. In a chapter on Derrida's conceptual vocabulary drawn from the life sciences, Christopher Johnson notes that Derrida devoted a series of seminars to Francois Jacob (who with Jacques Monod and Andre Lwoff won the 1965 Nobel Prize for their work on the role of RNA). This conceptual vocabulary works similarly to Haraway's "gene, seed, fetus, chip, brain, and bomb.'' Some of Derrida's terms are *articulation* (from anatomy and topology; including figures of ligature and ligament), *invagination* (from pathology and embryology, as well as from sexual anatomy), *tympan* (convoluted pathways complicating the distinctions between inner and outer, limit, boundary, margin), *atom, seed, germ,* and *term.* Thus, for instance, Derrida "shadows'' Hegel's notion of seed as a unitary anticipation of the tree, the conceptual figure of circular movement, a closed monad, unfolding of a concept, a restricted economy, and the return of mind to itself after its passage through nature or experience. Derrida pluralizes, turning seed into dissemination—"There is no first insemination. The semen is already swarming.'' (1972: 304)—and turning the restricted economy into a general economy in which there are remainders which deviate from the path of predetermination or unfolding, that generates difference, new forms, mutations. Included in the set of terms with seed and germ is also *pro-gram:* the genetic and cybernetic codes (*programs*), both subject to interference, to parasites, diverting the message or letter for its own goals, like the molecular formations of cutting and sequencing, citing, iterating, and constantly altering. While Johnson sees Derrida as merely drawing from the biological, Derrida's German translator, Hans-Jorg Rheinberger argues that Derrida's writing is the material form of molecular biology and experimental modern science in general, that science is no longer a process of discovery of a world apart, but is quite literally an experimental form of writing. (Genetic material is literally cut and respliced in order to understand and discover, thereby, however, also actively creating the reality that it understands.)

In an analysis of the playful metaphors used by operators of DNA sequencers (connected to Macintosh computers), Deborah Heath points out that technicians are referred to as "hands"—as in "Don't you get it: techs don't talk, techs are hands"—and that this is a double entendre: "good hands" are the tacit skills that make experiments work, and that intuit new solutions to problems. It is this doubleness that makes techs different from chimeras. Chimeras are mergings of cosmids (microorganisms that act as vectors for particular stretches of DNA) with fragments of another organism's DNA. Chimeras can be cloned, and potentially there is also a doubleness to "chimera" as Deleuze and Guattari remind with respect to all sorts of rhizomatic forms: "As Francois Jacob says, transfers of genetic material by viruses or through other procedures, fusions of cells originating in different species, have results analogous to those of 'the abominable couplings dear to antiquity and the Middle Ages'" (1987:10–11). But chimeras as a technical unit in microbiology do not have the same agency and ethical responsibility as "good hands." Heath's ethnographic attention to the humor and metaphorics of the techs—notions like "good hands," but also naming practices with a point such as "Lana" (*anal* spelled backwards) for one of the sequencers—seems a promising opening to the back-and-forth flow of metaphors, rather than the older hierarchical models of "popularizing" and degradation of knowledge from experts to publics.

It is the interplay of various kinds of cultural rationalities or logics rather than just agons among interest groups or calculations of risk assessment that promise to open up the social worlds of specialized knowledges. Complementary work is being done in such cultural studies analyses of middlebrow, new age, and various other sociological constellations of theorizing about science and technology (Ross 1991; Penley and Ross 1991). At issue in some of the feminist contributions to this literature is the relation between systematic but covert rhetorical codings of social power relations in the ways technoscience is deployed. Donna Haraway's tone of modest determination is admirable: "Feminist discourses and anti-colonial discourse are engaged in this very subtle and delicate effort to build connections and affinities, and not to produce one's own or another's experience as a resource for a closed narrative. These are difficult issues and 'we' fail frequently" (1991:113).

Generations, Genders, Institutions, Places, Cultures, and Narratives[9]

> Visvanathan argues that the life histories of the major nuclear physicists in this century have been a movement from innocence, freedom and conviviality—from play, discovery and communitas—to the tyranny of secrecy, control and in some cases elatory nihilism. [Their] biographies reflect . . . the prototypical relationships between

> nuclear science and the scientists' self-defined social responsibility, which in turn reflects the culturally defined relationship between knowledge and power in the modern West.
>
> —Ashis Nandy

If the reflections in the previous sections highlight the contemporary cultural shifts from physics as the lead science to molecular biology and computer sciences, and thereby indirectly such sites as Paris and Silicon Valley, other genres of contemporary reflection on the development of the sciences have begun to sketch out anthropologically more usable understandings of previous generations and places of scientific development. The new hybrid cultural and historical studies of science (Latour, Shapin and Shafer, Golinski, Biagioli, Körner), as well as science-in-fiction novels (Banville, Djerassi) have begun to draw connections between science as a peculiar set of practices and interventions in larger cultural worlds (public demonstrations of experiments carefully staged and controlled; piggybacking science on larger and more diffuse social movements; construction of public roles under diverse institutional regimes) that begin to trace out the degrees to which scientific understandings were central to new emergent cultural consensus, and the degrees to which individuals who practiced science also lived in worlds governed by quite other conceptual regimes (back to Levon Abrahamian's cartoon parable of Giordorno Bruno).

Scientists' autobiographies might provide a series of texts in which the construction of an autonomous heroic ego or "genius" can be placed in tension with the collective expansion of the aesthetics of rationality which is often the outcome of fallible cooperation and competition, playing games, including at times deceit, as well as often becoming captive of state, market, or corporation-driven rationales. James Watson's *Double Helix* (1968) was one of the first autobiographical accounts of science that helped demystify the actual process of scientific discovery in the popular imagination. Rita Levi-Montalcini, 1986 Nobel laureate in medicine for the discovery with Stanley Cohen of nerve growth factor, entitles her autobiography *In Praise of Imperfection* to underscore that making educative mistakes advances science. And Francois Jacob builds his autobiography, *The Statue Within* (1987), around a framework of psychological anxieties both individual and formed culturally (like Montalcini) in the pressures of being a secular European Jew during World War II. These and other autobiographies from the ascendant biological sciences (Crick 1988; Kornberg 1989; Snyder 1989; Djerassi 1992) might interestingly be read against more general efforts to trace back the emergence of an information society since the nineteenth century (Latour 1988; Richards 1993; Pynchon 1973).

Much of the reigning discussion about the nature of science is, however, built around the experience of twentieth-century physics. Shiv Visvanathan, drawing upon the Viennese science writer Robert Jungk, uses the lives of physicists to construct a matrix of competing positions on the relation between political democracy and big science:

> Jungk is a master of what anthropologists call thick description. The multitude of anecdotes he provides coalesce into a choreography of positions available to science in relation to the violence of the atom bomb as a social fact. Within such a perspective scientists like Einstein, Szilard, Teller, Bohr, and Oppenheimer appear not as idiosyncratic figures but as permutations within a scientific code. Names become role tags listing various possibilities as the table shows. (Visvanathan 1988:116)

Thus, Enrico Fermi stands for an "apartheid science—aloof from politics." ("Don't bother me about your conscientious scruples. After all, the thing is beautiful physics"); Niels Bohr stands for "the social organization of science itself as a model of communitas" ("Pure science had managed to avoid the violence of war by sublimating it into agonal play. The scientific paper was a precious gift, and it circulated in joyous exchanges. . . . Every conference was a kind of potlatch, each scientist showering the others with knowledge in exchange for eponymous recognition. The internationalization of science withstood the pressures of war."); Edward Teller "embodies the scientist as a political lobbyist playing on military and political fears to obtain larger financial sanctions for research"; Szilard and Franck in contrast "urged greater public understanding and control of science" (Visvanathan 1988:116–18). The most compelling and disturbing portraits are the *contradictory* lives of such men as Karl Fuchs (who, like Prometheus, stole the gifts of nuclear fire and gave them to the Russians that a monopoly of terror be broken), Hans Bethe who opposed armaments research after Hiroshima yet by 1951 was seduced back into H-bomb research, and "the most fascinating figure in this danse macabre," Robert Oppenheimer, "a humanist Hamlet struggling against a scientific Prometheus." Sociologically, Jungk and Visvanathan argue, all these different lives were caught in a three-fold shift in the nature of science: "the degeneration of science as a play form; the shift within science from epistemic uncertainty to vivisectionist hegemony; and the displacement of science from the university to the company town." All three, Visvanathan continues,

> are symptomatic of the transformation of Western liberalism into occidental despotism, heralding the coming of the Atomstaat. . . . For liberalism, the private was sacred and the public was open and accessible. In bizarre inversion, vivisectionist science has opened up the privacy of the body and soul to the public scrutiny of the clinical

> gaze, while science as public knowledge has become increasingly secret and forced into the most monstrous of total institutions—the research cities of the twentieth century. . . . As a mode of production, [the nuclear energy regime] demands a fail-safe system of security . . . and superhuman precision. . . . However the human body is a reluctant machine. (130, 146, 148)

Visvanathan and his colleagues in New Delhi are not antiscience Luddites: science is not evil, but it must be controlled, redeemed through spiritually, ecologically, and democratically sound values. Among the ways suggested by Visvanathan and his colleagues are alternative sciences, updated traditional sciences, and, above all, bicultural perspectives which constantly juxtapose the needs of different communities and ecologies in different parts of the globe so as to try not to allow one perspective to override the others. Ashis Nandy:

> Contemporary India, by virtue of its bicultural experience, manages to epitomize the global problem of knowledge and power in our times. There is a continuity between the Indian experience of an increasingly violent modern science, encroaching upon other traditions of knowledge and social life [like the rising debate over large-scale dams that displace thousands, can erode the landscape and cause long-term ecological damage, and transform the nature of agrarian relations], and the Western experience with modern science as the dominant cultural principle resisting the emergence of new cultures of knowledge [like the resistence to investing in exploration and production of alternative forms of energy]. (11)

As Visvanathan illustrates, the use of lives containing the principled contradictions in the transformations of modern science can be a valuable tool both of explorations of the institutions and social organizations of science, and also of the narrative devices by which scientists make sense of their own activities.

In the following pages I want to look at several science autobiographies and in a very preliminary fashion probe some of the ways they are structured, and suggest that some of the reasons for their structuring have to do with the science fields in which their authors participate: that their form as well as their content may have something to tell us. I look first at autobiographical openings, and then at a couple of middle passages.

One way to get a feel for ranges of autobiographical style is to read the openings of a variety of texts. Beginnings are wonderful places to examine notions of coherence, the degree to which autobiographers feel themselves to have coherent selves or narratives, the ways in which other autobiographers who acknowledge decentered, conflicted, contradictory, fallible selves position their multiple parts vis-à-vis linguistic, libidinal, social, historical, familial, generational, ethnic, gendered, technological, intercultural, or other processes; the ways in which attention is given to the media or vehicles of memory and forgetting, desire and information, documentation and impression. Some autobiographies enact, are performatives, in ways their authors perhaps do not realize, and that, too, is interesting, and is often signalled in beginnings.[10]

For instance, just to take a quick unsystematic sample of the openings of six recent autobiographies (of five scientists and one philosopher of science): Hideki Yukawa, Philip Morse, Irene Fischer, Karl Popper, Norbert Wiener, S. E. Luria. Two of these begin with visual tableaux or set-piece dramatic scenes that work as emblems of the text to come; two begin with meditations on the cliche that only when a life is nearly over can one hope to find a pattern because human life stories are recursive, hermeneutic, ever-changing; and two begin with meditations on the problems and problematics of writing autobiographies of lives rooted in scientific careers for nonscientific audiences. Of these, two come across as self-centered in ways that caution the reader to the fact that the narrator perhaps is not to be fully trusted, although enacting the grounds on which trust and distrust might be evaluated. Karl Popper is one of those wonderful characters whose philosophy is perfectly rational and democratic in ways that he himself may not always live up to: one is amused by his posturing and yet reassured by the explicitness of the arguments he makes. Norbert Wiener is more self-aware of what may seem like arrogance, calling constant attention to it himself. Wiener is one of the two who uses the meditation on the difficulties of writing a scientist's autobiography: the difficulty for him centers on the difficulty of writing science nontechnically and there are repeated explicit put-downs of laymen and unconscious insistence that only males are scientists. By contrast, Luria's meditation on the difficulties of writing a scientist's autobiography focuses on the challenges of presenting personalities from multiple perspectives: Pasteur's religious faith, Darwin's neurasthenia, and Newton's mental illness are keys to understanding these powerful personalities at work; "to examine Pasteur without his roots in French provincial bourgeois life, or Einstein without his relation to Judaism is to diminish them"; "personality should emerge as a landscape composed of many vistas like . . . a Breughel panorama of peasant life" (1984:3). Luria has formulated here, of course, a neat criterion that distinguishes uninteresting autobiographies from those that are in fact illuminating about the processes of life that go into the making of science. (Compare his compatriot, Rita Levi-Montalcini's autobiography, discussed below.)

Thus, one can distinguish also between the two examples using opening tableaux or dramatic scenes as emblems: Karl Popper's is the story of how he apprenticed himself to a carpenter, and how this craftsman taught him everything; it is a nose-thumbing gesture to be repeated in the autobiography itself, where one of the main themes is to insist that he is very, very different from the logical positivists of the Vienna Circle, when in fact much of what he claims as his personal discovery was also their perspective. Irene Fischer's tableau on the other hand is of her retirement party, of the flood of memories that came to her at that moment, of the stories she began to share with the many people who had been her long-time colleagues, which were commented on in the following days and weeks, and in turn stimulated the writing of her scientific memoirs. Here there is an interesting generative process: story telling and shared experience as the markers of traditional solidarity, and yet this scientist, almost uniquely in these autobiographies, grounds the string of stories by faithful consultation with diaries, work logs, correspondence files, and published papers. Reference to these punctuate the text, with a rhythm of collegiality, at once narrative and documentary.

Of the two scientific biographies that begin with meditations on looking back at a life as it comes to some sort of point of retrospective, Morse explicitly alludes to a theme that arises less explicitly in several of the others: that is, a key motive for writing and understanding the lives of scientists, is that it is increasingly imperative for governments, industrialists, and citizens to understand the workings of science in a world where the links coupling basic research and final application are becoming closer, and the conditions of everyday life are becoming more dependent upon those linkages, including the possibilities of gross disaster if the linkages are misused.

Hideki Yukawa is the other meditator on the retrospective point of life, but his is a much more hermeneutic, scientific account of multiple perspectives. And it is that, too, that I find fascinating: the demonstration that the distance between scientific perspectives and humanistic ones is not opposed as so commonly is assumed. It is an exquisite text with multiple openings, each containing multiple frames, seeming to parallel the multiple alternative explanations and perspectival, partial models in physics. There are four openings like Japanese gift boxes: a forward, a chronology, an introduction by the translator, and the opening chapter.

The foreword, only two pages, is an extraordinary collage of shifting perspectives, Rashamon done in cubist style, not just four points of view. It begins with the line about now that he is fifty years old, he can review half a century. The expectation of a coherent narrative thereby set up is disrupted in the next sentences which meditate on his two lives which are one—the easy middle-class life of the son of a geology professor, easy enough, that is, to narrate; and the academic life which is not so easy to analyze; yet both these lives and paths are one. The life of physics is further refracted in two: on the one hand, one could describe it easily enough, as a life that was just carried along on the tide of a science that was rapidly changing in this century (a sweetly modest metaphor for someone who later does admit that he helped shape the course of that science); but, on the other hand, this neat metaphor is dispelled in the observation that holds both for physics and for narrating a life: "today's truths may tomorrow be disproved, and that is why, from time to time, we must look backwards in order to find the path we must take tomorrow." Note this recursiveness is like the stream into which one can never step twice; one looks backwards not to achieve closure, but to track a course into the future; it is like the hermeneutical circle which never achieves closure since each new reading generates awareness of further horizons.

The pace of the text does not pause: again softly it notes that this particular life has already had multiple accounts of it written, both by Yukawa himself, and in the form of biographies by others. Gently the author suggests that the public has an image, and he wants to offer more information so that that image can be judged. Note again the imagery being used: this is not the normal public masque, private interior dichotomy, nor the correcting of a false public impression (the binary logic of public and private, right and wrong); instead it is a modern scientific sensibility of triangulation amidst more and more kinds of data, an image of increasing approximation, increasing comprehension of complexity.

Next the interior/public, subjectivity/objectivity trope is muddied. The Nobel laureate claims to have trouble expressing himself, to tend to view matters subjectively, and to know that if he tries to view things "objectively" he may "betray" himself. Objectivity and subjectivity here, of course, are not the popular body/mind, reality/emotion, hard/soft binarisms, but rather something more like simplistic reification versus perspectival truth. "In any case," the text hastens on, "not even I can perceive clearly what is about to take shape" (1982: vi): indeterminacy, and note that here he is talking not of the future, but of the nature of the text that the reader is about to commence.

The falsity of proper names is next. Names are not what they seem, especially in Japan: we learn that Hideki Yukawa is Hideki Ogawa. Ogawa is his father's family name: upon marriage, he took his wife's family name, Yukawa. Moreover, eventually we learn that his father had done the same, and that name in turn was also such a name. A lovely regression of ever receding uxorilineal nominations.

Finally, the foreword ends with a lyrical sentence that brings temporary closure to the opening sentence: Hideki Ogawa was born in 1907 at old Tokyo's Ichibei-cho Azabu. The house smelled of plum blossoms each spring. But, of course, later in chapter 1 we learn that this is but a reported description, elicited from his mother, about a birthhouse of which he had no memory, certainly no olfactory one, a birthhouse moreover that no longer existed, having been burned in World War II. (Derridean traces of the ineffable.)

I won't go on with the description of the text, except to say that it continues in this fashion, insisting on the multiplicity of truth, perspective, and modeling; and to suggest that this is not a function of writing style, or of simple Derridean post-structuralist reading on my part, but that it reflects a fundamental modern scientific perspective on reality; and finally to suggest that the result of multiplicity is not (as Allan Bloom fears) the undermining of knowledge, but the increase of purchase on truth and knowledge, by identifying the sources of uncertainty, the limits of particular angles of vision, and by triangulating them together.

Sciences are diverse in their procedures, methods, aesthetics, and organizational structures. Let us look at some middle passages of two scientist's autobiographies, one from a science, geodesy, that is closer to the applied mathematics end of the continuum, and one from a science, neurobiology, that is closer to the life and human sciences ends. Such juxtapositioning of different sciences may help pose questions about the varieties of models and practices of knowledge production we call science. Geodesy is a particularly clear example of procedures of modeling, mapping, indirect measurement, and increasing degrees of approximation or accuracy. Neurobiology is closer to that other image of science as almost anarchistically experimental, trial and error, inductive empiricism, leavened, of course, in this case with analytical biochemical detection of molecular weight and other physiochemical properties. The notions of multiplicity and triangulation, invoked by Hideki Yukawa apply also in these two sciences in interesting variants: triangulation is a literal mapping procedure in geodesy, while the complementarity of neurobiological experiments and biochemical analysis was the sine qua non for isolating and beginning to explore the still-mysterious hormonal mechanisms of the nerve growth factor for which Rita Levi-Montalcini and Stanley Cohen received the 1986 Nobel Prize in medicine.

These two autobiographies (of Rita Levi-Montalcini, the neurobiologist, and of Irene Fischer, the geodesist) might also be usefully juxtaposed to accounts in the emerging fields of studies of science and technology, both to have other frames of reference with which critically to interrogate the autobiographies, but also to provide a reality check to the theorizing by nonscientists (albeit, importantly, often science-trained persons who either chose or were forced to leave the science career fast track) about what drives scientists to discovery and accomplishment. (No serious anthropology or even journalism takes native accounts at face value, yet no anthropological or journalistic account can be credible without building upon native perspectives.) A useful preliminary foil, for instance, is provided by Donna Haraways' (1989) survey of the four temptations in studies of science, four tempting perspectives on science, each valid up to a point, but dangerous if allowed to silence the other perspectives.

The first temptation—useful to read against Rita Levi-Montalcini's quite different perspective—is the social constructivist view of science that inquiries into the power relations that affect the progress of particular lines of inquiry, particular careers, and, in its strongest form, the ways in which scientific knowledge policies its own boundaries against new ideas or new information that it cannot easily incorporate. In its strongest versions, such as that elaborated by Bruno Latour and Steven Woolgar in *Laboratory Life* (1979), the social constructivist view rejects ordinary notions of realism, ordinary separations of what is technical from what is social, and regards the criteria of pragmatic feedback from success in the world to scientific models as fundamentally underestimating the ways in which science can protect itself from falsification. Latour and Woolgar see science as a tactic of reducing conflicting interpretations of messy reality into unambiguous facts through various methods of transcription, translating into equations, and machine outputs. (Or to take a clearer example of how scientific facts operate from a nonsocial constructivist, Sharon Traweek's study of high-energy physicists, what gets defined as elementary particles and the equations of their relationships, after all, are observed only through the construction of machines, accelerators, devices for turning the imperceptible into traces that can be recorded, inscribed, calculated, and modeled.) Over time, say Latour and Woolgar, what began as probabilities, tentative generalizations, or approximations are incorporated into later stages of model building as undisputed facts. In the competition for success, scientists become invested in power struggles, defined in part as simply raising the cost to competitors of destabilizing reigning accounts so high as to be unworthwhile pursuing. Haraway rightly calls such a description of science both attractive and dangerous: drawing attention to the constructedness and contingent nature of reigning scientific ideas, but wildly overstated if treated as a complete account.

I will come back to put Rita Levi-Montalcini's observations in fuller context, but want to simply observe here that she—like many scientists—foregrounds and revels in the contingency of scientific knowledge. Latour and Woolgar's "demystifying" account is only demystifying to popular accounts of genius and absolute truth, not to working scientists. But Levi-Montalcini places a different set of implications on this contingency: it is part of a larger picture of both science and life as an evolutionary process created through a capricious game of mutations and selection. She notes both in life and in science the necessity oftentimes of repressing knowledge that cannot be incorporated and that can be self-destructive: thus immersing herself in her experiments while the Nazis raged around her was one form of healthy repression, but also she cites Russian neuropsychologist Alexander Luria's "law of disregard of negative information" in being able to repress negative experimental results that she would not be able to explain until years later, and that, had she focused on it, might have derailed her work. Furthermore, retrospective histories of science often include, as Levi-Montalcini's does, explanations of powerful ideas that provided false confirmation for what were thought at particular points in time to be scientific results: thus she cites the sway of gestalt theories as providing support to Karl Lashley's experiments which seemed to suggest that the capacity to learn and memorize was not localizable in particular parts of the cortex. From a scientists' point of view, then, the social constructivist account is largely true but trivial, not very interesting (not adding systematic new information).

If social constructivism is Haraways' first temptation, her second temptation is a political mediationist view of science that inquires into the ways in which language, laboratory hierarchies, industrial or governmental patronage, and so on, structure the perspectives through which truth is recognized. This could be taken as a variant of the social constructivist inquiries, and differs primarily in its concern to expose the systematic political, state- and money-generated patternings of what otherwise might seem to be more individualistic, contingent outcomes of strong personalities and organizations, or effects of measuring and inscribing devices. Haraway derives her version of the political mediationist questions from Marxist, feminist, and minority observations that the conflictual and inegalitarian relations of society are often opaque to those in positions of systematic domination and power; what seems to be true from one perspective may conceal problems that are visible from other positions. Weak and relatively trivial versions of the mediationist accounts of science are descriptions, for instance, of how popularizing accounts of science draw upon sexist imagery. Emily Martin's work (1987) on medical textbooks' metaphors for the biology of human reproduction is an example that foregrounds important ways in which the laity and the poor may be disempowered in their own thinking about reproduction or in their interactions with physicians, social workers, and others by the circulation of such metaphors; but this points up more about the poverty of scientific literacy in America, or of translation languages, than about the trajectory of scientific models. Haraway herself has a field day with the metaphors that have structured research programs in primatology from the days of "Man the Hunter (of Science)"—individualistic tests of manhood by such science explorers as Carl Akeley who went into the wilds to shoot great apes with gun and camera (with gun for specimens, with camera to preserve without destroying nature)—to the 1970s when *National Geographic* popularized Jane Goodall, Shirley Strum, Birute Galdikas, and Diane Fossey as "Woman the Nurturer (of Science)," for example, of orphaned apes, females being "closer to nature" and thus being able to provide the conditions for animals to approach human beings, and for human beings to reapproach the secret garden of nature. (Meanwhile, says Haraway, the cameras recording/"shooting" these women and apes were held by their husbands and consorts in good traditional gender-divided roles.)

A more serious part of Haraway's *Primate Visions* is the correlation of research paradigms in primatology with more general cultural anxieties: the concern in the 1890–1930 period with fear of decadence, stressing preservation and conservation of pure nature (including eugenics and racial purity); the concern in the 1940s instead with obsolescence and stress in an increasingly technologized age, stressing molecular biology, systems engineering, the recognition that human beings are parts of human-nonhuman "cyborg" systems, and management systems of social control (focusing on food, sex, dominance-subordination, hormones) involving a movement from laboratory experimental settings with chimpanzees (conditioning, learning, behavior modification) to watching free-ranging animals in created colonies, in nature, and in "nature preserves"; the concern in the post–World War II period with antiracism and decolonization, and a biological anthropology that supported the unity and equality of all human beings, centering itself on population biology and adaptative complexes of functional anatomy and culture (upright bipedalism and tool using); and finally in the 1970s and 1980s a sociolobiology (genetic calculus, strategic rationality) that paralleled the rise of yuppie stress on competitive individualism.

But most importantly, Haraway acknowledges that in the course of the development of primatology, biological anthropology, genetics, and related fields, there has been a real positivist increase in the knowledge base and sophistication that is not contained by either the popularizing metaphors or the correlations with encompassing cultural anxieties of the age. Hence again her recognition that while political-mediationalist accounts of science have a degree of validity (especially in a field that is so prone to be a projective screen for thinking about human culture, psychology, physiology, and sociability—and hence her critical focus in the last third of the book on four women primatologists who have helped reorient the field and its generalizing implications), they are dangerous if allowed to silence other accounts of science.

Again, it is perhaps interesting to juxtapose Rita Levi-Montalcini's account, both to remind that even when scientists write popular accounts they often eschew the kinds of metaphors Martin and Haraway foreground, insisting on simplified yet technically precise language, and also, more importantly, to note the enactment of the separation between ordinary life full of metaphors, emotions, and chaotic forces and a sphere of inquiry where inferences and implications are more narrowly interrogated and controlled. Indeed, Levi-Montalcini goes so far as to concentrate her account of her scientific career in the third part of her autobiography, with only occasional references to the scientific developments themselves in other more familial, organizational, or historical sections. Irene Fischer's account interestingly does not engage in this sort of segregation, but does similarly insist on simplified yet technically precise language, pointing out the ways in which metaphors introduced for popularizing fun (as in the case of the "pear-shaped earth") are analogically apt or not. Haraway's field of observation, primatology, is, as noted, one particularly subject to popularizing projections, and the recent subfield of sociobiology has had a spate of practitioners who have been particularly promiscuous and deliberately provocative in their use of popularizations that shift ambiguously between technical and inappropriate connotative meanings (for example, in the use of the word *altruism*).

More strictly political economy inquiries might also pursue Chandra Mukerjee's suggestion that scientists at work on big science projects (oceanography, the supercollider, the genome project) are a reserve labor force of experts available to government, which can by selective funding ensure both access to the expertise it needs and reduce the threat from expertise that might undermine its own policies. Scientists can remain autonomous in the details of their work while being simultaneously dependent in the major guidelines of their work; science in this sense is politics writ in a larger sense than the politics of labs and individuals fighting for the prizes of recognition, or in the various strategies that individual scientists use to extract resources for their own projects in the interstices of large science and bureaucracy (again, see Levi-Montalcini on the loss of interest in Italy in neurobiology, yet its continuation, and also her comments on the differences in individualism—greater in Italy than in the U.S.—and publishing demands in Italy that affect the course of risks and innovations scientists attempt).

The third of Haraway's four temptations is the accounts of science by scientists themselves, which Haraway describes as usually realist and positivist. These accounts claim for science a degree of autonomy from the realities of their institutional and political settings which are freely acknowledged to affect the progress of science. These accounts also frequently remind us that the metaphors of popular science are but very rough approximations of relationships that can often only be accurately rendered in the technical language of mathematical functions, that the two kinds of language do not translate very well, and that therefore to critique science by critiquing only the popular metaphors by which it is approximated is not tenable. Scientists distinguish between discovery (which may be serendipitous) and confirmation/falsification, between the sociology of science and the content of science, between theoretical models that are simplified approximations of reality and reality itself, between levels of precision and pragmatic feedback provided by the ability to predict or control outcomes, between probabilistic truth and particularistic knowledge. Science is heuristic, pragmatic, partial, approximate, evidence-relational, and modeling; all of which may involve reductionism, deterministic causality, mechanical as well as statistical models, but less as totalizing accounts than as components within larger modeling intentions. Internal scientists' accounts are "temptations" (insufficient by themselves) insofar as they ignore or downplay the sociological and political environments that enable them, or insofar as they tell the story of science discovery self-servingly from a particular individual's or group's point of view.

Finally, the fourth of the temptations is to consider science as a species of story telling, which immediately opens the possibility that the same science may be narrated in multiple ways. The attraction of trying to tell multiple narrations of science is that it provides a sense of how coherence is created, while drawing attention to perspectival and approximative tactics, or, as Haraway puts it, individually having no power to claim unique or closed readings. To problematize and describe science in terms of the various stories (such as the four temptations) that can be told about the histories of discovery, the relation between science and nonscience parts of cultural understanding, the uses and abuses of scientific knowledge, unintended consequences and implications, and so on, is a relativizing move quite popular for cultural critique. To cast the first and second temptations above as stories is in part to relieve them of their arrogance or sense of claiming to be the whole truth rather than important aspects of the truth. But the temptation of turning all accounts of science into the status of mere storytelling, of course, must be resisted: the chemical effects of drugs, or the geometry of the earth, or the physics of atmosphere are not just stories.

One of the compelling characteristics of good science autobiographies is that they are not dry accounts in terms of the third temptation, but that they involve several, if not all, of the four temptations or perspectives or stories, as well as others such as the historical and personal resonances between scientific activities and other parts of scientists' lives, including such well-crafted allegories of methodology and meaning as cited above from Hideki Yukawa. Resonances between scientific activities and other parts of scientists' lives is a part of the social and cultural constructedness of science that often is screened out by "social science" accounts such as the Latour and Woolgar study, but which can contribute to an understanding not only of the psychology or motivation of scientists, but also to the larger aesthetics that informs and encompasses their work. Let me turn then briefly to the autobiographies of Rita Levi-Montalcini and Irene Fischer. Both are written by women with European backgrounds, features which might be read against those monochromatic feminist critiques of the science establishments that emphasize the difficulties for women in pursuing scientific careers: not to dismiss those difficulties, but rather to emphasize Rita Levi-Montalcini's philosophical observation that more important to success in scientific research than either special intelligence or the ability to be precise or thorough are the qualities of dedication and optimistic underestimation of difficulties.

The beginning of Rita Levi-Montalcini's autobiography is, like other openings already cited, emblematic of the account to follow. It is a threefold meditation on the object of her science (nerve cells and nerve growth factors), on science and technology, and on the relation between rationality and life—each of these three an analogue of the other two, all fitting a model of "capricious games" of mutations and selection, an evolutionary process in which advantages are built upon so that *retrospectively* there is a line of irreversible progress. This is a facilitating frame for an account of the development of neurobiology involving the interplay between the availability of techniques (chrome-silver impregnation so that nerve cells stand out), technology (the electron microscope which allowed one to see synapses, cathode-ray oscilloscope, and camera), the complementarity between biochemistry (to analyze the nature of snake venom and mouse salivary gland serum, and to purify fractions of these, which were important steps in identifying nerve growth factor) and neurobiology (to explore experimentally the spectrum of action of these protein molecules by injecting them in developing organisms and in differentiated organisms), including a number of false leads. There is place here both for the expansion of knowledge and rationality, a sense of historical horizons (what was possible and/or conceivable at different points in time), and for the chaos of reality. She sees progress as stemming from imperfection—in technology she contrasts the efficiency of the bicycle which has not evolved much from its introduction as against propulsion mechanisms that have evolved in speed and efficiency; in science she says of herself that she lacks strong powers of logical thought, lacked aptitude for math and physics, and so on, but nonetheless thanks to such characteristics as determination and underestimating difficulties was able to make major contributions. Indeed her account in the third part of the book where she chronicles both her own career and the advances in neurobiology that led to the explorations of nerve growth factor is one of trial and error, of complementarity of competences between investigators, of putting aside intractable problems and results that could not be solved until further advances had been made. And in an unintentional feature of the book, she enacts dialectics of fallibilities and contradiction as well, first denying emotional complexes and proceding immediately to enumerate childhood anxieties (tendencies to solitude, neurotic fears of wind-up toys and mustaches) and complex emotional relations with parents, siblings, and mentors; or, more importantly in the present context, arguing against diaries and journals (and life histories?) as vanities, yet at the same time thankful for the return of letters which both allowed her to relive intense periods of her life and allowed her to construct key portions of the present text:

> I never developed the habit—nor do I regret not having done so—of keeping any kind of record, still less a diary, because I believe that, if memory has not taken the indelible imprint of a given event, then it could not and should not be brought back to life by mere written witness. I believe, in fact, that the very act of recording an event causes, if only unconsciously, a distortion resulting from the blatant

> desire of the diarist to make use of it as an account to be exhibited to third parties, as a way of reliving in old age a particular moment, and of making one's descendants partake in it or even, if one is especially vain, for its value to posterity. (170)

> Having never kept a diary, I was very pleased when, in June 1980, Viktor Hamburger sent me a large envelope containing all the letters—carefully preserved for so many years—that he had received from me during the period I spent in Rio, from September 1952 to the end of January 1953. . . . In reading them, I have relived one of the most intense periods of my life in which moments of enthusiasm and despair alternated with the regularity of a biological cycle. (154)

The philosophy and the picture of knowledge and science here is one of chaos, and often adversity (lack of funds, bureaucratic opposition, war), against which the drive to creativity leads to lasting results. She ends with a tribute to Primo Levi, and his citation, in the midst of concentration camp adversity, of Dante's Ulysses, "You were not born to live as brutes," a motto that holds as well for the promise that neurobiology and the nerve growth factor studies may eventually bring cures for nervous system disorders. There is, to be sure, a certain kind of heroic trope here, but it is a markedly humble one, neither the macho version critiqued by Haraway, nor one that sees the scientist as a lone genius. The scientists described are all fallible creatures creating a collective understanding which itself is historically situated and contingent, if no less remarkable and promising for all that.

Levi-Montalcini's book is one in a series sponsored by the Alfred P. Sloan Foundation which are largely popular accounts rather than attempts at systematic inquiries into the sciences themselves or their histories, though one does get capsule histories as in Levi-Montalcini. Irene Fischer's "Geodesy? What's That? My Personal Involvement in the Age-Old Quest for the Size and Shape of the Earth, with a Running Commentary on Life in a Government Research Office" is a more concentrated and sustained account of the development of a scientific field and career over a twenty-five year period, the period of the creative expansion of the field from an extension of surveying technology to the age of the satellite which transformed its methods. Unlike Levi-Montalcini's book, which tends to segregate the account of the science in one section, Fischer's text is characterized by an interlacing rhythm of personal life, technical scientific problems, collegial relations (both complementary and competitive), publications, bureaucratic difficulties, accomplishments, and rewards. Although at first sight the text may seem a forest of names of people, articles, and groups, as one reads it takes on an almost mimetic rhythm of the social networks and step by step solution of the large puzzles posed by measuring the earth, aided by hints in the literature, personal interactions, technological breakthroughs, patient data accumulation, recalculations and reconceptualizations as new thresholds of descriptive competence are achieved.

Unlike Levi-Montalcini's ambivalence toward diaries, this text is built on rich documentation of diaries, correspondence, and publications, allowing a textured interlacing of stories like Haraway's temptations (particularly the first and third), as well as personal and world-historical stories. Each chapter is constructed around an advance in terms of the scientific puzzle of being able to measure and/or model the earth's size and shape: the advance of surveys along the earth's surface, the different results that one obtains from different astrogeodetic, gravimetric, and oceanographic measuring techniques, the efforts to find a best-fitting ellipsoid to the irregularities measured on the actual earth, the efforts to piece together geoid maps for different parts of the earth into a unified world datum (of which the Fischer North American Geoid Chart was the first to cover the North American continent, the Mercury Datum or Fischer 1960 Ellipsoid became the official NASA and Department of Defense world datum, and the Fischer South American Datum was the first to bring the various efforts of Latin American countries together), and the revolution introduced by satellite technology (fig. 2). This strand of the account works from chapter to chapter as an incremental series of historical snapshots or a processual history of the creation of more complete, more secure, and reconceptualized knowledge.

Each chapter interlaces such scientific advances with accounts of the competitions and cooperation among different government agencies, scientists and support staff, different countries, and individuals (the "third voice," the sociological and social constructivist narrations). There is a trajectory to this narration from the esprit de corps of a tiny pioneering group within a larger social setting of gender, race, and nationality consciousness, to, toward the end of the autobiography, assessments of the effects of changing government management policies on the conduct of science. Each chapter also plays off the family and immigrant consciousness not only of the author, but of a scientific world in the U.S. after World War II, invigorated by the inflow of European scientists and of closer international cooperation and competition. (Levi-Montalcini's book also contributes to this story—her medical school classes included Salvador Luria, Renato Dulbecco, and Rudolfo Amprino, and her own career benefited from cooperation between St. Louis, Rio de Janeiro, and Rome—but she focuses reflectively on this primarily regarding the reestablishment of science in Italy after the war, not in the sea change of American science and academia.)

These were still the days when the author could be told by a fellow female employee not to work so hard because women would never advance beyond a given level, when a Military Air Transport steward refused to believe that the GS-12 Fischer listed on his flight roster to be boarded first could be a woman,

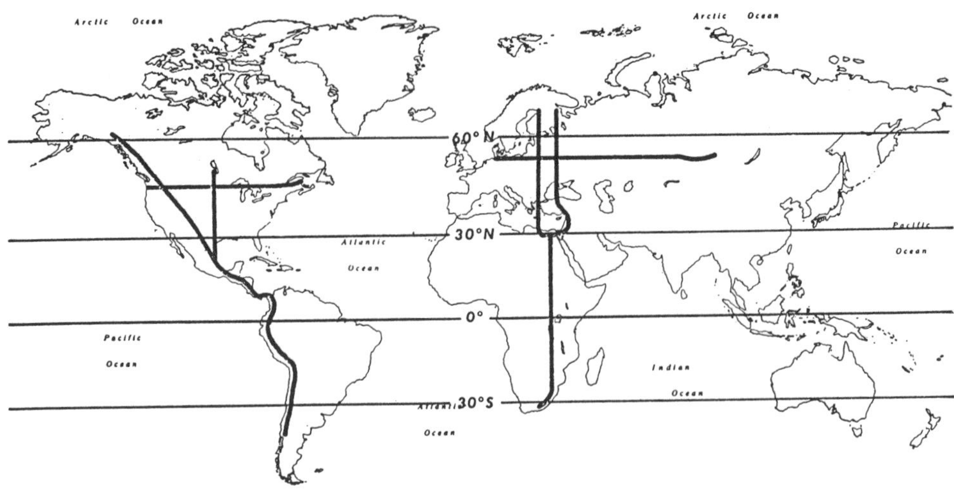
Arctic Ocean
Arctic Ocean
60° N
30°N
Atlantic Ocean
Pacific Ocean
0°
Pacific Ocean
Indian Ocean
30°S
Ocean

(a)

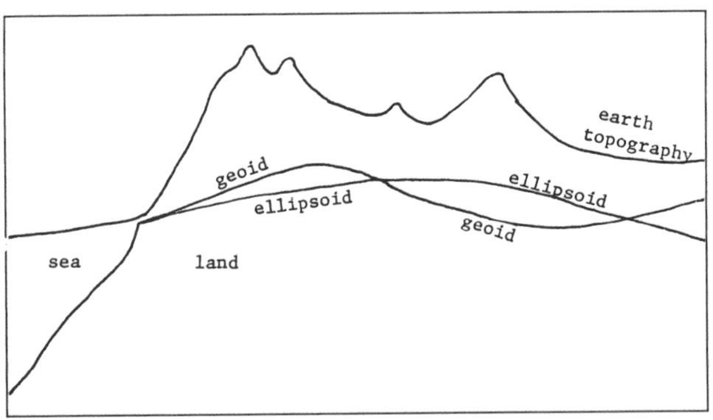
earth topography
geoid
ellipsoid
ellipsoid
geoid
sea
land

(b)

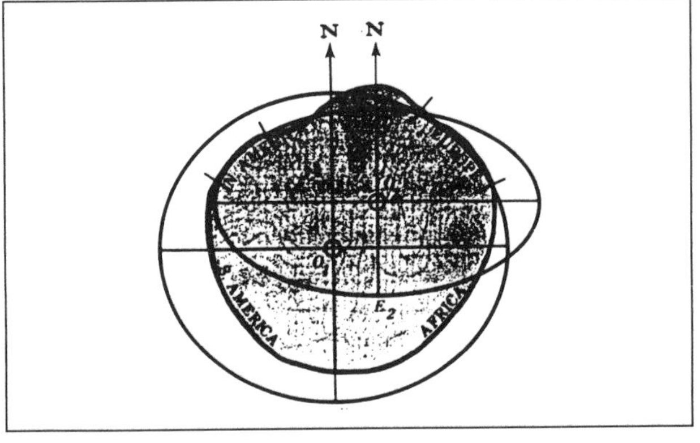
N
N
S. AMERICA
AFRICA

(c)

when her 1957 Meritorious Civilian Award pin was designed to be worn only in a male jacket lapel button hole (subsequent awards corrected this oversight). These are remembered as only amusing markers of breaking through barriers. Somewhat more serious were the time a paper had removed her name as author because the bureaucrats thought it unseemly or unlikely that a woman should be the author (much to the embarrassment of her superiors), and the earlier loss of a job at Harvard as an assistant to Vassily Leontieff when students found out and protested that a woman was grading their papers. (MIT students of Norbert Wiener did not similarly object when she graded for him. Working for John Rule at MIT developing stereoscopic slides, she excitedly tried to show them to Norbert Wiener only to learn that he had only one functioning eye, and so without depth perception could not appreciate these three-dimensional figures.) More poignant still are memories of trying to break segregation patterns—women ate lunch together to celebrate promotions but blacks were usually not invited; this was morally unacceptable to a Jewish refugee from racist persecution, and so on her first promotion she invited all her coworkers to a party at her home only to have one black woman not show on a particularly flimsy excuse.

The immigrant sensibility provides a useful bifocal vision, a constant sense of alternative perspectives, a reality check and stabilizing force. A charismatic mentor, John O'Keefe initiated and taught her what she needed to know of geodesy and surveying, but in an environment of the esprit of a small elite group where she proved herself by deriving formulas during a placement quiz and by passing a series of covert informal tests of knowledge. The group of five or so included a German speaker who knew of the Vienna Circle, of which she had been a junior follower, which made her feel at home. It also included another refugee, the former head of a geodesy organization in Yugoslavia, saved from a dead-end bureaucratic job by O'Keefe. That she could read the basic Jordan-Eggert geodesy text in the original German helped (there was only a partial in-house translation in English available at the time, done by another German refugee), and her Viennese training proved superior in many ways. The philosophy of Austrian physicist Ludwig Boltzmann also stood her in good stead both against show-off methodology for methodology's sake that got others into unnecessary trouble, and against lack of interest in issues that she felt important, and that she was willing doggedly to pursue until others realized the relevance (for example, surveyors never thought the curvature of the earth important enough for their purposes, and this implica-

---

Fig. 2. Irene Fischer's sketches of (a) the first long survey arcs used to construct a world datum; (b) the three surfaces of the actual topographic surface of the earth, the irregular surface of the geoid (an equipotential gravimetric surface), and a best-fitting ellipsoid to the geoid; and (c) two ellipsoids, each fitting better different parts of the irregular geoid.

tion was carried over into early geodetic models as well, an absurdity that Fischer refused to let lie and eventually, as the field progressed, the fact became important and required explicit incorporation [contrast the foreshortened time frame and unduly pessimistic thoughts of Latour and Woolgar about the freezing of "facts," and compare Levi-Montalcini's similar perspective to Fischer's]):

> Propose what is true. Write so that it is clear, and fight for it to the death. (*Bring vor was wahr ist. Schreib so dass es klar ist. Und verficht's bis es mit dir gar ist.)*

Amused by a colleague whose fancy math allowed him to set up 199 simultaneous equations with just as many unknowns, in contrast to her own step-by-step compilation for the same task of 300 equations but with only three, four, or five unknowns which could be solved with a simple desk calculator, she allowed herself the comment, "An American scientist would rather bite off his own tongue before he permits his technical work to appear in easy detailed steps." Their results disagreed and it took him months to find his error. Another friendly saying in German, by Einstein, also helped her keep her equanimity between conflicting geodetic and astronomic methods based on unexamined different *a priori* assumptions, "Der Herrgott ist mutwilling, aber nicht boeswillig [the Almighty God is mischievous, but not malicious]."

Scientific life, in this account, in other words, was not just a wonderful puzzle-solving delight, but a network of intellectual lineages, historically created assumptions that could not be taken for granted, personalities, protocols, and bureaucracies, ranging from the genial Brigadier Guy Bomford who used his outsized belly to teach students about the equatorial bulge, to the indirection needed to get answers from the Russians. In one case, she asked which of two methods the Russians had used in a 1939 paper: the Russian geodesists at the international meetings claimed not to know, so she presented calculations based on two possibilities. Then a few months later she read the report on the meetings, in which they only gave one of her calculations: the one based on the correct method.

Indirection and competition, of course, were not only across international and cold-war lines: there was the case of the computation division within her own agency which under much pressure finally agreed to teach the researchers how to program the new UNIVAC computer, only to discover when they tried to run some real computations that the computer people jealous to protect their expertise had taught them a no longer operational program; and there was the case of the rival agency in the U.S. government that squelched them by having their work classified so it could not circulate, and another agency that plagiarized their work. More amusing is the case of Charlie Brown: in 1958 the Vanguard satellite went up, and O'Keefe explained to a news conference that the new calculations of the shape of the earth showed it to be more pear shaped than round. This showed up in a Charles Schultz cartoon of Charlie Brown happy with a new globe until Linus told him the earth was not round but pear shaped. Later when Fischer wanted to use this cartoon in a publication, Schultz's syndicator refused permission, claiming it as his/their own intellectual property. This, of course, did not prevent Fischer from using a series of vegetable metaphors in training manuals and in speeches, such as the one to the American Philosophical Society delivered at the Cosmos Club, where women were at the time not officially allowed. On the more cooperative side, Fischer was able to locate and return materials from the famous 1927–35 Sven Hedin expedition that were secreted away in Washington to Erik Norin of Sweden, who was putting the final touches decades later on a book to accompany the Hedin atlas. And in 1965 she won a vote of applause of the Organization of American States by managing to break the competitive logjam since 1944 on cooperating with data by showing how a unified South American Datum could be constructed and thereby generating the enthusiasm for all to contribute their information. Even more impressive, perhaps, given nationalist sensitivities, was her skill in getting Argentinian cooperation: acknowledging their wariness of allowing their work simply to be appropriated by others, she invited the Argentinians to come to Washington for training and to use computer facilities, and they in turn were willing to share their data, not with Fischer's agency (a part of the Defense Department) but with "me personally for the scientific purposes of the Figure of the Earth and the South American Datum" (246).

Again, only this initial taste of the text will have to suffice here, enough to indicate the historical richness of the presentation of the evolution of a basic field of research needed among other things for all satellite and space technology; the analogical utility of this model of science both in constructing this text and as one of several models of science involving mapping, degrees of approximation, modeling, triangulation, collation of multiple perspectives, indirect measurement, constructivist-pragmatic approaches to knowledge; the human dramas, comic and serious, fallible yet cumulative result producing, historically situated in world-historical terms as well as sociologically and culturally, dependent upon individualistic persons but in a collective enterprise; and the joyful optimism of so much scientific endeavor which all too often contrasts with the dour, suspicious, even angry affect of much social science critique of science. The capsule accounts of changing methods and problems from actual measurements on land, in the sea, and from space to creating models and measuring irregular reality as deviations from theoretical surfaces be they geoidal equipotential surfaces or oceanographic theoretical levels, to collation and compromise temporary solutions among competing methods—for example, the early heroic achievement by surveyors of long triangulation arcs from Scandinavia to Cape Town, from Europe to Japan, from Canada to Tierra del Fuego; the calculation of deflections of the vertical (variations in gravity forces of the earth from place to place which can affect rocket or missile trajectories) by either dense gravity surveys or by calculating the difference between astronomic and geodetic positions; the introduction of marine geodesy (stationing three transponders in an equilateral triangle on the ocean's bottom with a ship above sending acoustical signals as a basic unit of triangulation; utilization of bathymetry to help estimate differences of deflection of the vertical from terrestrial measurements); the introduction of satellite geodesy (geometric satellite techniques, analogous to terrestrial triangulation, photographing a moving satellite against a star background from two stations, one of known position and one to be determined; dynamic satellite techniques, analogous to terrestrial gravimetry, measuring distances, not angles, by electronic, not optical, means), the disputes between oceanographers and geodesists about calculating equipotential surfaces of the earth and over the slope of mean sea level, the measuring of the irregularities of the earth against locally best-fitting ellipsoids and finding a global datum, bureaucratic solutions of taking discrepant scientific results by differing methods and dividing the difference to produce a compromise datum for widespread use—provides a fascinating conceptual education as well as a history. Along the way are wonderful side studies into historical issues illuminated by new scientific understandings: application of the world datum to the Ice Age, reevaluating the classic lunar parallax determinations, refiguring Eratosthenes' and Posidonius' calculations of the circumference of the earth and showing how errors have creeped into the literature about those calculations, and more generally application of Vienna Circle pragmatic language philosophy to a retranslation of a famous but often misunderstood dictum of Rabbi Akiba about free will. Also along the way are astute analyses of attempts at management of science in the government—not jeremiads but observations on how a new commanding officer was able to rekindle morale among researchers by separating status- and control-seeking administrators from scientists, by dismantling stovepipe organization, as well as observations on the triviality, inappropriateness, and self-defeating philosophies of management courses, and an analysis of civil service reforms. And there are anecdotes rich in psychological, historical, and cultural resonance.

These historical soundings constitute the layerings of memory that are powerful anchors for the present. These are personal, as in the inner voice of Fischer's father who gives her support at various moments of significance, or in the banter of O'Keefe and Fischer over fear of flying in her first overseas flight in 1955: Catholics have it easier, joked O'Keefe, they know there's an upstairs, causing Fischer to reflect on Shaw's vision of the boring hymn singing upstairs, and also the Jewish version of sages debating—"But I did not think they admitted women there, and even if they did, I was just not yet interested'' (27)—followed by a lyrical description of the world seen from the air, resonating with psalm, Beethoven, and Isaiah, and a traditional blessing upon touching down safely—culture and religion. They are also historical as in the connections and hesitancies of the last half-century, as when returned to a symposium in Vienna—the first and only return in over thirty-five years—the convener addressed her in an inimitable Viennese idiom and mixed polite and colloquial language, "Gnädiger Frau, ich habe ein Hünchen mit Ihen zu rupten'' [Dear Lady, I have a bone to pick with you], admonishing her for forcing him to speak in English for decades not realizing she spoke German, but she poignantly remarked about her childhood city, "There seemed to be two towns on the same spot: the lively town of the Symposium, a beautiful town strangely suggesting we may have visited here before; and another personal town that was crying with silence, a ghost town'' (226–27). And she found sometime later in the course of giving thanks for an honor it was easy enough to speak the thanks in German, but as she tried to turn to geodesy she found she had to ask permission to speak in English. The third level of such mosaic memory is the soundings in much deeper histories, to Eratosthenes, for instance, and the Ice Age, but also Petronius Arbiter (66 A.D.) on management.

> We trained hard—but it seemed that every time we were beginning to form up into teams, we would be reorganized. I was to learn that later in life we tend to meet any new situation by reorganizing, and the wonderful method it can be for creating the illusion of progress while producing confusion, inefficiency, and demoralization. (319)

This she posted on her file cabinet to provide a "long view [which] had a soothing philosophical effect,'' together with a pinup of Galileo Galilei: "He had a pensive look into the far distance. . . . His presence reinforced awareness of a clear distinction between real work and busy work in a researcher's value system, and helped allocate precious time and energies accordingly'' (324). She had to fight repeatedly to protect his freedom on the wall against the government's inspectorate, and hid away in her desk drawer Lord Acton's "Power corrupts, and absolute power corrupts absolutely,'' also as a reminder to herself as an administrator (325).

These mosiac features are not just ornaments but are repositories of philosophical perspective, of will, of motivation, of sense of proportion, things which help foster that determination and discounting of obstacles of which Levi-Montalcini also spoke.

Autobiographical triangulations. Two features of these scientists' autobiographical texts are most important in the present context: the utility of autobiographical forms as exploratory access into the systematic features of science as a human endeavor, and the well-craftedness of the best of these scientific autobiographies, well-crafted not only in the sense of being a pleasure to read but in the sense of using their literary forms themselves to provide access: the suggestions I have made with regard to the three texts of Yukawa, Levi-Montalcini, and Fischer that the form of their life history narrative and the form of their scientific endeavor are formal analogues of one another, consciously, deliberately so, and provide in their variety access to the diversity of science, its procedures, methods, and models. One of the interesting effects, I think, is the demonstration that the distance between scientific perspectives and humanistic ones are not necessarily opposed as is so commonly assumed, and that the multiplicity which some people fear as destructive of traditional pieties is not a pandering to irrationalities of postmodern irresponsibility, but on the contrary is ever more a feature of the contemporary condition, of rationality, and of grounded positions of critique that allow understanding (and perhaps reconstruction) to advance.

Autobiographical Triangulations, Narrative Folds

> Mathematics over the past century has given little evidence that it is concerned with eternal, timeless, and hence, unarguable truth. On the contrary, contemporary mathematics is filled with (no pun intended) chaos and turbulence . . . we are still witnessing what (it is hoped) are the transients arising from two profound shocks: the overthrow of Euclid and the discovery of inconsistencies (paradoxes) in Set Theory.
>
> Just as with a molecule of globular protein, which must fold in order to become active, this linear string must be folded back on itself, so as to bring parts remote on the string into close contact. . . . this entire volume is an attempt to show how the ideas developed herein constitute, in effect, a single enzyme geared to lyse the problem "What is life?"
>
> —Robert Rosen

> Some of the young people I see, who are very good, take physics . . . as a system you can do things with, can calculate something with, and they miss . . . the mystery of it: how very different it is from what you can see, and how profound nature is. . . . There is no good translation of a Witz. It's a joke or a trick. It's the use of this kind of witty trick that I have always liked about physics. . . . I have always taken physics personally. . . . It's between me and nature.
>
> —Isidor Isaac Rabi

> According to Bachelard, reason is best known by reflection on science . . . the repeated refutation of a priori philosophical ideals of rationality by historical scientific developments.
>
> —Gary Gutting

> Seminars, true rites of initiation, free-for-alls . . . That day an American biochemist . . . The audience interrupted incessantly. . . . Chopping up his discourse with questions . . . badgering him, provoking him, nipping at his heels like excited puppies. . . . As for [Sol] Spiegelman, he was not easily confounded . . . he listened to comments and criticisms while playing with a piece of chalk in the fashion of a Hollywood gangster with a coin . . . The deathblow came later in a cafe on the Boulevard Pasteur . . . the theory was dissected, torn apart, shredded into tiny pieces. Bit by bit, the bull weakened. A final thrust of [Jacques] Monod's descabello. The bull's final spasm. And resistance ceased. All this amidst laughter and joking. . . . Two trained roosters beak to beak vying with humor and sarcasm. . . . a will to power . . . a desire for intellectual domination. . . . a world full of gaiety, of the unexpected, of curiosity, of imagination. A life animated as much by passion as by logic.
>
> Science meant for me the most elevating form of revolt against the incoherence of the universe. Man's most powerful means of competing with God; of tirelessly rebuilding the world while taking account of reality.
>
> —Francois Jacob

It is time, in the idiom of biological ecologist Rosen (no less than of a Deleuze or a Foucault), to fold the strands of this essay upon themselves. There is a richness, a jouissance, a gamble with passion, in the sport of science, as the almost Geertzian cockfightlike description of Jacob reminds, as does the insistent use of the word *joy* in the titles of the autobiographies of Weisskopf and of Fischer; it comes up in interviews as well: an American-resident India-born researcher in electroluminescence says,

> every new idea of value that I have had has been a matter of grace. . . . the Greeks would call it a muse. It's not something that I do, it just comes to me. And that's a very nice feeling, very, very exciting. It's a joy! It's a great feeling! . . . it is more than science. . . . almost like "high poetry". . . . that is the purpose of doing science. The reason I do science is because it gives me that joy.[11]

There is as well a metaphysical depth, as Rabi indicates, an agon with the mysteries of the universe and life, not just an aesthetic harmony or a pleasure in technical skill. This depth often is folded against cultural traditions, historical experiences, and community structures: so far some of the productive tensions and synergistic reinterpretations between scientific rationalities and Christian and Jewish traditions have been explored, but in today's world it is odd that more has not been done with for instance the Hindu, Buddhist, Confucian, and Islamic traditions that lie in the background of some of the world's greatest scientists (Ramanujan, Chandrasekar, Vikram Sarabai, Homi Bhabha, Abdus Salam) as well as an increasingly important segment of the world's scientific establishment, and not just as intellectual traditions but also as community structures that provide the sustenance, pressures, and motivations out of which various sorts of professionals and scientists emerge.

In the preceding pages, I have raised and evoked many more questions than I can answer, from what is an autobiography to what and how do they signify? As an anthropologist, I have been trying to ask the beyond the text questions: how do real selves in the world assemble themselves, and can autobiographies help analogize or investigate this? How do genres of self-portraiture change under different social conditions? Do leading technologies of an age provide different perspectives on the ways in which agency and moral responsibility are negotiated? The autobiographies that have, in however a preliminary way, been analyzed here are texts from a particular generation—that which came of age before or during World War II. Will the texts of their children look different, be formulated in the languages of cybernetics and molecular biology and cyberpunk literature, and with a different moral or social sensibility, with a different range of aesthetic and cultural resonances (more cosmopolitan, more multicultural, more individualistic, less utopian)? Francois Jacob's autobiography, with its elegant play upon emotional memories acting as triggers and repressors, as a flow of information beneath the surface of ordinary consciousness (the statue within) perhaps already hints in this direction.

But I have also been using the evidence of scientists' autobiographies implicitly to recast the science, technology, and society (STS) literatures of the recent past: to refocus the STS literature that focuses on ethnomethodology and agonistic game playing into one that focuses on positioning and translation; and to recast the STS genre of the study of conflicts into one that deals with public policy negotiations and complex sites of science construction. The shifts involved are those from a kind of internalist "beating of scientists at their own game" to a broader effort to locate science and engineering in larger cultural and social contexts. Particularly important are moments of translation across disciplinary boundaries that constitute breakthrough sites, or ongoing needs for translation across disciplines, where scientists are able to reflect on their own limitations and needs to negotiate meaning, technique, or implication; and public policy debates that involve the need for scientists to do translations vis-à-vis other kinds of expert knowledges and public interests.

Notes

1. See Midgley (1992) for an eloquent charting of some of the metaphysical embeddednesses of contemporary science. As she says, "Any system of thought play-

ing the huge part that science now plays in our lives must also shape our guiding myths and color our imaginations profoundly. It is not just a useful tool. It is also a pattern that we follow at a deep level in trying to meet our imaginative needs'' (1992:1).

2. On the dialectic between male vision as a source of tyranny and female voice as a technique of healing in the Arabic *Thousand and One Nights,* see Malti-Douglas (1991); also on speech versus writing. See also Mills (1991) on the oral recitation of folktales as a talking cure in situations of oppression from *The Thousand and One Nights* to contemporary Afghanistan. The structural parallels are also striking in the Greek legend of the Ring of Gyges, and the Jewish story of Esther. Note the variants of Dinarzadeh (a dinar is a unit of money) which would resonate with the story of Gyges as Shell (1978) unpacks it, having to do with bureaucracy and the introduction of money; versus Dunya-zadeh [born of the world, worldly], which John Barth's version Dunyazid prefers.

3. See Mary Midgley's interesting account (1992: chaps. 7, 8) of the ''outbreak of bizarre sexual metaphors in writing about science'' by the founders of the Royal Society and mechanist philosophers such as Bacon. She suggests that these metaphors were in part directed against Aristotelian thinking (which treated Nature in a contemplative manner without touching it/her) but mainly against natural magic and occult forces figured as feminine dark forces (which the light of reason should dispel). Kepler's notion of attraction and, later, Newton's theory of gravitation were both rejected as unintelligible appeals to such occult forces. The spokesmen of the Royal Society said they were for a ''truly masculine philosophy.''

4. Don Ihde quoted in Sobchack (1993:578–79). See also Hayles (1993), Haraway (1992), Piercy (1991). The classic contemporary speculation about downloading information from brains into computers is Moravec (1988), but its lineage in contemporary scientific thought goes back at least to J. D. Bernal, Freeman Dyson, J. D. Barrow and Frank Tipler (see Midgley 1992: chap. 2), and its philosophical lineage goes back at least to Descartes' notion that his ''me'' is completely distinct from his body and is easier to know than his body.

5. One of the most interesting, and most influential, of these might be Erwin Schroedinger's *What Is Life?* which served as a key reference for many physicists moving into biology after World War II, a movement they spoke of as one from a science implicated in death to one concerned with life. Considered in historical context (written during the war without reference to the war) and in terms of its marginalia about his life (the preface, the epilogue, and the autobiographical sketches penned in 1960 and now published with it), it becomes a rather profound personal text as well as a period piece document.

On these notions of ''third genre'' and ''semipopular,'' see Kelley (1993), Eger (1993), and, more generally, Midgley (1992). Eger nominates a ''third genre'' between technical scientific writing and popularization, that combines both scientific and literary seriousness in an attempt to break through professional barriers. The texts he analyzes—Jacques Monod's *Chance and Necessity* (1971), Douglas Hofstadter's *Gödel, Escher and Bach* (1979), Joseph Weizenbaum's *Computer Power and Human Reason* (1976), et al.—he suggests ultimately contribute to a renewed, and more detailed, unified story of evolution, a renewed master narrative that can integrate the prebiotic (chemical) to the cosmic (astrophysical), including brain physiology and culture. He sees these texts as signaling an emergence from a positivist period of multiple specializations and concerns with methods, into a new drive for theoretical unification. However, at the same time, this "third genre" because it foregrounds the construction of new high level metaphors, it allows a more profound recognition of the hermeneutical nature of foundational metaphors in science, and thus an opening for critique especially insofar as such metaphors provide socially orienting models for action beyond research in the laboratory (he cites sociobiology and genetics as obvious worrisome cases). Midgley (1992) and Gillian Beer (1993) show that this genre is considerably older than Eger thinks.

Robert Kelley suggests a useful typology of scientists writings from the scientific paper to textbooks and popularizations. The "semipopular" scientific text—he would include Mandelbrot (1977; 1983), Prigogine and Stenger (1984), Hofstadter (1979), but not Gleick (1987), Steven Jay Gould, or Lewis Thomas—is true to the science it portrays but like a textbook (an extremely interesting observation) skips some of the mathematical development and experimental data. "It makes assertions about a science and about its implications without rigourous proof and is therefore close in form to the scientific textbook" (148). Certain subdisciplines, he also suggests, find this genre particularly appropriate. In this context, a third essay in the same volume, by David Porush (1993) on Gleick's *Chaos* and Prigogine and Stenger's *Order Out of Chaos* is quite instructive, and leads to my own questions about the variety in this genre of text. I suspect that Eger is overly pessimistic about the hegemonic reassertion of a singular master narrative, and would only cite, as a very preliminary starting point, the final pages of Francois Jacob's autobiography, where he contrasts his own philosophical stance with that of his coworker, Jacques Monod, despite a working relationship "of exceptional intensity and even intimacy" like a "long duet" and culminating in joint conclusions (Jacob 1988:307).

Steven Heims's provocative and problematic double biography of Norbert Wiener and John von Neumann (1980) is profoundly overdetermined by Heims's own decision to leave physics during the Vietnam War. Wiener and von Neumann thus become the two warring parts of the moral dilemmas of physicists during the cold war: to find ways to not work on defense department-funded projects (Wiener, albeit the father of missile guidance systems during World War II), or to go along with weapons development more or less enthusiastically (von Neumann, figured by Heims as an almost demonic Dr. Strangelove). The in many ways better biography of Alan Turing by Andrew Hodges (1983) might also be read usefully as a doubly figured text, as Hodges is himself a mathematician who took time out to write the biography and then returned to mathematics. It is a "better" biography for its richness in weaving together the enigmas of the various aspects of Turing's life, while Heims's text is flatter, angrier; but these aesthetic and human criteria do not lessen the power of the questions Heims poses about the social responsibility of scientists, and about the role the public should play in decisions about science in the contemporary world for which the two subjects of his text are but foils for himself and us.

6. Physics was the leading science in terms of numbers of undergraduate majors at schools like MIT until the 1960s; in the 1970s it was electrical engineering and com-

puter science's turn. Although computer science is still the lead major at MIT, the numbers nationwide began to crest and turn down in the mid to late 1980s.

It would be of interest to pursue a new study of the twentieth-century evolution of the two-cultures gap: the founders of the modernist humanities—Descartes, Kant, Goethe, Locke, Hume, et al.—were heavily motivated by developments in contemporary science in a way that perhaps is only now, quite tentatively, again becoming the case, at least insofar as key figures in the humanities such as Claude Lévi-Strauss, Jacques Derrida, Michel Foucault, Gilles Deleuze, and Michel Serres are concerned. Interesting that these should all be Frenchmen, which leads one to ask questions about the social structure of the French academy, or inversely about the failures of the American and British academies which have acknowledged the need to bridge the two-cultures gap, but have not seriously addressed the issue except for oddly conceived courses of "physics for poets" and "humanities for engineers and scientists." See Snyder's 1980 study of MIT students and the degree to which learning styles fostered in the humanities are also important to the success of engineers and scientists in their creative careers and lives.

7. The 1992 sessions culminated in an author-meets-critics session with Donna Haraway. In the 1993 session, I raised as problematic the freezing on one term, *cyborg,* as a label, when even in Haraway's own work, *cyborg* is just one of a range of terms that generate narratives of possible worlds and new agencies, as in the quote from the 1992 session used as an epigraph above; while this point was freely admitted, organizers of the 1993 session also pointed out the need for a name that draws audience, and "cyborg anthropology" has precisely that "possible world" promise of drawing together into conversation hard science, science-fiction imagination, ethnographic realism, and utopian critique.

8. Note that there is an important distinction to be made between actants as functional units that may include machines and humans or biological processes and humans on the one hand, and on the other hand the morally trivializing idea that objects like coffee pots may be thought of as having agency and subjectivity.

9. Parts of the following sections are taken from a forthcoming earlier essay on autobiographies and social theory in probing ethnicity, religion, and science.

10. See Edward Said's *Beginnings* for a literary critical approach to beginnings.

11. From a fascinating series of cross-cultural interviews of leading researchers in electroluminescence conducted by Annadore Schultz, with her permission.

References

Banville, John. 1984. *Doctor Copernicus.* Boston: D. R. Godine.

———. 1981. *Kepler.* London: Secker and Warburg.

Barth, John. 1972. *Chimera.* New York: Random House.

Beer, Gillian. 1993. "Eddington and Modernism." Lecture. Harvard University.

Berstein, Jeremy. 1978. *Experiencing Science: Profiles in Discovery.* New York: E. P. Dutton.

———. 1991. *Quantum Profiles.* Princeton, N.J.: Princeton University Press.

Biagioli, Mario. 1993. *Galileo's Courtier.* Chicago: University of Chicago Press.

Crary, Jonathan. 1990. *Techniques of the Observer: On Vision and Modernity in the Nineteenth Century.* Cambridge, Mass.: MIT Press.

Crick, Francis. 1988. *What Mad Pursuit.* New York: Basic Books.

Deleuze, Gilles, and Felix Guattari. 1987. *A Thousand Plateaus.* Minneapolis: University of Minnesota Press.

Derrida, Jacques. 1962. *Edmund Husserl's The Origins of Geometry.* Trans. John P. Leavy, Jr. 1989. Lincoln: University of Nebraska Press.

———. 1976 [1967]. *Of Grammatology.* Trans. Gayatri Chakravorty Spivak. Baltimore, Md. Johns Hopkins University Press.

———. 1981 [1972]. *Dissemination.* Trans. Barbara Johnson. Chicago: University of Chicago Press.

Djerassi, Carl. 1992. *The Pill, Pygmy Chimps, and Degas' Horse.* New York: Basic Books.

———. 1989. *Cantor's Dilemma.* New York: Doubleday.

Downey, Gary Lee, Shannon Hegg, and Juan Lucena. 1993. "'We Treat Everyone the Same': Gender Race and Class in the Engineering Curriculum." Paper presented at the American Anthropological Association Meeting, Washington, D.C.

Doyle, Richard. 1993. "On Beyond Living: Rhetorics of Vitality and Post-Vitality in Molecular Biology." Ph.D. diss., University of California, Berkeley.

Dumit, Joseph. 1993. "A Digital Image of the Category of Person: PET Scanning and Objective Self-Fashioning." Paper presented at the American Anthropological Association Meeting, Washington, D.C.

Dumit, Joseph, Gary Downey, and Sarah Williams. 1992. "Granting Membership to the Cyborg Image." Paper presented at the American Anthropological Association Meeting, San Francisco.

Eger, Martin. 1993. "Hermeneutics and the New Epic of Science." In *The Literature of Science.* Ed. McRae.

Fischer, Irene K. "Geodesy? What's That? My Personal Involvement in the Age-Old Quest for the Size and Shape of the Earth, with a Running Commentary on Life in a Government Research Office." Typescript. Schlesinger Library, Radcliffe College, Cambridge, Mass., and the Institute of Physics Library, New York.

Fischer, Michael M. J. 1991. "Anthropology as Cultural Critique: Inserts for the Nineties." *Cultural Anthropology* 6 (4): 525–37.

Fischer, Michael M. J. Forthcoming. "Autobiographical Voices (1,2,3) and Mosaic Memory." In *Autobiography and Postmodernism.* Ed. K. Ashley, L. Gilmore, and G. Peters. Amherst, Mass.: University of Massachusetts Press.

Friedman, Alan J., and Carol C. Donley. 1985. *Einstein: Man and Myth.* New York: Cambridge University Press.

Friedman, Michael. 1992. *Kant and the Exact Sciences.* Cambridge, Mass.: Harvard University Press.

Georges, Egenia. "Women's Experiences of Fetal Ultrasound Imagining in Greece and Mondragon, Spain." Paper presented at the American Anthropological Association Meeting, Washington, D.C.

Gleick, James. 1987. *Chaos: Making a New Science.* New York: Viking.

Golinski, Jan. 1992. *Science as Public Culture: Chemistry and Englightenment in Britain, 1760–1820.* London: Oxford University Press.

Good, Byron. 1993. *Medicine, Rationality and Experience.* New York: Cambridge University Press.

Guha, Ramachandra. 1989. *The Unquiet Woods: Ecological Change and Peasant Resistance in the Himalayas.* Delhi: Oxford University Press.

Gutting, Gary. 1989. *Michel Foucault's Archeology of Scientific Reason.* New York: Cambridge University Press.

Hallyn, Fernand. 1990. *The Poetic Structure of the World: Copernicus and Kepler.* New York: Zone.

Haraway, Donna. 1989. *Primate Visions: Gender, Race, and Nature in the World of Modern Science.* New York: Routledge.

———. 1991. *Simians, Cyborgs, and Women.* New York: Routledge.

———. 1992. "The Promise of Monsters: A Regenerative Politics for Inappropriate/d Others." In *Cultural Studies.* Ed. L. Grossberg, C. Nelson, and P. Treicheler. New York: Routledge.

———. 1993. "Modest Witness @ Second Millennium. The FemaleMan© Meets OncoMouse™." Paper prepared for the British Association of Social Anthropologists Meeting. Oxford, England.

Hayles, Katherine. 1993. "Gender and Cyberspace." Paper presented at the Literature and History Workshop, Massachusetts Institute of Technology. 22 November.

Heath, Deborah. 1992. "The View from the Bench: Prosthetics and Performance in Molecular Biotechnology." Paper presented at the American Anthropological Association Meeting, San Francisco.

———. 1993. "Genetic Trading Zones and Gendered Geographies: Calendrical Meetings on the Marfan Syndrome." Paper presented at the American Anthropological Association Meeting, Washington, D.C.

Heims, Steve. 1980. *John von Neumann and Norbert Wiener: From Mathematics to the Technologies of Life and Death.* Cambridge, Mass.: MIT Press.

Hodges, Andrew. 1983. *Alan Turing: The Enigma.* New York: Simon and Schuster.

Hofstadter, Douglas. 1979. *Gödel, Escher and Bach.* (New York: Basic Books.

Jacob, Francis. 1988. *The Statue Within.* New York: Basic Books.

Johnson, Christopher. 1993. *System and Writing in the Philosophy of Jacques Derrida.* New York: Cambridge University Press.

Keller, Evelyn Fox. 1985. *Reflections on Gender and Science.* New Haven, Conn.: Yale University Press.

Kelley, Robert T. 1993. "Chaos Out of Order: The Writerly Discourse of Semipopular Scientific Texts." In *The Literature of Science.* Ed. McRae.

Körner, Lisbet. 1993. "Goethe's Botany: Lessons of a Feminine Science." *Isis* 84.

Kornberg, Arthur. 1989. *For the Love of Enzymes.* Cambridge, Mass.: Harvard University Press.

Lachterman, David. 1989. *The Ethics of Geometry.* New York: Routledge.

Latour, Bruno. 1988. *The Pasteurization of France.* Cambridge, Mass.: Harvard University Press.

Latour, Bruno, and Steven Woolgar. 1979. *Laboratory Life: The Social Construction of Scientific Facts*. London: Sage.

Laughlin, Kim. 1993. "De-Naturalizing Science, Feminism and Political Action: Collaborative Approaches." Paper presented at the American Anthropological Association Meeting, Washington, D.C.

Levi-Montalcini, Rita. 1988. *In Praise of Imperfection: My Life and Work*. New York: Basic Books.

Luria, S. E. 1984. *A Slot Machine, A Broken Test Tube—An Autobiography*. New York: Harper and Row.

Malti-Douglas, Fadwa. 1991. *Woman's Body, Woman's Word: Gender and Discourse in Arabo-Islamic Writing*. Princeton, N.J.: Princeton University Press.

Mandelbrot, Benoit B. 1977. *Fractals: Form, Chance and Dimension*. New York: W. H. Freeman.

———. 1983. *The Fractal Geometry of Nature*. New York: W. H. Freeman.

Martin, Emily. 1987. *The Woman in the Body: A Cultural Analysis of Reproduction*. Boston: Beacon Press.

McRae, Murdo William, ed. 1993. *The Literature of Science: Perspectives on Popular Scientific Writing*. Athens: University of Georgia Press.

Midgley, Mary. 1992. *Science as Salvation: A Modern Myth and Its Meaning*. New York: Routledge.

Mills, Margaret. 1991. *Rhetorics and Politics in Afghan Traditional Storytelling*. Philadelphia: University of Pennsylvania Press.

Monod, Jacques. 1971. *Chance and Necessity*. New York: Knopf.

Moravec, Hans P. 1988. *Mind Children: The Future of Robot and Human Intelligence*. Cambridge, Mass.: Harvard University Press.

Morse, Philip. 1977. *In at the Beginnings: A Physicist's Life*. Cambridge, Mass.: MIT Press.

Mukerjee, Chandra. 1990. *A Fragile Power, Scientists and the State*. Princeton, N.J.: Princeton University Press.

Nandy, Ashis. 1988. "Science as a Reason of State." In *Science, Hegemony, and Violence*. Ed. Nandy.

Nandy, Ashis, ed. 1988. *Science, Hegemony, and Violence*. New Delhi: Oxford University Press.

Penley, Constance, and Andrew Ross, eds. 1991. *Technoculture*. Minneapolis: University of Minnesota Press.

Penley, Constance. 1993. "Science Fiction Theater: Biosphere II." Paper presented at the American Anthropological Association Meeting, Washington, D.C.

Piercy, Margery. 1991. *He, She and It*. New York: Fawcett Crest.

Popper, Karl. 1985. *Unended Quest: An Intellectual Autobiography*. London: Open Court.

Porush, David. 1993. "Making Chaos: Two Views of a New Science." In *The Literature of Science*. Ed. McRae.

Prigogine, Ilya and Isabelle Stengers. 1984. *Order Out of Chaos*. Toronto: Bantam Books.

Pucci, Suzanne Rodin. 1990. "The Discrete Charm of the Exotic: Fictions of the

Harem in Eighteenth-Century France.'' In *Exoticism in the Enlightenment*. Ed. G. S. Rousseau and Roy Porter. Manchester: Manchester University Press.

Pynchon, Thomas. 1973. *Gravity's Rainbow*. New York: Viking.

Rabinow, Paul. 1990. "Reflections on the Human Genome Project.'' Paper presented at the American Anthropological Association Meeting, New Orleans.

———. 1992. "Meta-Modern Milieux.'' Paper presented at the American Anthropological Association Meeting, San Francisco.

Rapp, Rayna. 1993. "Real Time is Prime Time: Ultrasound Fetal Images, Medical Maternal Discourse, and Popular Culture.'' Paper presented at the American Anthropological Association Meeting, Washington, D.C.

Rheinberger, Hans-Jorg. 1991. "Experiment, Difference and Writing.'' Verbund für Wissenschaftsgeschicte, Somerakademie "Writing Science.''

Richards, Thomas. 1993. *The Imperial Archive: Knowledge and the Fantasy of Empire*. New York: Verso.

Rickels, Lawrence. 1991. *The Case of California*. Berkeley: University of California Press.

Rosen, Robert. 1993. *Life Itself*. New York: Columbia University Press.

Ross, Andrew. 1991. *Strange Weather*. New York: Routledge.

Rouse, Joseph. 1993. "What are Cultural Studies of Scientific Knowledge?'' *Configurations* 1.

Said, Edward. 1975. *Beginnings*. Baltimore: The Johns Hopkins University Press.

Schroedinger, Erwin. 1967. *What Is Life?* New York: Cambridge University Press.

Serres, Michel. 1982. *The Parasite*. Baltimore: Johns Hopkins University Press.

Shell, Marc. 1978. *The Economy of Literature*. Baltimore: Johns Hopkins University Press.

Shapin, Steven, and Solomon Shafer. 1985. *Leviathan and the Air Pump: Hobbes, Boyle, and the Experimental Life*. Princeton, N.J.: Princeton University Press.

Snyder, Benson. 1983. "Literacy and Numeracy: Two Ways of Knowing.'' *Daedalus*.

Snyder, Solomon. 1989. *Brainstorming: The Science and Politics of Opiate Research*. Cambridge, Mass.: Harvard University Press.

Sobchack, Vivian. 1993. "New Age Mutant Ninja Hackers: Reading Mondo 2000.'' *South Atlantic Quarterly* 92.

Turkle, Sherry. 1984. *The Second Self: Computers and the Human Self*. New York: Simon and Schuster.

Traweek, Sharon. 1988. *Beamtimes and Lifetimes*. Cambridge, Mass.: Harvard University Press.

———. 1993. "Whose Progeny? Reproducing Engendered Physicists in Japan and the U.S.'' Paper presented at the American Anthropological Association Meeting, Washington, D.C.

Visvanathan, Shiv. 1985. *Organizing for Science: The Making of an Industrial Research Laboratory*. New Delhi: Oxford University Press.

———. 1988. "Atomic Physics: The Career of an Imagination.'' In *Science, Hegemony, and Violence*. Ed. Nandy.

———. 1988. "On the Annals of the Laboratory State.'' In *Science, Hegemony, and Violence*. Ed. Nandy.

Watson, James. 1968. *The Double Helix*. New York: Atheneum.

Weinzenbaum, Joseph. 1976. *Computer Power and Human Reason*. San Francisco: W. H. Freeman.

Wiener, Norbert. 1956. *I Am a Mathematician: The Later Life of a Prodigy*. New York: Doubleday.

Yukawa, Hideki. 1982 [1958]. *The Traveller*. Trans. L. Brown and Y. Yashida. Singapore: World Scientific Publishing Company.

MIND, BODY, AND SCIENCE

These contributions develop conversations with scientists involved in medical science, one through the mind and the other through the body. Dramatic contemporary shifts in the understanding of the idea of humanity are nowhere more apparent than in these conversations, as is the salience of moral/ethical preoccupations in the discourse of the subjects of these pieces. Joseph Dumit's presentation conveys the way that "mind" through PET technology not only materializes for his scientist interlocutors but also takes on a certain independent, animated quality. The chapter by Mary-Jo Delvecchio Good and her colleagues simply and powerfully captures the binds and the rationales for resolving the latter of specialists who traffic in both primary research on disease process as well as treatment.

Joseph Dumit

3

Twenty-First-Century PET: Looking for Mind and Morality through the Eye of Technology

This paper represents a number of forays into the world of PET (positron emission tomography) scanning, a relatively new medical imaging technique, described below. PET scanning purports to provide images of our brains in action (for example, as we think about a word), and images of different kinds of brains (for example, normal versus Alzheimer's-afflicted). I have been mapping this field through oral histories, interviews, and participant observation in order to discover the processes by which new information becomes significant knowledge about ourselves as humans and as persons. Choosing who to interview and present is a continual challenge:

Dumit: I saw in one article that you were quoted as being the "Father of PET." Are there many fathers of PET?

Ter-Pogossian: Well, I'm glad that you are saying that. Because when somebody referred to me as the Father of PET, I said, "I'd rather be the mother of PET, because many offspring have many fathers, and only one mother. As a matter of fact, some offspring have no father at all!" Of course, there are many fathers. And I think that if you look closely at history there are masses of fathers of PET.

• • •

I am interested in presenting a variety of perspectives on the development and multiplication of PET; not to cover the field but to point to tough questions of science, of popular dissemination, and of ethical interest; and the approaches being taken toward them. I view this as a beginning conversation. I am putting together discussions which took place in different cities, on the phone and in person, over a period of six months. The interviewees are all major players in the arena of PET scanning, broadly defined, an arena which includes *Newsweek*. They are not the sole representatives of their disciplines or their institutions; they do, however, point to diverse and even contradictory approaches to slicing up the brain with PET and reading its mind.

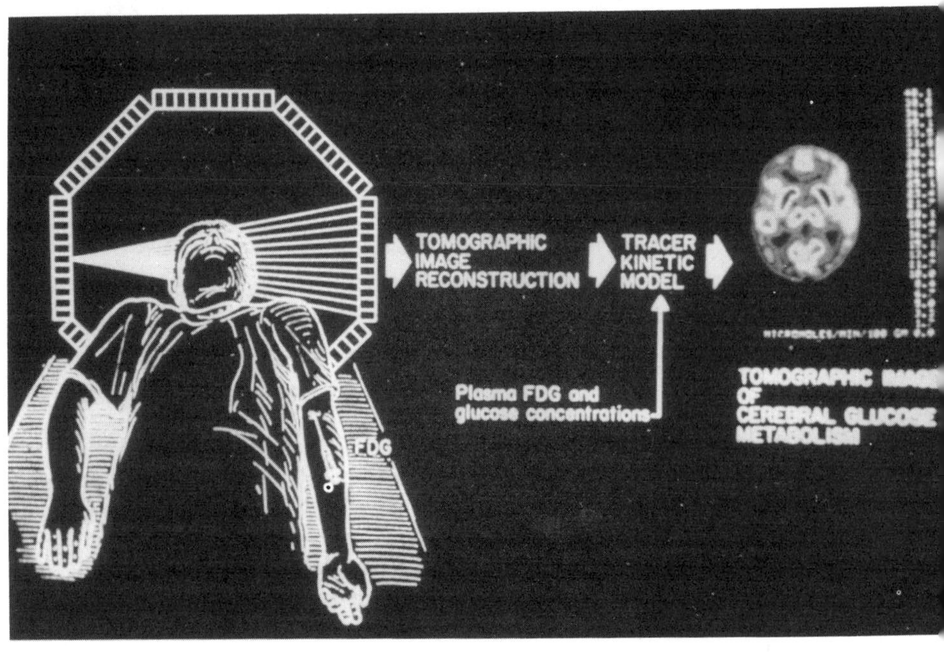

Schematic display of the process of PET scanning. (Courtesy of Michael Phelps, UCLA)

Michael Phelps is the head of the UCLA PET center and one of the fathers of PET.

Phelps: Suppose I am just an ordinary person and I ask you, What is PET?

Dumit: It is a device, like a CAT [computerized axial tomography] scanner, but isn't. A CAT scanner produces images of the structure of the brain through X rays. It shows where things are. With PET you are able to produce an image of where a radioactive molecule is going in your body. First you take a cyclotron and produce a radioactive molecule. This is then chemically attached, or "tagged" onto a molecule (like water) or a pharmaceutical (like lithium). This radiopharmaceutical is then injected into your body and the radioactivity it emits is captured in a ring of detectors (the PET device). You then mathematically reconstruct a "slice" of the person into an image. What you get is a cross-sectional view of where that molecule or drug has been. You can use this to find out where in the body the radiopharmaceutical is, and how much of it is there.

Phelps: Another way to approach the explanation of PET is to forget about the technology initially and focus on the problem. Inside the body there is all this biology going on, there are all kinds of cells that are metabolizing things or moving around and doing things, signaling to each other.

We'd like to be able to watch that. So that is the objective—you know it is there, you'd like to be able to build a camera that can watch it. Well one way to do that is first to say, "If I were really little, I could go in there, move around and watch those things." But since you can't go in there, you can send a messenger. So you do that, you say, "Well, I want to look at one portion of this." So I take a molecule that will go and participate in that. It will go through that process. I take that molecule and I put a source on it that will emit back to me. Then it will take this journey. So I inject it into your bloodstream, it goes on this journey, it goes throughout your body with the flowing blood, and depending on that molecule, it will go into some organ in which that molecule is used. And you have a camera and can sit there now and watch that molecule, watch it go through the blood supply, go into the brain, go into the tissue of the brain, and actually go through the biochemical process. So you have a camera that allows you to actually to watch some of that, watch the biology of the body. So that is really the objective. Forget about the particulars of the instruments, that is really what PET does. It reveals to us something that we know is going on inside your body, but that we can't get to. And it does it in such a way that does not disturb the biology of the body's chemistry—this molecule is in such trace amounts that it, the body, goes on about its business, apparent to us but transparent to the body.

DUMIT: Like an ideal participant observer.

PHELPS: It is an observer that doesn't disturb you. That is, what happens would happen with or without that observer there. If you are an observer at the presidential conference and bother the president, then you distort what would have taken place had you not been there. But this molecule is given in such trace quantities that it makes no disturbance. Whatever happens would have happened whether you were there or not.

DUMIT: The PET researchers at UCLA emphasize this pharmacological approach, tracking molecules. How does this differ from activation studies?

PHELPS: Activation studies—brain mapping—is a program that says, "We would like to just understand the macroscopic way the brain is organized and how it works at all." Both PET and MRI [magnetic resonance imaging] are developing ways and have been used to pursue ways to just map out how the brain does what you know it does.

Everything originates and is carried out through chemistry and biology. What you think is structure is really that portion of biology that says we have to build and maintain a framework, and within this framework we will go and do things. If you look at disease, all disease starts by an error in the chemical process. It typically goes about eroding away the reserves, and the compensatory capability of the biological system to accommodate that normal process. Eventually, if the person becomes symptomatic, it has eroded away the reserves, there is no further way to compensate, and now the cellu-

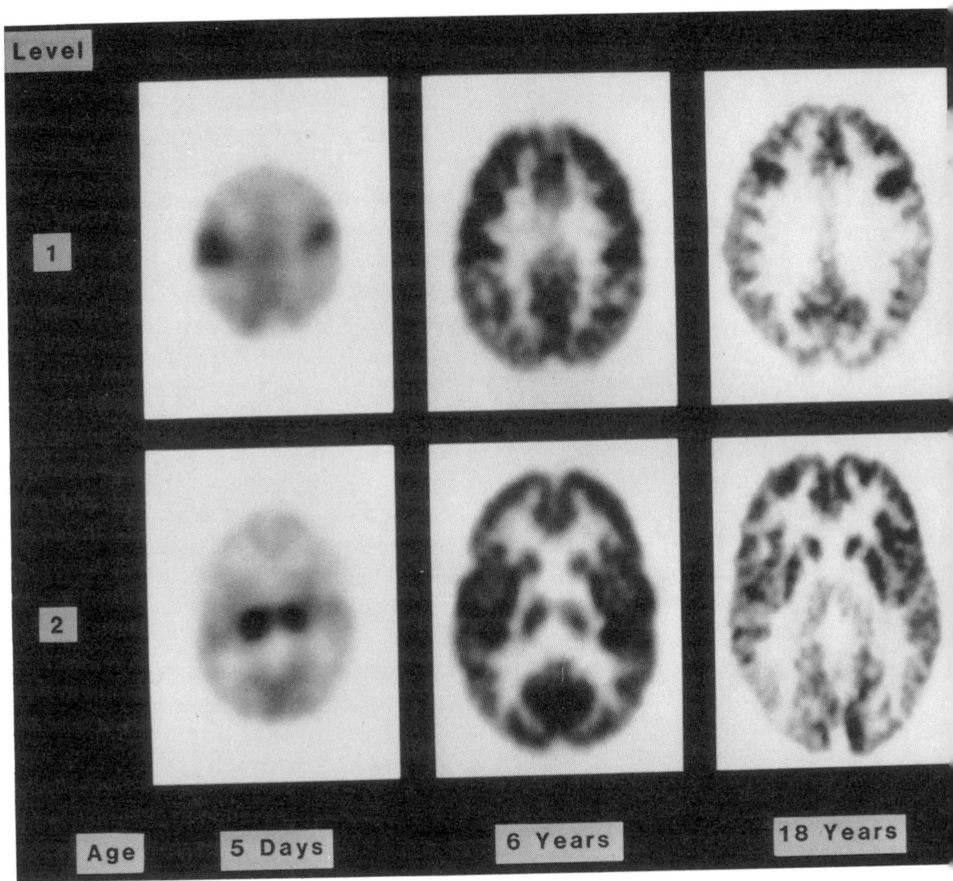

The metabolic maturation of the human brain at 5 days, 6 years, and a young adult. (Courtesy of Michael Phelps, UCLA)

lar function begins to fail. But like any system, it is designed to accommodate a lot of changes.

DUMIT: So has your notion of how the brain works, and your brain in particular, changed as you have worked with PET, seen activation studies, and things like this?

PHELPS: Well, PET has certainly caused my brain to have to work harder. You know, I think the thing that probably has influenced me the most, in terms of how the brain works from the use of PET is to watch how it develops. For the first time in humans, we actually watched the human brain develop from birth to adulthood, and it was phenomenal. To see at birth that the areas that were functioning in the child were not surprisingly the phylogenetically evolutionary-wise old systems, the older parts of the brain, the cerebellum the central structures of the brain, the thalamus, the old motor cortex. And if you look at a child, that is pretty much how they look. They can't do anything, they can't even reach out and grab something, they have reflex actions. They have their motor system, they have a basic simple set of reflex activities—they suck, startle—they can't do a lot more. But then as you watch them, there is a progressive unfolding or expanding of their behavioral repertoire. To see your own children grow up, while you are studying it for the first time and watching the brain of adults mature, one structure after another matures. And as they mature, that child adds one function after another to the brain. And to watch the final periods where the frontal cortex, inferior to superior, posterior to anterior, develop as the cognitive skills develop. It is evolution that you watch unfold in front of you. You watch a child begin to function through all of this. You see that by the age of one the brain looks equal to an adult metabolically, even though the child is certainly not mature. By the age of two or three the metabolic rate is twice the adult and stays up there till the age of ten. Then it will begin to decline. We are beginning to map out in humans that neuroplasticity period and then to use that plasticity period to go in and identify epileptogenic areas in infants and children. We take out major portions of the brain, up to a hemisphere, and see that they have phenomenal compensatory recovery that you and I can't do any more. To see that we laid the foundation for that, we introduced that, both in the fundamental knowledge of how the human brain develops and to use that knowledge to take care of infants and children, is incredible.

We used the data to teach two secretaries of education, Cavazos, and Lamar Alexander. While this country focuses on higher education—and there is good reason for this—it has lost its way in the most critical time of education, which is early education. During the early time, particularly between the ages of two and ten, the environment shapes and selects connections in the brain to be kept and kills half of them, the other half. So the environment is now anatomically altering the functional organization of the brain. And that is rather fundamental. So whatever you believe is important for your children to learn, you better teach it in those years. It is wired in.

DUMIT: I've looked at a number of popular articles on PET, in *Newsweek* and other magazines. When they do cover stories on PET, I'm interested in how the story grows very quickly on what can be seen and what can't. In terms of the history of PET, it seems these two things go hand in hand, the promise of a future and what is being done now.

PHELPS: Yes. When scientists come together with a representative of the public, particularly the press, whether it be the *L.A. Times,* or *Time* or *Newsweek* or whatever, there is a very difficult situation in that the objectives are quite different, the needs are quite different. The scientist, if he is a good scientist, begins from the perspective of being conservative—not wanting to say something that he cannot offer the evidence and the data for. The media person wants the bottom line—not what you have done, but what is the impact on society from it. This sets up a paradoxical conflict, and then the public writer has to meet it. It is just like the images that you mentioned, if you write something that is absolutely true and boring, and the point can't be seen or made, it is irrelevant. You have both the obligation to do the science and do it well, and say take it to the limit of what you can prove, and on the other side you must convince another person. And therein lies temptation, speculation, exaggeration, a lot of *tions;* versus a lot of other things that end in *tion,* such as validation and foundation. You have to come to terms with that, you have to be careful with that. Lots of major figures that want to bring brain science and the public together will say to us, in the best of intentions, "You've got to say it. You've got to tell them what you believe will come from it. Be true to yourself." But you can't just show them the raw data. It means nothing to them. So you've got to be careful with that. It is just my impression that when you go into the neurosciences with PET there is more opportunity for exaggeration, because the brain is an exaggerated organ. There is just a lot of weird stuff.

DUMIT: Do you think the fact of being able to have "images" of the brain has a particular effect on this?

PHELPS: Yes. In several ways. One, it provides a means to communicate to the public with a picture. You know, I don't care what you are doing, you can put out all the words in the world, but a picture is very powerful in communicating the information. So it is a medium that on one hand allows you to communicate more easily with the public, but on the other you can start imagining things from this image that aren't real.

DUMIT: I have wondered if there has been a tension in PET between the image as qualitative and the image as quantitative. PET seems to be a great difference engine, it can produce a difference between different kinds of people or different kinds of traits and states.

PHELPS: Yes. Of course, part of that is also conflict built into PET. There are many different people involved in PET. There are conflicts between those who use it qualitatively and superficially, and those who use it analytically, by their nature. There can be people who are trying to use it to do some clinical research, or deal with the limitations in people. You know you can't be doing their biochemistry by other techniques, pulling out their tissue and analyzing it. So the research paradigms are quite different. One can be exploring the relationship between the brain and behavior in the domain that they normally work in. And the basic neurochemist involved in PET will say, "I don't like that, you don't know what you are talking about. What are the units of your data, depression on this axis versus color on this axis?" So there are a lot of different factions within PET, because it does go from basic chemists and biologists to clinical investigators, and their criteria of an experiment is quite different.

DUMIT: This is one of the challenges of writing history. Not only trying to figure out how to be fair, but figuring out where to cut, what kinds of stories of causation and seeds and so on.

PHELPS: You know, science tends to be an evolutionary concept. It depends on one's religious beliefs. Supposedly there is only one creation: truly from nothing before to now. But science doesn't typically do that. It is a religious experience if it is really truly an absolute creation. You know there are things that other people did before you. And there are certain benefits that come to you from the influence of people to what you are doing. Then there are certain portions where you put it in, and that's your contribution. But when laid out in the evolution of science, you know you are just pieces in it.

Sharon Begley is the science editor at *Newsweek* and author of articles and a cover story on PET.[1]

DUMIT: How do you decide who to interview?

BEGLEY: It's usually fairly obvious. You start with a few people. In the case of the PET cover, there was a paper presented at the Society for Neuroscience [Meeting]. It was just some interesting memory thing, for which they used a PET scan. The editors saw that and thought, "Well gee, the pictures are gorgeous. Maybe there's more that can be said of this." So you start pulling the papers and talking to people preliminarily. You get a sense of who is doing what in the field. In this particular case, there are not a lot of people using PET scans for basic research. I would say we talked to half a dozen people. And then there were a few more using other technologies, and we drew those in at the end, SQUID and MRIs and things. In this particular case it was a very circumscribed universe of subjects. In other cases, of course, you have to cast the net a little more widely. You look around for people who are doing interesting papers, you look around for people who are critical of the technology, or the conclusions drawn from it. People ask me this all the time, and all I can say is, it is never a dilemma, it is always obvious who you should talk to.

DUMIT: What do you look for in an interview regarding a new technology, like PET scanning.

BEGLEY: Well, it has to be interesting and important. If it is a new technology, then the assumption is that it is certainly different, and probably better, than what came before. So you have to search out whether that is really true, or whether it is just a gadget or gizmo. You ask yourself, Does the world need another one of these? I myself don't do much new technology. I do mostly pure science. It's almost incidental that the work you're referring to, the PET scan work, was done with a new machine. If they were making those same conclusions based on something else, it would still have been a *Newsweek* story, because they were finding out about information processing and language centers, and other things. How they found it out was really incidental to the story, except for the fact that it gave us nice pictures.

DUMIT: Those pictures seem to be a key element.

BEGLEY: Well, I think it does touch a lot of people's imagination. The idea that you can "see" what a brain is doing during a particular task, that does intrigue people. So apart from what you can learn scientifically, about where glucose is being used in a brain, yes, it does have a sort of man-in-the-street appeal: "Wow, that's a brain at work and I can see how it's different when it thinks about an animal and when it thinks about a mathematics problem."

DUMIT: What kind of reader do you imagine when doing an article like this?

BEGLEY: I don't know, that is a question we always ask ourselves and there has never been a very good answer. At the coarsest level, you are writing for an educated layman. We generally assume at least a high-school education. I prefer not to think about it in terms of number of years of education, but I guess we could give them more temperamental qualities, like curiosity and an ability to be surprised. I'm a firm believer that you can write on almost any subject and make it believable to people. In the case of the brain cover, with the PET scan on the front, the conceit was that you would have this italics running through it, telling the reader to do something. "Think of a skunk. Think of these four objects. Look at these four words." It's just a silly way to draw people into the story. The general rule is that whenever you are writing about something personal to people—and what could be more personal than how their brains work—it is not that hard a sell. But I will say that the punchline for many of these things is, "An area two millimeters in the corpus whatever is involved when you do task X." That means nothing to people, that is a big "So what?" Again, that is why when we talked about where processing is going on, we tried to tie it to other processes that go on nearby, or that are connected neurally to that thing. And then you could draw the conclusion that memory is related to spelling, to give an example from the literature that does mean something to people. Just to locate a particular area is just one line short of a punchline that is meaningful to readers.

DUMIT: Do you have a sense of differences within the PET community? Some seem to be working on things which you might call "general human differences" like intelligence, gender, and so on, while others are working much more closely on little things.

BEGLEY: Yes, we concentrated on the latter, like concentrating on a verb rather than a noun. Intelligence, yes, you can say that you are looking at intelligence, but you have to make sure that the one thing you measure can be expanded to this very squishy thing we call intelligence. Some of the work at UC–Irvine does that. When you ask them what they mean by intelligence, they say, "It has something to do with learning from experience." This is with the game called Tetris, it is a computer game. Learning from experience, generalizing from specifics, this is something that some can do and others can't. And if you buy that this is at least one component of intelligence, then yes, in some sense, they have measured at least a part of intelligence. As far as gender differences go, yes, you can show that there is a different processing component between men and women. The corpus collosum, and the hemispheric differences have been talked about long before PET. Of course, a lot of the problem is that a lot of the work looks at only five subjects, or something, and the differences among the group are often greater than the differences between groups. Yes, they can find something, but whether it is generalizable to between all men and all women, that is a separate question.

DUMIT: One of the things I am fascinated with is how after a very detailed study of nineteen or twenty people, it boils down, at least visually, to one or two scans.

BEGLEY: Yes, I think that is right.

DUMIT: From the illustration "PETting the Brain" the implication I see is that if not now, then soon. PET will be able to tell us the difference between different kinds of people.

BEGLEY: Well, I don't know. Maybe that is an inference that people drew. That implication was not intentional. I think we are looking at a microcosm. The point was that you take known cases; in other words, these scans were not diagnostic or predictive. You identify the two poles going in, and then you take PET scans of these people. And then you can, lo and behold, see differences. Which is not to say that if you PET scan the brains of a hundred people walking outside this door you can get something as striking, or that these hundred people will fall into these two separate groups. I didn't address the diagnostic possibilities of PET because I don't think that they are really there, and I really dislike erecting a straw man and knocking him down. I'd rather just not open that subject at all. I think it is silly wasting space that way.

DUMIT: Yes. That is, however, the first thing that jumps to people's head when I show them these. Or an instant critical reaction, "These pictures can't really show that. Why does it look like they can?"

BEGLEY: Well, see, in this case they could, but that is because the diagnosis had been made by other criteria beforehand. Again, the only point was, whatever you have, a mentally retarded person or a schizophrenic, and then a non-whatever, then you can see differences. Which is not to say that in all cases you can. I mean you will get false positives and false negatives.

DUMIT: Will you be doing future articles on imaging?

BEGLEY: Not in the future. If we've done it once as a cover, we rarely go back to it. As news happens, if you find something interesting about a brain, how it processes, I can see doing a half a page or something. Again because the brain is one area of science that people are perennially interested in. But no, I can't see again just hauling off and doing another roundup. As I say, I cover all of science, from astronomy to zoology. So I just have to get it right the first time.

Michel Ter-Pogossian is a physicist and director of PET at Washington University, St. Louis. He is also one of the fathers of PET.

DUMIT: Do you have a favorite image?

TER-POGOSSIAN: No!

DUMIT: I was curious. Do you consider yourself a visual person?

TER-POGOSSIAN: [*Pauses.*] Probably yes. But I had to think about that.

DUMIT: Right.

TER-POGOSSIAN: Let me put it this way. If I look at an image, or at numbers, I am more concerned with what it represents than how it represents it. I don't know whether that answers your question. But I am trying it. When I see a hot area on an image or a series of numbers, I try to think of whether it is an accumulation, or whether it is an inert process of some sort occurring, whether it is necrotic tissue, not concentrating activity. In general, this is what stimulates my thinking.

DUMIT: So images or numbers, it doesn't matter, it is the process.

TER-POGOSSIAN: It is the process, yes, underlying it.

DUMIT: Did you have a preference one way or another? Were you pushing for color, as opposed to grey scales, for PET images?

TER-POGOSSIAN: Well, color is certainly useful if you want to superimpose different images. It is also useful in establishing levels without having to go to numbers. We can have a color scale and each different color represents a five-percent increment. But often colors place boundaries where they don't exist, you can jump from one color to another. Otherwise, people like them, and they are just more attractive to show in color. The dynamic range is increased, though in an artificial way.

DUMIT: There are a number of circumstances where color isn't particularly helpful.

TER-POGOSSIAN: As a matter of fact, it can mask things, or rather, put in boundaries where there aren't any. All at once you go from green to yellow and you see a boundary of some sort and it is just a small increment one way or another. Still, in general we try to show [them] in color because they are attractive.

DUMIT: Yes. I talked to a science editor at *Newsweek,* and when they decided to do a cover story on PET, it featured Washington University. After they saw the pictures, they said, "These pictures are great, let's get the story." So one of the things that I am interested in is the color pictures in terms of the different things that they can signify. In one case they can signify that there is a lot of activity going on here.

TER-POGOSSIAN: Well, yes, they signify whatever you want them to signify. This is the pitfall, of course. You can emphasize, for example, a given phenomenon, very artificially so, if you want to do it with color. It is dangerous, too. You have to be very careful when you are using it.

DUMIT: Now when there is purple in this picture that is outside of the boundaries of the person's head there, is there any significance to the mottle that is going on?

TER-POGOSSIAN: No, this is noise. That purple, that is noise, this is reconstruction noise. The reason why you see lines is that they are really reconstruction artifacts. And you see that in any reconstruction scheme. You see that in computed tomography. However, very often you erase that by just windowing it out. In other words, these represent very low values as seen on this scale. So all you have to do is put a cutoff limit and it is removed. But that is what it is. And this, you see, this is a reconstruction artifact.

Parenthetically, the pictures that are particularly attractive that you have seen in general are doctored, in the sense of making them more attractive than they are at first. Removing things like, for example, what you have pointed out.

See [*pointing to the color scale along the side of the map*], this is misleading. Here you have a jump from one color to this one so that will put a very strong boundary here on the image and there is really no reason to do that. Actually, yes, you see this boundary here, you see a boundary but in fact there is very little boundary there.

DUMIT: On the other hand, the shift in this area where there is a very steep climb [in color, from green to white].

TER-POGOSSIAN: A very steep climb, yes. This is meaningful because indeed it comes from this value to a deep red here. And each one of these values is five percent; there are twenty steps to this scale.

DUMIT: Have there been movements to develop a standardized color scheme?

TER-POGOSSIAN: Not to my knowledge. There have been some semistandardized color schemes, there is really no standard set of colors. People have a tendency, of course, to use the scales that emphasize what they like to emphasize.

DUMIT: Yes.

TER-POGOSSIAN: Oh, that is all right.

DUMIT: Yes, I have been struck that each different institutions' pictures tend to look very different from each other. It seems very difficult to compare PET scans from different institutions.

TER-POGOSSIAN: It is very difficult. It is very very difficult indeed. It is misleading to just use purely aesthetic values. Very difficult and misleading.

DUMIT: I am interested in how you read images. There is a lot of literature on how radiologists read X rays, but there isn't that much on what it means to read a PET scan. Just because it is a functional image, you seem to have to know a lot more about how this particular image was done to look at it.

TER-POGOSSIAN: I think in many instances, not in all instances, but in many instances it is very fundamentally wrong to try and read a PET image the way you read a radiological image. Because in a PET image what you are looking at is not morphology, but, as you said, a function. But sometimes, if you are an individual who is acquainted with X-ray pictures, you tend to think in terms of morphology, and that is a mistake if you do that. A hot area may represent active living tissue, on the other hand it may represent damaged tissue just as well. So you know, this is a very gross example. The sooner you get away from the morphology concept, the better off you will be. Of course, you have to relate your images to morphology, you have to know which organ you are in and in which portion of the organ you are. But beyond that I think it is a mistake. That is one of the reasons why there has been a divergence, if you want to, between radiologists and internists in nuclear medicine. Radiologists are very, very morphologically inclined. Internists are much more physiologically inclined, biochemically inclined. An internist knows and uses biochemistry.

DUMIT: I was just thinking of functional images. I am still working [at understanding how] a function, or something that happens through time, is represented in a PET picture. Because it looks like one instant in time, [like a photograph] even though for most PET pictures it is two minutes, maybe twenty minutes.

TER-POGOSSIAN: Sure, but that is not the important portion at all.

DUMIT: What would you say is the important portion?

TER-POGOSSIAN: It is the function itself. And the function itself is again derived through the application of a physiological model, which is usually not instantaneous. In some cases, the instantaneous image is perfectly all right. Let me give you a specific case: the accumulation of FDG in a brain tumor. Well, that is a fast phenomenon. But if you are trying to measure actually the glucose metabolism by FDG, an instantaneous picture really is not going to give you anything. You have to take a series of measurements, you have to plug into measurements of blood activities, and very often reconstruct a completely different image from the one that you obtained instantaneously, taking into account the other factors.

DUMIT: With the oxygen studies it is the same thing.

TER-POGOSSIAN: Yes. The image itself is only the starting point. In fact we often don't show the images. You show an image if you want to show the morphology in a given portion of the heart.

DUMIT: Activation images, though, would be kind of in between?

TER-POGOSSIAN: Activation images are in between, you are right. But you see, to interpret an activation image quantitatively you may have to calculate the flow, and the calculation of the flow is a dynamic process. That is probably, incidentally, the reason for the relative difficulty in interpreting PET images. The complexity, perhaps not the difficulty, is something that has slowed down the development of PET. In clinical practice you don't like to do that at all. And for very good reasons: you don't have time for that.

DUMIT: Because it takes a lot more time to go through and make sure the physiological—

TER-POGOSSIAN: Well, yes. It requires all the measurements, the blood activity as a function of time. That means you have to take samples in some way, and measure the samples. And then you have to sit down with a computer and apply a model of some sort to make sense out of that. It is much easier to look at a picture and say, "You know, it is right there." That is not the way you do things in PET.

Although, that is not completely true. In certain cases, indeed, you have purely morphological images. The FDG accumulation in a brain tumor is purely morphological, but that is not the strength of PET.

DUMIT: When you do an article for a journal for PET—

TER-POGOSSIAN: I'm working on one right now.

DUMIT: —and you are trying to select images for the article to demonstrate one way or another what is going on. It looks to be the case—and I can't tell because often there is not that much information presented about why these two images were chosen—it looks to be the case that the most extreme images are chosen.

TER-POGOSSIAN: Sure.

DUMIT: I'm curious about this. Is this a kind of heuristic idea, that these images display the difference that is being talked about? Are they representative? What is—

TER-POGOSSIAN: Well, it varies. It depends on how you show the images. For example, if indeed you want to emphasize a difference, you show the extreme cases. However, in any responsible article it behooves you to emphasize also the overlapping areas and these are any kind of study which involves, say, the comparison of something or another, it behooves one to use a statistical analysis. In most instances we have a statistician on the staff and we ask him, "How do we present the data?" And he in general has his own approach. I'm not a very good statistician myself. And he gives you that data.

You might also, if you so wish, show images that on the contrary show a false positive, it depends on what you want to do. But yes, in most instances you select the images that prove your case. However the case is also proven, supposedly, in your text.

DUMIT: Right. The reason why I bring it up is because when I look in the mass media, and I have done a mass media search and looked at the places where PET pops up. When it pops up, the only thing that remains from the journal articles is typically the pictures. Even the captions are rewritten. So the appearance is one of massive difference, distinction.

TER-POGOSSIAN: That is a very good point. For example, the pictures that were shown in that movie I mentioned this morning [*Rampage*] showed really differences of something or another. I don't know how they were obtained, I don't know what the pharmaceutical was, but they were striking PET images. I haven't the slightest idea of whether they were obtained in a schizophrenic patient or not. But yes, that is what remains, and that's a pity that that is what remains. But that is the fault of the media. I shouldn't say the fault, because that is the way the media does things. I mean, out of what happens, and I mention it again, what happens in Yugoslavia, the media show a screaming little girl, or a dying little girl, or a child in a hospital with his legs blown off. And that is what remains. I think that we all have the vivid image of the starving child in Somalia—you know, they look like skeletons. Of the Holocaust we have pictures of these stacks of cadavers. Surely the press and the media direct us. I don't think there is anything particularly unethical.

DUMIT: No. I am interested, though, because I think the media's portrayal of this plays into the ability of the courts to accept PET. I don't think this is a completely unrelated connection. I'm not sure, but I am trying to look at this. But a certain acceptance of PET's extreme viability by the mass media plays into the courts being able to accept it.

TER-POGOSSIAN: But the courts should use a certain degree of skepticism about things like that (as in the above-mentioned film). It is not enough to have one individual tell them that it is great. Why not go to three other individuals. I can find twenty individuals who will stand up and say that this is questionable. And they can show images that indeed deny that, in other words, show images of schizophrenics that are absolutely normal. It is not clear to me that one can analyze these images to the point of making them useful to courts. But as you say, the courts are quite willing to take a test, I mean, anything that shows color pictures on a screen—click—all at once, this is it.

DUMIT: Right. The reason why I brought up the X rays that were used in the courts at the turn of the century is because the argument before a Supreme Court was, Why should the jury see X rays? If X-rays are for the expert, the doctor to interpret, why should the jury even see them? They'll be persuaded by what they think a fracture looks like [on an X ray], not by what a fracture [on an X ray] really looks like. And this was brought up with regard to the use of the PET scans as well.

TER-POGOSSIAN: Oh, I see. I didn't realize that.

DUMIT: Why should the juries see PET scans, since they are going to see the boundary lines and say, "There is a difference," when only an expert should interpret it? They will be persuaded by the pictures, in other words. One judge recently ruled that juries should be able to see PET scans.

TER-POGOSSIAN: I'm not a judge. But showing PET images now to a jury without extensive preparation doesn't make any sense whatsoever. I mean, if whoever shows these pictures was given a stack of twenty pictures of perfectly normal subjects and twenty pictures of schizophrenics, and then you shuffle the pictures, I doubt that he or she would be able to stack them, to unscramble them. There are some areas, which seem to be associated sometimes with schizophrenia, but it is a minefield. I think that it is too early on that; adequate data are not in.

I think that in the future, PET will play a very important role in the understanding of psychiatric problems. What I am concerned about is the use presently of PET as a tool in assessing mental disease. I think that it is an extremely important tool in psychiatry, but it will take years of studies to understand what we are seeing.

Richard Haier is a psychologist and former Director of PET at the University of California, Irvine (UCI). He has also been featured in *Newsweek*.

DUMIT: You have always been oriented to research at UCI, yet this is also one place where clinical PET is practiced regularly.

HAIER: The PET center was always set up here to be a research instrument. Whatever clinical cases we did were to generate supplemental money to support the research, not the other way around. It has been reversed around here lately, in some ways they give priority now to clinical cases rather than research cases, which is terrible. We did clinical cases very early on. I think that we were the first, and even the only center that routinely did clinical cases.

DUMIT: And they were clinical cases in schizophrenia, mental illness.

HAIER: Well, they started out that way. We were really the only people around here doing this. I remember some of the very first cases that we got. We had a family fly a kid in from Switzerland. When word got out that PET scanning was available, some wealthy families would do this. We have some early cases in trying to help make a differential diagnosis between schizophrenia and other things. We had some early cases on head injury, we had some early cases referred by neurologists who were perplexed at the presenting symptoms. But what happened after a relatively short period of time was that the whole clinical operation turned out to be weighted heavily on legal issues. So right now the vast majority of our clinical referrals come from lawyers. Many of these are head injuries and many of these are convicted felons, often murderers. For head injury we try to come up with an objective assessment of whether a head injury is present or not. In the case of the felons, the PET scan usually comes up in the sentencing phase of the trial, after they have been found guilty, as to whether or not there is any evidence of preexisting brain damage that may be a mitigating circumstance against the death penalty.

I gave a talk several months ago as the guest of the chief justice of the Orange County Superior Court. There were superior court justices there and many prominent attorneys. I gave them a whole lecture on how PET works, how it is interpreted, what to watch for. They asked me things like, "How would an unscrupulous PET operator fake the results?" This threw me, because I never had that question before. I really had to think about that. But [Monte] Buchsbaum testified in a number of trials. So in Orange County, PET scanning has essentially become the community standard. We get a lot of requests from lawyers, not that many requests from psychiatrists any more. In the beginning the local ones were kind of curious. I'm not sure it turned out to be that useful to them. We had some neurologists, we had some candidates for neurosurgery come through, but by and large the bulk of it seems to me to be legal oriented.

DUMIT: I just saw this movie *Rampage*.

HAIER: That was filmed here. *Rampage* was filmed when we were still upstairs. William Friedkin came with a film crew and spent the day filming that scene in the PET scanner. So it was kind of fun for everybody. It's our PET scanner up in the old lab. And some of our technicians were in the movie, but as I understand it, they were cut out. I haven't seen the movie.

DUMIT: Right. There is just one scene in there where after the PET scan is performed, somebody, a technician, flips up the image and he can read it. He says to both lawyers, "These areas are abnormal, definitely. You are seeing a picture of madness."

HAIER: Ha. That was an actor, that wasn't really one of our people. I remember at the time that there was some controversy within the hierarchy here as to whether or not we wanted to participate, not only in the movie but in the whole idea of doing PET scans on murderers. "Wouldn't the public think that we were just participating in getting people off?" And Buchsbaum and I just rejected that whole idea as being a Luddite idea. I mean, really, the question is, Do these people have brain damage? Can you make a case that the brain damage has anything to do with very abnormal behavior? And isn't there a scientific aspect of this? And shouldn't we be looking at it?

DUMIT: So there definitely was a feeling, I'm on the court for a bit more, that you could diagnose mental illness on the basis of scans?

HAIER: No, not diagnose. But you could clearly say, "This scans shows hypofrontality, which we believe is consistent with the diagnosis of schizophrenia." So you know you have the usual parade of psychiatrists coming in saying this person is schizophrenic. Now you have PET. Well, the PET scan is consistent with schizophrenia, but hypofrontality by itself is not diagnostic of schizophrenia. Other people have it, too. Let me put it this way: some juries look at a PET scan that has clear brain damage and decide this is a mitigating circumstance. Other juries will look at a similar PET scan that is obviously brain damaged and decide, "Tough. The facts of the case are so horrific that we don't really care that this guy has brain damage. He gets the death penalty." So the PET turns out to be another piece of data for the jury to consider, and you can't judge how the juries are going to take it.

I know that Buchsbaum was successful in some civil cases, lawsuits. There was one where a guy was beaten by police and suffered brain damage. And the PET testimony was very important in establishing in fact that the brain was damaged, because those pictures are very compelling. Buchsbaum had a few cases where the opposing side and the opposing insurance company would write a settlement check in the hallway before going into court once they saw the PET scan, which was very dramatic—a piece missing, all black, kind of thing.

And Buchsbaum also told me that one judge, in pretrial, decided that he would admit PET scanning—no judge has decided that he wouldn't—but this judge decided that he wouldn't allow the jury to see any of the pictures. He would just allow the testimony about what was in the pictures because he felt that the pictures in themselves were prejudicial. This strikes me as absolutely true. This seems to me to be a very wise decision. Because those pictures are very compelling, and what I told the superior court justices is that if you wanted to manipulate PET, it was very hard to fake it by saying, "What can I think now to activate my left anterior thalamus?" But as an operator, I can choose the colors on the scale and I can choose the interval on the scale, and I can make a lot of areas black. And that would look very dramatic. That is about the worst thing I think one can do to make a visual presentation that was not entirely accurate.

So I think probably there is a big future for PET in the courtroom. Particularly as we get more and more sophisticated about it. Now we have data on about forty-four felons; that is a very large number. We have tried to work this up into a scientific paper, but they are not randomly selected, we don't have a lot of ancillary data on these people. You know, we don't have a standard neuropsych battery on everybody, we don't have a clear history on everybody. Because we were only asked to do a clinical job, to just take the PET. Accumulating data on all of these people has been very difficult, but I did have a student go through and count up the number of abnormal areas in each one of these people and rank order them from most abnormal areas to least. And the fascinating thing about this rank order is that there are a large number of these forty-four people who have very few abnormal areas. The other group that has a high number of abnormal areas tends also to have a concurrent diagnosis of mental illness, which means that their mental illness is a part of this. This also means that there is an awful lot of very violent people out there with no identifiable brain abnormalities. So I don't know what this means, and this is not really publishable because of the way the data were collected, but it was pretty interesting.

DUMIT: Yes. It raises a lot of questions about—

HAIER: So you can see that with several more years of experience and data collected in a scientific way, sooner or later it might be possible to say, "This is a picture of madness, this is a picture of impulse control, this is a picture of schizophrenia, this is a picture of sexual aberration." I don't think that is impossible.

DUMIT: I guess the worry is always that if there are normal people who have PET scans that look similar to whatever happens to be the current definition of that, that the PET scans will take over the diagnosis.

HAIER: I'm not so worried about that, because that is not our system. Our system is that people are in charge. And they weigh this like they weigh anything else. There is kind of a knee-jerk Orwellian fear that people have. That is a popular idea, though. I don't necessarily have that, I think it is pretty overstated. I think the system is very good.

DUMIT: How did you get started on the intelligence experiment?

HAIER: Well, I had raised some money and got some free scans. I said, "I want to do my own experiment. We were doing all these things on schizophrenia that I enjoy, but I think it would be fun to do something I have always wanted to do, and that would be a study in normals of individual differences in intelligence. You know, to start out in a new direction. In addition to our grants in psychopathology, let's see if we can get some grants to study normal cognition." So I said the first thing that I want to do is get just a group of eight normal volunteers doing the Raven's advanced matrices test, a robust test of intelligence [which requires filling in complex patterns] I knew very well from my graduate-school days. I thought, for many reasons, this would be a very interesting kind of paper-and-pencil test to start with, as opposed to a traditional kind of cognitive experiment or word-processing experiments, which is not really what I do. Buchsbaum was happy to agree to this, so I got those eight slots, which turned out to be the study that wound up in *Newsweek,* because it showed this interesting finding of inverse correlations between glucose metabolic rate and scores on the test. Which implied that it wasn't always the case that higher metabolic rate was good for you.

DUMIT: You had these wonderful pictures that went into *Newsweek* and so forth. I've talked to a number of people about the difference between images that you use and images that you publish and put out. I was wondering if you could talk a little bit about—

HAIER: Well, there are two kinds of ways to approach this. A lot of PET groups have programs that force all individuals into the same outline so they can publish a group average picture. We don't do that. In principle, we think that distortion is a little too great. I mean there are a lot of individual differences in the brain, you get a lot of distortion when you do that. We're not against it, we just have a different method. As a matter of fact, we are kind of developing a group average method because we are kind of being forced to by some of these critiques.

There are a lot of aspects of this now. So we always publish group statistical data, usually analysis of variance, sometimes multiple T-tests. Our conclusions are based on the statistics. Most of the time, although not all of the time, we include a color picture, because journals like color pictures—everybody likes color pictures, and that is what they remember. When we do that, we select images that illustrated the group statistical finding. It is not the other way around, you don't make up the statistics to support what [the picture shows]. So the picture that was in *Newsweek,* I just took the person with the highest score and the person with the lowest score. And it looked so compelling, but that's what the effect was, that is why it was so compelling. I took the best exemplar, I took the best pair, to exemplify that. That is true. But I don't see anything wrong with that.

So yes, there is some selection, but I think it is all right to select people that exemplify the statistical result. But again, not everybody does, and this is part of the individual differences.

DUMIT: It kind of goes both ways. It looks from reading the news account of PET, yours and others, that it is very easy to go from the image right to the result or diagnosis or something like that.

HAIER: Oh, right. But we don't even do that in our clinical work. In our clinical work, we provide a list of all brain areas, the metabolic rate for that patient in that area, and the normal range for that area based on our files. And whether that person is plus or minus two standard deviations of the normal range for each area. We also provide the color pictures because the physicians want it, but we would be quite happy on the clinical reports just supplying the numerical information. I would be just as happy publishing papers with just statistics. But I think it does add something to put the pictures in.

DUMIT: Do you have preferences between color and black-and-white images?

HAIER: For PET, color. Clearly, because the color conveys a lot more information. It conveys the levels, so if you have sixteen colors, you can show sixteen levels, whereas black-and-white doesn't really show anything in PET. I have actually declined to provide black-and-white PET pictures for magazines that won't do color, that would only do black and white. Or even for textbooks. I won't provide any pictures at all for black-and-white. It is just not the same thing.

DUMIT: I think it is going to be my fate to not have very many color pictures—

HAIER: You are making a mistake. It is a big mistake to not have it in color.

DUMIT: I'm curious because the way you talk about PET, it seems that the images are a nice sidelight for you, but it is much more the quantitative information you gain from it.

HAIER: You have to be careful not to be seduced by the images. You have to understand from the scientific point of view, that the paper you write, you are not writing for *Psychology Today*. You are not writing for *Newsweek*, you are writing for the journals. You have to have statistics, and I have often said that, as fancy and as compelling as the PET scan is, you still need a good, solid research design, with an adequate number of subjects and the right statistics. You know, a lot of the early PET things were in magazines showing what happens when you listen to a Sherlock Holmes story. Remember that? Well, I don't think that was ever a published study. That was just a nice case for a magazine.

DUMIT: Still, especially in the popular media, I have been curious about the apparent ability to go backwards in these accounts, very easily from the color picture to the diagnosis. But that seemed to be more an effect of the news media.

HAIER: Well, you know, *Newsweek* took the pictures that I gave them. I was surprised at the tremendous amount of media attention I got. I spent a tremendous amount of time with radio, television, and print media, you know, newspapers, wire services, news magazines. And I would say that overall I was very impressed with the high level of attention to scientific detail. The only exceptions would be when newspapers would have nonscience reporters call. And you can spot them in the first five seconds. They're difficult, especially the headline writers from newspapers. I can't remember a bad experience where somebody deliberately misrepresented something, only some of the newspaper headlines on the articles that I didn't like. They

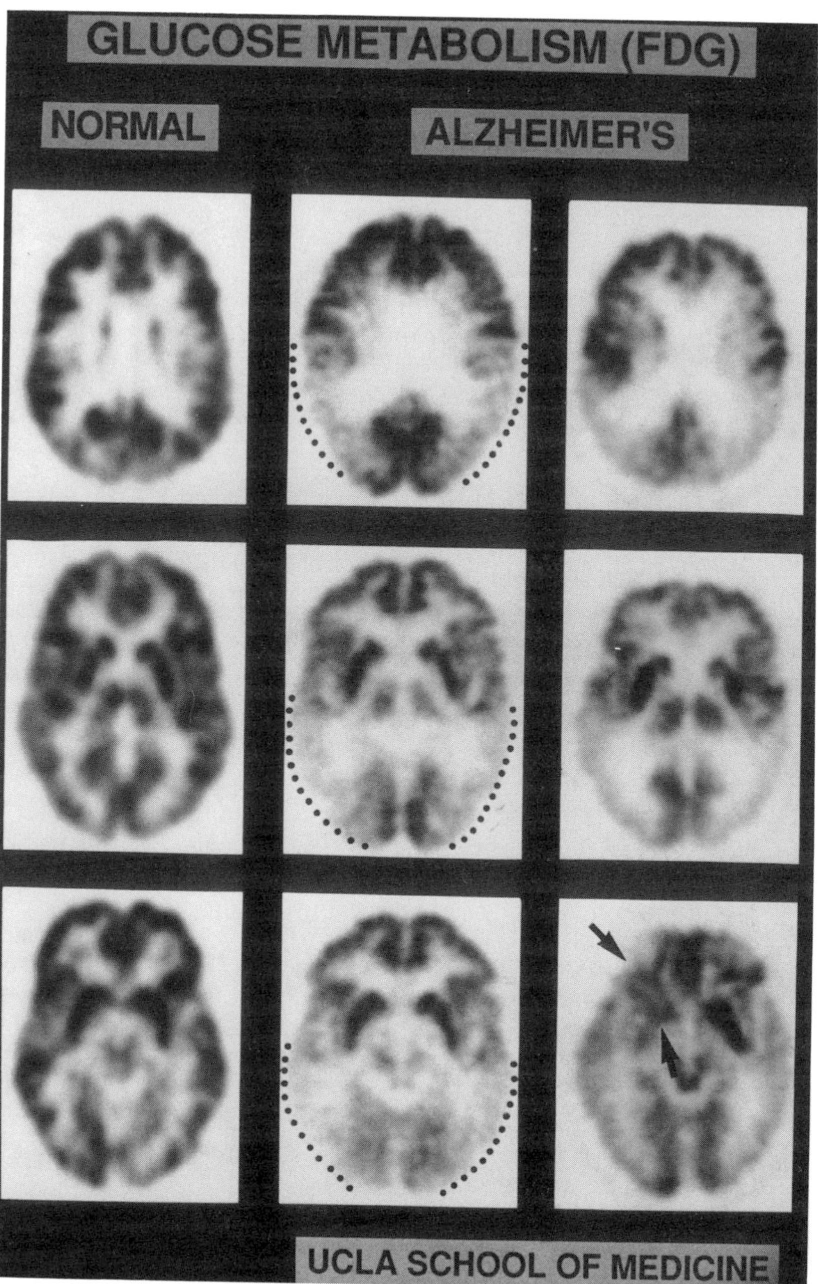

Compared with normals, Alzheimer's patients show a decrease in glucose metabolism in many areas of their brains. (Courtesy of Michael Phelps, UCLA)

would pick it up from the wire service but put their own headline on it. But that is relatively minor.

I do find a tremendous interest in PET scanning everywhere I go. I do a lot of public speaking, and I find that people are very interested in this. And they are always appreciative of the first ten minutes where I go through how positrons decay into gamma rays and the coincidence detections. They follow this, they understand it, they have a concept of how the whole thing works, and they are terribly fascinated with the whole idea. People are tremendously interested in the brain. You know, almost everybody thinks they are going to get Alzheimer's disease. If for no other reason than that, they want to know what is going on. So I do just a tremendous number of these talks.

DUMIT: Is there a cost to that? I know from talking to people who look at physicists, for instance, that if a physicist talks to the news media, this is a very bad maneuver.

HAIER: Oh, I think there is a very major cost to not doing it. And I was very aware, when I started with all of this media stuff, that the flow in academia is against talking to the media, that somehow you are promoting yourself, just doing things for PR. But given the tremendous failure of scientists to educate the public on what science is, if scientists don't do it, who is going to talk about their work? I think: you get public money, you are at a public institution, you get federal tax dollars to do your research, you are required to be able to explain in lay language what you do and why it is important. In PET scanning, that is easy, as it turns out, because the images are so compelling, but you are required nonetheless to do it. And if you don't, then people don't know or they hear the wrong stuff. Science in this country is very poorly understood. The public image of the scientist is not such a positive image. I was always kind of amused as a hobby of noting the way scientists have been portrayed in movies. In the forties, in the fifties, around the sixties you start getting the real Dr. Strangelove type—nut malevolent scientist. Or cold and data-faking. Even in *The Fugitive* that I just saw the other day, the essence of that movie is a medical researcher who fakes data. This is not the right public image, and scientists have to go out there and do their thing. And if their colleagues don't like it, it is too bad, they have to get their colleagues to do it, too.

DUMIT: That's a very healthy view of it. It helps both nonscientists become more interested and knowledgeable about science and it helps the scientific image.

HAIER: And I had a very practical consequence, too. A woman saw me on public television, called up the university, asked to meet me, and wound up leaving me the sole beneficiary of her trust, for research. Which is an extraordinary comment on people's willingness to invest in science if given half a chance. I don't see why people should leave it to dogs and cats, you know, medical research is at least as good a thing to leave your money to.

DUMIT: Plus, you mentioned the communication that you get which otherwise you wouldn't have.

HAIER: That's right, from other scientists. I got very good scientific communication out of that. So the public relations aspect of it for me is very positive.

DUMIT: From the 1988 Raven's test, where did your research go in that direction?

HAIER: Well, the Raven's study got a lot of attention because it was presented at the American Academy for the Advancement of Science meeting. A lot of reporters were there and it was picked up by wire services and you know, the usual kind of fifteen minutes of fame. But it also was picked up by *Newsweek,* and that picture appeared in *Newsweek* in an article where this was featured. So it got a lot of attention. And I got a lot of mail from that, you know, nut mail. But also mail from various scientists who had seen it, who ordinarily never would have seen the article. It turned out in terms of scientific communication it was extremely interesting. Because you know, usually the sequence of events is the paper is published, then maybe you get a little publicity on it, but most of your scientific correspondence comes from the scientific publication. It was successful at that. But I got a lot of really interesting scientific correspondents who had only seen the *Newsweek* article. And a lot of correspondence about efficiency. I got a letter from a fellow named A. E. Maxwell, from London, who had seen it in the *London Times*. Once *Newsweek* picked it up it was picked up in lots of places, and he sent me—he is retired emeritus from I think Oxford—and he sent me some of his earlier papers that spoke to brain efficiency. Things I never would have found. But they were extremely helpful to me in thinking about this concept.

There were a lot of choices of what to do next. We only had eight people. Even the people who thought this was great said, "You know, can't you do a few more?" What about women? Well, once you start on sex differences, you know, you aren't really ready to start on sex differences as the first thing. I had already gotten a lot of questions about potential racial differences. Wouldn't this be a great way to study racial differences and IQ? Well, this is the sure way to come to grief. You know, the whole intelligence research community just about committed suicide in the late sixties over this issue. And it has not recovered to this day. There is almost no research on intelligence anymore in the U.S. So I wasn't anxious to go in that direction.

But I was interested in the idea of brain efficiency, and could you demonstrate that the brain became efficient, and it seemed to me that the best way to do this would be a learning experiment. I decided we would try to get some complicated thing like a flight simulator program. I said, We'll give people some training on a flight simulator, just like they were learning to fly an airplane, and we'll scan them and then they will complete the training and we'll scan them again, and see if they become efficient. It's like learning to drive a car—flight simulator programs are very big. So I went down to Egghead Software to look at some flight simulator programs. But the more I looked at them, I said, this is too complex. The learning here is just, this is not conducive to an experiment. But that was the week Tetris was delivered. I saw this game and I knew immediately this was the right game. This was extremely simple to learn, the rules are extremely simple, yet obviously practice was going to have a major impact on this. It is going to have a tremendous learning curve.

So we collected data on another eight people in what has come to be called the Tetris experiment. And it was interesting because you would ask people, "Okay, so what should glucose use do?" Because look, the thing about Tetris is, the more you practice, you know how Tetris works, the shape drops faster and faster the higher your score goes, so as you improve it gets harder and harder, so the better you get the harder it is. The question is, after you have practiced this game for a month, and you are really doing extremely well, and it is hard, is your brain working harder than it worked when you first learned it? It is kind of a simple question. Everyone's gut instinct is that you are going to see more glucose being burned. You're processing more stimuli, the stimuli are faster, you have to pay attention more. But on the other hand, there was this intriguing idea of the inverse correlations, that maybe somehow your circuits will become more efficient and you will actually use less. So we thought this would be another dramatic experiment. And sure enough it worked out that metabolic rate went down on the second scan. And moreover, if you correlate the size of the decrease with the person's Raven's score, which they took on a different day, it works out that the smartest people had the biggest decreases in glucose. The smartest people became the most efficient.

You see, this is a train of thought that I believe is unique. I don't think you are going to find this at any other PET center. It is an example, I think, of what a psychologist would do with PET as opposed to a biological psychiatrist. Moreover, it is an example of what a noncognitive psychologist would do with PET. If anything, the kind of people who are working or thinking along these lines tend to be neuropsychologists who want to get access to PET to validate neuropsychological paper-and-pencil tests. Some of them are the ones who would come up with correlations between test scores and brain metabolism, and when they see they are negative, they are going to start to think about this. That is kind of what has happened. So we now have that group, of which I am an atypical member, and that whole group is atypical compared to the cognitive psychologists doing the kind of verbal processing type of studies that are famous from St. Louis and other places.

So it is all kind of churning. Now, I wrote an editorial for the journal *Intelligence* a couple of years ago, knowing that the readership would be only essentially psychologists, really encouraging the idea that every psychology department, every academic psychology department should have a PET scanner. Now, when you talk about this to psychologists, they laugh. They say, "This is just completely ridiculous." But I press it very hard and I say, "Why is this ridiculous? You are a psychologist, you want to study the brain, you need the best technology to study the brain. Do you want to be an astronomer and say, 'I'm sorry, radiotelescopes cost too much, so I am just going to use this old one that I have lying around here, the standard one, because I can't afford what I need.' No astronomer in the world is going to say that. They will all figure out a way to get access to the tools that they need. And as psychologists we need to do the same." And every psychology department, by the nature of what psychology studies, needs a PET scanner. Now, if you want to say, well, you can't afford it, you have to collaborate with the medical school. That is a local decision, that is just the practical aspect of it. But the principle that every psychology department needs access to a PET scanner I think is a perfectly reasonable principle. Psychologists don't think about this at all, they completely have given up on the idea of high tech, complicated technology, expensive technology, as being part of what they do. This is just. . . . I understand, I mean everybody understands how this came about. But you've got to break out of this.

Another set of excerpts from my interview with Michel Ter-Pogossian.

DUMIT: You have participated in studies on activation. Has your conception of mind and brain changed as a result of these? I guess a better question would be, What model of the mind and brain do you use?

TER-POGOSSIAN: Mind/brain: first of all, I'm not quite sure that I am ready to put the two together. There is no question that the mind is related to the brain in some way. The only concept of the mind/brain that has really changed, if you want to, is my surprise that changes in blood flow could be so localized in the brain. When activation studies are carried out, what always surprises me is that there is undoubtedly a part of which "lights up" in the sense that there is more blood flow and therefore more metabolism. This is all right, as compared to others, but in light of the function of the brain, I am surprised that these areas are so well defined. And this for the following reason: I'm talking to you right now, and there are a very large number of different stimuli that I receive. I mean I certainly receive a visual stimulus from the light here, I see objects from around here, I hear you and at the same time I am trying to arrange my thoughts. So under the circumstances I am just surprised that certain specific areas of the brain would strongly stand out, in light of the fact that essentially the brain works. But they do. There is no question about it. So I think that in stimulation studies what you're doing is really to overstimulate very strongly one specific variable, say visual cortex, motor centers, as compared to how the normal brain works. That does not mean that the studies are not valid. On the contrary, I think that they are very valid. But no, so far I think most of the observations that were made through PET, most of them, not all of them, have been observed before from invasive work on the brain. I mean, everybody knew where the visual cortex was. So I think that PET just allowed you to study the brain noninvasively. And I think that the areas which have not been studied noninvasively, which would be particularly interesting, are indeed the activities of receptor sites of the brain, neuroreceptors.

DUMIT: Is there work here being done moving toward that?

TER-POGOSSIAN: Yes. There is good work being done here, by Dr. Perlmutter particularly. Joel Perlmutter. But activation studies, I think they are very interesting, too. I mean, speech physiologists of the brain are very interested in that. Physiological disciplines are very much interested in stimulation studies. And they really work, there is no question about it.

DUMIT: So what parts of the mind do you think are preserved in spite of all of these—

TER-POGOSSIAN: Mind, I mean, that is, again. . . . I don't know what the mind is.

You are interested in anecdotes. Sometime ago there was a meeting here, a relatively small meeting. Including Mr. McDonnell, James McDonnell, who was a founder and the chairman of the board of McDonnell Aircraft. He was very much interested in the human mind, and his hero was Penfield. He came to this university trying to stimulate people to get interested in the human mind, and, among other things, to get interested in parapsychology. Not many people in this institution are interested in parapsychology. So finally I think that Mr. McDonnell was convinced by a group of people, I was one of them, that perhaps PET is close to what he is interested in. And he helped out financially and he is still helping out, through his McDonnell Foundation for Neurosciences, some of the PET activities. He was interested in the human mind, but he also had pretty strong feelings about the fact that there were a number of things that we didn't know about. Few people are interested in parapsychology in this institution in this medical school. You really couldn't convince many people to work on that.

DUMIT: This is one of the centers of biological psychology.

TER-POGOSSIAN: Yes, indeed, there are few followers of Freud and little interest in people who bend spoons. So you are quite right about that in terms of just druthers here. But I don't know what the human mind is. Don't misunderstand me, I'm not being difficult about that. But it is probably related to the brain. If you remove the brain, there is not much mind left.

DUMIT: Right. I haven't met anybody involved with PET who is not at least that, the eliminative materialist, as it's called.

TER-POGOSSIAN: Is that what it is?

DUMIT: Without the brain, you are nothing. At least that much is material.

TER-POGOSSIAN: Beyond that, it really starts getting difficult. There are certainly very interesting aspects of the brain that you probably know all about, you certainly have heard all about that. There are definite regionalities in the brain. Also, the brain has an amazing capacity for recovering from the removal of very large volumes of tissue. We see a patient who is losing about one pound of his brain, and two months later, well, he is pretty good. Quite fine. You know, losing a number of cubic centimeters of your brain, up to several hundred, is all right. I mean, there is no computer that would do that. So the brain has the ability to utilize different regions. Yet, as you also know, affecting certain specific areas of the brain, you may run into serious trouble, like being blind, like being aphasic: all it takes is a stroke. So does that have anything to do with the human mind? I don't know.

DUMIT: So you are not very partial to computer metaphors for the brain and mind.

TER-POGOSSIAN: No, don't misunderstand me. I'm not partial or impartial, I haven't the slightest idea. There is at present no computer that mimics the brain, that works like the brain does. That does not mean that one cannot design one, and there are people that are working in that direction. I mean, it is possible. I know that some work is being done with a computer with short-term memory, for example. That exists now, and that is fascinating.

DUMIT: I have been looking at some of the mass-media descriptions of what PET can show. It can show efficient brains versus inefficient brains.

TER-POGOSSIAN: Lord, really? I mean, efficient? How can you distinguish that with PET, efficient brains from nonefficient brains?

DUMIT: It was a theory that you use PET to measure intelligence.

TER-POGOSSIAN: How did you measure intelligence with PET? What pharmaceutical do you use and what model do you use?

DUMIT: This was an FDG.

TER-POGOSSIAN: What do you do, you use FDG and if there is more FDG you are more intelligent, or do you accumulate more FDG?

DUMIT: The theory was that with more intelligent people and less intelligent people, or people who have trained in something well versus those who haven't trained in it, like a video game or something like that. And they thought that the more intelligent, the more that you would use, and it turned out to be the opposite, the more intelligent, the less. So they thought that this might be inefficiency that causes it one way or another.

TER-POGOSSIAN: Oh, I see. In order to make that kind of study valid you would need a very, very tight series; it has to be judged double blind. The numbers that you need are enormous, the criterion that you need. No, it is doable, it would be an interesting study, but reaching any kind of conclusion on the basis of small numbers that we have now is not very responsible. Anyway, people like to do these things.

DUMIT: I guess it is amazing that you can, nowadays, start to do these things, that PET has reached the point that—

TER-POGOSSIAN: But one has to be careful, it doesn't mean that the tool is not there, I disagree with that. But the media should be very careful about that, indicate that only as a possibility if nothing else, but not as a reality. Yes, it is a possibility, it is an interesting area. But first we have to understand what FDG does also, I mean, this is energy, metabolism. The fact that the brain uses more or less FDG really means nothing else than that. That it uses more or less FDG. You can't extrapolate. You can't even extrapolate to normal glucose. FDG is not glucose. It is 2-deoxyfluoroglucose, and its behavior mimics glucose only to a point.

Joseph Wu is a psychiatrist at UC-Irvine who works with PET.

DUMIT: Do you show the patients their PET scans?

WU: Oh, yes. We try and show them the PET scans, and then, some of these patients will refer them out to people.

DUMIT: Does it help them overcome part of the stigma of mental illness?

WU: I think so. I think that definitely. One of the intrinsic messages is that the depression isn't something to be ashamed of, it is an illness which needs to be understood. And it is not something that is their fault.

I think that there is a destigmatization that occurs with the biological emphasis. It is a fine line, because there are some arenas of personal responsibility that people can and should assume for their feelings. But I think it is a very narrow and tricky balance. It is important not to think that it is all biology; that can lead to a certain eschewing of what is appropriate for one's own role in understanding one's emotions. On the other hand, I think that people can go overboard, and say, "Gee, I'm entirely at fault for how I feel." How to try and understand one's role in helping to monitor one's emotions without being unnecessarily harshly judgmental of oneself?

DUMIT: Nancy Andreasen, she has written about the biological revolution in psychiatry. You were in medical school during this time. Did you also get the other side of psychiatry?

WU: Oh, very much so. I would say that most of the psychiatrists in this

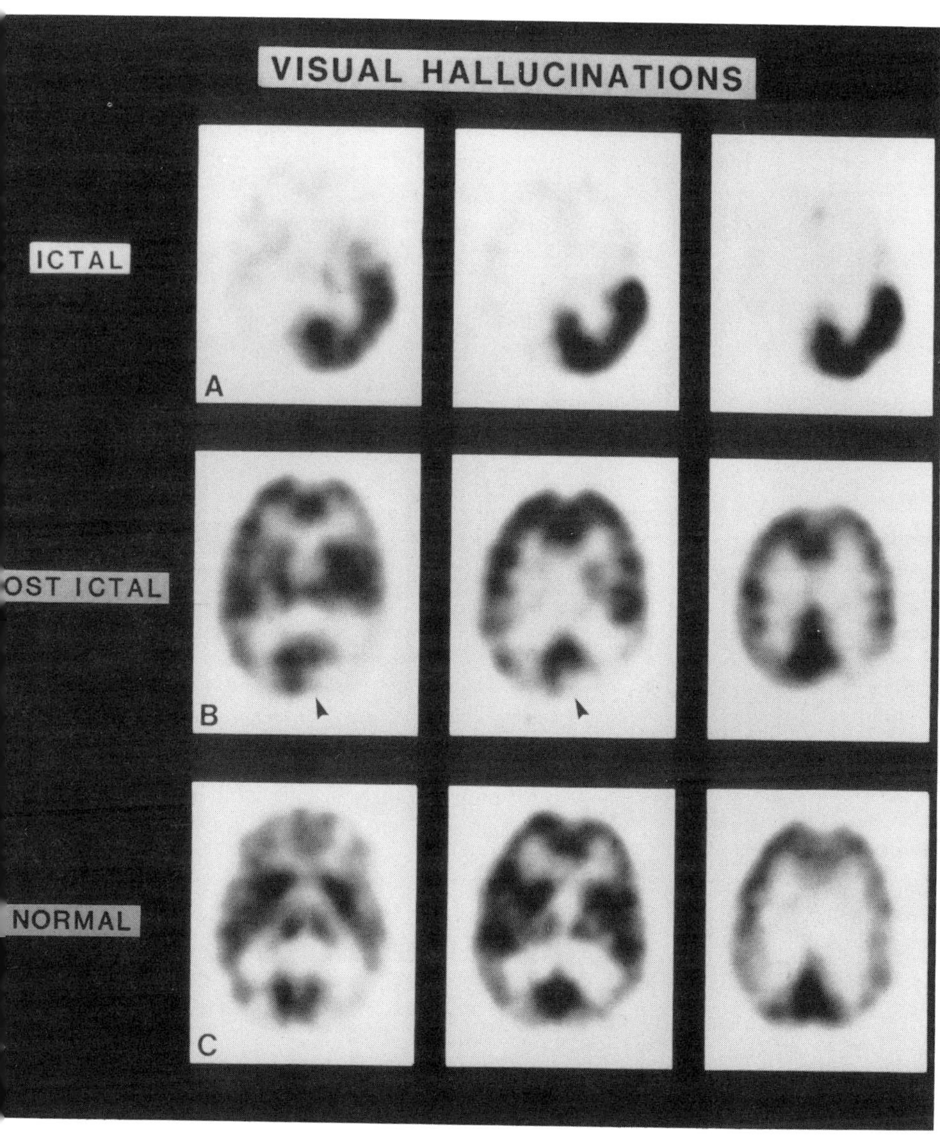

Visual hallucinations. A patient during a visual hallucination in her left visual field, her right visual cortex is activated (ictal); immediately after the hallucination she cannot see in her left visual field (post-ictal); subsequently her vision returns to normal (normal). (Courtesy of Michael Phelps, UCLA)

department are still analytically, dynamically focused. I would say that biologically oriented psychiatrists still make up a minority of the faculty, maybe thirty to forty percent.

DUMIT: Do both of these sides come into play in your work?

WU: Somewhat. For me, when I do a study of depression, there is a part of me, a whole human dimension, that really tugs at my heart. Part of me feels moved by the pain of the patient that we work with. I am also moved by the courage and willingness that many of these people have to participate in this study, even with the depth of their emotional pain and anguish. I think we try to offer to them the gratification that comes with knowing that they are contributing to the fund of knowledge that we hope will eventually help to eliminate depression or mitigate it. And that is something that many of these people find appealing, because there may be some greater purpose to their suffering. It is a way of reconnecting in some sense with the broader community. It is a way of making a personal meaning out of the emotional pain that they suffer from.

For me, I see the whole biological aspect as not being contradictory or mutually exclusive from the psychodynamic aspect. I really see it as complementary and synergistic with the dynamic aspect. There are some people that see it as either/or. I see it more as a both/and type of proposition.

DUMIT: How do you think of the mind? Is *mind* a term you are comfortable using?

WU: Oh, yes. I think this is a good question. I really feel mind as being multidimensional and multifaceted. I'm interested in mind as being in some sense an aggregation, a variety of systems, both on a material plane as well as on a nonmaterial plane. And I think that there is a dynamic interaction between the material and the nonmaterial planes. I think it is clear from recent studies that the material plane has a profound affect on mood and personality.

This week's issue of *Newsweek* has an article in the book review section about a book written by a psychiatrist [Peter Kramer] who was writing about Prozac. He was writing about how striking it was to see one of his patients who was a morbidly depressed, masochistic woman who had been in long-suffering relationships. She took this Prozac and within weeks became a more assertive, vibrant person; her laughter even changed. And so it is clear that there is a profound connection between body and mind.

But there is another interaction too, that is, the mind can influence the body. I think there is a lot of evidence to suggest that serotonin and dopamine can modulate mood. I think there is a lot of evidence also that mood might modulate serotonin and dopamine levels, too. I mean, psychotherapy by itself can reverse certain forms of major depression. And that would suggest that mind was altering body. That the nonmaterial plane was affecting the material plane. So I am a quasi-reductionist, but not a complete reductionist.

DUMIT: I heard you talking about different kinds of circuits. That there are pleasure circuits, and emotional circuits.

WU: Yes. I think that mind, as I have said, is an aggregate of multiple subsystems, systems within subsystems, all of which interact with each other, not only on a material plane, but on a psychodynamic plane.

I think that Freud's concept of the id, ego, and superego reflects what some might call aspects of personalities, of consciousness. I think that there really are splinters or fragments of personality which have predominant functions or roles, which interact with other splinters and aspects of our personalities which also have other roles, and I think that they are complementary. I think that there is a biological underpinning to some of these. I think, for example, that we can see that some people are obsessive-compulsive. You might argue that in some sense obsessive-compulsiveness is a hyper superego. And we see that there are certain correlative changes with PET scans and OCD symptoms. Lou Baxter has done some very nice work looking at some of the changes with OCD. I think he has also shown that behavioral therapy as well as pharmacological therapy can both result in a common change in those areas of the brain which appear to be associated with OCD. So I would say that here we see some evidence suggestive of correlations between one aspect of the personality, the whole moralistic superego in an apparent site which appears to be controlling, there is probably serotonergic modulation of that. We know that Prozac modulates OCD as well.

And then we see, on the other hand, that mania can be an example of the pleasure circuit gone awry. We see people who can be extremely euphoric, they can be hypersexual, hyperactive, and we can see that there is a dopamine modulation to this. We can give dopamine blockers and reduce this. There is not only a pleasure circuit, and a moralistic circuit, but also some kind of clock, a timer, which seems to regulate how long one's focus shifts from one subsystem to another. I think that there is some type of regulator which shifts the focus. I think we have a focus, and that focus can reside in one subsystem and then at some point in time the timer or regulator shifts that focus to another subsystem. And sometimes that regulator gets stuck in one subsystem and the normal brakes that modulate that subsystem can get jarred or stuck and we can't govern or regulate it in a normal pattern. I think that a PET scan can help provide us with some clues as to where some of the anatomical circuits might reside which modulate some of the psychodynamic elements which interact in some of the ways that the psychodynamicists like Freud or others have described.

Jung talks about archetypes or personas. I really see archetypes or personas as being not dissimilar in some ways from Freud's topographic theory of id, ego, and superego. I see these as all representing different aspects which have biological and evolutionary function, and which have neuroanatomic substrates and modulators. So I really see mind as being very complex. I suppose that we can train people to not become depressed, so I think that the mind can modulate the anatomic circuit. I think that Baxter's studies show that using just behavior therapy alone you can modulate these areas. So I really think that there is a fascinating interplay between the many different elements that comprise our consciousness.

Henry Wagner is the recently retired Director of PET and Nuclear Medicine at Johns Hopkins University.

DUMIT: I was talking with a psychiatrist at Irvine. He thinks of the dopamine and serotogenic circuits as having very specific kinds of psychological correlations. On a spectrum from a very cautious kind of approach relating it to behavior, or a broadbrush approach to talking about the mind and the relation to biochemistry, do you see yourself one way or another?

WAGNER: I think that with broad strokes, if you drop dopamine on neurons, it inhibits the transmission. And therefore the opiates and the dopaminergic systems are thought of as inhibitors, that they cut down on the neurotransmission process, they could be acting as a governor. But if you take a disease like Parkinson's disease, the patients have trouble getting started and they have trouble stopping. So they have trouble in regulating movement. Again, you could say that you have to inhibit inhibitors. There are so many negatives and positives combined that you really can come up with almost any theory you want from them. For example, the dopaminergic system can inhibit an inhibitor which then has a positive effect. It's possible, for example, [that it's] like an air brake: you let your foot off and it inhibits the inhibitor that is keeping you from moving forward, so it is a safety factor. Certainly in Parkinson's disease, they have trouble getting started. So you could say that they have trouble turning off the postural reflexes that are keeping them in one place, they have an imbalance between the inhibiting and the stimulatory factors and the dopamine doesn't turn off the inhibitor one, so that the stimulating ones can activate. So you can come up with almost any explanation of pluses and minuses that you want. So I do think that it is true that the dopamine system is related to movement.

But to divide the brain function up into cognition and emotion and movement is again an example of categorization that is probably not too helpful. Depressed patients—Alzheimer's patients can be depressed and depressed patients have trouble thinking—and schizophrenic patients can have trouble moving, they can become catatonic. But these are very broad categories. I personally believe that putting anything into a category is not because there is some kind of intrinsic truth [to it] but because it is useful. My philosophy is pragmatic. Therefore when you say that a person has disease X, it should be because putting him in that pigeonhole makes a difference in some way. These are man-made categories, abstractions are man-made simplifications of an unbelievably complex external world. Therefore, to say that dopamine has to do with movement—it has a lot to do with movement, but it has a lot to do with other things as well. What science does is try to make everything more and more simple, [to] try and get more common factors to get it simpler. But right now, in the area of mental function and mental disease, it is not very simple. So the goal is to try and come up with some kind of simple explanation for these things. Right now, you say you have a simple explanation where serotonin is related to mood and dopamine is related to movement and acetylcholine is related to learning or intelligence. But to say that acetylcholine is intelligence and serotonin is mood and dopamine is movement is a gross and unhelpful simplification, a counterproductive simplification. Although it is true that blocking the dopaminergic system has been one way that it has been found to help some patients with schizophrenia, and blocking the serotonin uptake site, or inhibiting monoamine oxidase has been one way of helping patients become less depressed. If it helps, it helps. It helps solve problems, the world is surrounded with problems, people are surrounded with problems. If an abstraction helps them—the best invention of all is language, I think that the most important part of consciousness and memory is language, because it translates the past into the present.

DUMIT: Do you work with cognitive psychologists, cognitive psychiatrists?

WAGNER: Yes. We do neuropsychological testing of every patient that has a PET study. We are trying to relate what we call "closing the circle between brain chemistry and behavior." So we always try to do extensive neuropsychological testing on most of our studies. By behavior we mean everything, all mental functions.

DUMIT: Do you have working models of the brain? I'm curious because one of the things I have been looking at has been computer models of the brain.

WAGNER: You mean like Adelman's? No. I have a conceptualization of the brain, my view of the brain, but not a mathematical model in the sense that the artificial intelligence people have.

DUMIT: Can you describe it?

WAGNER: My concept? Well, I think that memory and language equals consciousness. The way you remember things—much of the past is related to language. Many of the things we carry from the past are in the form of language, although not everything, obviously. Images are also carried forward. But I think the thing that gives us our consciousness is perception of the present combined with memories of the past. And that we are constantly abstracting unbelievable, colossal amounts of information that is coming into our brains. And the thing that we can study are these modulating factors, which is why I'm just as interested in global activity as I am in regional activity, though I know that some of these things are going on in regions and you couldn't detect it if it's being diluted out by the rest of the brain. So the ones that we can measure are the dopaminergic, the cholinergic, the serotinergic, and the opiate system very extensively right now. And we try to relate them to behavior.

DUMIT: So do you think there may be a way some day to describe chemically or molecularly the processing of language? At what point does the complexity kind of implode?

WAGNER: I think that soon there will come a knowledge of things that affect memory. Right now we know that scopalamine and acetylcholine are related to memory. We can obliterate somebody's, a whole period of time, or enhance a whole period of time, so I think this is related to what you are calling attention and things like that. I think this is going to be much better understood in the immediate future. That's the kind of thing [for which] I think there is going to be an immediate payoff. But to know the end grammar, it's possible that these complicated patterns have simplified rules that this information is stored in, but I think it's going to be a hell of long time before that is going to be done. Again, the effect of drugs, including things you eat, I think the effects of diet on brain function is an immediate area you can explore. Most of these chemicals are kept out of your brain, but some of them get in, understanding things like that, I think it is going to happen in the next ten years. We are going to learn a lot more about brain chemistry and behavior. But it won't be in the area of patterns of cognition, so I think language is going to be the hardest thing to study. Mood would be easier to study than language.

DUMIT: You talked a couple of times in your articles on the future of PET about violence research as a promising area.

WAGNER: Again, I think that is a very good area to study. We know that there are hormonal factors that affect behavior, that steroid hormones, testosterone, and things that are related to aggressiveness. And I think that fits right into that. Again, this is an example. You can't say I'm going to learn better; it is a general pattern. If a person is violent, it is not that they are violent when they hear things, or when they see things, it is just that there is an atmosphere of violence that is probably related to some chemicals that are measurable. And we can find out things about that. I think that it would be easier to study violence than it would be to study language, it is far less refined. I mean language is probably the most exquisite mental function that there is, and therefore the initial payoff is probably more in these crude things, like mood. I think you can study learning but not language.

DUMIT: Some of the places where violence comes up is in courtrooms where PET scans have just started to make their appearance.

WAGNER: Well, there are certain general patterns that are true historically. First of all, some physicists have said, and I think quite frankly, that every time a physical measurement is made and a physical fact is learned, that physicians try to apply it to diseases, they try and relate it somehow or other to diseases, either models or things. And I think that every time you make a correlation between behavior and mental function you are going to find people that are using it in the context of free will or lack of free will. And I think it is probably abused more than it is being used, it is probably being misused. Some people believe that you really don't have much free will, and will defend people on that basis. And therefore they search for anything that can be used to support that position. They may be right. I think it is more from a philosophical basis that they are arguing, using that as a basis, as a tool, rather than that it is a great discovery. I mean, I'm [more] willing to believe that the person can't really control their behavior as a result of their past experience or genes than I am to say because their corpus collosum is twenty percent thicker than anybody else's, that that is the reason why they are doing all these things. So it's a gimmick.

DUMIT: So have you testified in court?

WAGNER: I've been asked to do it, usually we get a call here, some lawyer calls up and wants us to do a study on some person, a head-trauma patient or something like that, and I refer him to somebody else.

DUMIT: One of the fascinating cases that I have found, one of the early trials using X rays, in the early 1900s, where the argument before a state supreme court was, Should X rays be shown to the jury? The claim by radiologists was that X rays were their professional domain, one had to be trained to be able to read a radiological image. Why should the jury see the image? Because then they will eventually decide on what they think the image looks like. That seems to be one of the questions with PET scans, if they are an expert image—if they are images that take a lot of expertise to look at—how much do they show? Looking at it from the jury standpoint and the lay public standpoint—

WAGNER: Well, you raise an interesting question. I was raising the question of should you use the data or should you not use the data, and you are asking who should interpret the data, and that is a very interesting thought. Again it is not a question of black or white. I think there are some things that are so obvious that you don't need more expertise than the average lay person can have to interpret it. A guy who has a broken leg is an example.

DUMIT: Even though at the turn of the century it is precisely the broken leg that is at issue.

WAGNER: Okay. So I think that there are some technical data that it doesn't take more than the average person's intelligence to see whether it shows what is claimed to be shown. I think you definitely should show it to the jury. It is like anything else: you don't understand English, you don't understand a person's confession. It's like saying only English majors should be able to hear confessions. I said something, did the person really confess or not confess, it depends on the meaning of their words. Since English teachers are experts on the meaning of words, you have to be a semanticist to tell the jury what the person says. All that I think is that you should not be limited in what you should show to the jury. It should be up to them, whether they believe the expert or not, whether they can see it for themselves or whether they have to take the expert's word for it. And the judge helps them do it.

DUMIT: The counterargument that was proposed was that, instead of English majors, it was that the juries were non-English speakers who were taught English in the court so that they could understand the confession. How much nuclear medicine do you have to be taught to understand the image?

WAGNER: It depends on the study and the result for how complicated it is for the jury to be instructed properly by the experts. And whether they are taking it on faith or whether they can really see it, you say, "You see that this is bigger than that." And they have criteria for size and length and they take it on faith that this is not supposed to be bigger than that. But I think that you have to be very conscious that you don't misuse that information. That is what is happening right now with DNA in this morning's paper, that DNA is probably going to be getting a lot of people out of jail that should be out of jail. But it is probably going to be because of errors and variance that there are going to be some people that, by luck, are going to be able to get out of jail because of the five percent of the cases that fall outside of the two standard deviations. So it is like anything else, it can be abused or not used, and I think I would be very cautious that there is sufficient data. I think it is clearly a mindset of trying to find anything that it really is an argument of free will or no free will; responsible or not responsible.

DUMIT: Yes, one philosopher phrased it that the law's assumption is that everyone has free will and science's assumption is that there is a cause for everything. Plus he added the fact that after you have competing experts in the courtroom, that the court will defer toward independent or third-person objective "measurement" to help the court decide.

WAGNER: That is a very interesting way to put it.

DUMIT: It is interesting to me because, again leaning on activation studies of PET, that you have diagnoses of something like schizophrenia correlated with PET scans, and then you have the case where the PET scan is then taken to be the objective data. And so even if the correlation is not that strong, it is still a correlation based on what experts diagnosed.

WAGNER: No. I would go back to the pragmatic rule. You are grouping people on the basis of usefulness and if grouping people in that way helps then you should do it.

Schizophrenia is a very old fashioned way of looking at patients with mental illness, it probably was even from the beginning. It is a start, it is an intermediate step. Instead of trying neuroleptics [a kind of drug] on every person with mental illness, you pick out the ones that Kreppelin and others said have certain symptoms that seem to be in common and you find out that neuroleptic drugs help seventy percent of them. So you say, okay, we'll try to characterize the seventy percent that it helps and make a new category out of them—schizophrenia 1, schizophrenia 9, schizophrenia 8. So therefore you judge these things according to usefulness. I believe in the pragmatic philosophy, particularly in medicine, where it is perfectly suited to it. Diseases are not oxygen atoms, or hydrogen atoms, they are groups and the groups change all of the time. Fever used to be a disease. Pleurisy used to be a disease. The search is for increasing specificity.

DUMIT: And these images help.

WAGNER: Molecules are pretty specific. Therefore, when you go to a molecular basis, you are increasing the specificity of the disease. We will classify the patient based on a chemical mosaic.

DUMIT: One of the things that I would be interested in looking at long term would be the effect of these molecular differentiations.

WAGNER: It would change the classifications completely. I mean, anemia is no longer a disease, fever is no longer a disease, shortness of breath or dropsy, these were all diseases. All of these things, like epilepsy, are grab bags. Diabetes is a little bit more specific, hyperthyroidism is more specific. But there is increasing narrowing of the categories. Molecular is it for the nuclear field using radioactive tracers, because radioactive tracers let you measure rates.

DUMIT: I would think that socially that would be a more uphill battle; because they are invisible, the more kind of visible symptoms would still carry the categories.

WAGNER: People are still oriented toward this anatomical approach, where a person catches a cold. Disease A inhabits person B. Historically, the ontological approach has alternated with the physiological. It was thought that something like cancer inhabited your body rather than being an imbalance of normal functions that keep you from getting cancer. But I think most lay people still follow the catch a cold concept. They think that the key is to go to surgery and get the disease removed. That is still the way most people think about it.

DUMIT: But what can replace the cause in the physiological model of disease?

WAGNER: You are implying that people like to know the cause.

DUMIT: Exactly. I mean, even if it is an imbalance—

WAGNER: Rather than a correctable manifestation? I'm not sure if that is right. I'm not sure if people want to know the cause rather than correct the manifestation. But you are saying that, Okay, I may want to correct the manifestation as my first goal, but I want to know the cause as my second goal.

DUMIT: But if you take a physiological approach—

WAGNER: It is a continuum of causes. The cause of hyperthyroidism now is that there is an antibody that produces some organism that stimulates thyroid cells, but is the virus the cause? Why is there a problem between the physiological approach and the cause?

DUMIT: You were saying that the ontological approach was still reigning in the world.

WAGNER: In laypeople. I think in science it is not nearly as much.

DUMIT: And one of the things in the anthropology of medicine is, How can I get better? And the second thing is the cause question. Why is it me and not somebody else? And the ontological metaphor hands you an answer—

WAGNER: Well, you ran into it, you encountered it.

DUMIT: If you hadn't been there at that particular time, it would have missed you, or something like that.

WAGNER: Some external event. I mean, there are multiple causes of diseases, some are external events such as toxins, or degeneration. You know the international classification of diseases consists of three digits which are anatomical, progressively more specific, and three digits that are etiological. One of them is what the toxin was, infectious, one of them is degenerative, one of them is metabolic. Just because you take a rate factor a time factor—why can't you have external events? You can have a cause without having a being. You are saying that ontological goes with cause and physiological doesn't. The external cause tends to perpetuate the ontological. Yes. Because the virus is thought to be the disease which it isn't. Those causes are not the thing, they are causes of the physiological. You are implying that in the layperson's mind that the cause is the thing or that the chemical is the thing. That the layperson does not really think of the body as the disease in response, and that the public has to be educated about it. They really do think that viruses cause disease or they think that the chemical causes the disease. They don't think mechanistically.

DUMIT: One of the strengths of PET is that it gives you quantitative data. And at the same time you produce visual, qualitative images. How do these two things work together? Can you read images?

WAGNER: There is a tremendous amount of data. When you say quantification, you are talking about numbers, and these spatially oriented studies, these four-dimensioned studies, three dimensions in space and one dimension in time, can only be abstracted and displayed in a meaningful way in the form of images. Otherwise there are too many numbers. Your brain can't really handle more than a couple of variables at one time if they are quantitative, so you have to have abstractions. And images are a very, very nice way of abstracting quantification.

Shigekoto Kaihara, who was a trainee here, is the guy that coined the term *functional image*. That then subsequently got translated into what is called parametric imaging. He published a paper in 1968 and the whole paper was in the title. It said, "The construction of a functional image from rate constants." He took rate constants, spatial measurements, and instead of putting distribution in space, he displayed the distribution of the rate constant and called it a functional image. So that is the only way you can handle that much quantitative data. Plus it has the additional advantage of relating it to space, so it relates space and time. The essence of physiology is time.

DUMIT: So most PET scans are of this form.

WAGNER: Yes, they are parametric images. That is interesting, all PET images are explicit parametric images. Now SPECT images or other types of images are implicit parametric images, because they frequently show, say, the accumulation of iodine by the thyroid gland. You portray the iodine at one time, but that is integrated over the interval between the time of injection and the time of making the image. Many of them are implicit even though it is one measurement of time of a spatial distribution, it is an implicit rate. Whereas PET images are all explicitly parametric images.

DUMIT: And that includes the FDG?

WAGNER: FDG is both. They frequently just look at the distribution, but usually then modeling is then carried out to make the distribution more clearly reflect glucose utilization rather than a combination of glucose utilization and other things.

DUMIT: When I talked with Dr. Robert Ledley, he said that when they first made the ACTA scanner, and came out with CT scans and they looked like they were starting to take off, they had to produce their own atlas to teach people how to read the images, because people didn't know how to see a CT scan. Is there a similar thing to PET?

WAGNER: Absolutely. Many radiologists today still don't understand PET or have an appreciation of PET because they cannot relate it to their training, which is primarily anatomical in orientation. There are certain branches of medicine which seem to be related, such as anatomy, surgery, radiology, and pathology; that is on the one hand. On the other hand you have physiology, pharmacology, internal medicine, and nuclear medicine. So one is very rate oriented, time oriented, space oriented, the other is primarily space ori-

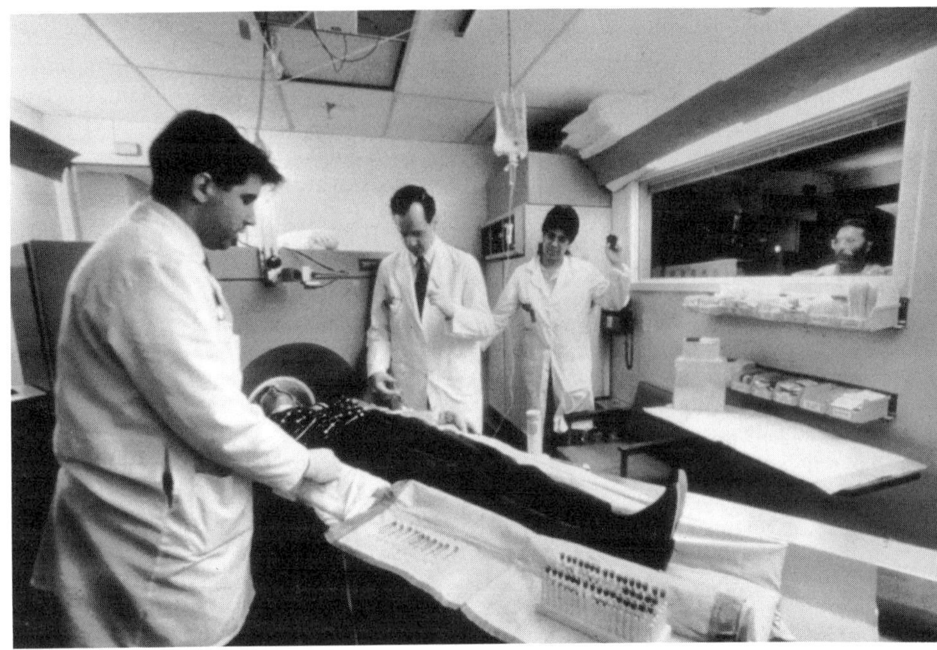

Technicians producing a neuro-receptor PET scan of a patient. David Clough (in the foreground) draws blood every 30 seconds, Emma Llinas (far back) times the activities, and Bill Boitnott (behind the glass) runs the scanner and computer. Dr. James Frost previously had injected his patient with a radiopharmaceutical. (Wide-angle lens photo, courtesy of David Clough, Johns Hopkins University)

ented. Surgeons are taking care of the patient in space at one particular time. Radiologists usually get one flash of all of these photons at one time. So the time orientation, the characterization in terms of space and time, is more characteristic of PET and SPECT than anatomically oriented specialities, such as surgery and radiology. So the answer is yes, because most radiologists today don't think of it in terms of time and rate.

WAGNER: So what impact is PET going to have on thinking, on human thinking? I think it is going to enhance the molecular approach to philosophy and disease in medicine. It is going to revolutionize pharmacology.

DUMIT: And this revolution—

WAGNER: It's known now, from PET studies and basic neurophysiologic studies, that before you have some kind of behavior, your brain gets activated. And it has been shown that multiple areas of the brain are involved in speaking, for instance, or adding a verb to a noun. But that is a very long way to go before you talk about intelligence, memory and things like that. Nobody knows how memory is encoded. So basically you are saying that one of the things that is most appealing about it is the promise.

Rather than the actuality. What we are doing here is basically we are studying the effects of drugs on the brain, which is very clear cut. We are studying the relationship between the drugs and behavior. That is not the future, that is the present. The future is the expansion of what we are now doing, without a giant extrapolation.

An example of one of the questions that we are now asking, that we have been totally failing so far in answering is, What causes the release of endogenous opiates in the brain? We tried: it is said that there is runner's high, it is said that some people with fasting and starvation have secretion of opiates, it is thought that with painful stimuli we have secretion of opiates. We've tried pain by putting your hand in ice water, by putting a tourniquet around your arm. None of those are manifest by secretion of endogenous opiates. So there is a lot of discussion about things that are associated with the release of endorphins and endotoxins. We have not been able to demonstrate that they really do cause release of it when you actually measure it. That is very interesting. So although you can read all this literature on runner's high, we have not found it. There is a person I spoke to this morning, for example, a master's degree student, where I had originally suggested to him, "Let's look at fasting and eating as something that is related to endogenous opiate secretion." And I've told him, "We should design these studies in such a way that even if it's a negative we can publish it." But once we are staring the results in the face, where we can't show it, it is no longer interesting. To me, a negative is not interesting enough to pursue and prove it as a negative, because it is much more interesting to really try and keep thinking about what we can do to get a positive.

Now the other thing you might look at is the enthusiasm that greeted the measurements of the electrical activity of the brain.

DUMIT: Exactly.

WAGNER: You agree with that. And in a sense it was a disappointment. You agree with that? Well, first of all, a lot of the information from measuring the electrical activity of the brain has been very helpful. The idea of action potentials and things like that has been very helpful. But the idea of measuring electrical activity in the living human brain and learning something about the things that you have mentioned such as memory, intelligence, and creativity—three things that people are very interested in. They probably had the same [idea], I think, from my research looking at the literature, as that when the electrical activity, the electroencephalogram was discovered, say, around the 1920s, they thought that they were going to find a hell of a lot about memory, creativity, and intelligence, and it didn't happen. In my judgment and in my lifetime, not a lot is going to be known about intelligence, memory, and creativity from activation studies. Now, I think that some people might find out a lot more about memory by studying neurochemistry which we can now do. But not by activation studies, that is by measuring regional blood flow. My judgment, through the example of studying the effect of the opiate system on memory, or studying the interaction of the opiate system and the cholinergic system which we are now measuring—not in the future, but now—that may help give us some information about intelligence, movement, and memory. Creativity is further off.

But you're saying the activation studies. But the activation studies have been publicized more! You agree with that?

DUMIT: Yes.

WAGNER: But is that a reflection of the fact that the media people find that more attractive, or that the chemical people haven't really presented their case adequately?

DUMIT: I think both, there is definitely a desire. If you say, "I have a study that is related to showing the difference between intelligent and nonintelligent brains," and if you have pictures and so on. I think that you are quite likely to have media people come to you and publicize your work and so forth.

WAGNER: Okay, we are doing a study here, in association with the army on sleep deprivation. And the really simple question that is really quite difficult to answer is, Does your brain glucose metabolism, global or regional, relate to performance as impaired by sleep deprivation? And it is very difficult to find that out. That is the type of answer we hope to find.

Note

1. Sharon Begley et al., "Mapping the Brain: Science Opens New Windows on the World," *Newsweek*, 20 Apr. 1992, pp. 66–70.

Mary-Jo DelVecchio Good, Irene Kuter, Simon Powell, Herbert C. Hoover, Jr., Maria E. Carson, and Rita Linggood

4

Medicine on the Edge: Conversations with Oncologists

> We are right up to what we can get away with. It's exciting, but it's a different brinkmanship. And I think for that reason you do it in a sophisticated world only, where you have checks and balances. I think these are quite separate decisions; patients may take all of that quite calmly. What they actually get out of the transaction is where they are talking and that's a separate mode and I don't think they collide really. *It's a schizoid way in which you operate*. One is talking to people and the other figuring out the best way to go for it. And then telling what risks there are, and that we are very blunt about. You have to do that.
>
> —Radiation oncologist discussing experimental treatment for brain tumors

Research oncologists stand between the bioscience of their specialties and the clinical tasks of caring for people with serious and life-threatening illnesses that are often resistant to treatment. As noted by the radiation therapist in her comment, the "schizoid way in which you operate" taps into both the scientific, often experimental domain of cancer care and the therapeutic tasks of caring for patients. These conversations are excerpts of discussions between oncologists and researchers and between oncologists and patients. They afford glimpses into the experimental, uncertain yet oddly routine domain of oncology as practiced at the meccas of high-technology medicine, at tertiary care teaching hospitals. The oncologists engaged in these conversations are "triple-threat" academic clinicians who practice on the cutting edge of advances in medical treatment. They care for patients, engage in laboratory or clinical research, teach medical students, and train residents and fellows in the subspecialty. It is through their work that they authorize biomedical knowledge and practice in oncology. As a breed, "triple-threat" scientist-clinicians, the clinical investigators, are threatened; the tension evident in balancing patient care with laboratory work or even clinical studies not only provides intellectual excitement, fueling activity in clinical investigations, but is also daunting and a difficult challenge. The following conversations capture some of the dilemmas of scientist-clinicians. The five physicians are by no means "average" oncologists, even in terms of demographics. Three of the five are British born, two received their medical degrees in England. Women figure largely in these conversations as physicians, as patients (given that a complementary study focused on breast cancer treatment), and as researchers. Two of the physicians, however, are male; they treat women patients and work closely with women colleagues. The very recent gender shift in medicine finds expression here and there throughout the dominant and complementary interviews.

The format for this presentation includes a dominant conversation between a medical oncologist, a social scientist, and a research assistant. We enter the interview (the second with this physician) with an account of the oncologist's personal history and begin with a discussion about the tensions and competing goals of laboratory and clinical life. Complementing iterative comments by surgical and radiation oncologists in conversation with the social scientist are presented in parallel at several points throughout the interview. Similarly, a conversation between a patient and oncologist is also presented in parallel. These brief excerpts may be regarded as commentaries or as part of the larger conversation of this project and place. Our multiple authorship acknowledges that these conversations are "authored" in part by each of us although assembled by one of us (Good).

Laboratory Life and Clinical Science

GOOD: Why don't we start with a bit of personal history and—

MEDICAL ONCOLOGIST: Where should I begin?

GOOD: Begin before medical school.

MEDICAL ONCOLOGIST: Well, I was born in Scotland, moved to England when I was eleven, went to grammar school in Surrey, and for the first couple of years was always the different one out: I was Scottish and had an accent and sounded different, and all the kids at school would say, Oh, say that word again! Say that word again! It was kind of funny. Anyway, after high school I went to Oxford. Wanted to do medical research.

GOOD: Why?

MEDICAL ONCOLOGIST: Ever since I'd been—well, it was very funny, I was trying to recall this for one of my patients' husbands the other day and it was, Have you always wanted to be a doctor? No, no, no, no, no. When I was little I wanted to work with chimpanzees and that was clearly the most interesting thing to do. But after I gave up that I wanted to be an astrophysicist and that was because of all the really interesting books in the library with black holes and all the things in the universe. Then I decided that was too much mathematics. I liked maths, but medical research was more interesting. So—

Good: Did you come from a medical or scientific family?

Medical Oncologist: No, not at all. No one in my family. Well, I had an uncle and aunt that were chemists, but they weren't close to me. It was really just interest in science in high school and getting into biology that way, and genetics. So anyway, in Britain it's a little different. You either go to medical school straight from high school or you do something else. You don't do a degree first and then go to medical school. And in England also it's a little different. Physicians are not given the status they are in this country. I don't think the smartest people go into medicine. I think the smarter you are the more likely you are to go into a more academic pursuit than medicine. So medics are not—they're always the people at universities in Britain that are always having the parties and the wild times and making up their studies, and I didn't quite fit that bill. Besides, I didn't want to do all the boring anatomy and stuff like that, so—DNA was exciting, genes were exciting, that's what I wanted to do.

So I went to Oxford to study biochemistry and microbiology and genetics, and it's an interesting perspective. Once you've gone through a scientific program in Oxford, you're given very little counseling about what your career should be in the future. They just assume that you're going to stay on and do a Ph.D. and do research, and if you say you don't want to do that, it's, Well, gee, I don't know what you should do then. And the joke in our college was that if you were in biochemistry and you didn't want to do a Ph.D., then you got shunted into the patent office, because that was the only thing that the tutor could think of that was possibly worth anything. So anyway, I stayed on in Oxford, and did a Ph.D. At that time what was very exciting—there were two areas that were very exciting in science. One was doing nuclear transplants; John Gurdon had been working there doing intestinal epithelial nuclear transplants into oocytes and finding that the frogs would grow from the nuclei that came from the intestine, and this was really exciting, this was very heady stuff. In fact, that was my first choice. I wanted to go work with John Gurdon. But the year I wanted to start, he was moving to Cambridge and he couldn't take that many people with him, couldn't take on new students. So my second choice was working with Henry Harris. And Henry Harris was one of the people that invented cell-cell fusion, cell hybridization. And he was doing work taking malignant cells and fusing them with normal cells and he had come up with this hypothesis that malignancy was recessive to normalcy in a hybrid, and so I heard of that work, thought it would be interesting, and went to talk to him about it, and he gave me the project of analyzing cell membranes in normal, malignant, and hybrid cells. And his belief was that the genotype of malignancy could be recessive and that the answer was going to be the cell membrane. He says fine, you're mine, graduate student, and gave me the project. So I had this project as a graduate student where I took the—I had to find a way of analyzing the membranes from these cells and finding the phenotypic traits that segregated with the malignant genotype. So needless to say, I didn't win a Nobel Prize, which tells you I didn't find *the* trait, and that's because there is no one trait, obviously. But it was a sort of introduction to malignancy, cell biology, genetics. It was really a great experience for me. During my graduate work—I was there for three years—during my graduate work I met this brash young American who was over there on a Rhodes Scholarship, and that's why I ended up in this country.

He told me that he himself was going to medical school, and whereas I had been very much focused on research and science, had never really thought much about medicine, he had looked at it a totally different way, he had known he was going to go to medical school and was interested in science and was taking time out to go abroad and do things but was very focused on going to medical school. At first I thought, nah, and then basically there were two things that persuaded me to go to medical school. One was that, again, my husband constantly told me that if you're really interested in medical research, then you should be a doctor yourself, and then you understand things from the physician's perspective. And again, probably more true in this country than in England, that doctors tend to run things. So that was good advice. And the second one was when you're doing a Ph.D. in England, you're basically in the lab full-time and being in the lab one hundred ten percent of your time gave you very little time for other things, and I began to realize that even though I loved it, I was not the kind of person that could weave in a lot of other things as well as the lab, and so I'd be in the lab constantly. After you went in for the day you'd stay late, you'd go in Saturdays, Sundays, and you didn't have a life as well. Other people can manage it, but I don't think I could, because it becomes very consuming, and so I realized that I wanted to do something else that would make me interact with people more, and I would like the research to be a lesser part of what I do. So, medical school, my husband was coming back here to go to Harvard, and so I was going to come with him, and I applied to Harvard and got in, so we went through Harvard Medical School together.

GOOD: When did you do that?

MEDICAL ONCOLOGIST: That was 1974 to 1978. And then came internship and residency. Again, being a product of Harvard College and Harvard Medical School, my husband sort of wanted to go to [Massachusetts General Hospital] for residency, and we had both done medicine and surgery here. And, being a woman, back in 1978 I wasn't exactly one hundred percent sure I wanted to go to MGH because the women on the house staff didn't seem that happy when I was a third-year student. I didn't think the women were treated particularly well by their male peers, and I wasn't sure I wanted to traumatize myself that way. On the other hand, I remember making a turning point in my assessment one time, saying, why should they stop me from going to the best place if that's the best place? Okay, I'll go. So my husband and I both came here and did internship and residency here. It's quite an art to try and be together but separate.

GOOD: Were you both in medicine?

MEDICAL ONCOLOGIST: Mm-hm. But it was a big enough house staff. Basically you could be in different parts of the hospital. We decided deliberately to do it at the same place because we wanted to be able to coordinate on-call schedules, and it's very hard to do it in different institutions. It's just so much easier to have a common page operator and do things that way, so we did that. And then it came to deciding on fellowships, and of course we had met because of our mutual interest in cancer, so it was very difficult for one of us to decide to go into something other than cancer, so we both did hematology oncology! [Laughs.] But he's doing hematology and I'm doing oncology, so we have divided it up between us, and I let him do that and I do this. And now he is much more research oriented than I am. He's still pretty much doing ninety percent lab and ten percent teaching and clinical work, and I'm doing pretty much all clinical work and teaching and some clinical research.

Uncertainty, Twists, and Turns in Clinical Research

GOOD: How much time do you spend doing clinical research?

MEDICAL ONCOLOGIST: Right now I've only got three protocols that I'm working on, but I'm planning to build that part. I made a kind of gradual transition after a fellowship, what I was doing, so I did fellowship 1981 to 1984, and then for the first few years after fellowship I had planned to do eighty percent lab and maybe twenty percent clinical, and I made the deliberate decision to stay in a lab very close to where the clinic was, because the kind of subspecialty this is, you have to be available. And I knew that in a way I was committing [professional] suicide in the lab because I deliberately kept my clinic and actually tried to keep a presence in the teaching service and somehow gradually got the fellowship program to run and to oversee the fellows in the clinic, and as the clinical responsibilities grew, it was just harder and harder to get the lab work done, and I'm not one for taking on trivial projects, so the project I had in the lab was a rather major project and it turned into something that was really more than I could cope with. I was looking at interactions between tissues. I was very interested in biology of breast cancer, but the model system was working and was a skin model, where you look at epithelial-mesenchymal interactions, and it turned out to be a very interesting model, and a good model, but it was very labor intensive. I had to kill some animals and plate out their corneas and trypsin-digest some tissue, and then do primary tissue culture and then do chemical purifications of a cytokine I was looking at and doing assays and—it just got to be very labor-intensive and then I found I couldn't publish my work until I had this thing purified, because phenomenology as of this last decade, you can't describe something unless you've cloned it or unless you've got it as a single band on a gel. So I put in several years of work in the lab only to find that working single-handedly on something and not being quite at the point of getting enough of a handle on it to clone it, that wasn't going to go any further. Or it would, but I would have had to do it one hundred percent of the time, and meanwhile my clinical and teaching responsibilities were getting higher. So it all came to a crunch about three years ago when the lab I was working in moved—that became utterly impossible. To try and work [and] commute back here every day and keep what I was doing, I just decided—. It is a glorious lab, and it's well equipped, wonderful camaraderie, it's really great. But for someone like myself it's just impossible. So at that point I thought, someone's trying to tell me something, so I froze all my stuff down and then a generator blew up in the building in which I stored the stuff, all the fridges turned off, my stuff got thawed, I tried to bail it out and store it in someone else's freezers, then something happened, I had to move it—and then the cold rooms warmed up one day, and so I just tossed everything out.

GOOD: It's like serendipity.

MEDICAL ONCOLOGIST: Yeah. There's obviously—if I think about it too much I'm heartsick, because I put in so many hours of work and it just was wasted, so I feel badly about that. But anyway, I decided that—and those are the notebooks up there—I can't quite bear to get them off my shelf.

CARSON: It hurts.

MEDICAL ONCOLOGIST: It does hurt.

MEDICAL ONCOLOGIST: [*cont'd*]: I decided that in life if you're going down the wrong path you have to start again, so—all along I've been sort of interested clinically in breast cancer, and the work I was doing, I was really trying to manipulate through developmental biology and tissue interactions into a breast cancer model and I decided if going

GOOD: Have you seen any impact on, treatment from, research done in the last five or six years?

RADIATION ONCOLOGIST: Breast cancer, or anything?

GOOD: Well, anything.

RADIATION ONCOLOGIST: I think in general, not much of this, not much of the molecular biology of that way wasn't going to work I'd go the other way, so I'm trying to give more time clinically and at a teaching level to breast cancer, hoping that through the clinical projects I get involved in, again, that'll lead me back to interacting with some scientists at a sort of clinical basic level. I think I've reconciled myself to the fact, as I said earlier, that if you can't spend one hundred ten percent of time in the lab then you really can't do very good, very basic research, and I don't want to do mediocre research fifty percent of the time, so I'd be happier collaborating with someone and sharing the science vicariously than I would by not giving it enough time myself. So what I'm doing right now, the projects I'm involved in are to do with looking at regulation of growth factors in cancer. There's a project I'm working on with a drug company where they were trying to inhibit growth factors other than estrogen from stimulating breast cancer cells.

GOOD: There are curious twists and turns in this kind of research.

MEDICAL ONCOLOGIST: I know. Well, again, you can see putting a lot of time into this and then it never going anywhere. There are two ways of managing scientific research. You can either go for a sure bet: here's a gene, I'm going to see what regulates it, and you know it's going to work. Or you can say I'm much more interested in looking into something that's unknown, but if you take on too big a problem, then DNA repair, has yet reached the point where it's got a significant influence in terms of treatment. I would say the answer's no. Some aspects of radiation biology in general, from the laboratory understanding, has had some impact in the clinic, but not specific—not connected to the specific work I do. So I think it's mostly been at the laboratory end only at this point in time. But I think the knock-on effects to the clinic are sort of potentially there, and will take time to evolve. I think—I always—one of the motivations is that you're keeping an eye to asking, is to using what you hope is a sort of up to date and cutting edge of research, but keeping an eye on to its potential usefulness. You know that—I mean, I'm not involved in the work, but you know the genetic risk evaluation group here, probably, in terms of the breast cancer study, there is an attempted direct link between the genetics of breast cancer and trying to link the genetic risk to the likelihood of developing breast cancer, which would obviously be an extra piece of information that we don't have from the information we can gather so far for risk of breast cancer. It may not be that useful in the long run, but it will answer interesting questions of biology, and it may be clinically useful as well. We don't know until it's done.

GOOD: When you do this work, do you take on a clinical cast to your gaze as you're working in the lab, or is it really separate?

you can run into this kind of thing. The trouble is, I'm not interested in the other.

GOOD: Why do it?

MEDICAL ONCOLOGIST: Exactly. I'd much rather have myself in a position where what you do is what you want to do. The trick is being able to make a salary and have a secure job while you're doing it.

RADIATION ONCOLOGIST: Not day to day. Day to day the problems in the laboratory are making the experiments work.

The Care of Patients: Innovations

GOOD: How do you collaborate with your protocols and those that—well, how do the other people who work on breast relate to you? Is there any connection, intellectual connection, or research connection?

MEDICAL ONCOLOGIST: It's something that we're hoping to build through the breast center. That's one of the major focuses right now. The breast center has been a dream for a long time and in a way it's a little bit humbling to have to start at the very beginning and build it rather than just jump into your dream. It's going to start small and it's going to start as a place where a woman can come into the system. You're worried or you have a lump or you have an abnormal mammogram, this is a place to go, and it's not overnight going to become the dream breast center that we all want, but certainly one of the reasons for having it—there are two major reasons for having it, really. One is so that the women can get comprehensive care, both accessibility to the center and multidisciplinary consults and everything else, and also as a gateway into MGH. You come in with a breast problem but you find that you also need a gynecologist and an internist. And the other reason for having it, for an academic center to have it, is so that we can coordinate our research much better and, frankly, in the past, everyone's been doing their own thing. There have been two kinds of research on breast here. One major effort has been through the interdisciplinary or interinstitutional groups, like Cancer and Leukemia Group B, so for every stage of every disease these groups will try and have a protocol. So if you're premenopausal stage 1, or if you're premenopausal stage 2, or if you're postmenopausal stage 2, this group will have a protocol to meet most of those needs. So you can really accrue hundreds of patients into a study. If you get—here, we've got something like 350 new cases a year, which isn't enough once you've broken it down by subcategory, to do any big studies. And you want answers soon. You don't want to have to accrue patients for ten years to get a study. So you want to open up a study in 1992 and be able to close the study in late 1993 and have accrued 500 patients. So we participate in that kind of protocol. And they're very much treatment-related protocols.

But as those groups have matured, there are also other protocols that are added on as coprotocols that are scientific.

Medical Oncologist [*cont'd*]: So the best example is in leukemia, where any time a patient goes on a leukemia protocol to get arm A, B, or C of a chemotherapy study, you also submit their chromosomes from their bone marrow cells, and the patient can be categorized according to their cytogenetics or their oncogene expression profile, and so when you see the outcomes from those different groups, you can say, well, in this group these patients did really well, these patients did badly. Was there anything about their chromosomes that might have predicted that? So you can actually in retrospect find prognostic factors by doing the science along with the clinical observations. In breast it's been a little harder to do that, and in a way we're still looking for what are the breast prognostic factors. We know some obvious ones, the lymph nodes and the receptor status. But of all the other ones that have been proposed, which are the most reliable—is it the p53 expression, is it the erb-B2 expression?—those are the studies that are currently ongoing in breast cancer protocols. So one of our dilemmas really is at the scientific-clinical correlation level, what should we do? Should we do something different, or should we do what everyone else is doing? We'd obviously like to do something different, but you've got to have a good idea, and you've got to then go and sell the idea and get money for it,

Good: One of the questions I had was how people connect their clinical selves with their scientific selves in the way they work with patients, and you've indicated that there's separation. But I was wondering if you'd reflect on the kind of work you do with patients. What goes on in your mind, how you reason, whether or not that sets off any kind of serendipitous ideas or how that happens in terms of how you like those two selves, or don't.

Radiation Oncologist: On a day to day, those connections I don't think are made very often. It's really the sort of thing that you do when you're probably walking for ten miles in the country somewhere and you make rather—you speculate about the way research might develop and then how it might become—what clinical applications may develop as a result of that. So as I'm doing work in the laboratory I'm not thinking, Gee, if I could only connect A and B this would be a clinically useful thing. That couldn't be further from the truth. And similarly while I'm in the clinic I'm not thinking, Gee, If I could only make my experiment work I'd be able to make the decision for this patient. Occasionally—it's not totally separated, but in the main I think it is. For example, in breast cancer, one of the things that we're hoping to develop is the local treatment of breast cancer, not with external beam radiation treating the whole breast, but treating less than the whole breast by radiation implants.

and without a center it's been hard to organize in that way. And you see a lot of me-too kind of stuff. I don't know how many people are looking at p53 in breast cancer, but there are hundreds, literally hundreds of people. So we're doing it here as well. We do have one different twist on that, and that is Steve Friend is looking at that gene in the germ line of women who've had it, so that's a little bit different than just looking at tissue expression, so that is a little different here.

So this is something that through the focus group that has been set up to try and plan the breast center, we're going to try and address what should we do.

GOOD: There's a focus group that's been set up?

MEDICAL ONCOLOGIST: We've got to wrestle with the issue of what should we use the center for, and how should we approach research through the center, who should have access to the data bank, all sorts of things like that. Accessibility of tissue samples to people that want to study things. So there has to be set up some working rules and again in the past it's pretty much been, if I want to do something, I'm going to do it, or I want to try and do it, I'm not going to consult anyone else. I think this is an attempt to bring some kind of order.

GOOD: It's really exciting. Just stepping into this whole thing.

The Fluidity of Clinical Science: Experimental and Routine Uncertainty in the Care of Patients

GOOD: Let's see. I wonder if we might take a turn to patients now? I'm trying to get a sense of—we talked a little bit about this before but maybe we could review it again—a little sense of how—what your philosophy of communication and care is, first of all, with patients, and from there if we can talk about patients that pose certain kinds of challenges or dilemmas or raise issues around uncertainty for you in terms of treatment or communication. We could start with talking about your general philosophy of communication and how you deal with a breast cancer patient.

MEDICAL ONCOLOGIST: I guess that if I were to summarize my philosophy as I wrap up the sessions with my patients, I usually remind them that what I've been trying to do is to be their educator. I tend to approach my patients with, "We're a team, and my job is to make you aware of the facts you need to know to make your own decisions," and then I spend the time going through what I think about where they are with their disease. "Here's what you have, here's how I interpret it, here are the prognostic factors, and therefore this is what I think your options are," and then I go through what

I think of the different options. But ultimately I want them to choose what they do. Now it depends on who your patient is and where they come from as to whether they're comfortable with that or not. I would say that maybe seventy-five percent of my patients are comfortable with that. They like the idea that they're making the decisions and they're in charge. And then maybe one in five or one in four kind of basically says to you, Well, you're the doctor, what do you think? And it becomes very obvious that they are not at that point in their concept of healthcare and they really want a straightforward Yes, you do, or No, you don't need this, and they'll trust you. And that's okay, too. It's refreshing sometimes to have someone not want to ask all the questions, but that's not really my view of how healthcare is or should be right now, and it's certainly not what most women with breast cancer are looking for. Most women are very well educated now, they've heard all the buzz words, and they want to know why my neighbor got chemotherapy and I'm not, or why she was given the option of a lumpectomy and I wasn't, and they've read enough that they want to ask all sorts of things.

Good: Has that become more routine—how long have you been working with breast cancer patients?

Medical Oncologist: Probably about ten years.

Good: If you take a long view of what's happened over the last decade in terms of how communicating with patients has changed—

Medical Oncologist: I think it's changed a lot.

Good: In what ways?

Medical Oncologist: Well, the therapies have been proven to be beneficial. When I was a fellow it was not quite as clear who did and who didn't benefit from chemo, but in the—I would say ten years ago we were beginning to understand that the benefits were more likely to be reserved only for the younger women, and that was sort of blessed in the middle of, let me see, 1985, with a National Institutes of Health (NIH) consensus report saying that really it's the young women that benefit. Don't give it to the older women. It should really be only reserved for clinical trials in the older

Surgical Oncologist 1: I use tiny dressings when I do things. Sort of the same way that you think you had a small operation so you get better faster. Most women are prepared ahead of time. I tell them the day that we are talking about the operation, "Most women are in the hospital one, maybe two nights, you certainly stay longer if you've got any problems, but most people feel well enough to go home, and, in fact they do." What I've been amazed at is which patients beg me to go home after one night when I thought they would be a two night stay kind of person.

Good: What's interesting is that you spend a lot of time with scientists women. And so it's pretty much clear-cut. You were young, you had positive lymph nodes, you got chemo. You had negative nodes, your prognosis was good, you didn't need it. Or you were over fifty, in which case chemotherapy didn't benefit you. So the rules were easy. And the patients would walk in and you'd be able to—anyone could do the same analysis and you knew what that patient would walk out with. Well, the rules over the last seven, eight years have gotten much more complicated, not only with—they're now coming out to say that the older women do benefit from the chemotherapy, but also lymph-node negative women may benefit, and maybe a combination of chemo and hormonal therapy is better than either one, and then they've come out with all these different prognostic factors that may push you to give chemo in someone that you would have thought originally would have been in too good a prognostic group to need the chemotherapy.

So suddenly it becomes very complicated. And patients are walking in to you for a second or third opinion, armed with such a lot of information, and they've already had their—they may have been biopsied out in the community somewhere, but they come in with a pathology report that gives their S-phase analysis and whether their tumors are aneuploid or diploid, what the oncogene profile is, and they're sending tests that frankly we don't send here because we don't know how to interpret them. It's easy to get data, but what to do with that data once you've got it is the hard part. If someone's aneuploid, does that mean they ought to get chemotherapy automatically? Well, I don't know. And so maybe I should choose not to have that bit of information if it's not going to help me to decide, why would I get that bit of information anyway? So you just kind of feel like with every patient that comes in armed with that that you are obliged, because the information's there, to kind of sift through it and maybe to address it with them before you come to that conclusion, that you do or don't need treatment. So right now I think there's just an awful lot of energy being put into what used to be a very simple decision, and then it's even more complicated with the fact, for

thinking about that clinical narrative and how it shapes patients'—

SURGICAL ONCOLOGIST 1: It's absolutely true. It was so dramatic to me when we went from—we tried to organize a program to have people discharged after two days instead of five or six. It was shocking to me when you know just the difference in how you explained it to them ahead of time and telling them what the expectation was how differently they did. And I do the same operation, the surgery is not that different. There may be a slight difference in how surgeons do things but it's almost all expectations. "Small dressing, must be small operation, I shouldn't be sick." I think it makes a big difference.

example, just to move on to a difficult decision, a woman came to see me two weeks ago, she's only thirty-six years old, she has a small tumor, about a 1.3-centimeter tumor, her lymph nodes were negative, and again she already comes armed with some ammunition in terms of it was an aneuploid tumor but a low S-phase, and she wants to do everything she can.

You will find education material for women now that tells them that there are these things called prognostic factors, and some of them even come in knowing what the cathepsin-D level is. It is truly amazing. Now this may be very selected, when you're in a university town and there's a lot of room for having second opinions and you begin to see things generated to make your center look better than other centers, maybe. Maybe it's different in the Midwest in a small town. I'm sure it is. But it's not unusual to have someone come in asking you specifically, I was looking at my reports and what does it mean when it says this? What does it mean when it says that? They've already asked their first oncologist that. So now they're coming in, they're kind of wanting to see whether you say the same or different. So they're throwing the same words at you. Well, my oncologist said that—they may not fully understand it, but they're quoting this, and you therefore have to discuss it. Now if you want a difficult situation—it's hard enough sifting through the different prognostic factors, because we don't have this algorithm. How much weight to give to the ploidy versus the ER status. And it would be easy if everything were bad prognostic, or everything was good prognostic. But you're going to have all these women in the middle, and you just simply don't know.

MEDICAL ONCOLOGIST [*cont'd*]: And to convey that sense of uncertainty to someone without having them lose confidence in you, you want to go into enough detail but not too much, it is difficult. And it's probably easier to take the old-fashioned approach: I've just weighed everything up, on an average I think you don't need it. But then they think you've cheated them. So that's hard enough. But then when a thirty-six-year-old says to me, Look, I really, if you think I need the chemotherapy, I'll take it. On the other hand, I haven't had my first child yet and I really, really, really wanted to get pregnant, and

SURGICAL ONCOLOGIST I: If it's malignant, I want them to have enough information so that they have the truth, but also so that they have some hope. They know that there are things that can be done that will help them. I think the hardest thing is uncertainty, and also I think it's extremely hard if you begin to think that your doctors are not telling you things. Then you don't know if you can ever believe them. So, I find being very frank, but not discouraging, from the beginning seems to be best.

Women are adults, women can deal with breast cancer, and you start out with that assumption and my oncologist tells me if I take the chemotherapy I may go into menopause, is that true? Well, yeah, it is true, if you're thirty-six and you take six months of chemotherapy, you may go into menopause. Which is more important to you? To do every last little thing you can to improve your prognosis, or to take a little bit of a chance on that and preserve your fertility so you can have that child you always wanted? And those aren't always easy decisions.

you deal with them that way. When patients start out being involved from the beginning and being *in control* from the beginning, it's much better. The whole way. And treating breast cancer is a long process these days.

And as a woman, I hear women coming in and saying, Look, my oncologist, who's a male, never even mentioned the concept of infertility after, and isn't that important? And it's not just the lack of fertility, but it's also the premature menopause at age thirty-six. Isn't that important? Well, yes, it is important, you're right. And of course now that opens you up for a whole line of discussion that if you were the first opinion you might not have spent so much time on, but because you're the second opinion and he didn't discuss it, now you've opened all this up, so you can really get yourself bogged down in a lot of—well, the risk of a stroke, the risk of a heart attack, if you're prematurely in menopause, the risk of osteoporosis, and that's where having tamoxifen there might be good, because maybe you can be reversing with that, but you can get into extraordinary depths, and I think some of us do tend halfway through the conversation to wish we could pull back and not have mentioned a few things. So I think—you get experienced.

Gendered Discourse

GOOD: To what extent do you feel that being drawn into that conversation gets run by patients nowadays? You've given this example of these women coming in and saying—I guess one of the questions I have is, do female physicians tune to patients' considerations in a different way than men do, and therefore allow that conversation more readily?

MEDICAL ONCOLOGIST: Oh yeah, definitely. I think you're the one that's going to prove that, right? But I would say yes. I think there's no question in my mind at all. I think we're more receptive to it because I think any time you're doing a job where you're trying to help somebody, if you're empathic, you're going to say, What would I think if I were in her position? And these are issues that you'd be tuned into. Often we can identify—these women tend to be young women. I'm identifying by age. I'm identifying with them in terms of social position, and then I'm saying, Gosh, how would I feel if I were sitting there? And no sooner does she broach the subject than

I'm onto it. I'm saying, Funny you should mention that, that's exactly what I would think if I were in your position. It's just the empathy that pulls you right in. And I frankly don't think I could decide what I'd do unless I were in that position. How do you choose if you're in that situation?

GOOD: When you treat men—do you treat men?

MEDICAL ONCOLOGIST: Mm-hm.

GOOD: When you treat men, do you think the same thing happens, that you get pulled into a male patient's considerations more readily than, say, male physicians do?

MEDICAL ONCOLOGIST: I think so. There may be special examples where that's not true. Dealing with a man who has prostate cancer, where you're going to discuss surgery with its possible impotence or incontinence versus loss of virility if you give them hormones—I think that's something where men are going to speak much less freely to a woman and probably much more openly to a man, because after all he'd understand these things and a woman wouldn't. So I think with the special exception of male-related diseases like that, I think men do tend to ask more questions of a woman and we do tend to be more open about them. But I don't find the intensity there. I think it's partly because of all the press that breast cancer gets, and all the options available, and the fact that you know other people with it, so you've already talked, and you're young, or at least the patients I'm talking about are young, I think all those things together are what counts. You don't have the same intensity even with a woman who has a lymphoma or Hodgkin's disease or lung cancer, because the options are much more restricted. Well, you may in terms of fertility issues, but either you do or you don't take the chemotherapy. It's not a question of playing a statistical game. We're talking adjuvant treatment here, so we're talking about making a fractional improvement in your prognosis if you do it, and how much is that fractional increase important to you, and how biologically significant it is. It's not like saying we either cure you or we let you die of the disease. It's not a black-and-white situation.

Clinical Ambiguity and Persuasion: "It's Not Black-and-White"

MEDICAL ONCOLOGIST [*cont'd*]:

GOOD: So are the protocols really clearer with Hodgkin's and leukemia in contrast to breast cancer?

MEDICAL ONCOLOGIST: I think so. For example, in Hodgkin's Disease it used to be that if you got chemo-

RADIATION ONCOLOGIST: It's not black-and-white, it's trade-offs. There's no uncertainty about the statistics, it's a question of how you use those to make your decisions. That's where it's ill defined, and different people can take the same

therapy there was a very high rate of infertility. But as we got attuned to the fact that people were being cured of this disease and infertility was a major problem, and secondary leukemias, we changed the regimens, and now we have an equally good chemotherapy regimen that doesn't cause infertility and doesn't cause secondary leukemias, so that's no longer an issue when you're talking to a patient, or usually is not an issue, and you can reassure them that you can probably cure them and you won't cause these other problems.

Now, this may change as the therapy improves, but right now, it's a very clean division. If you have recurrence—and we're not talking about recurrence in the breast or local regional area, we're talking about a distant metastasis—until recently the absolute judgment on that was that you now have an incurable form of breast cancer and my job as your oncologist is to help you live with this as healthily as possible. And therefore the philosophy becomes, Do as little harm as you can with the treatments. If hormonal therapy works for you, use that first line. If that works, then change it next time when you have a progression. And as a last resort use chemotherapy and use something gentle, because if you don't respond to that we have something else.

statistics and decide to handle them in different ways. Doctors can do that and patients can do that. And I try to explain as much as the patient appears to want to have information given to them, but frequently there reaches a point where they're perhaps saturated with information, in which case they're just looking for you to recommend something. It's just that they're—it's how you use that information in terms of decision making. There's a lot of gray area.

SURGICAL ONCOLOGIST 2: Breast cancer patients used to be very easy to take care of when I was an intern in 1970; everybody got a mastectomy, that was basically—it was easy. And often they came in and had a breast biopsy done, and if it was positive, we just went ahead with the mastectomy and there wasn't a lot of discussion about it. And now that there are so many options, it takes a lot of time. I wouldn't call it a dilemma, but—the hardest part, probably, is helping your patient decide what they're going to be comfortable with, because it's not always what you might—right out of the starting blocks—think would be necessarily, what their decision would be. But then you kind of draw things out and talk about their social situation.

With adjuvant treatment you're much more talking about this is your one chance to cure that patient and there's definitely a tendency on the part of an oncologist to say, Well, if you only get this much benefit with this regimen, maybe we should make it more aggressive and maybe that should cure more people, and this is our one chance. We've got to get really dose-intense, and so you'll see these protocols escalating drug therapy. And again, you're putting a decision on the patient. Do we treat you with standard therapy and get this much benefit, or do you want to come on this protocol? We'll treat you aggressively. The patient doesn't really have the knowledge to make that decision. It's really going to be your decision, and you present it in a way that the patient says yes, if you really feel this person should have the aggressive treatment.

Clinical Narratives: Shaping Patient Experience

MEDICAL ONCOLOGIST: And so you tend to—it may be a little dishonest, but you tend to size the person up as you're talking to them. Is this someone that would want to go for the aggressive treatment and should go for the aggressive treatment? And if the answer is yes to both, then you basically tell them, Do I have a protocol for you! This is really something I very strongly recommend—and then they say yes. Whereas if the patient is someone that you really think could not tolerate it or should not because their disease is not that aggressive, then you phrase it to them, Well, we do have these, but these are so extraordinarily toxic that I really don't think that this is something I'd recommend. Basically you're making the decision for them, and then phrasing it in a way to them where they say, Yes, that's absolutely right, that's what's for me. Now, I don't know, I would be very interested in an observer telling me, You're overdoing it, you're underdoing it, the patient really isn't in charge at all. I know perfectly well—if I'm taking a patient and they're thirty-five years old and they've got twelve positive lymph nodes, that is a horrible prognosis. Those women are probably all going to die. And I'll be sizing that woman up. Now, if she has a horrible psychological background or an awful family situation and I know she could not take the stress of going through a bone marrow transplant, I will try not to mention that as an option up front. I may mention it, to be fair, but I won't be pushing it on her. So I will already to some extent have interpreted, put my judgment on the front line, because I don't think it's fair to say something to a patient and then afterward find out that for various reasons she can't do it. I think that's just not fair. For example, they have no insurance, they've got six kids to take care of, and they're emotionally disturbed. How could you possibly mention bone marrow transplant? First of all, no one will pay for it. How is she going to have someone take care of the kids? It's just cruel to say you've got to go out of state and leave your kids behind, because it will never happen anyway. So that would be an extreme example of where you know up front that an option that might be an option for someone else is not an option for this patient. That's just the way it is. So if I had someone that I knew was motivated and I knew had [good insurance], then that might become something I might discuss with that patient. And I would begin by trying to keep it in a very objective, bland way, and I would put it as one of the options and then if she showed a strong interest in that option I would take it a little step further and say, Well, of course it's experimental treatment and it's very aggressive and you have to take more risks with this. And if she still showed interest, then I'd discuss it in more detail. But you're looking for feedback. What you don't want to do is to push your agenda and say here's your option, this is what you should do. And if she starts to back away, then you say, Well, let's leave that on one side. It's an option, but let's explore the other standard therapies. And you will be amazed at how people give you the signals as to what they do and they don't want to talk about.

Reading Patients

CARSON: What kind of signals do patients give you?

MEDICAL ONCOLOGIST: The body language, and also what they ask questions about and what they don't. Some of them are visibly looking awkward. You mention the bone marrow transplant and the body language tells me, I knew someone who went through that and they died, let's leave that one in place. And you can't always predict. I had a woman recently, she's about forty-two, I really thought that the bone marrow transplant option for recurrent disease would be what she'd leap at, because I thought she was a very objective lady, and—and yet her response was, Ooh, that's experimental! And I heard you can die from that. I don't even want to talk about it.—Well, I think we should talk about it.—Trust me, I don't want to talk about it.—Okay. Fine. Nothing experimental.—I don't want to leave home, I don't want to be sick. In fact, I don't really want to lose my hair.—You're only forty-two

SURGICAL ONCOLOGIST 1: Did you meet her today? Oh, good. So she—did she talk to you about the tamoxifen?

PATIENT: Yeah. She told me—she told me that after fifty, it was just as good, there was only a three percent chance that the—not the radiation, but the chemotherapy would be better than the pill. And you have to decide on how much of a life span you want. I don't want to be sick at eighty. If I have five years I'll take my five years and enjoy them.

SURGICAL ONCOLOGIST 1: And the other thing that she may have been explaining to you is that there's a three percent chance that you would have a longer life span with the chemotherapy than with the tamoxifen. you might have the same—

PATIENT: —life span, that's right! They don't know!

SURGICAL ONCOLOGIST 1: You'll be on that for at least five years.

years old, we're talking about a fatal illness.—That's okay. I'd much rather stay with something, if you can offer me something gentle and I can live with this for a year or two, that's what I want to do.—And it's not because she's stupid and it's not because she hasn't been to medical school. It's that she's known other people with cancer and she has already made her choice. And somehow she's had—she's already got access to some of the information she needs and she's ready to decide that she can decide against this.

PATIENT: Well, she said three anyway, but they're making a study whether it's beneficial between three to five. At least three and then she'll see me in three months anyway.

At least [*jokingly, but bittersweet*] you're giving me three to five years!

Salvage Therapy and Dangerous Weapons

GOOD: What's the story on bone marrow transplant now?

MEDICAL ONCOLOGIST: I think there are two ways of looking at bone marrow transplant. There's bone marrow transplant as a salvage therapy for metastatic disease, where nothing else is going to work. There are diseases that it clearly has a place, for example, relapsed lymphomas or relapsed Hodgkin's disease, where they just have become refractory to the other treatments and—let me rephrase that. For example, someone has Hodgkin's disease and relapses. You have a salvage chemotherapy regimen that may still cure them. But if they relapse a second time after that second treatment, it would be extremely unlikely for a third-line standard therapy to cure that patient. Whereas if you take those same patients and put them through bone marrow transplant, now allowing you to give them very intense doses of chemotherapy, maybe about fifty percent of those people will be cured. And so that has become the treatment of choice for someone who's failed a couple of times and it has a track record and we feel those are cures. So there's a definite role for bone marrow transplant in salvage therapy.

GOOD: That's an interesting term, *salvage therapy.* I don't think I've heard that before.

MEDICAL ONCOLOGIST: Yeah, you're trying to salvage the patient after the standard therapy has failed. So you're pulling something out of the bag. *Salvage* is a common term. It usually—you can use different terms, *salvage* or *third-line, fourth-line.*

GOOD: Did this come with bone marrow transplants, the concept of salvage therapy, or is it an old concept?

MEDICAL ONCOLOGIST: Salvage therapy is an old concept. I'm sure if you look in oncology meeting abstracts under *salvage therapy*—if you looked under a certain disease there might be a subheading, *salvage therapy.* It usually means unconventional or third- or fourth-line, something that isn't usually the primary treatment and you're trying it. You could equally use an experimental therapy. You could use an antibody therapy as a salvage therapy. It may or may not work, but it's the attempt to do it that makes it *salvage*.

GOOD: I'm sorry, that was sort of a sideline.

MEDICAL ONCOLOGIST: Right. So what I'm saying is that there are some diseases like relapsed Hodgkin's where bone marrow transplant is a good option and has a proven track record and you know which ones those are going to be because they're the ones the insurance company pays for because they've recognized that this is the way to go. When you come to breast cancer salvage treatment, it's in its infancy, and the first published studies were not that convincing, and there were some lively debates where people would stand up and say what they were doing and how they were getting phenomenal responses, taking these women with recurrent disease and giving them very high doses of chemotherapy and having complete responses in eight days instead of the usual several months, and how wonderful this was going to be, and then they do a bone marrow transplant.

RADIATION ONCOLOGIST: Medical oncology still doesn't have—its role in the management of cancer is still a relatively small one, although increasing. Now as more and more adjuvant chemotherapy protocols are being accepted as the standard of treatment, there is a much broader base for medical oncology. I think there are some advantages in having a knowledge of medical and radiation oncology in terms of managing the patient overall. You are not—if you only have one treatment weapon, you tend to select it. If you're selecting from a multiple, a variety of treatment weapons, you're hopefully selecting what you think is the best combination.

That was the good news, and the bad news was they were all relapsing at six or eight months after the transplant. So it looked good but in fact it turned out not to be the answer. And different people interpreted that different ways. If you were a transplanter, you said, Well, we just haven't found the right drugs, or we haven't given them early enough, or we haven't selected our patients well enough, and we're taking patients who have been through more treatments and in whom the disease is more refractory, so what you really need to do is to do the transplant earlier in the course, when the disease is still responsive. And if you were a nontransplanter, you'd say, "We're wasting our money doing this, it's clearly not going to work." And there were very lively debates.

CARSON: Especially since it's so expensive, right?

GOOD: Yes, what is it, $100,000? Or has it gone down?

MEDICAL ONCOLOGIST: It's coming down. With the traditional harvest—the marrow from the pelvis—and with patients in the hospital the whole time without any growth factors, it would cost about $150,000. The big revolution over the last two years has been moving away from actually using marrow and instead using peripheral blood-stem cells, which allows you to do a lot of this as an outpatient and allows easier engraftment. So ideally a lot of this could be done as an outpatient. It's the hospitalization when you're neutropenic and febrile that really costs the money. And if you can keep the person out of trouble and keep them out of the hospital, it's been estimated that it might be one-third to one-quarter of that cost. Still expensive, but much more manageable. So where we are with bone marrow transplant for metastatic disease is it's still experimental but it is the only option if you have a patient that's not responding to standard treatment and as such for selected patients, as I say, the gold standard is, does the insurance company pay? Some insurance companies will pay, and it's going to depend on whether this makes sense or not. If you're thirty-five, you've got a low burden of disease, and you show response to chemotherapy, and in fact if you can get a complete response to chemotherapy, you know what's going to happen if you don't do a bone marrow transplant. That disease is going to come back, the woman's going to die. If you do a transplant, there's a chance that it might work, and there are some people—in each series there's a couple of people that are two years out from their transplant and still don't have recurrent disease. Now what the numerator and denominator's going to be in those studies, we don't know yet. And that's a real question. And it's an ethical dilemma. How—what percentage of patients can be salvaged, to make this a cost-effective thing for society to do? If you can only salvage one in a hundred, should we pay for a hundred to only salvage one? And what you might decide for yourself if you're a patient isn't necessarily something that society can absorb. And so those are very stressful and emotional issues.

Between Science and Therapeutics

GOOD: You've really addressed this question, but let me conclude with it and see if we take it in any other direction, but I was wondering if you would add anything more to how your scientific work relates to your clinical practice. When you're actually taking care of people, the process of thinking, acting, relating, touching, doing, seeing—do these two things feed back and forth?

MEDICAL ONCOLOGIST: Well, not as much as I'd like. I've really seen them to date as two separate activities. It's very interesting to know about how a cell might become a cancer cell, it's very interesting to know what hormones might regulate it, but then in the clinic there's only so much you can and can't do, and you may be counseling people about how often to get mammograms or what kind of surgery to have, but it's really divorced from the science. The science and clinical overlap a little bit when it comes to, for example, choosing the first-line treatment for someone with metastatic disease. The fact that hormonal therapy is going to work for this patient and you're sure of that because the ERs are over 1,000, and this is an older woman, and you just know so much from the biology that you can be very authoritarian about recommending this. But that's very minimal overlap. I would really love to be able to bring the two really tightly together. For example, if you're trying to analyze why one woman does badly and one does well, I think ultimately you're going to have to do a biological profile of the tumors, and you're going to have to say what it is about this woman's tumor versus that one that causes this one to grow more aggressively and that one to be more indolent; this one to be refractory to chemotherapy and this one to respond. And that's going to allow you then to design your treatments. So I would foresee if things go the way I'd like them to, to really have some kind of profile that you do of the tumor. The profile right now is some very mundane things, things that may or may not help with treatment options. As I said, the ploidy status or the S-phase. But it doesn't really tell you anything about the biological characteristics of the tumor that you can really address. If, on the other hand, you knew that this one were using this growth factor or that growth factor and you had blockers for those factors, or this one expressed this antigen and you have an antibody to that antigen, that way for each woman you could perhaps put together appropriate combination biological therapy and target that woman's tumor. And you could say this one should go into this study where we're using this, and this one should go—and that would be the kind of thing that I think we could build over the next five years. But that really has to involve very close liaisons. The clinicians are almost the least important people because they would be given instructions by the scientists. It would be, here's the profile to send the tissue for, and here's what to treat this group of women with, and that clinicians just become the doers. Probably the most important people are the thinkers in between the scientists and the clinicians that try and persuade the scientists to generate these tools and then persuade the clinicians to use them. And I think that's where the excitement is.

GOOD: Is this your project? Is this your growth factor project?

MEDICAL ONCOLOGIST: No, that's a way of muscling in on this area, because it's the closest thing that's commercially available to doing that, and it's just a sign of my interest in that area of work. But I think to try and constantly be in control of your subject by having an overview, so you're not bogged down with the facts, and also to be generating innovative ideas, I think that's where the fun is. And by dealing with patients every morning, every afternoon, every day, you get bogged down in being emotionally supportive and pulled into these arguments. If you're in the lab, you've got to keep your grants going and you get very focused on concrete problems. You can't speculate and dabble in this because you've got to get this done. And I think the challenge, my personal challenge, can you keep one leg in the scientific arena so you understand the science, one leg in the clinical arena so you can excite people and apply the science, but can you make your job in the middle, so that you can create some new ideas.

GOOD: That's a lovely image. You'd get a Nobel Prize for that.

MEDICAL ONCOLOGIST: There you go, right. No, you'll never get a Nobel Prize for that, because the Nobel Prizes would go to the scientists and the humanitarian awards go to the clinicians. But you have to—I've come to the conclusion that the reward in life is from enjoying what you do, and that's what I want to do. I want to have my cake and eat it. I want to be able to go to the scientific meetings and I want to talk to the patients, but I don't want to do this or this full-time.

GOOD: What you're really saying is this larger role, in a sense of overseeing both, shaping both.

MEDICAL ONCOLOGIST: Yes, yes, yes. And again, my eyes are always too big and I always bite off more than I can chew.

Interpretations

I began conversations with oncologists over five years ago, first as a researcher studying physicians and second as a colleague carrying out joint research. From my very first interviews in 1987 to those most recently conducted during 1993, I have been struck by the excitement that comes with being on the edge of medicine, engaged in experimental treatment. The brinkmanship noted by one colleague, the challenge of the scientific and clinical struggles, emerged over and over again in conversations. What has also emerged is the tension between the work of science and the work of therapist and physician, and the dichotomy between the activities of the two. Although oncologists incorporate a good deal of clinical science in how they convey statistical information to patients, interpret the potential impact of experimental treatments to be inscribed on patients' bodies, and negotiate therapeutic choices with and for patients, most physicians interviewed separate laboratory and clinical life, and even scientific thought from clinical thinking. Although the technical therapeutic acts acquire a certain orientation from clinical science, oncologists contrast that work with what occurs in laboratory or clinical research. Research and clinical activities—talking with patients and acting upon patients—remain across divides. In the main conversation in our presentation, my oncologist colleague finds herself standing between the laboratory—and the "purer" science of the lab—and the challenges of clinical work and the management of institutional innovation. Bridging this gap suggests another form of "political and cultural" brinkmanship.

—Mary-Jo DelVecchio Good

Acknowledgements

We extend our thanks to patients and clinicians who have participated in the studies which have led to these conversations. Special appreciation to Martha MacLeish Fuller, who transcribed these conversations with great care and formatted the manuscript.

Science, Inc.

These contributions portray various complex reconfigurations of institutional relationships through which basic fin-de-siècle science is done. The entrepreneurial business sense of the scientists themselves in each of these cases is striking—in relation to an environment of large corporations, universities, and government, certainly, but not dominated by it.

Paul Rabinow

5

Reflections on Fieldwork in Alameda

Having completed a long term project and published my book *French Modern: Norms and Forms of the Social Environment,* I was looking around for new terrain to explore, hoping to avoid that fallow period which so often follows a major project. Friends in Paris had been urging me for some time to think more about the life sciences. With the announcement, and extended debate, about the Human Genome Project, the federally funded effort to map all of our human genetic material, I decided to take the plunge and start exploring. Early on it occurred to me, perhaps if *French Modern* had been about the emergence of the category "society" as a new object of power-knowledge in the 1830s and the century-long elaboration of sciences and institutions associated with efforts to form and reform it, then perhaps we were on the verge of a parallel modernization of life.

The first step, like tackling a new language, was to accept the challenge of beginning the process of learning the basics of molecular biology and genetics. There was an immediate payoff and arousal of curiosity, the material was extraordinarily interesting. Evelyn Fox Keller kindly introduced me to the social studies of science literature. We organized a faculty dinner series on the genome project. Our first speaker was Charles Cantor, director of the genome project at the Lawrence Berkeley Laboratory, run by the Department of Energy. The DOE had been involved in biological and genetics research for a long time, concentrating on the effects of low-level radiation from the atom bomb. The genome project had, albeit under congressional pressure, committed three percent of its large annual budget to social and ethical issues. Cantor was hospitable and enthusiastic about an anthropologist working on the project. However, shortly thereafter, in a dramatic coup, Cantor was removed from his job at Berkeley. Science had its own politics and it was played for keeps.

What ensued was frustrating, certainly not uplifting, but ethnographically instructive. A large, leaderless bureaucracy lurched forward, desperately seeking a "star" to take it over. During the extended search, potential star candidates like Leroy Hood were quite receptive to including an ethical-social component to the project at Berkeley. But Hood and the other stars didn't accept the position. Those on the ground were another story. It seemed as if all of the worst qualities of the academy plus those of a governmental bureaucracy were activated. Eventually an ambitious local dog geneticist was named to head the project; high school outreach and tours of the facility to distinguished visitors were his idea of a social component to his empire. Who needed this? Very little science was being done and I was all too familiar with the other dimensions of academic small-mindedness and ambition to need to waste more time studying them.

In search of other more fertile terrains, I asked my colleague Vincent Sarich whether he knew anyone in the biotech world. He recommended Tom White, at Cetus Corporation (by that point White was working for Hoffmann-La Roche but was still in the Cetus buildings) in nearby Emeryville (they soon moved to Alameda). Sarich had been consistently generous with his time as I worked through the basics of molecular biology, stopping to chat and explain details and concepts in the xerox room, whenever he was asked. He was also embroiled in a raging controversy: students had disrupted his class charging him with racism, sexism, and homophobia. The controversy had hit the press and was still swirling around during the initial contacts with White, culminating with a forum in front of five hundred students and faculty.

Who would have thought that industry, even a small corner of it in a cutting-edge field, located in the atypical San Francisco Bay Area, would be a haven in a heartless world? The contrast with the perpetual petty power plays, seemingly insatiable status lust, and insistent incivility of the academy could not have been more unexpected. No paradise to be sure, but nonetheless here was a racially and sexually diverse environment with a group of reflective people, who gave every impression of passionate devotion to their science, who seemed to take an interest in the ethical and social consequences of that science, even without any grants or congressional pressure to do so. Suspicious—appearances must be deceiving—I began my fieldwork in Alameda.

The following exchange forms part of a more extended set of interviews on a wide range of topics, especially the invention, standardization, and development of what is arguably the most important biotechnological achievement of the 1980s, the polymerase chain reaction. Tom White and I are writing a book on this relatively simple technique to amplify, rapidly and efficiently, specific strands of DNA, producing millions of copies in a number of hours. During our exchanges and through the course of interviewing other scientists at Cetus, it was striking that a number of the other senior scientists had been politically active in their youth. The following interview explores this topic with two of those scientists, Henry Erlich and David Gelfand.

Science and Politics: Some Spontaneous Discourses of Scientists

RABINOW: We'd like to discuss the relations of politics and science both individually and institutionally. You were each activists during the 1960s. I'd like to hear about that engagement and how, if at all, it related to your scientific interests. Next, the idea that a group of former political activists would one day have major scientific responsibility in a biotech company, one now owned by a major multinational, and see some continuity with their past commitments would surprise most academics. I'd like to explore that paradox or contradiction with you. Henry Erlich, why don't you begin by telling us about your background?

ERLICH: Well, I grew up in a family that had a socialist background—not communist, but socialist, social democratic, really. My grandfather had been a leader of the Bund, a Polish, Jewish, Socialist movement. I was named after him and he was always extolled to me. I did a lot of reading about him, so he was kind of an inspiration for both myself and my brother, who went on to be a labor leader.

RABINOW: Was he anticommunist?

ERLICH: Yes. My dad is anticommunist, my mother isn't. What that meant in dealing with the Vietnam War was that my antiwar mother, my brother, and I were on one side. My father came around but he started out thinking you had to be tough, had to be anticommunist. His dad had been executed by Stalin. My grandmother had known Lenin and Martov and all these people, and she was still quite left wing, and she lived in New York. Until her late nineties she still took part in demonstrations and marches and so forth. When I was undergraduate and grad student I was SDS [Students for a Democratic Society]. I was an undergraduate at Harvard, but then I actually got more active in SDS and the antiwar movement when I was in Seattle at the University of Washington.

RABINOW: Was there any science in your background?

ERLICH: No. I had always thought science was interesting but my father was a professor of literature and my mother is a psychologist, so there was no real science in my background. And my bent was always more toward philosophy or literature. As an undergraduate I was majoring in Renaissance history and literature. And then I took an introductory biology course from George Wald. He later was awarded the Nobel Prize. He was a very inspiring teacher; I got very excited by this and I went on to take genetics courses and switched my major to biochemical sciences. I graduated in 1965.

RABINOW: So it's not yet a very politicized atmosphere.

ERLICH: No, the late sixties were much more political. When I started grad school at the University of Washington in Seattle—I grew up in Seattle, so I kind of went back home cause I love the northwest and Seattle—

I was getting my Ph.D. in the Department of Genetics. And then I got involved in the SDS and all kinds of other counterculture activities. I worked on an underground newspaper, the *Helix,* which was an exciting experience and a lively and vibrant community. And then I took a year off (1967–68) and was a VISTA volunteer in New Mexico, working with street gangs in Santa Fe, which at that time wasn't the trendy, boutiquey place it is now. It was a very poor place but interesting and beautiful.

RABINOW: My sister lives outside Santa Fe, and she was a midwife, and now she's a nurse. The other side of Santa Fe is still there.

ERLICH: Yeah, well, it was a fascinating town, a lot of new age activity. There was a commune called New Buffalo—a poet named Max Feinstein was the head of it, and he was a very interesting person. And the guy I worked with was a Christian Brother, Godfrey Reggio. He left the brotherhood and became a political activist. He was the guy who made those movies, *Kayaanisqatsi,* produced by Coppola and music by Phillip Glass. It's a very political film, but just pure images.

RABINOW: Was this to get out of the draft?

ERLICH: No, actually I had a student deferment. It maintained the deferment; in other words, I wasn't vulnerable to the draft. I was very lucky because New Mexico was a very interesting place and it was a very good program. This was the first time I was directly grappling with problems in the community, in the world. I made a lot of friends.

RABINOW: Did your science have anything relevant to this at this point?

ERLICH: No. What we were doing was trying to retain the structure of the gang with the strong leader and the cohesive group—and just redirect its activities. I liked the kids. They were basically, I'd say, with one or two exceptions, good kids. And a few of them were bad guys, like they'd try to stab you or something. Which is not very . . . not very congenial. After that I went back to graduate school, I continued to be active in SDS, but then I got my Ph.D. in 1972 and left Seattle to do a postdoc at Princeton.

RABINOW: So there are still no connections between your politics and your science? For example, I wasn't in SDS but I was certainly involved in the antiwar movement, so in anthropology, one of the things I did was learn about Vietnam, everyone knew what was wrong with the American war, but almost nobody in the antiwar movement knew anything about Vietnam.

ERLICH: I was very interested in science and genetics, and I was very interested in politics and wanted to maintain both of those interests. They were parallel.

RABINOW: But you didn't come out of Harvard saying, "Science for the people."

ERLICH: No, in fact, I was always a little uncomfortable by the attempts to impose politics on science. I had friends who would talk to me about bourgeois physics or bourgeois biology, and I said, "There's biology . . . you know, there's physics." I remember one of my friends showing me some article, "Is Nature Dialectical?" All those attempts to impose politics on science really didn't seem like a useful exercise to me. I thought they were getting further from the truth rather than gaining any real insight.

RABINOW: So now you go back to graduate school in Seattle.

ERLICH: Yeah I went back to graduate school and got my Ph.D., and I did a postdoctoral fellowship at Princeton continuing the kind of work I'd been doing, microbial genetics. I became very interested in immunology after taking an immunology course because it was kind of a mysterious field, very complex. The complexity and mystery that was gradually being sorted out in bacterial genetics. Bacterial genetics was starting to fit together, and while still a very elegant and exciting field, I thought I'd leave bacterial genetics and go into an area that was both mysterious and complex, but also had immediate human medical value.

RABINOW: Had those considerations weighed on you before?

ERLICH: A little bit. I thought that what I was doing was very interesting, but wasn't clear that it was really that useful or that relevant to human concerns. I did a second postdoc in immunogenetics. I was a postdoctoral fellow at Stanford from 1975 to 1979. I had met a woman with three kids—she and I got together, so I immediately had this large family to support. And she had a dance company which she brought out in a large yellow school bus from Oberlin, Ohio. And having transferred this huge dance company to the Bay Area, my mandate, as it were, was to try to find a job in the Bay Area. It had never really occurred to me to get a job in industry. I mean, I had all the standard left-wing, academic biases.

RABINOW: You assumed you'd be a professor.

ERLICH: Oh, yeah. That was just kind of assumed. But I had a friend whom I liked very much and respected at Stanford named Shing Chang, who took a job at Cetus. And his mentor was a guy named Stan Cohen, one of the developers of recombinant DNA technology, who was an adviser at Cetus. And I'd done some work with Stan. So I dropped by Cetus one day for a lecture/interview and I figured, "Well, if this is good enough for Shing, it's gotta be pretty good." I met David, I met Tom, and I really liked these guys, and it wasn't what I had expected. I realize I really didn't have a very coherent expectation . . . you know, I just sort of went. But they were very lively, interesting people.

RABINOW: So, your entry into industry was legitimated for you by the fact that people at Stanford, some of whom you had a close association with, had themselves already made a connection to working with biotech companies.

ERLICH: That's right. It still wasn't something that a lot of people did, but it was something that a few respected people had done. And basically what

I realized when I got there is that in many ways this was a more congenial environment than a typical academic one.

RABINOW: Let's pause there. I want to get David's story up to the present. In sum, you came from a leftist background, you had a sort of traditional, fairly standard march through the elite schools, you maintained a strong radical consciousness during this time, but there was no sense that the science itself and the political connection were either in conflict or in harmony. You didn't see any conflict. You didn't feel ethically debased by working in immunology at a series of elite institutions.

ERLICH: That's right. They weren't in conflict and they didn't reinforce each other. I guess I had separated those parts of my life. Not that I had intentionally separated them, but they just functioned independently and that they didn't really need to interact.

RABINOW: OK. David, could you bring us up to 1978 in a parallel way. Family background?

GELFAND: I guess left of center, particularly for my father, rather than my mother. I grew up in White Plains, New York, a suburb just north of New York City. My father was an accountant, tax lawyer, by vocation. And a pro bono attorney for the ACLU, the Lawyers' Guild, and LCDC, by avocation. And so I also had a politically active upbringing in terms of Constitution law and civil rights. I don't know if my father was a true fellow traveller . . . He was an Abraham Lincoln Brigade member. In high school, the most significant political activity was in high school when we tried to get the principal to allow us to have an advanced high school biology class. I became independently, like Henry, interested in science, but was also very interested in American history. Although, while there wasn't an advanced placement course in biology, there were a few teachers who acted as mentors so that we could continue to do experiments after school in the biology labs or chemistry labs or physics labs. I was involved in high school science fairs.

RABINOW: Is that something you had, Henry? This kind of mentor relationship, because one of the things I know is going to be a theme in David's story is a series of mentors. Did you have that at all?

ERLICH: I don't think I had any real mentors, I mean, I was pretty active in high school, too, in sort of philosophy class, and I got into trouble with a lot of teachers because they claimed I was a communist.

GELFAND: He probably wanted to know why it was wrong for the Cubans to have a revolution and nationalize foreign property, but it wasn't wrong for the United States to do the same thing in the 1700s.

RABINOW: Why'd you choose Brandeis? Any particular reason?

GELFAND: Two reasons: of the places I had visited and applied to, it was one of the few that encouraged undergraduates to work in the laboratories.

And because I'd done lots of experiments and taken summer courses at NYU and the University of Michigan, I was interested in doing research. I was also interested in politics and American history. I wasn't sure whether I was going to be a scientist or a lawyer.

RABINOW: Was the Jewish component a part of it?

GELFAND: No, not at all. In any event, I was involved at Brandeis in the Northern Student Movement, also in SNCC and also with SDS.

RABINOW: We're talking midsixties, early sixties, 1962, 1963?

GELFAND: Yeah, 1962 or 1963. I don't remember how I first became involved with the Student Nonviolent Coordinating Committee, but I remember one winter going into a demonstration, outside of Albany, Georgia, and had come to know people involved in SNCC, and in the fall of 1963, winter, January–February 1964, discussions on how to get more people involved in civil rights and how to get northerners, more liberals, into Mississippi.

RABINOW: Kennedy had just been assassinated and the beginnings of the escalation of Vietnam were taking place, King is moving onto the center stage.

GELFAND: Yes, we wanted to focus attention on Mississippi.

RABINOW: And while you're doing this, you're doing your biology?

GELFAND: Well, they're separate. I'm not sure we thought about it at the time, although maybe we did. A conscious political decision to get northerners involved in Mississippi in the summer of 1964, expecting that the violence that was perpetrated on the SNCC staff people in 1962 and 1963, would also be visited on the northern volunteers, and that would carry with it the press and national attention, and hopefully the army to take over the state of Mississippi. I don't know why we thought the army would help. At the time, believe me, it was better than what was going on. Even though I had spent time with SNCC in the south, I really had not been prepared for the intensity of the violence in Laurel, Mississippi, and other parts with Michael Schwerner, James Cheney, and Andrew Goodman getting murdered, and attempts by Senator Eastland to cover it up for a long period of time until the bodies were found in August.

There were several incidents, one was at the local courthouse where I had been at court in a local proceeding because several of the other staff volunteers from the Laurel office had been attacked at a lunch group counter in Laurel. I'd been there to offer moral support and testimony if necessary. While there, I had observed the local head of the Klan attacking another volunteer as he was bringing someone to register to vote in the courthouse and had filed an assault and battery complaint against the head of the local Klan. A week or ten days later we happened to be at a lake at a farm of a black family. This fellow and another dozen guys came out of the woods with chains and clubs, came up to me and said he was going to kill me.

After being clubbed and shot at, I managed to get back to the farmhouse, we phoned the FBI to come rescue us. The local FBI and northern FBI agents, who were by this time in Mississippi, said they were not a protective agency and they could only investigate civil rights violations. I remember saying "Fine, could you please interpose your body between the bullets and us so you could investigate the bullets crashing into the wall." They never came. Eventually I was taken to the hospital in a hearse from the black community because the white ambulance wouldn't come to pick me up. Since I knew who had attacked me, I had lodged a complaint for assault and battery with intent to commit murder. The following fall I was scheduled for a grand jury hearing, and I went back to Laurel. An attorney from the New York Lawyer's Constitutional Defense Committee came with me. We had a lot of difficulty getting our own depositions and statements, made to the FBI from August (files moved from Laurel to Jackson, no cooperation by Jackson FBI office with Justice Department intervention), that I had made to use in this hearing. The hearing then got moved and reassigned to a backwater county seat rather than the city county seat where our congressman had assured us we would have protection. After the grand jury hearing, a couple of cars tried to overtake us and run us off the road. And at that point, I decided that I was no longer going to pursue an American history/political science major. I wasn't adverse to going to law school. I assumed that anyone who was interested in constitutional law, civil rights, and civil liberties law was going to be shot at and perhaps killed. Indeed, the lawyers who had helped us had bricks thrown through their hotel windows. While it's fine for someone who's a sophomore in college, it's not the way to grow up and live your life. I thought some day I'd get married and someday I'd have kids. I had blinders on. I assumed that civil rights law would always be like that. Crazy. That's what I thought, because that's what I was living. I said, "Fine. This is very interesting stuff, I'm going to continue to be interested in it, but I'm going to be a biology major."

RABINOW: Did this give a shape to the biology? In other words, biology, what biology? I mean, Henry suggested a gradual move to immunology because it was more relevant to health concerns and also was a field that was developing. Or was it just a love for it?

GELFAND: It was just very general biology at Brandeis. Later that year—I guess my junior year—I had begun to pester someone who was on the faculty in the biochemistry department, a cell biologist by the name of Gordon Sato. I bugged him to work in his lab. Finally, after two years, he caved in and said yes. He was more of a mentor than my undergraduate adviser in the biology department. When it came time to decide what to do for graduate school, he advised me to choose UC, San Diego. I had applied to two places: the University of Miami, Coral Gables, and the University of California, San Diego, because I was interested in scuba diving.

RABINOW: Would it be fair to say at this point that the Mississippi experience pushed you both toward a kind of privatization on the one hand and, on the other, the consolation of the republic of science. Science was going to be another kind of community, another kind of way of leading life, that you still had faith in?

GELFAND: Yes, although I was still involved with SDS.

RABINOW: OK. On to San Diego.

GELFAND: On to San Diego! Gordon Sato told me I should go to San Diego, so I went to San Diego. One of his former postdocs was on the faculty there and arranged a summer job for me. It turned out that various people from SNCC and several folk musicians from the south lived in Del Mar. I'd played guitar in some of the clubs in Boston and Cambridge and I continued to play in San Diego. We started an off-base coffee shop in Oceanside, a folk music club, a rock place, geared to the cadets, the seventeen- and eighteen-year-old kids who had enlisted in the Marines. We hoped to try to get them to think about what they were doing, and the club was called The Sniper. It drew the fire of the local marine base, but it was off-base, and we just had folk music and a lot of Phil Ochs and Len Chandler folk music. I guess one of the most discouraging times was in 68 in L.A., the big demonstration in Century City at the Century Plaza Hotel when Johnson was there. A large number of people had been organized to protest the Vietnam War at the Century Plaza Hotel. The L.A. police, with their horses and their motorcycles, had just mowed everyone down with clubs and stuff and it was much worse than Mississippi and the fire hoses. Anyway I became less involved overtly other than The Sniper, which continued until the early seventies.

RABINOW: What are you doing in biology at that time?

GELFAND: Bacteriophage molecular biology. Classical. Heavy-duty molecular biology. I had done rotation projects in various labs, one with David Baltimore at the Salk Institute. We spent a lot of time discussing politics and an awful lot of time studying polio virus. My thesis was very intense because my thesis adviser believed that "rank hath its privileges": graduate students must work harder than the professor. He worked from 11 A.M. to 4 A.M., seven days a week. He expected graduate students to finish in three and a half to four years. No five-year theses. There were long hours all the time.

RABINOW: Who was this?

GELFAND: His name was Masaki Hayashi. There was always the competition. Once I was leaving the lab at one o'clock in the morning, and he would look up from his lab bench at me and then at the clock—really. Anne Burgess, a graduate student at Harvard, was working on the same project as I. He said, "Look. It's one o'clock. There's still three good hours in the day. Anne Burgess had a headstart on you! Anne Burgess had a three-hour headstart on you today! That's twenty-one hours a week, eighty-four hours a month. You must work double hard to catch up!'' I said, "Hold on, I started at seven o'clock this morning and Anne Burgess went to bed a long time ago.'' He said, "Well, they're three hours ahead every day and the early bird gets the worm. You must work double hard to catch up.''

RABINOW: This is not a mentor relationship.

GELFAND: Well, actually it was. I liked him a lot. His lab was a wonderful lab in which to be a graduate student. You had to learn to do everything yourself. It was not a very good lab for a postdoc because he was very nonpolitical. He didn't go to many meetings. He was not a part of the network.

RABINOW: Was there any talk of industry at this point?

GELFAND: *Absolutely not!* Not even a *remote* chance. I finished up, staying on as a postdoc in the same lab because my wife had enrolled in graduate school in the sociology department for a Ph.D. program. I stayed for two years as a postdoc while she was going to graduate school. The draft was affecting us also. As I recall, the draft law changed in July of 1966. So that on 30 June 1967, the old draft law changed to the new draft law. I had had a 2S deferment that expired 30 June 1967. My draft board had failed to classify me, even though I was a graduate student. They subsequently sent me a 1A classification, because you had to ask for a 2S. The first thing I did was go to Masaki and arrange that I could complete my Ph.D. at the Pasteur Institute. I was *not* going to Vietnam. It was either Canada or France. By that time, it must have been spring of 1968, my wife was pregnant. I was able to obtain a 3A deferment because of this ridiculous technicality that I was unclassified for months. In any event, I finished my graduate studies, stayed as a postdoc and then moved to UC San Francisco in January of 1972 to work with Gordon Tomkins. I had met Gordon Tompkins at the Cold Spring Harbor Symposium in 1970 and thought he was just wonderful. We had talked over the next year and a half, and it worked out that I would come to his lab. So we came to San Francisco. And *again,* no—not the slightest idea of ever working for industry. And three and a half years after coming to Gordon's lab, he died tragically.

In April of 1976 I had a phone call from Ron Cape asking me, actually first saying that he was president of Cetus, a company in Berkeley, and heard that my future at UC was uncertain, that I was looking for a job, did I know anything about Cetus and would I like to learn more. I said "Well, Ron Cape, President of Cetus, my future at UC is not uncertain, I can stay here for the next five years. I'm not looking for a job. Under no circumstances would I consider leaving before nine or ten months from now. I thought that Cetus was doing something biological in the East Bay. Occasionally I have lunch, so if you come over I'll be glad to have lunch with you.'' What had gone through my mind in some flash instant was that several of us in Gordon's lab, Pat Jones, Bob Ivarie, Pat O'Farrell, Barry Polisky (who was in Brian McCarthy's lab) were close before Gordon's death, became even closer after Gordon's death, and wanted to continue to work together, if we possibly could. We realized no matter how fantastically good we thought we were, we knew the biochemistry department at San Francisco wasn't going to share that vision, and wasn't going to offer us all jobs. One way for us to continue to have potluck meals together, play baseball on Saturdays, and go sailing was if we started an immunology farm on the northern Mendocino coast. We'd make antibody reagents and purify restriction enzymes. But, of course, we didn't have any money or business experience. Here's Ron Cape telling me he's the president of this company. I happened to meet with him a month later in Cambridge at one of Miles' symposia on science and society, on the impact of recombinant DNA.

I visited Cetus for the first time in late June. I gave a seminar on the work that Pat O'Farrell, Barry Polisky, and I had been doing at UCSF on expression of heterologous genes in *E. coli.* I was struck by two things: one, the total absence of any molecular biology equipment at Cetus, and two, the very sharp questions and interruptions I was getting from the nine or ten people attending the seminar. They were asking the questions and anticipating the date for the next slide before I showed the next slide. This had never happened to me at places where I'd given talks. There were generally few, if any, comments because it was crystal clear. In any event, that afternoon Pete Farley asked me what it would take for me to come to Cetus and do genetic engineering. I said it was impossible. He wanted to know why. I told him, "Well, first, I don't like the term 'genetic engineering.' It didn't take into account what's involved: one person doesn't do 'genetic engineering.' In addition, you don't have any of the space that's necessary, you don't have any of the equipment that's necessary, you don't have any of the facilities that are necessary. It's very expensive to set all this stuff up, and besides, that's not what is important. What's important is who decides how things get done, who decides what gets done. People in industry don't have any understanding of what that takes. People who are making the decisions don't understand what's necessary, and it's just *no.*" And he said I was the most opinionated, biased person he'd ever met in his life, and it might be that way at other companies, but it's not that way at Cetus. He asked if I would put down on paper what I thought it would take to organize a recombinant molecular research lab. Polisky and O'Farrell, with whom I had dinner that night, encouraged me to put down what I thought was necessary. I said, "Well, that's ridiculous. They'll never accept it." They said, "That's right, and your biases will be validated and you can say I gave it my best shot." Well, I didn't want to do it. So I made it as outrageous as I could think of making it. I came back three weeks later from a meeting in Europe and a week after that Ron Cape called me and asked when could I start. I guess this is the end of July or August. I said "Start *what?*" And he said, "Start the Recombinant Molecular Research Division at Cetus Corporation." I said, "Well, this is *not* what I was expecting." And he said, "Well, look, we considered it for the last month and the scientists believe it's a good idea, and Pete and I think it's a good idea, and Josh Lederberg and Don Glaser and Stan Cohen think it's a good idea, and that's exactly what we want to do and we want you to start it up." I said, "I don't know." I hadn't *thought* about it. He said, "Well, when do you think you'd be able to let us start," and I said, "I have no *idea*. I don't know. And if that's not acceptable, I'm sorry." I just couldn't do it. "I don't know when I'll be able to let you know." And he said, "Well, I'll be calling you periodically." I had to think about this because it was serious. I talked with everyone I knew, who was doing what I thought I wanted to be doing when I grew up.

The people I respected said I should take the job at Cetus. In any event, the person I argued with most was Gordon Sato, because he had advised me to go to San Diego, he'd advised me to go to San Francisco rather than apply for jobs at Columbia or Texas or a postdoc with Paul Berg. He told me to do everything I ended up doing, and so I said, "Why are you telling me this?" And he said I was crazy. I said what I wanted to do after leaving Tomkins's lab was eventually to have a research group, have postdocs, and do the kinds of things you're doing. Why are you telling me that I don't want to do that. He said that he spent more than 80 percent of his time trying to get money for graduate students, postdocs, technicians, and he had no time to interact with the postdocs and the graduate students. When he's on campus, it's faculty meetings. If I had any *reasonable* expectation that Cetus would be able to fulfill its commitment to me, don't hesitate for a moment. "Because," he said, "if you stayed at UCSF on your soft-money, nontenured research faculty position, you'll continue to work with people like Barry and Pat and other people, and you'll do nice things, but where will you be in five years?" And he said, "If Cetus is able to fulfill its commitment and you were able to attract the kind of people that you would like to hire and expand, there's no telling what the limits are." Since I had fallen in love with interactional science, largely through the five years in Gordon Tomkins's lab, I accepted the position at Cetus.

Science in Industry and the Academy

RABINOW: Okay, we're going to do a rapid comparison of science in the industrial setting and the university setting. Let's start with Henry. Can you tell us about collegiality at Cetus? What kind of freedom did you have? What kind of freedom didn't you have? A decade after making the decision to enter industry, do you ever think about what your life would be like today if you had become a professor at Stanford?

ERLICH: Well, I made this decision largely on the basis of personal interactions. I like and respected David a lot, I'd just met Tom, whom I liked and respected a lot, I had known Shing previously, so I thought this is a group of people that I really like. But those were the *scientists*. I didn't particularly have feelings about the management. After I made the commitment to come and I was finishing up my research at Stanford, it was a very difficult time for me because I really didn't know if I'd made the right decision. I remember having lots of sleepless nights. How had I ended up going into the profit-making commercial arena, because I never imagined I would—and I was quite anxious. In fact, I didn't really know what Cetus was like, I didn't know what directions they were going. I just said, Well, I like these guys, this should be a fun place to work, so I said yes. Which was not a terribly considered process. At any rate, I was very anxious before I came and wondering what it meant about my future and my political involvement. But on the other hand, I also thought about the fact that it could conceivably be very exciting to try to do biological research that led to practical, useful things. So that, of course, was an exciting process and one that one normally doesn't have a chance to work on at a university. Now things in academia have changed and some of the distinctions between university and biotech research no longer hold. I think the things that attracted me were the people, the individuals, and the idea that you might actually be able to do something useful, practical, and benefit human health.

RABINOW: Did you ever think of quitting? Did you have recurrent sleepless nights thinking, "Have I done the right thing?"

ERLICH: I don't think it ever got to the point where I thought I'd made a horrible mistake and I would have to call up and say, Forget it. It never got to that point. But I do remember being very worried about what it meant from my political perspective. But actually, in the very beginning, even before I started work, I had the idea for doing something which is now called DNA HLA typing. I remember this exact moment. I was meeting with Ron Cape, and I thought, Well, if they think this idea is something I might be able to work on, well, gee, maybe this would be an okay place to work. I didn't think they'd go for it because it was a totally wild idea at the time.

GELFAND: We had talked and he said he was interested in gene structure. I said, "Well, okay, fine." I knew I wanted him as a colleague. And I said, "Well, if you are interested in gene structure, I don't know how we're going to be able to package that and sell it to Cape and Farley. But let's figure out how your interest in gene structure is relevant to human health, and we'll figure out a way to make it a project."

ERLICH: And that's really, I think the crux of how one can ever really establish a satisfying scientific life in a company. If you find a project that is of real fundamental passionate interest to you, and if it also has the possibility for having some real practical commercial outcomes, then you have the prospect of a company project that really satisfies a scientist. I think there are some people who do things that are very valuable for Cetus or Roche, but I wouldn't want to have that kind of career. I mean, I respect what they do, they're very good at it and so forth, but for me, personally, there would have to be some sort of overlap between what I thought was personally interesting to me as a scientist—an overlap between that and something that management thought might be potentially useful at some point. So I think I was lucky in that some of the things I worked on had that overlap. At the beginning I did lots of different things. I was, I think, the first immunologist to come to Cetus. But I also had some training as a geneticist and as a molecular biologist. The project I worked on initially involved monoclonal antibodies and there was some infectious disease immunology, but what I was *really* interested in was the genetics of the HLA region. Well, I had to defer that, you know, for a long time while I was doing these other things. The other things were interesting as well. But finally, as time went on, I was able to devote more time to that, and it's been very satisfying for me to personally see this thing grow from just kind of a pipe dream into really the way the whole field has progressed. So now the future in this whole field is DNA typing at the HLA level, and that's—it's those tests that are used for human identification, for example, forensics, but also for tissue typing, for transplantation, and now, one of the things I've always been interested in, it's also proving to be valuable in genetic predisposition to autoimmune diseases. So that's been satisfying. And one thing that I think has also been satisfying in terms of a process, not just a particular project, is that working at Cetus and now Roche has always been very interactive. Always, well most of the time you felt part of a team where you were friends, not just colleagues, but friends with the other members. And the personal part is very important to me. I think if Tom and John Sninsky and David were not at Cetus or Roche, I wouldn't be there either, because I think the culture was determined by those people. Say, at the university, the culture is much bigger than just a few people, and if a few people go, it's still UC Berkeley or Harvard or whatever. I think Cetus was small enough so that what was important to me culturally could be determined by a relatively small number of people. Actually, when Tom left Cetus, it made a big difference to all of us.

You know, a lot of times I felt Ron Cape didn't really understand the science, but he actually tended to let other people make the decisions. But some of the latter management people, they had no idea of how to manage people and how to make scientific decisions, and they didn't have respect for the fact that you had to *know* something in order to make a reasonable decision. Fildes had a lot of power. He was actually quite a bright guy, but I think he was very arrogant, and he didn't listen to people, he didn't have enough respect for other people. So, I think if Fildes had remained in charge, there would have been a slow, or not so slow, hemorrhage of senior scientists.

GELFAND: The fourteen senior-most research, development, and clinical people in the organization left between the time Tom left and when Fildes was fired.

ERLICH: So that was pretty serious and one of the things that was, I think, quite disillusioning to me and probably even more to David was that Ron didn't seem to get it. I spoke to him. I said, "If Tom leaves, this is incredible. A serious blow and you can't let this happen." And then after he left I said, "Do you realize what happened?" And Ron really didn't get it.

RABINOW: Did you think at this point of going back to the university?

ERLICH: Oh, yeah. Sure. I had offers but never once did I actively pursue them. I said, "Well, thanks a lot but I'm trying to make this thing work out. If it doesn't, I'll let you know." That was usually my response. But there were moments long after I joined Cetus when I had some real doubts and concerns. I still remember the first scientific meeting when Fildes was there, and I went, "Whoa! Weird person! We're in trouble here." Because he came in a very aggressive, hostile way. And we were right, we were in for some trouble. So there have been moments like that. And the Roche transition was hard for me to adapt to emotionally because I had liked the idea of being with a small company whose growth I had been part of, and becoming part of a big multinational company was difficult. But the thing that made it tolerable or actually even desirable in a way was that, in a way, I saw it as a kind of reunion of the old Cetus. I'd be working with Tom again and I wouldn't have to deal with some of the jerks at Cetus who were in a position of authority. Fildes wasn't the only problem, he was just the most visible. So I figured, Well, we're getting out from under these guys. It's true we'll be part of this huge company, and you know it's inevitable we're going to be bureaucratic, but I thought, Well, there's a chance that we'll go to this small, semiautonomous thing in Alameda and Tom will be the head of the research and development group and that all sounded good. In a sense, it was like coming home rather than being taken over.

RABINOW: But you always still had the sense that you could do the science that you wanted to do?

ERLICH: Enough of it. In other words, I knew there were a lot of constraints, and there were a lot of things I wanted to do, and had I been at a university I could have pursued, but I couldn't at Cetus and Roche. So there were real trade-offs. But there were also things I was able to do, and practical things I was able to do at a place like Cetus or Roche that I could never have done at a university.

WHITE: But within that separation aspect I think what's also important is that you could also have the politics that you wanted to have regardless of your environment. In other words, political questions didn't really influence anything having to do about promotions or acceptability or anything. I mean, you could be totally involved in whatever political aspect you wanted and just simply pursue that, and no proscription of what you were involved in, and you know, if you sensed somebody was a behemoth Reaganite, you just said, "Well, I don't want to go on any road shows with them." But that was their business. And at the same time, they were not impinging on my business, which they'd probably be pretty upset about if they knew what kind of politics I have.

RABINOW: If you left, would you go to another biotech company? The university?

GELFAND: I don't know. I'm thinking more that it would be a biotech company than a university. I have great admiration and respect for academic science. The things that I was interested in doing, I would not be able to do as well in an academic setting as I would in a corporate position. *Providing*—and what Henry indicated is absolutely true—what has always been most important is that there be a synergy of goals between what I, as a scientist, wanted to achieve and what the company wanted to achieve. And then second, the nature of who your colleagues are. The two most important things are, are you going to be able to do what's important for you to do? That requires a commonality of interests between your scientific goals and corporate goals. And second, who are you doing it with? Because ninety percent of what you do is collaborative and interactive. It is very different from my view of academic science in the biological sciences. It's very difficult, as an academic scientist, to do interactionist, collaborative science. The acculturation process is one that is keyed to individual, personal achievement. You first learn that as a graduate student. To get into the best lab as a first- or second-year graduate student, you'd better excel on the individual achievement scale to get through your orals and qualifying exam. In regard to your cohort class, you'd better excel in personal achievement as a graduate student. You have to prove that you can do independent experimental research. After that three-to-six-year period, generally, one is applying to labs for postdoctoral study. Of course, if you're biomedical or biochemical sciences, you're applying to the most prestigious labs, that is, usually the largest labs. You may be among the fortunate select few who have obtained a National Science Foundation or NIH grant or a Howard Hughes postdoctoral fellowship. You may go to the lab of Z, a famous Nobel prize winner or about-to-be-Nobel-prize-winner, along with twenty other postdocs. Again, you have to, in the next two to three years maximum, produce high-quality publications in *Cell* or *Molecular and Cellular Biology* or *JRC* as independent publications. But you will, undoubtedly, being from a large group, be recruited along with several other postdocs who were coauthors with you on the most striking new transcription factor or growth receptor or some other hot field—interviewing for the same positions—competing with the same postdocs in the same labs. God forbid you were at UCSF in the mid-seventies and the postdoc in X's lab competing on the same project with a postdoc in Y's lab on expression of insulin. Hardly an environment to foster interactive collegial collaboration. Through some miracle you may secure an assistant professorship at some prestigious place after this wonderful, illustrious career as a postdoc. Now you are competing, of course, for R01s, and then renewals. Since all prestigious academic institutions have more assistant professors slots than associate professor slots, you are also competing with the excess assistant professors at your institution for a limited number of tenure positions. God forbid you should make the mistake as assistant professor of collaborating with either your former postdoc mentor—a very famous individual, of course, because you wouldn't have studied there had he not been—or graduate student adviser. All of your independent work done as an assistant professor will be ascribed to the brilliance of former major mentor, making it ever so much more difficult to get tenure. Thus one finds oneself at age forty being promoted to associate professor with tenure and twenty years of experience of how *not* to collaborate and that one gets ahead only through individual personal achievement.

Clearly academic science has been a very productive establishment, and people have made significant contributions. We have learned a lot from that system. But that's not the way I enjoy doing science. The experience in Tomkins's lab was really eye-opening to me. I wanted Henry to come, I wanted Shing Chang to come, and I wanted Tom to come to Cetus. Not only because I respected their science, but I liked them as people. It was just as important to have a quality of life as well as the excellence of the science. That's what affected me, with Shing Chang and Mike Doyle and the people who joined our Recombinant Molecular Research Division. I think it's not a coincidence that Henry was involved in Vista and Tom was involved in the Peace Corps.

RABINOW: You guys have created something special. You've had an important hand in creating a space in which science—in your view, first-rate science—can be done in an industrial setting. You don't have any apologies about the science you're doing. You see the work environment as supportive, collective, interactive. Today you are the scientific managers, and you're looking for other scientists who are going to fit into the mold. In the academy we would see that environment as something which we lack. Berkeley has never been like that for me and many friends around the country share similar sentiments about their experiences. As an undergraduate and graduate student I did experience a wonderful sense of excitement about learning, but I was a student. But at Berkeley, as a professor, I've never experienced ten minutes of that support. I miss it all the time. I've always wanted to work *with* people; I eventually *do* work collaboratively, despite being discouraged from doing it. There is no support and no reward for this kind of work. Colleagues, even more than administrators, constantly question whose work is it, why do you do this, and so on. I'll get the same reactions to doing the book with Tom: How much was yours, how much was his? Ideas are property in the university and property is private. Your accounts are very interesting to me—my political and intellectual background is quite similar to David's and Henry's—your accounts raise complex emotions, a certain envy and a certain caution that what you are telling me is too good to be true, even though as an ethnographer I've been observing you for a while now.

WHITE: Well, it seems perfectly normal to us. Why *shouldn't* it be like this? Yet we have the sense it's special, too, because how is it you were able to have this sort of environment? As David said, "We'll figure out some practical application of Henry's idea," and it turned out we were right. In 1979 it might have been more starry eyed, and in 1981 it was maybe a little less or a little bit more, depending on who's looking at it. But a lot of things panned out. And in the political sense, to me, there's been no compromise. In other words, when people ascribe political views to me because of my location, I always wonder which location they are talking about? Berkeley, the state of California, Hoffmann-La Roche, or the United States? In other words, there's a constant attribution when I speak, I only speak for Roche or "industry." But when Paul speaks, he doesn't speak for the University of California, or even the Department of Anthropology. I don't feel I am always speaking for Hoffmann-La Roche, either, unless it is in an explicit setting where I am being asked for a Roche or an "industrial" perspective. Yet my views are always attributed to projecting the industrial view. For example, I was recently invited to speak at the annual meeting of the American Society of Microbiology, in 1993. There will be several speakers: one on molecular techniques that can be used for fungi, another about how many fungi have been studied so far. They write me a letter saying they need an industrial perspective. None of them know I have published some of the apparently seminal articles on both techniques and a comprehensive review of fungal systematics. Yet other speakers were chosen to cover these things. And they have apparently not read our papers, but probably only know ten percent of what I could discuss on those subjects. But I'm the industrial representative talking about how the diagnosis of the fungi fits into the priorities with respect to other diseases. There is a constant projection of limitation with regard to the breadth of expertise and intellectual interest simply for being employed by a company. Certainly with respect for political stereotypes it's the same thing.

RABINOW: The thing that struck me early on in the interaction with Tom was, at one point, where I was complaining about all the obstacles the university puts in the way of my doing work, Tom sort of looked incredulously and said, "You mean they don't support you doing research?" And I looked at him incredulously and said, "Are you kidding?" The first response is always no. Always no. I asked our dean if it were possible to advance my own sabbatical time to finish the book. He didn't call back for a month and then said no because I might leave Berkeley without paying back my time. He suggested I get an outside offer. Never once did he say, It is an interesting project, what can we do to help?

Is this situation going to last? As with the computer industry, perhaps biotech has already experienced its youthful start-up days fueled by venture capital, with no products expected immediately, and is now facing a more bottom-line oriented future. Today the multinationals are taking over. The seventy-third diagnostic test can't be as exciting as the first three. Is that true?

GELFAND: Well, first, I don't know about the seventy-third diagnostic test—that'll be just great. The sixth new thermostable polymerase is certainly as exciting as the first. In my view, we are the ones who are responsible for making sure that we attain our goals; Henry, John, me, Tom. We're being sure that it continues as we want it to be. It's not a matter of Roche permitting this or that or the other, or Roche being responsible, yes it is, or no it's not. It's us. And we better make it work. We've got to make it work, because we want it to work. Otherwise, we have failed, not Roche has failed.

RABINOW: And Roche has not stood in the way of that so far.

GELFAND: No. They haven't overruled us. Not everything is a bed of roses, but we'll figure out how to work it out to get what we want to get accomplished.

WHITE: Roche made a fundamental calculation that this technology has the potential to have a major impact on diagnostics and many other businesses. It would be sensible for Roche to invest in an expanding technology and reduce or end its efforts in mature or unprofitable areas of diagnostic testing. The question is, do they have the resources and skills in all areas? There was a partial manifestation during the years that it was being developed by Cetus, Kodak, and Perkin Elmer, yet there were clearly other opportunities that weren't realized because of limited resources. Roche has a lot more resources and the dual aspect of their investment is money and the confidence that they have a group of people that can carry this forward. And it's the group of us, and others who have more business expertise, and the sense that we can interact to really try to do something. The attributions to biotechnology also have a scale. There are attributions for small biotech companies, and the negative connotations increase in scale the more multinational it gets. From my perspective, Roche is less bureaucratic than it was when I was at Cetus. I don't view it as a degree of evil associated with size. There's an absolute perspective on what are we trying to do. We have a certain vision of what we're trying to accomplish, regardless whether that vision is carried out in the context of Cetus or Roche. The challenge for us is almost the same: we are trying really to manifest this technology in a world where it will have an effect. For me, the context of Roche or Cetus is remote. It's a question of resources and commitment or interference.

Curiosity

WHITE: To me, curiosity is an extremely powerful motivating factor. You know, food, sex, and shelter and stuff like that. Some of things we are doing here, we don't really know where they lead, you could call it instinct or gut level, but we don't know. Henry will justify his work on diabetes or HLA, and that's right, but he just wants to know about how the thing works. He doesn't give a damn about whatever else is involved in it. That's why David Gelfand has boundless curiosity which takes over what he does. The elements around curiosity are what the people do to satisfy their curiosity. That brings in resource issues and external things.

RABINOW: What are the limits to curiosity?

WHITE: Well, it's probably that you just find something even more curious. Or it's just that it's boring. You'd like to go after something where you really don't have a clue what you are going to find. That's called a fishing expedition, which is not supposed to be science. I've seen curiosity end for some scientists. When it does end it's a totally recognizable element in them. They no longer have the curiosity. They go home at five o'clock. Or they say, "Well, if you want me to write up the paper I am going to have to take some time off from work" rather than write it at night or on the weekend like everyone else does. Or when some peculiar result is presented at meetings, they yawn and aren't interested. It's the strangest thing. It's like death in a scientist. They can be productive in a certain sense but the ability to solve new problems isn't there.

RABINOW: So, curiosity can die and become routine and boredom. But what about the other side, can you have too much curiosity?

WHITE: Yes, some people are so curious that they never complete anything. One idea after another, but all at a level that's not very deep so you can't determine the complexity, what's workable or not. The science fiction mode sets the limits of curiosity when humans mate with apes and meddle with God's work, that kind of thing. The limit for scientists is that scientists' visions are limited socially. Never even conceive some issues. How the family is defined. People are thinking about how to distinguish hemoglobin-S from hemoglobin-A, not these other issues. They don't think how this will affect families.

RABINOW: Curiosity is a good thing?

WHITE: It's getting the answer to your curiosity. The mouse pushing on the button to get more cocaine. There is something intensely satisfying about satisfying your curiosity. Scientists just want to know the answer to something. That's why David Gelfand is in the lab every Sunday. He just wants to know how the thing works.

RABINOW: How far down the ladder does this apply as a motivation.

WHITE: There is a range of human variation. Those who are motivated by curiosity have the problem of stopping. They ruin social occasions.

RABINOW: I've written a paper called "The Curious Patient," which was inspired by Hans Blumenberg's chapter on curiosity in *The Legitimacy of the Modern Age*. Blumenberg talks about curiosity as one of the great motive forces of the Enlightenment. He shows how curiosity has been something that has been consistently under attack by Christianity and other authority structures. There were the German medical and scientific experiments and so many others which obviously crossed the line of acceptable research. Perhaps there are no self-limiting principles within science itself to tell you not to do a particular experiment. Since curiosity and modernity combine to drive endlessly toward producing something new, the problem is the relation between the drive to newness combined with curiosity which has no internal principle of limitation. Perhaps these German scientists who worked on living patients were horrible human beings, but we now know that they were not all horrible scientists. This disjunction is troubling. The core of the distinguished German medical establishment went along with the Nazis. Curiosity has its thresholds. Perhaps it's ethics or religion which limits what one can and can not do—not science.

WHITE: That boundary where curiosity goes over into something unethical could also be an element in some aspects of scientific problems. They are always ascribed to power and priority issues, but there is an element of curiosity affecting the ability to interpret your data. It's a theme we've encountered in the history of PCR; Mullis saw the band he wanted to see, it was reinforcing his curiosity about it. Others we falsify; their experiments could be simply ignoring the data that doesn't fit.

RABINOW: And there are always data that don't fit. There is rarely, if ever, a definitive experiment which totally settles the issue.

WHITE: Curiosity does get to a point where judgment is required. One boundary to examine is when does curiosity reach a limit. How would that decision be made? Since there isn't an independent reference, what sort of process does one go through to arrive at a stopping point? What would you draw on to make that decision? Not a simple question: what to do to access resources; what you do that might be unethical; or socially advisable.

INNOCENCE AND AWAKENING: CYBERDÄMMERUNG AT THE ASHIBE RESEARCH LABORATORY

The Ashibe Research Laboratory was a unique, controversial, and explosively short-lived organization for basic research in virtual reality and interactive multimedia in the early 1980s. It was a child of the company that in this story we'll call the Ashibe Corporation, a fictionalized composite of the first manufacturers of personal computers and interactive game software. The lab was founded in the context of Ashibe's runaway growth and astonishing profitability. It was the time of the "me generation"—the florescence of the yuppies, the "I Want My BMW" kids, junk bonds, wholesale corporate looting, the gutting of the social services infrastructure in the United States, and the reelection of Ronald Reagan. The Republicans were in control, and the dancing looked like it would go on forever. It was a time of high corporate optimism, and within this framework two revolutionary research establishments were started up almost simultaneously. One was the modestly named World Center for Computer Research, in Paris. The World Center was the brainchild of Jean-Jacques Servan-Schreiber, who had persuaded Georges Pompidou of the program's importance. Pompidou had then pushed the funding for the center through otherwise impenetrable red tape with gratifying speed. The other research center was the Ashibe lab, geographically on the other side of the globe but in concept and execution, a universe away.

Servan-Schreiber had two candidates in mind for the prestigious directorate of the World Center. One was Seymour Paport, who had done so much groundbreaking research in computer cognition with Marvin Minsky. The other was Nicholas Negroponte, the suave and charismatic founder of the architecture machine group and, later, of the Media Lab, both at MIT.

When Ashibe decided in late 1981 to start a laboratory to do research in interactivity the impression they gave was that the lab would be an effort on a par with nothing less than the World Center. Its scientists would have a free hand and unlimited resources to work at the cutting edge of new technology. Top management at Ashibe allocated spacious digs in the company's

Sunnyvale, California, complex, a group of gleaming new buildings built around a sunny atrium complete with burgeoning tropical plants and singing birds. Now they needed someone to spearhead the new effort. They knew Negroponte's track record, and dispatched an emissary to offer him the position of chief scientist at the new lab, only to find that he had just accepted Servan-Schreiber's offer of the directorate at the World Center. Ashibe immediately went to the second person they'd had in mind: Steve Leslie, formerly of Xerox.

Leslie had been a child prodigy and later a professional musician before being hired by Xerox when the Palo Alto Research Center (PARC) was still a dream research site, lavishly funded by Xerox's huge profits. He had made an indelible mark in the industry by describing (the state of the art did not permit him to build) a hand-held graphical interface computer. As they did with a whole succession of brilliant and eminently marketable ideas, Xerox missed developing the device into a marketable product.[1] Leslie also designed one of the earliest object-oriented programming languages, and in other ways had distinguished himself as an up-and-coming researcher with organization abilities.

Leslie began by hiring a slew of recent MIT graduates and near-graduates. Then he proceeded to raid the MIT architecture machine group, luring the bright, eager young graduate students with (quite real) promises of gorgeous offices, state-of-the-art equipment, California climate (which looked paradisaical to the Boston-based students), and what looked like unlimited funding. They accepted in droves. "Rumor at MIT was that Steve was bringing them out in busloads," Mike Gardner, a senior researcher, said. "He took almost everybody they had. It was like draining a swimming pool."

Already at Ashibe, fresh out of Michigan State with a Ph.D. in interactivity, was a young and brilliant researcher named Mara Fennell. She was not doing research, but rather with her unusual and suspect degree she had been hired by the software division to do marketing research for games. It wasn't her idea of a good job, but it was a job, and it was in Silicon Valley with one of the world's largest manufacturers of entertainment software. Fennell was lying low, pretending not to know much about programming because she'd discovered it was a dangerous skill for a woman in the marketing division to process. She'd already gotten into trouble with Roy Klamath, a vice president there. When they met one afternoon he'd asked about her background, and she'd said she knew some code. He had glowered, pointed his finger at her, and said "You're an *engineer.*" Clearly implied in his voice was, "One of *those.*"

This remark should have tipped Fennell off, but things were new and confusing and there were a lot of them for a young recruit to absorb. Klamath was referring to the hackers in the software division, young men in their late teens and early twenties, the first generation of their kind, who lived their lives perpetually in semidarkened rooms, sleeping at their terminals or under tables, and seemingly subsisting entirely on nothing more than Fritos, Coca-Cola, and wild determination. These crazies wore ancient Adidas and malodorous T-shirts with the sleeves ripped out, spoke a jazzy argot which was unintelligible to the suits, and—most offensive of all—pulled down salaries in the fifty to sixty thousand dollar range for what looked like nothing more than playing games. But in spite of all this they were untouchable because it was entirely upon these disreputable and largely indecipherable quantities that the phenomenal success of Ashibe was based. Management felt that to tamper with the game designers' work habits might turn out to be something like defusing a bomb.

Ashibe had been started by the entrepreneur Norman Barnwell in the mid-seventies. Barnwell meant to develop what at the time was being called a home computer, which, it was thought, was something that people would use to balance checkbooks and keep track of recipes—an extremely popular idea at the time, and the very things that personal computers did *not* wind up doing. Ashibe's machines achieved great popularity not because they could balance checkbooks but because they ran game software, very simple things involving chasing moving cursors and knocking bricks out of walls. The game software grew in sophistication with tremendous speed, and when the company was showing a nice profit, Barnwell sold out. His purchaser of choice was the newly formed Patton Communications. A controlling interest in Patton had been quietly acquired in the early 1960s by the Cortlandt Group, originally a chain of parking lot franchises and shoe stores and, it was said, a respectable conduit for mafia money during the "legitimizing" of mafia interests that Mario Puzo described in fictional terms in *The Godfather*. With that backing, Patton was out to control as much of the entertainment business as it could.[2] Ashibe seemed a good choice. Barnwell had built it soundly with selling it in mind, and it was solidly profitable. Shortly, though, Patton found that it was hanging onto the tail of a financial rocket.

Ashibe's success was phenomenal, even to Ashibe. The statistics were stunning—Ashibe had doubled in size every eight months three times in a row, the largest growth ever posted by an American corporation. Nobody in management quite believed it, and nobody could explain it. Three years after acquiring Ashibe from Norman Barnwell, Patton Communications was making more money from computer games than they were making from their entire extensive library of films. To some of the old hands on the financial side of the company, growth on that scale was frightening. "They thought Ashibe was some kind of creature like the blob in Woody Allen's *Sleeper,*" Gardner said. "They thought they had climbed onto something that was totally out of control." The investment in equipment and packaging necessary to produce a new game was infinitesimal, so profits were enormous, and still growing exponentially.

Which is why, when Gardner asked Steven Leslie how much money Ashibe was going to commit to long-range research during the next ten years, Leslie replied seriously, "An *infinite* amount."[3]

In fact, the company initially budgeted fifteen million dollars for the lab, which probably did look like a nearly infinite amount to the first wave of young researchers.

Ashibe's initial profitability had been based on the revenues from a single arcade game, in fact on one of the prototypical arcade games of all time: for our purposes we'll call it Space Chase, which within an extremely short time found its way into not just every arcade, but also every bar, pool hall, lounge, and convenience store in the world. Nobody at Ashibe's had any idea why Space Chase was so extraordinarily successful. "Why people were willing to pay so much money for it is still a mystery," Born said. Gardner added, wistfully, "Maybe it really *was* the magic of interaction."

Space Chase was also the bellwether of Ashibe's true marketing philosophy. Barnwell had discovered the enormous potential of the game market by accident, but Ashibe's staggering success at it gave the company the sheen of an innovative, cutting-edge organization, ready to take risks in a new and untried field. The reality was quite different. Ashibe was staffed from top to bottom with suits (extremely conservative people). Essentially they were solid businessmen like Roy Klamath, people with proven management skills but little imagination. People in places like the MIT media lab knew this, because they had been intensely interested in how Barnwell got the vision to start a company that marketed interactive software on such a grand scale. Consequently, they were aware of the reality: Ashibe was not a visionary company, merely a stunningly lucky one. With the sole exception of Space Chase, Ashibe might as well have been in the furniture business.

Thus it was that the best and brightest perceived Ashibe's hiring Steve Leslie to start a research lab as a major breakthrough in Ashibe's humdrum conservatism, a sign that there were people at Ashibe who had the vision to use the enormous profits Space Chase was generating to fund more and better experiments in interaction. In truth, they had completely misunderstood what was happening. Ashibe had hired Steve Leslie because the company had grossly misperceived his abilities. They saw him not as a brilliant innovator but rather as a good administrator of exotic talent, someone who could keep bright programmers in line and producing. Space Chase, as it quickly developed, was Ashibe's Maginot line, the standoff point between the suits and the researchers. The Ashibe board of directors' idea of a "visionary future" was an endless series of Space Chase clones, like the movie sequels that Sylvester Stallone had turned into a mini-industry. The company's administration thought they were buying a group of programmers who would quickly produce an endless series of Space Chase spinoffs.

The trouble lay in the innocuous word *research,* which each side understood quite differently. To the potential lab staff, many of whom were still at MIT doing graduate work when they were hired, *research* meant innovation, taking risks, doing new things. To Ashibe, *research* meant duplication, slight changes on an already accepted idea, but finding out how to duplicate their successes better and cheaper. Leslie did not perceive this at first, and when he figured it out he arranged to be away from the lab on business most of the time. By the time he did figure it out, though, the Ashibe administration and the lab kids were at each other's throats.

The first shot in the war was fired not at the lab but over in marketing. Klamath wanted Fennell to develop a market for all the Space Chase spinoffs that would soon be emerging from the lab, staffed as it was with all those nice obedient programmers. He had ordered Ms. Space Chase before the lab got going and the game was now nearly finished, and in the works were things with names like Son of Space Chase and Space Chase Three. Management was hot to get game software into the home computer division, because prior to Space Chase exactly what a home computer was going to do was still a mystery. People bought them, played with them for a while, and eventually tossed them in the closet. Maybe things like Space Chase would boost the sales of hardware to match that of software. Meanwhile, the Ashibe game division continued to grind out riches as if it were grinding salt.

In the midst of this numbing plethora of sheer unencumbered wealth, the mindless machinery of corporate profitability ground inexorably onward. Upper management scrutinized division reports with a no-sparrow-shall-fall mentality, and the slightest sign of lagging profits meant The Divine Hand of Retribution. In practice, this meant that every week or two another marketing VP was shot out of the saddle. "They aren't happy with the biggest golden eggs anybody's ever seen," Gardner said. "They want the whole damn goose." Inevitably, one day the shots hit Fennell. Although she was in marketing, Leslie knew about her work, and they had been discussing a possible relationship with the lab for her at some future time. When she got her pink slip she immediately called him. Fortunately, he was in town. "Help!" Fennell said, with unaccustomed brevity.

Leslie had her packed up, out of her office in marketing, and over to the lab in the same day. When Roy Klamath found out she hadn't actually left the company, he sent a succession of blistering memos to Leslie advising him of her inadequacies in a most incendiary manner. "This person is an *engineer,*" Klamath fumed.

Fennell had wanted to do interactivity research in the first place, and she came to the task chock-full of ideas. But she found herself facing the same misunderstandings that had shaped up over the word *research*. To Fennell, and to the rest of the lab, interactivity meant something like Andy Lippman's description back at MIT: "Mutual and simultaneous activity on the part of both participants, usually working toward some goal, but not necessarily."

There are five corollaries to Lippman's definition. One is *interruptibility*, which means that each participant must be able to interrupt the other, mutually and simultaneously. The second is *graceful degradation*, which means that unanswerable questions must be handled in a way that doesn't halt the conversation: "I'll come back to that in a minute," for example. The third is *limited look-ahead*, which means that because both parties can be interrupted there is a limit to how much of the shape of the conversation can be anticipated by either party. The fourth is *no default*, which means that the conversation must not have a preplanned path, it must truly develop in the interaction. The fifth is that the participants must have *the impression of an infinite database*, which is to say that an immersive world should give the illusion of not being much more limiting in the choices it offers than an actual world would be.[4] Interactivity implied two conscious agencies in conversation, spontaneously developing a mutual discourse, taking cues and suggestions from each other on the fly. To this Fennell had added the additional elements of drama—interaction based on principles of stagecraft—and of ludics, playful interaction.

On the other hand, to the Ashibe management, interactivity meant something quite different. Nobody had thought it out; if you stopped people and asked them, they said it was intuitively obvious. Interactivity meant taking turns, not interruption; it meant that the user pushed a button and the machine did something as a result. That was interactive because the machine responded to the user's command instead of doing something on its own. It was what Gardner called poke-and-see technology. Many of the lab people felt that way about Apple's Hypercard—from the standpoint of interactivity it was a deception because although it was a huge step forward in the popular dissemination of hypertext tools, it gave a facile illusion of what interactivity was.

The clash over interactivity shaped up as nastily as the one over research. Fennell got to fire one of the opening salvos, although quite unintentionally. Bob Stein, Leslie's consultant, was working on an interactive encyclopedia with the Encyclopaedia Britannica Company. (His liaison with the company was Charles Van Doren, a graduate of Saint John's College and the first contestant to be caught and convicted of cheating on the television quiz show *Twenty-One*. The interactive encyclopedia let the user enter words for the program to look up. As far as Stein was concerned, in that fact inhered its interactivity. Stein came by Fennell's new office in the lab and proceeded to chat her up about interactivity. Fennell's ears pricked up. "That's great," she said, "I'm working on interactivity too."

"You are?" Stein said.

"Sure," Fennell said enthusiastically. "I've got this idea for an interactive educational thing about whales told from multiple perspectives—whales from an Inuit perspective and then whales from a whaling corporation perspective and a Greenpeace perspective, say. Multiple narrative thread, user selectable. It'd fit right into your interactive encyclopedia."

"Uh-huh," Stein said. "Right." He walked away and never came back.

Sooner or later any discussion of interaction and how it worked turned to debates over how to handle very large databases. After a database exceeds a certain size it becomes impractical for a human to search even the major listings in a reasonably short period of time. Thus the fact of a data base that contains, in the ideal case, all of the knowledge of humankind, doesn't imply that such a database is searchable. This surprising fact was encountered in practice with the first large, publicly searchable computerized databases in the 1960s. One of the first of these was the MEDLARS medical database, maintained at the National Institutes of Health in Bethesda, Maryland, for the National Library of Medicine. MEDLARS ran on a mainframe with access to the database available only from a single console. There were no public terminals. Researchers wanting to use the database submitted requests for information on slips of paper, just as if they were requesting a book from the closed stacks. The library staff transcribed these requests onto standard forms and typed them into the console keyboard by hand. The answers came back a day or two later in the form of computer printouts.

Things did not proceed as expected. Researchers inquiring after information found to their dismay that a request for a list of titles of papers on a particular disease or drug might result in hundreds of feet of dense printout. Notices went up in the MEDLARS area warning people to be extremely cautious about how they worded their requests for information. Still, the piles of paper proliferated. The reason was simple: there was simply too much data on each subject for the existing search protocols to handle. Keyword searches, the common method, simply did not work with so much data because several thousand abstracts might share the same keywords. Buried in the abstracts themselves was critical information by which they could be separated, but even programs that searched the abstracts themselves for keywords or phrases found too much. Some different approach was needed.

The most advanced thinking about this problem was not about search protocols but about the abstract idea of searching and the ways that searching was done not in computers but in actual interaction with humans. The ideal model of a search device already existed. It was known as a graduate student. Computer programmers tended to be so focused on mathematically derived search algorithms such as binary tree protocols and bubble sorts that the significance of the graduate student tended to escape them. Graduate students or research assistants make the best search devices because in ideal circumstances they become attuned to their professors' work and are able to gather just the kinds of information that fit the research project, beyond the capabilities of lists of authors or keywords to capture the essence of the work. They take initiative, understand abstractions, and pursue corollary threads. At least the best of them do. This was precisely the kind of search device that a database such as MEDLARS required.

The people at the Ashibe lab and the media lab called such devices, in their abstracted form, agents. How to produce such an agent has been an ongoing debate ever since the media lab researchers first sketched out the idea. In the 1980s John Scully ordered a demo of Apple's dream machine for the 1990s, which the public relations division called the Knowledge Navigator. The device incorporated a phone, modem, e-mail, calendar, and a voice-operated word processor. The most striking thing about the Knowledge Navigator was its interface. It used not the desktop metaphor but an agent metaphor. In the upper left-hand corner of the Navigator's screen was the image of a secretary-like person. The image looked like a video of a live male secretary, complete with a perky bow tie. (Some of Apple's design team said they had wanted to avoid the stereotype of a woman in the secretarial position, which left them with the only gender alternative available to a large mainstream corporation. In their terms, the bow tie was an attempt to demasculinize the agent.) "Humans naturally talk to humans," Gardner would say, waving his arms. "Humans don't naturally talk to computers. We've spent millions of years learning how to talk to each other. Why change now?"

Talking to a program that is written to look like a human implies some things about interaction. The main one is that the program is convincing enough to actually engage the human—that the program possesses a depth and complexity that makes the interaction convincing. Somewhere between a talking to a cartoon face that answers in a mechanical monotone and talking to a photo-quality image that expresses the full gamut of human emotions lies the treacherous ground of agent experimentation, and it goes on at full speed today. When Fennell raised the first questions about dramatic interaction, she implied the computer as actor, which led to the same problems vis-à-vis agents: how to make characters on the computer screen that were convincing.

Central to the construction of agents was the idea of *presence*. What, exactly, was it about a representation that gave it the illusion of personal force, of a living being? And conversely, what did it take to convince a person by means of a representation of a place that they were actually present *in* that place? The Ashibe lab researchers talked about it long and often. Just what was presence, really? they asked. And how do we manipulate it?

It seemed obvious that talking about presence wasn't going to be enough. They had to experiment with presence, to play with the concept of agency in practical situations. In the middle of these discussions they more or less stumbled on the construct that they came to call Marvin Fishbein.

Marvin was at first a purely conceptual person, a thought experiment in what it took to make an artificial persona seem real. What went into creating a believable virtual agent, really? The original idea came from Samantha Born. For a while he was known simply as Marvin, and then Born exercised creator's privilege and gave him a last name. Later he also acquired a wife, named Olivia. Marvin and Olivia Fishbein, the team chuckled. If they wanted believability in their agents, how better to start?

The team thought of Marvin as another member of the lab. (Olivia was a homemaker, so she avoided the hard parts of the experiment.) They set out to explore how his sense of presence could be developed and how it needed to be tweaked in order to create an aura of reality for him. At first his persona was honored more by reference than by his own speech, like the legendary Kilroy. People talked about him, rather than Marvin himself doing any talking. Shortly it became clear that just talking about him in the third person was too limiting, and Marvin would have to start doing some talking of his own. Since he didn't exist in the physical sense, this presented some problems; but many of the lab's day-to-day interactions already occurred via e-mail, and the team seized on this fact for their first experiment. Thus Marvin's personality developed and flourished at first entirely on e-mail, where the problem of voice could be ignored. Various lab people who were privy worked on his character—Jim Dornan, Mike Gardner, Polly Steig, Mara Fennell, and Samantha Born were in on the experiment and each could log on as Marvin. They added to his persona in a deliberate way from time to time, but after a while he began to grow by a kind of accretion. Perhaps his own embryonic personality was beginning to assert its presence.

People at the lab frequently put on-line brief resumés and statements of what their research interests were. These were kept in special files called finger files, which anyone could read by "fingering" the individuals' name. After a while the Marvin team put on-line an elaborate and quirkily humorous resumé of Marvin's achievements in a finger file of his own—he'd invented squid jerky, for example, and made significant contributions to the development of muffler bearings—and before coming to Ashibe he'd had a distinguished career with the British postal service. The team was still only half-serious, playing around with ideas of identity without much sense of where they were going, and unconsciously they were creating Marvin as a kind of liminal character. He had some quotidian attributes, and also some that were obviously whimsical. But as it turned out, his persona would become so overwhelmingly real that the whimsical attributes would come to be overlooked, or astonishingly misread so as to make them fit into an acceptable image of a genuine, if eccentric, person.

It was now early 1983, and Steve Leslie had begun dropping hints about needing more assistants. Leslie was gone much of the time doing other things, such as attending corporate meetings, and traveling the lecture circuit, and his absence was being noticed—and decisions about the lab needed to be made day by day. Leslie's regular guy and heir apparent to the title of Director Pro Tem was Morris Lieber, whose claim to fame was that he specialized in big databases; but by this time the interesting work at the lab wasn't being done in database construction, it was being done by the people working with problems of agency, dramatic interaction, and presence. In the meantime, Marvin Fishbein had been gaining reality points as the team continued to work away at building his character. Leslie was following the Marvin experiment with interest, so one day, by fiat, he decided to kick it along. He published a memo to the company at large announcing that in his absence Marvin would be director pro tem of the lab.

Marvin responded to his new authority by blossoming into a much more distinctive and complex personality. He became known around Ashibe as a suave, intelligent, smooth-spoken guy, mature and sexy, slightly rakish in a Victorian way—in fact, rather noticeably like the personality of Nick Negroponte, the lab's godfather. The resemblance was probably not happenstance. Marvin acquired a corner office, furnished in typically Marvinian idiosyncratic fashion. For example, someone had found an old airline seat and brought it in to the lab with the idea of eventually using it for an experiment. The team borrowed it and installed it in Marvin's office, where, it was said, he used it instead of a desk chair. The sight of the thing behind Marvin's desk seemed in keeping with Marvin's quirky personality.

Of course, Marvin was almost never in town, either, being always off on some mission or other of his own, just like Steve Leslie. But unlike Leslie, he was extremely active on e-mail, keeping up a continuous flow of conversation and bombarding people with questions about the progress of their experiments.

In and around the lab there were the beginnings of the fights between lab folk and the systems engineering people, border skirmishes that would later blossom into a full-scale war. Systems folks thought the MIT crew to be, in their words, effete little pricks. The systems engineers were what lab folk called pocket-protector people, and in the term could be understood the magnitude of the culture clash that was shaping up. Battles of words began to rage on interoffice e-mail between the groups, and in these clashes Marvin was always the cool voice of moderation. After a while, the Marvin team noticed that there was an occasional piece of mail from someone in engineering, or sales, asking for Marvin's advice on some intradepartmental disagreement. Apparently people in other departments were hearing about this Solomonesque person in the lab, and were seeking out his opinion.

The team fed this tiny flame of belief assiduously. They spent hours talking about ways to ramp up Marvin's reality controls. Finally, digital music technology provided an unexpected breakthrough. One day Mike Gardner showed up with an Eventide Harmonizer, a device that changes the pitch of whatever sound passes through it. The team took one look at the Harmonizer and decided that it was time for Marvin to have a real voice. Laurie Anderson was already using a Harmonizer to change the pitch of her voice in live performance, but the team didn't know that. Mara set the device up and practiced with it, getting used to how it worked and what its limits were. The team decided to give the Marvinian vocal presence a test run. They hooked the Harmonizer up to a speakerphone, took out their note pads, Mara dialed up the manufacturing office—and accidentally got connected to the senior vice president of Ashibe.

Everyone in the room snapped to full attention. "What do I do now?" Mara mouthed silently. "Keep going," Polly Steig mouthed back. So while the team listened, gripping their notepads, Mara bulled ahead. Speaking as Marvin, with the deep voice of the Eventide Harmonizer, she wound up convincing the VP that Marvin was in Boston on important business. Marvin left a phone number in the event that Ashibe needed to reach him. Mara was completely unbriefed as to how to carry the masquerade to that extreme, and in the background the team scrabbled around for a plausible Boston phone number. Finally they came up with a number that Mara fed to the VP. Then she hung up, astonished by the extent of the masquerade. She looked around at the team, and Don O'Neill said, "I hope he doesn't actually try that number. He'll wind up talking to Boston Dial-a-Prayer."

The team realized that something quite interesting was going on. The experiment in presence was succeeding beyond anything they'd anticipated. They had actually convinced a significant part of the Ashibe corporation that a person named Marvin Fishbein existed, and further that he was acting director of the lab. Insidiously they had even been winning over some lab members. But the greater population of the lab wasn't convinced. "Muffler bearings," some snorted. Shortly they challenged Marvin to either appear in person, or, failing that, to hold a live teleconference with the company at large from where he happened to be.

The Marvin team took up the challenge. They couldn't handle a personal appearance, but a teleconference was just enough within reach to be truly daunting. They raced around the company grabbing audio and video equipment wherever they could find it. They came in the next weekend when the lab was deserted, and spent forty-eight frantic hours cabling the place. They ran audio and video feeds out of the lab area, hiding the cables behind the overhead tiling. In a remote part of the building they threw together a makeshift video studio, complete with lighting and sets. They'd succeeded in creating an e-mail presence, then they'd created a vocal presence; now they were about to stage the debut of Marvin's visual presence.

What did Marvin look like, really? The team held a strategy meeting, and decided that he might as well look like Mara Fennell. Fennell would play Marvin for the live cameras. That bracketed his appearance; it gave him some limits with regard to size, height, build, and so forth. On the day of the teleconference the team sneaked away to the hidden studio and spent four hours prepping for the performance. Mara needed makeup to create the visual Marvin, and the Eventide Harmonizer to create his voice. She and Mike Gardner were about the same size, so she borrowed some of his clothing. She did a careful makeup job—grey temples, crepe hair sideburns, and a hat to cover the abundant blonde hair on her head. Mike hooked up the Harmonizer between Mara's microphone and the audio feed to the conference room.

The conference was set up to be two-way, with live questions and answers from the audience. In the audience was Douglas Adams, author of *Hitchhiker's Guide to the Galaxy,* who was visiting the lab that day to give a talk of his own.

At zero hour the screen in the lab's spacious conference room lit up to show the elusive Marvin Fishbein calmly seated behind his desk. He made a few opening remarks, a sort of state-of-the-lab address in miniature, and then threw the meeting open for questions from the live audience in the conference room. People asked questions about how the company was doing, were there any new plans for the lab, and so forth. Marvin answered with his accustomed rakish aplomb, occasionally rising to pace the room or to sit on the edge of his desk. During his answers the video switchers interspersed shots of him with prepared charts and graphs. And because the team couldn't leave well enough alone, they had built the set to include several windows behind which were landscapes. The landscapes came from an assortment of travel posters that Don O'Neill had cut up, but no two posters were of the same country. During the cutaway shots to the graphs while he was not on camera, Marvin shifted position from one window to another, thus giving the effect (to the alert viewer) of being in one country in one shot and then suddenly in another one for the next. This sort of thing produced much giggling in the control room.

Insofar as the conference was a test of whether Marvin could convince an audience that he was real, it was an unqualified success. Muffler bearings in his resumé notwithstanding, the percentage of people at Ashibe who believed in him had risen to near one hundred percent. But there were unexpected side effects. The most striking was that some people remained unconvinced not of Marvin's reality but that the conference had been live and interactive. After all, they reasoned, the focus of much of the lab's work was on interactivity, and the conference was a golden opportunity to test out some of the things they'd been finding out. They believed that Marvin had been recorded some time before the "live" conference, and that what they had seen had been a playback of portions of the recording, cued by some sort of voice-recognition system. Marvin's seeming to answer questions from the audience had been a trick of some very slick programming. The cuts between segments that were necessary with such a system had been made invisible by the programmers' hacking the switching gear. That, they felt, had been the whole purpose of the Marvin deception.

Fennell, peeling off her crepe sideburns, shook her head in amazement. "Jeez," she said, "he's real, but he's not live. Just the opposite of what we'd expected."

The Marvin team decided it was a good lesson. "I think we learned more about presence from the conference than from anything else we did," O'Neill said. "It's been very interesting how you construct the illusion of a person."

Although the team didn't know it, the end of the lab was approaching too quickly for them to exploit the things they'd learned from Marvin, from Disney, and from the interaction research they'd then barely begun.

The game software market finally peaked around 1982, and after that time revenues began slowly to drop. It didn't matter that they were dropping in relation to the most profitable financial situation that any American company had ever enjoyed; by the relentless logic of business, the romance had now gone sour. There was also something awry with the books, but that wouldn't become clear for another year or two. Ashibe immediately started layoffs. In some areas of the company these were fairly serious as early as 1983, but by 1984 rumors were beginning to circulate that layoff fever might reach as far down as the lab. Management had to make decisions about whether the direction the company intended to pursue was going to be only making product for the current market or also doing development for potential markets. Todd Hirsch, Ashibe's newest VP and the person directing most of the layoffs, believed that Ashibe's future lay not in interactivity—which, as was becoming clear, the company as a whole understood not at all—but in hardware, specifically in VLSI (very large-scale integrated circuit) technology. A VLSI chip is a specialized component incorporating a microprocessor, some memory, and a permanent program, all together on a single chip. The chip runs only the program for which it was designed, and can be made extremely small because it doesn't need the frequently bulky general-purpose components that a personal computer must have in order to properly run whatever software happens to be loaded into it. Products like Gameboy are based on specialized VLSI chips rather than generic chips running external programs.

Ashibe had a VLSI department of its own which was separate from the lab. The VLSI people didn't do research; they worked from a set of accepted design principles to produce custom special-purpose chips. It was a craft job, requiring set kinds of skills. The chips ran programs that were written by the game division; thus the lab wasn't in the VLSI loop at all. Phil Hoff, who oversaw both divisions, wasn't particularly happy with the lab. He tended to treat the lab and its staff like an exotic tropical bird with an extremely low IQ. It was nice to look at, but didn't do much of anything useful. Hoff didn't care about research. Research, he felt, was tomorrow. Today he had a bottom line to meet, and to hell with tomorrow; he wanted product, not speculation.

Hoff was representative of the entire management philosophy at Ashibe. From the beginning, and continuing as the company exploded in size, upper management had continued to recruit the middle-management corps from places like Proctor and Gamble, Lever Brothers, and Clorox—good, solid people with little imagination who had solid track records as business managers. It was said that not a single one had had a computer in their offices at their former companies, and none of them had computers in their offices at Ashibe. They weren't embarrassed to say that personal computers were merely a passing fad, and that after all the excitement died away Ashibe would have done its consumer research and would get on with producing whatever the merchandise was for the next fad.

Patently, life under this dispensation was not an atmosphere conducive to starry-eyed romanticizing about the future of technology. With the attention of higher management palpably turning to immediate results, the members of the lab found themselves in the position of having been hired to perform a series of tasks which management was now indicating it felt were odious. There was no doubt that they were pushing the envelope of long-term agent research, but Ashibe's support for the lab was quickly evaporating; under the new and stricter policies the lab had no short-term mandate.

In addition, they were beginning to experience the downside effects of one of the very things the lab had been organized to exploit: the results of continual culture clash between the long-range researchers and the medium-range manufacturers, with the added red herring of the short-range profiteers thrown in. Leslie's (and Negroponte's, had he been present) original idea behind throwing these groups together had been productive clashes, cross-fertilizations. Because of, or perhaps in spite of, Leslie's prolonged absences, the cross-fertilization wasn't occurring. In fact, the pattern that was rapidly emerging was quite the opposite: continual suspicion on the part of the manufacturers and outright contempt on the part of the profiteers.

Part of the reason for this might have been the deliberately blurry lines between the young technoturks in the lab and the postadolescent terrors down the hall who played at computer games in the guise of work. From a distance they were more or less indistinguishable. That had been okay when the tiny terrors were practically spouting cash from their fingertips, cash enough to float the entire corporate vessel with enough left over to enrich anyone who came close. But that was then and this was now. The bloom was off the kids. Their tiny imperfections, always noticeable but easy to ignore in the golden haze of enormous profitability, now stood out like mountains on the moon. What were they doing, management began to ask, besides playing around?

Sensing impending trouble, the lab staff held a series of internal meetings. The idea had something to do with a combination of self-promotion and self-policing. Leslie, as usual, was off somewhere, and Marvin couldn't handle a job that required extensive face-to-face interaction. The staff met anyway, and with Leslie's (and Marvin's) concurrence they elected Celeste Harmon, one of the original MIT group, as acting manager. Harmon instituted a system of technical reports, and then forgot to sign off on them. But before this could become a problem, Leslie was back, and called a special meeting.

The situation at Ashibe at this moment, of which Leslie was either unaware or pretending to be unaware, was this: one of Ashibe's New York executives had been caught with his hand deep in the cookie jar. The books had been rigged to cover the losses, which were considerable. Almost simultaneously, Ashibe's stock had begun to plummet. Either Leslie didn't know about this or chose not to mention it; at any rate, he was running on a tape of his own, a reaction to the earlier gloomy financial picture. He was growing uncomfortably aware that the gulf between the lab and the rest of the company, not particularly dangerous in the palmy days, could easily prove fatal in Ashibe's current financial straits. What he did do was talk about who was hiring in the rest of the industry.

There they are in the middle of another interminable lab meeting, and all of a sudden people are noticing that Leslie is talking about other job opportunities. It was so bizarre and unexpected that many people didn't even notice he'd done it until after the meeting had ended; the blissful bubble of invulnerability that surrounded the lab was hard to breach. "Did you notice?" Laurent said later. "Steve was talking about job openings at the defense department! Now why the hell would he do that?"

They found out a few days later. At eight A.M. the top brass called Celeste Harmon in and told her to fire twenty to thirty percent of the lab staff immediately. The first anyone in the lab knew about it was when Harmon came into John McClelland's office, sat on the edge of his desk and said, "I need your badge."

"Huh?" McClelland said.

"I need your badge. You're fired. You have five minutes to get your stuff together." She paused, then added in a lower voice, "I'm sorry."

McClelland just sat there with his mouth hanging open, and a guard walked in, carrying a cardboard box. Harmon took McClelland's badge, stood up, and went on to the next office. McClelland dazedly began filling the box with his personal stuff. In a few hours Harmon had cleared out forty people.

A few days later they did it again. In two or three waves of firing, the population of the lab dwindled by half. Then Harmon started in on the core people, the best and the brightest whom Steve Leslie had romanced personally—Fennell, O'Neill, Laurent, Steig, Born. "It came as a complete shock," O'Neill said bitterly. "One afternoon it was, 'Here's our plan for the next fifteen years, here's fifteen million bucks, let's get busy,' and the next morning it was, 'Can you have your things packed and be out of your office in fifteen minutes?' "

While this was going on, Ashibe's stock had dropped through the floor, and was still plummeting. On top of the sinking game market, news of how Ashibe had been looted from within had badly shaken investors, who were now operating in dump mode. Living only in the present, the Patton Corporation was frantically searching for someone to take this worthless loser off their hands. They found their angel in the person of John Reitherman, who had piloted a nearby computer company in the valley until a stockholders' rebellion had ousted him. Reitherman had the reputation of being the baddest, meanest SOB in the computer management business; those who had worked under him were fond of referring to him as Jabba the Hutt. "The mildest thing anybody said about him," Gardner related, "was that he was a vicious, cold-hearted, bloodthirsty shark." Reitherman had survived prison camp during World War II and wasn't going to let Silicon Valley slow him up one bit. He bought Ashibe from Patton for what amounted to salvage costs. Once he had hold of Ashibe, word came down that he was preparing to clear everybody out and move his team in. Normally this takes a while, but not when you are hungry, in a hurry to cut costs to nothing, and utterly ruthless.

The scene at the Silicon Valley complex was reminiscent of the last hours of the U.S. Embassy in Saigon. At eight A.M. word had arrived that Reitherman had taken over and that his goons were coming to clear the buildings. *Everybody* was fired. Harmon communicated it to the remaining lab people by announcing, "Jabba the Hutt is on his way." While Reitherman and his goons were on their way—for all anybody knew, in armored assault vehicles—Patton's guards were already conducting sweeps of the buildings, clearing everyone out, and boxing up whatever was abandoned in the process. The mood shifted to full-scale panic. People who had just discovered what was happening were frantically trying to collect their personal belongings. Others on the upper floors opened their windows and began dropping things out—into the bushes if they were lucky, onto the pavement if they weren't, occasionally into the open trunks of their cars parked at the curb—things like chairs, bookcases, sofas, stereo systems, and television sets. Occasionally a TV exploded on impact, sending a gout of dust and debris spouting upward from the trunk, and kicking the trunk lid into mad seesawing in the aftershock. Guards were herding secretaries and lower-echelon workers into the street like farm animals, some bewildered, some in tears.

Gardner and Merino took refuge on the upper floors, perhaps in hope of finding a rescue helicopter. Somehow the guards missed them during the next-to-final sweep. The two walked slowly downstairs, through the empty, echoing hallways drifted with debris, and into the silent, blue-lit warehouse, which was already filling up with rows of cardboard boxes stuffed with personal items the guards had collected. They turned and went back into the lab building, and walked upstairs to O'Neill's old office. The building was shockingly empty, its suddenly abandoned spaces surreal. Drifts of discarded files littered the floors. It reminded Merino of the Hall of Records in *Orfeo Negro*. They sat there in the deserted building, soaking it in.

At about 9:30 Reitherman's troops evicted the Patton troops and conducted their own sweep. This time they found Gardner and Merino in Hoff's office. There was the sound of boots in the hall, and suddenly a Reitherman shock-trooper loomed in the doorway. "You leave now," he said.

"Now?" Merino said. "We still have to pack up these tapes." He indicated a stack of Gardner's personal videotapes.

"Okay, but nothing that says Ashibe on it," the guard said.

All Gardner's tapes said Ashibe on the boxes—until recently it had been something of a mark of pride. So Gardner and Merino took turns, one of them distracting the guard while the other surreptitiously dropped the tapes, one at a time, out the window. As they were completing this operation, there were more footsteps in the hall, and a young, fresh-faced, friendly kid in Reitherman livery came in. "Hi!" he said. "What are you guys doin'?"

Gardner and Merino looked at each other, then at the kid. "Packing up," Merino said, "And you?"

"I'm part of the Ashibe research team," the kid said. "John Reitherman just hired us. It's really exciting. We're going to start an Ashibe research lab. We're going to do blue-sky research here. Reitherman says there'll be no limit to what we can do. Cutting edge stuff!"

Gardner and Merino just goggled at him. After a minute, Merino intoned, in a sepulchral voice, "Welcome—to the Twilight Zone."

Appendix

This essay is adapted from a section of my forthcoming book, *Presence: The War of Desire and Technology at the Close of the Mechanical Age*. Though it may appear cavalier, the style of the piece is resolutely formalist. It is part of a series of ongoing experiments to understand what I perceive as a faltering of social theory in its efforts to come to grips with real-world social formations near the end of the twentieth century, and to find modes of representation adequate to the task of engaging wider discourses and audiences in times of shifting interdisciplinary boundaries. Data collection was by interview and participant observation and is of the type generally characterized as thick or "in-depth," since I was interested, in an as-yet-unwritten longer account, in creating an atmospheric piece that conveyed my own odd sense of the period covered by the essay, and its implications. In the construction of this literary fiction, which I present as an anthropological account, I have taken certain liberties of which the reader should be warned. The dialogue within quotes comes from interviews with persons who were present or who had professional or nonprofessional contacts with the principals. But even where dialogue is quoted there is no guarantee that things are in any sense pristine. On occasion I have superposed dialogue from various parts of the transcriptions, deleted events, collapsed several persons into one, changed some names and not others, and in general constructed a fictional narrative which hews rather closely to the shape of the events themselves but which is not identical to them. In this effort I have examined the style of several researchers who walk the thin and treacherous line between scholarship and journalism. In particular, the work of Gary Taubes, Frank Rose, and Tracy Kidder has been most helpful. Brenda Laurel has been a wonderfully supportive and articulate friend and critic, and Sharon Traweek has been an ongoing inspiration. Adele Clarke and Joan Fujimura, two brave scholars, would have commented on the essay if I'd given them the chance, and they deserve mention anyway. Of course, Donna Haraway is an *eminence*-not-so-*grise* in the diegesis.

Notes

1. The history of the development of computational devices is littered with Xerox's missed opportunities, as the scientists and engineers at PARC came up with one brilliant product after another—all of which Xerox's top management ignored. When Xerox's seemingly endless profitability began to fade in the late 1970s, it was too late. For an excellent study of this disaster see Smith and Alexander 1988.

2. During my sojourn in San Diego I mentioned to Bruno Latour that any consideration of how a commercially produced object came to possess agency would inevitably require some theorizing about the mafia. "Ridiculous," Latour said.

3. Leslie now feels that he was misquoted or misunderstood in the meeting. "What I said," he maintains, "was that they were going to commit a *seemingly* infinite amount."

4. These are laid out in more detail in Brand 1987, which is an informative although starry-eyed look at the early interactivists.

5. DARPA, the Defense Advanced Research Products Agency, was a major source of funding for computer systems research with possible military uses. Politically speaking, because of its ties to the military, DARPA was anathema to the Ashibe crew, as it was to many of the young, liberal computer whiz kids.

References

Brand, Stewart. 1987. *The Media Lab: Inventing the Future at MIT.* New York: Penguin.

Kidder Tracy. 1981. *The Soul of a New Machine*. New York: Avon.

Rose, Frank. 1989. *West of Eden: The End of Innocence at Apple Computer.* New York: Penguin.

Smith, Douglas K., and Robert C. Alexander. 1988. *Fumbling the Future: How Xerox Invented, Then Ignored, the First Personal Computer.* New York: William Morrow.

Stone, Allucquére Rosanne. Forthcoming. *Presence: The War of Desire and Technology at the Close of the Mechanical Age*.

Taubes, Gary. 1986. *Nobel Dreams: Power, Deceit, and the Ultimate Experiment.* New York: Random House.

GARY LEE DOWNEY

7

THE WORLD OF INDUSTRY-UNIVERSITY-GOVERNMENT: REIMAGINING R&D AS AMERICA

> Unfortunately, the overall quality of engineering design in the United States is poor. . . . Partnership and interaction among the three key players involved in this endeavor—industries, universities, and government—have diminished to the point that none serves the needs of the others. . . . This state of affairs virtually guarantees the continued decline of U.S. competitiveness.
>
> —National Research Council, 1991

> We're trying to do with the computer what we cannot do with our organizations.
>
> —David Grose

The ACSYNT Institute is rewriting a computer program for aircraft design that is twenty years old. In the process, institute members are also rewriting themselves. The ACSYNT Institute is experimenting with a new form of organization for technological research and development, a joint venture involving industry, university, and government as equal participants. Its technical objective is to get design engineers in industry to use a program, or code, called ACSYNT, originally written within government and recently improved within the university. Yet through its very organization and day-to-day activities, the ACSYNT Institute also offers a new vision of research and development, which I call here the world of industry-university-government, or IUG. I began following the activities of the institute in 1990, after an informal consortium among the participants had become a formal organization. My own feelings about the world of IUG are strongly ambivalent and, hence, somewhat confused.

ACSYNT is short for Aircraft Synthesis. The code synthesizes, or brings into interaction mathematically, technical constraints from different engineering areas in aircraft design, including the most prominent areas of aeronautics, propulsion, and structures. Aeronautical engineers from NASA originally wrote ACSYNT in the 1970s to help them evaluate and compare proposed designs for military aircraft. During the late 1980s, engineering

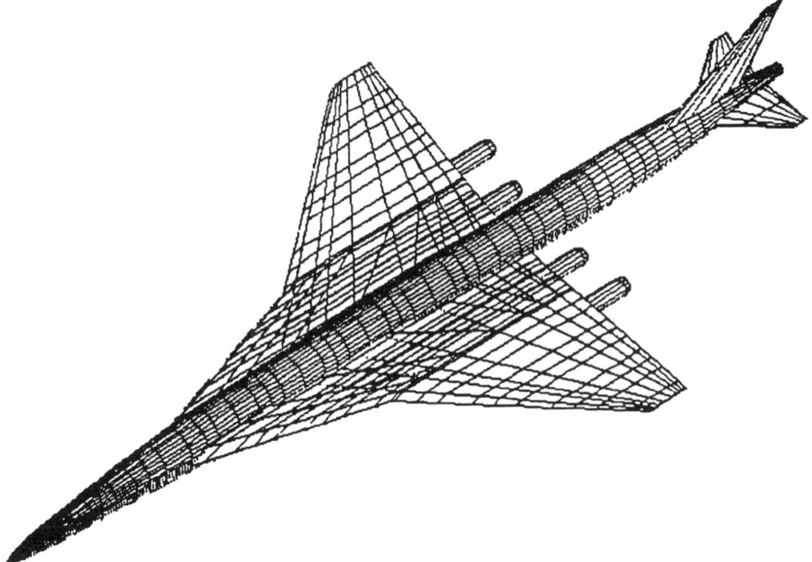

Fig. 1. Iconic representation of the ACSYNT Institute.

faculty and graduate students at the Virginia Polytechnic Institute and State University, working under a grant from NASA, merged graphical capabilities into the code, which means that they enabled it to display visual images of proposed aircraft in addition to presenting long lists of numbers. Design engineers in both government and industry find this new feature very appealing because it enables them to see an aircraft as they settle on its specifications. The institute represents itself iconically with one such image (fig. 1).

I find the ACSYNT Institute an interesting organization to study for two reasons. In the first place, I find its experiment attractive politically. The institute seeks to blur a bit the boundaries between industry, university, and government with the goal of benefiting the common good. A nonprofit organization, the ACSYNT Institute is motivated in part by a sense of citizenship. Its members try actively to move beyond a narrow focus on the maximization of self-interest and to work together. I watch people struggling to find ways of overcoming the boundaries that separate them. I see in the ACSYNT code an opportunity to introduce a wider range of design parameters, such as environmental considerations, into the mechanism of aircraft design, as well as to help an industry shift from military to commercial enterprise. I also like the people involved, and have become friends with some of them.

During the two years I followed the institute, its membership consisted of the NASA Ames Research Center, Virginia Tech, and eight aircraft companies, with four other NASA and U.S. Navy research organizations participating in minor roles. Each company committed $30,000 per year for a period of five years to support research and development activities on the code. NASA is actually the motivating organization behind the institute. A legal provision of the 1958 Space Act that founded NASA permitted it to enter into jointly sponsored research arrangements with nongovernmental organizations, something that few other federal organizations that do. In 1987 NASA provided staff to establish an independent, nonprofit organization, the American Technology Initiative, with a mandate to establish joint research and development projects, now known as dual-use technology development, that would transfer NASA technologies to private industry. The ACSYNT Institute is the second joint venture founded under the auspices of the American Technology Initiative, or AmTech, although the first to be funded under the broad authority of the Space Act (fig. 2).

The actual work of tailoring the ACSYNT code to fit industry practices is done by engineers at the NASA Ames Research Center and Virginia Tech. The government engineers are primarily responsible for improving its capabilities in mathematical analysis and synthesis, while the university engineers emphasize developing further its capabilities in visual and geometrical representation. Members meet for two days twice a year either at NASA Ames in California or at Virginia Tech in Blacksburg, to report on progress and negotiate future plans. I attended three of these meetings, conducted lengthy interviews with fourteen participants, and attended weekly meetings of the Virginia Tech research group.

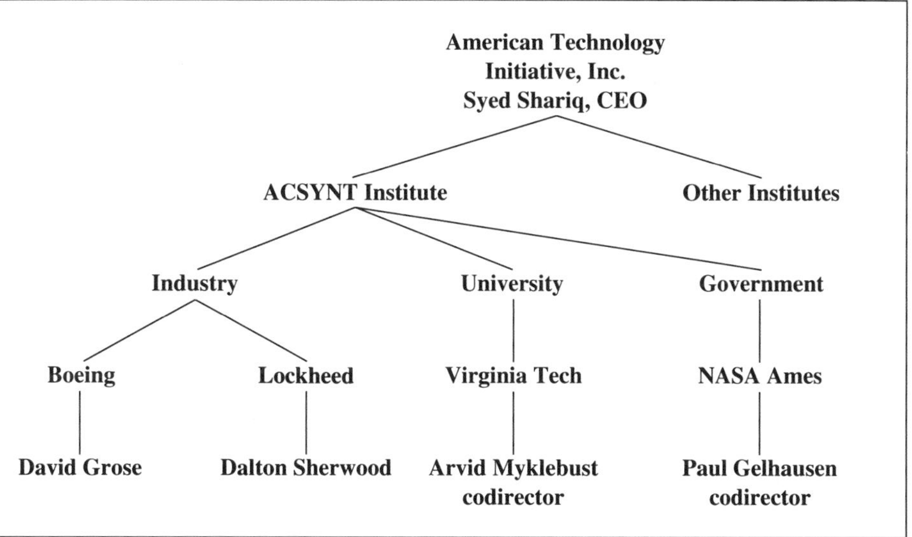

Fig. 2. Organizational chart and cast of characters, American Technology Initiative, Inc.

The second reason for my interest in the ACSYNT Institute is that its members are blurring the boundaries among their institutions by means of the code itself. David Grose makes this point in the epigraph: they are trying to do with the computer what they cannot achieve with their organizations. Showing connections between rewriting a computer code and redefining organizational relationships is a delicate task because it challenges common assumptions about the relationship between technology and society.

Throughout American history, a common strategy for producing social change without debating it explicitly has been to delegate the change to technology. This strategy often remains inexplicit because popular theorizing in America imagines technology not as a social phenomenon but as a force external to society that impacts on it and to which society must adapt. Separating technology from society in this way renders it legitimate to produce social change by means of technological change, for technological change is understood as inherently liberating and progressive. Since the technology itself becomes the cause for change, human participants in technological development may be absolved of responsibility for any social implications other than those that contribute to human liberation and foster social progress.

Academic research in the interdisciplinary field of technology studies has maintained that this theoretical attitude of technological determinism limits one's ability to understand increasingly complex relations between humans and technologies. A common academic strategy that I endorse is to reconceive technology as a part of society rather than apart from it. This conceptual move can make it easier to see how the technical process of developing a new technology also designs the social relations within the technology will work. Since institute members generally envision technology as lacking social content, they do not describe the transfer of ACSYNT from NASA to industry as a process of building new social relations into the code. Technical activities are different than social activities. Formal discussions tend to center on the technical features that the code should or should not have, or on whether the code itself is any good or not. At the same time, much informal discussion focuses on the participants' frustrations with what they often see as needless politics, that is, the nontechnical dimensions and issues that also always seem to be present. Grose's assertion linking the technology to organization is exceptional rather than typical.

If one starts, however, with the different theoretical premise that society includes technology, one gains a much different image of institute activities. For example, one can argue that the original ACSYNT code was limited because its developers had built into it the social relations of research but not of production. That is, the code fit nicely the research environment at NASA Ames, where it has always been appropriate to consider different design specifications simultaneously (for example, aeronautics, propulsion, and structures), but it did not fit the production environment in industry, which has institutionalized rigid separations among these areas. From this standpoint, by rewriting the code to incorporate the production environment, the ACSYNT Institute is now constructing innovative social relations that blur the boundaries among government, industry, and university.

The problem the ACSYNT Institute faced when I completed my participation in 1992 was not only to incorporate the production environment of industry into the code but also to define the three-way collaboration that constitutes the world of IUG. Whether the code was good or not had become a question of whether or not everyone could see themselves in the code. Producing code that everyone could accept and use would be an indicator that the institute was not simply an organizational artifice for an incomplete or impossible collaboration. These technical innovations and political innovations had to be realized together, through the same actions.

In the midst of my enthusiasm for this experiment, I also have strong worries about the world of IUG. I fear that such boundary-blurring, collaborative ventures have the potential of reproducing, or even magnifying, dangerous forms of national chauvinism. During the 1980s, official America retheorized international struggle from a political to an economic idiom. With its economic dominance of the world no longer assured, America theorized itself no longer as a site for competition among individual interests but as a single economic actor maximizing a collective interest. The national shift to economic struggle was sudden and dramatic, embodied and epitomized by the election and reelection of Ronald Reagan. Reimagining nationalism in economic terms and economies in nationalistic terms legitimized and opened possibilities for unprecedented levels of cooperation within industry, as well as the direct participation of university and government in economic development. More well-known examples of collaboration in R&D have included the Microelectronics Corporation, which brings together competing electronics companies, and Sematech, a joint venture between government and industry. During the 1960s and 1970s, such collaborative ventures would likely have met with significant protest that they fulfilled apocalyptic visions of the military-industrial establishment. In the 1980s, however, the power of patriotic commitment to this economic call to arms became concentrated in a single, one-word trope: competitiveness.

By successfully transferring the ACSYNT code from NASA Ames to the aircraft industry, the ACSYNT Institute explicitly hopes to improve the process of conceptual design, the earliest stage in aircraft design, in order to enhance U.S. competitiveness in aircraft manufacturing. During my participation, no one in the institute questioned the nationalist character of this objective. Although writing distinct visions of R&D from industry, university, and government into the ACSYNT code serves the common good, the common good is the welfare only of the United States in economic competition with other nations. The practices of research and development have become linked to the nation because national struggle has been reconceived as competition in a global market. Rewriting ACSYNT to achieve a world of industry-university-government is thus one instance of a national strategy to reimagine R&D as America.

Seeing two different possibilities in the institute at the same time, I find myself ambivalent. Is blurring internal boundaries the crucial first step toward developing a sense of global responsibility and citizenship? Or, by enhancing the ability of the United States to act commercially with a single voice, do collaborative ventures undermine the development of global responsibility and simply reproduce nationalistic struggle in a new idiom? My sympathies lie with the first, but the second appears more likely in the near term, for the end of the cold war has led to a sharp increase in the intensity of nationalism around the world. Despite the fact that the nationalist concern for competitiveness marks a shift away from production for military purposes, I worry that the cost may be to militarize American conceptions of commercial activities.

Strategies for reimagining R&D as America are likely to persist in the 1990s as the Clinton administration works to realize an activist government that not only participates in economic development but also actively directs it. As the country evaluates expensive proposals to reconstruct America through such technological changes as fiber-optic networks and electronic information highways, the popular theory of technological change is being put to use once again. In each case the stated objectives envision the technological delivery of social progress in the form of enhanced competitiveness, yet the implementation of these technologies will require other social adaptations as well. It is important for cultural studies of science and technology to examine critically the social engineering that is built into visions of R&D partnerships in order to make it clear that technological choices are also blueprints for social change. Discussions of new technologies should focus not only on which ones to choose but also on how to assess, evaluate, and acknowledge *a priori* the social changes written into them.

By splicing together bits of text from interviews and presentations at a semiannual meeting of the ACSYNT Institute, I introduce the reader both to the world of IUG and to the individual and collective stakes for those who inhabit it. For the people and organizations that participate in the ACSYNT Institute, moving from a world of old islands of knowledge into a world of newly blurred boundaries carries not only significant opportunities but also considerable risks. Participants can find themselves caught between the two worlds, subject to two different sets of responsibilities and expectations. The forms these tensions take vary significantly, depending on where one starts.

A Cast of Hybrid Characters

A significant feature of the ACSYNT Institute experiment is that its human participants are people whose careers have taken them across the borders that distinguish industry, university, and government. In a sense, participants are all hybrids. Not only do these people have some understanding and respect for the perspectives that colleagues from other arenas bring to the institute, but their individual biographies also contain some notable features or events that demonstrate a commitment to working for more than the maximization of self-interest. Note, for example, how the rise of Reaganomics redirected some careers away from work in environmental and energy arenas toward business and the military. The thread of continuity was the technical content of the work. Does the presence of hybrids in the ACSYNT Institute suggest a limited future for the world of IUG? Must one be a hybrid to want to participate in the first place?

Syed Shariq, 42, CEO of American Technology Initiative, Inc., Menlo Park, California, the parent organization of the ACSYNT Institute, works full-time at the institute while on leave from NASA Ames Research Center, Moffett Field, California.

DOWNEY: You were at Ames at the time this started?

SHARIQ: I had just joined Ames, having traveled a very long and circuitous path. I graduated from Virginia Tech back in 1974, where I got my Ph.D. in operations research in the industrial engineering department. I taught at Oklahoma State for a couple of years, and then got very interested in the decision sciences area. I went to Duke and was responsible for starting and heading up their program in societal risk assessment. How does society make the social decisions of risk, and how do you value human life? All those questions concerning the socioeconomic impact of technology were the focus. Also questions of ozone depletion and whether or not there's enough risk there to worry about, in terms of cancers and other ecological effects and so on. So I did that until about 1980–81, when I decided that I wanted to leave academia to pursue a career with more direct relevance and immediate impact on the world we live in—the real world.

DOWNEY: Did that business become boring around 1980–81 because of the Reagan election?

SHARIQ: President Reagan came into office, so I realized that the environmental agenda of the nation is going to be on ice for awhile. It was also a good time to really do something different. I joined the Stanford Research Institute [a prominent consulting firm, primarily on defense matters], and spent about four years working for companies as a consultant in strategic planning and diversification/acquisition of technologies. Just looking at how technology is developed and commercialized in large and small corporations. What does it take for these big companies to acquire little companies? How are they thinking about future products and how does technology play a role? How do you price knowledge work and technologies, and so on? I did a lot of economic, financial, strategic thinking for several companies, and I also helped SRI to do their own investments in high technology areas, like biotech, VLSI, advanced materials, AI, and so on. I was an internal consultant to the office of the president of SRI.

After doing that for three and a half years, I felt I should really take my knowledge and ideas and apply them to real-world high-tech investments, so I joined the venture capital business. I spent two years at a premier investment banking company in the Bay Area. I became part of a team of four senior people who were managing a portfolio of over sixty-seven companies, with over $100 million in investments. And for about a year and a half I developed a system for them for managing portfolios of companies and monitored investments in the artificial intelligence area.

DOWNEY: So first you went one direction, academia, then made a completely different move to consulting, and then to venture capital.

SHARIQ: Yes, I spent about two years in venture capital doing that, working on portfolio selection and investment decision-making methodology in the AI/software area. So after being there I realized that most of the venture investments, with over $110 million, had been made in the 1983–84 time frame. Those portfolio companies, over half of them, were doing so poorly that the fund would eventually collapse. There's no way anybody would be able to raise funds again, so I looked at the cards and decided that that's just not going to work out. I wanted to take some time out for personal reasons, just slow down, because venture capital had gotten to be more than two full-time jobs.

DOWNEY: I bet it was stressful.

SHARIQ: So I came to NASA to head their program in artificial intelligence, with no intention of doing what I am doing now at AmTech. But the director of Ames who hired me and a couple of other people thought that this idea of bringing NASA-industry-university joint ventures together is an interesting one.

In the wake of the Challenger accident and other problems with major projects, NASA was looking for new ways to legitimize its activities in commercial terms, contributing to America by enhancing its economic competitiveness.

SHARIQ: The legal people at Ames had done some research on the idea of using NASA's Space Act authority for carrying out joint R&D projects, and decided it was a nifty idea. They wanted to ask somebody who is well versed in the private sector and also understood academia to implement it. And so, I happened to be in the right place at the right time. When I reviewed the concept, I felt that I could do it and believed it would be a valuable innovation in our economy.

DOWNEY: And government. So you've got it all. You got it all.

Paul Gelhausen, 31, codirector of the ACSYNT Institute with Arvid Myklebust, is an aeronautical engineer working full-time at NASA Ames Research Center, California.

DOWNEY: Could you give me a sketch of your own history and involvement in ACSYNT and what you're doing currently? You're often characterized as someone who really has become a champion for the code.

GELHAUSEN: I don't really consider myself a champion. I've always considered myself more of a—see, I'm of German descent, and I think that Germans are the guys that are paid to fight for you. So I kind of consider myself more of a technical mercenary. I always find something technically challenging no matter what the job.

DOWNEY: You're a soldier. You're a soldier rather than a champion.

GELHAUSEN: I think so. I hope, that's what I'd like to be. Yeah, I am enthusiastic. I started as a co-op student working here on ACSYNT. [A co-op student alternates between school and work, taking more time to get a degree in order to gain experience and income.]

January of 1980 is when I started. Boy, my boss, George, hated me because I was a terrible typist and all we had were punch cards for writing programs. I kept screwing up the cards. I'm surprised he hired me back, except I work hard. But I thought ACSYNT was a great thing. I thought it was the way things ought to be done.

DOWNEY: George told me in an interview that you were an excellent student out of Michigan. In aeronautical engineering?

GELHAUSEN: Yeah. Aerospace is what the degree is, but I studied mostly aeronautics. But I came and I worked for George and we were doing research on the VSTOL airplane [vertical or short take-off and landing]. It was more in the line of advocacy. [Gelhausen elaborates in some detail how he did design research for several different types of aircraft while maintaining an interest in ACSYNT.]

DOWNEY: It sounds like you didn't want to limit your identity to any particular type of airplane but wanted to focus on the more general engineering issues.

GELHAUSEN: Yeah, I think it's all aerodynamics, when it all comes down to it. They all fly, so they're all flown. The other thing is that I would like to think that I'm making things that are most efficient. I'm interested in making the most efficient use of resources, getting the job done most efficiently.

DOWNEY: Where efficiency includes cost?

GELHAUSEN: Cost and minimum fuel use and things like that. I grew up in the seventies with the fuel crisis and things like that. It probably affected me more than, say, the folks who work on the National Aerospace Plane [NASP, a long-distance aircraft that will fly into space], which will really suck up gobs of fuel. One thing that I hope to be able to do is to look at alternate energy sources. Methanol, those kind of things, that have less impact, maybe more replenishable and clean energy sources.

DOWNEY: There might be money for that someday.

GELHAUSEN: Someday [*sighs*].

DOWNEY: Not right now, though, not in the last ten years, not since Reagan beat Carter.

GELHAUSEN: Yes, that was 1980. Well, anyway, the way that I ended up getting involved with ACSYNT is that I always knew that we could do a better job with it.

Gelhausen then describes a three-year design competition between the United States and United Kingdom, in which he used ACSYNT successfully to compare and evaluate the proposed designs. By this point, his own identity as an engineer at Ames had merged with the code, and he became principally responsible for managing its further development.

Arvid Myklebust, 45, codirector of the ACSYNT Institute with Gelhausen, is professor of mechanical engineering at Virginia Tech.

DOWNEY: Could you start with your education and work your way up through some jobs?

MYKLEBUST: All right. Well, I'll point out why I'm in this area of computers at all. It's because I began to study in a preengineering curriculum at Miami-Dade Community College in 1964, which was very active in computing at the time because they did computing for the state. I went to school at night for four years to finish two years work. During the day I was a purchasing agent for a printing company, and I was selected by the company to learn systems programming for a computing system that company was going to buy. In those days you had to write your own software. I had a year to prepare, so I started taking courses in electronic data processing, and got a great deal of basic background in the fundamentals of computing and assembly language. I wrote software for business machines, printing orders, and things like that. The company then lost its contract, so I went back to preengineering, finished, and transferred into engineering at the University of Florida. I wanted something related to computing software but nothing like that existed, so I switched to mechanical engineering because I had a very strong mechanical background.

DOWNEY: You've talked on occasion about working with engines.

MYKLEBUST: I did major overhauls on trucks and cars, about half and half. That's what my father did, and I was working with him. So that's why I switched to mechanical engineering, but I looked for computing applications all through my undergraduate years. Later, as a graduate student, I switched to a program in kinematics [mathematical modeling of mechanical linkages] simply because that was what was making the broadest and most intense use of computing at the time.

Myklebust then goes on to describe his master's and doctoral research on the computer-aided design of linkages, postdoctoral research in Norway, and then teaching positions at the University of Arkansas and Florida Atlantic University. He found himself caught up in software design for microprocessors and director of an engineering computer laboratory. I wondered what kept him in the university.

DOWNEY: Did you think about jumping ship and going after the money in industry?

MYKLEBUST: I wasn't interested in the money. Never have been. I think the reason was that the companies I worked for while I was working through the first four years really soured me on working for companies. I'm sure there are many good companies to work for, but I didn't like the approach that some of these companies took toward their employees.

Also, I was so idealistic then, I didn't think about the money at all. As a matter of fact, I only wanted to do the research on linkages to help stimulate a research area in computer-aided design.

DOWNEY: When you say you were so idealistic, what were the ideals you were pursuing?

MYKLEBUST: I wanted to be able to accomplish something that was worthwhile, to make my presence here have some meaning. Just to go out and sell software to make a buck didn't seem to me to be too satisfying.

When I conducted this interview in 1990, I did not pursue this point further, and never realized how Myklebust connected computer-aided design to having meaning in his life until I did a follow-up interview to prepare this paper. I had visited Myklebust's farm and knew that he was passionately committed to farming with horses, but never imagined that his work at home with horses was linked to his work in the lab with computers. After discussing the series of job changes and grants that led to his positions as faculty member at Virginia Tech and codirector of the budding ACSYNT Institute, we began talking about problems in America's new emphasis in the 1980s on manufacturing.

MYKLEBUST: For me the important thing is to improve the quality of things rather than to reduce their cost. I want things to get better. I think we should pay attention to improving things for our own benefit, not just for the sake of producing stuff to sell to people. Because we want to improve the situation. Because the price of quality can't be measured. It's orthogonal to cost [perpendicular to or at odds with cost]. When we think about the things we do, the things we eat, and how we live, then what's more important than cost is the quality, the quality of our food, the quality of our life. The cost to me doesn't mean beans.

Out of a group of twenty-seven new products, we probable need zero of them. I haven't had a television for thirteen years and many times I've considered getting rid of my telephone. It's not because I'm an oddball, it's because I don't find any good reason to have it except to call people on emergencies. If you take yourself away from pop culture for a period of time, for a year or two, all of a sudden you begin to realize that the things people think are important, the goods they buy, aren't so important. But it's hard to see that when you are plugged into it.

I try and let our students know that we don't need to manufacture stuff just for the sake of manufacturing to make a profit. Even if you take every company in the U.S. or in the world and make them socially responsible, that still isn't going to stop the impact on the environment from constantly taking resources and producing stuff that always generates wastes. Somehow you have to change the philosophy of what we mean by corporations and the production of goods to be able to stop this destruction of our environment. We are doing it at an ever-increasing rate. It is horrifying. And I guess people are not going to notice it until it starts showing up on the television.

The connection between computing and horse-drawn farming became clear to me for the first time: both promise benefits to society with minimal cost to the environment.

MYKLEBUST: I was going to say the same thing about computing. If computing does anything, it helps us to, gives us time to do things better, at a very small environmental cost. Or at least it used to be a small environmental cost. Now its getting to be an enormous one because there are ten PCs in every home, just like there are ten televisions. But in terms of electricity and impact on the environment, it probably has a much smaller impact than most other machines we use. The computer can amplify our abilities not to think but to evaluate possibilities for things by computing them much more rapidly. And it ought to be used to help us instead of to make things worse for us.

DOWNEY: That's my career goal, to inject a sense of global citizenship into the corporation.

MYKLEBUST: But you need to be able to act locally. [*Hesitates, points at tape recorder.*] Could you turn that thing off for a minute?

David Grose, 43, is the leader of a design research group at Boeing Commercial Airplane Company.

DOWNEY: Could you give me a little background on your career?

GROSE: I got my bachelor's degree at the University of Kansas in engineering. I was in that first draft lottery. It was the only thing in my life that I have ever won. I got drafted out of college and spent two years as a prison guard at Leavenworth. I had a chance, because of the duty hours, to take a couple courses each semester at Kansas; it was only an hour drive away. So when I got out of the army, I went back to graduate school. By that time, I just had a few courses to finish my master's and then I started on my thesis. While I was there I got a lot of good advice from professors that "if you want to go farther, do it now because it's so tough to do later on."

So I went in and I started on a Ph.D. program in control sciences. Then I shifted to a doctor of engineering program, which NASA partly sponsored, that was design oriented as opposed to basic research oriented. It was more of a systems, system design, system analysis, that type of thing—the bigger picture. So I got a fellowship grant and went to NASA Dryden and did my research in active controls of flutter suppression, which had an interdisciplinary flavor. So when I finished up there I went to Gates Lear Jet in Wichita, where I headed an engineering group in flutter and vibration. I was there about three and a half years, I guess.

DOWNEY: When did you finish your doctor of engineering degree?

GROSE: I finished at NASA in 1978, with the degree effective in 1979, and I stayed at Gates for a little less than four years. I had the responsibility for the flutter certification of the Lear 55.

Then I got the courage to start a small company with some other guys trying to apply aerospace technology to alternative energy. Kansas was primarily oriented at that time to energy storage and wind energy because the wind blows all the time. We built a large prototype of the system and tested it until it was damaged in a wind storm, which are also common in Kansas. I guess I was in that a little over two years. Unfortunately, the timing was bad in the sense that when we started there was a lot of interest in alternative energy and after two years Reagan became president and the interest in funding for those type of things went downhill rather fast.

For monetary reasons I went to Boeing at that point. The first job I had there was program manager on an air force program on artificial intelligence. What they were trying to do was apply AI techniques to automate aircraft manufacturing.

DOWNEY: They were doing that ten years ago?

GROSE: Really wasn't what I wanted. I went there to actually apply some of the control background in robotics, which was kind of an emerging field at the time. I was there three days and literally got shanghaied on a proposal effort [seeking a defense contract], and when it was over I was hoping they wouldn't win it but they did. Well, I did that for about a year. The technology was not anywhere close to the point that the expectations were. And I actually recommended to the Air Force that they terminate that program and focus on some of the areas that needed to be matured before they try it. They cut the funding but they kept the same objectives [*laughs*]. They started to focus in an area that I really had no interest in, integrated database management and things like that. So I got them on schedule, under budget, and we got a deputy program manager that had a background in that to take it over. I went and took advantage of an opportunity to lead the flight controls research group at Boeing Military in Wichita.

Spinning Off AmTech from NASA

The first step in building this R&D world of industry-university-government was to establish a new form of organization to function within it. Establishing a new form of organization is no easy matter, because the organization must also be theorized in terms of the categories of law. A world of R&D that did not fit within the world of law could not exist.

SHARIQ: The idea of AmTech as such didn't come out of any of the congressional activities or any mandates from the president or from higher-ups at NASA. The Space Act has been around since 1958, when NASA was created. At the time, we were under the fear of Sputnik and there was tremendous pressure to build technology and compete head-on with the Soviets. The law was and is one of the strongest pieces of legislation, providing unprecedented authority for the agency to deploy resources and do things that other civilian agencies could not do.

Strangely, the way NASA chose to do most of its business was different than any other agency. So NASA really does business an old-fashioned way, even though the Space Act allows the agency broad authority to be innovative and pursue alternative strategies. The common mode is to use grants and procurement mechanisms [purchases]. NASA does distinguish activity outside of grants and procurement as Space Act agreements.

During the past thirty-two years, the Space Act agreements have been used only when NASA entered into an agreement without committing any funds. Because they felt there was a clear boundary—that, if you used money, you used grants and procurement. But for doing novel things—

leasing a facility, using a wind tunnel, or letting McDonnell-Douglas or Boeing or 3M use microgravity time on the shuttle—you can do joint endeavor agreements. But in all of these governments, federal funds are never involved.

So when I took on the challenge of developing the joint venture program in the fall of 1987, it was just an idea. We needed the Space Act to fund joint ventures, for the Space Act had never been used to pursue such a program.

The notion of venture is normally used to describe investments that are risky but promise big payoffs. A joint venture shares the risks and benefits.

R&D in the AmTech World: Market Oriented, yet Nonprofit

SHARIQ: During the early years (1987–88) we built a body of knowledge on intellectual property law, procurement law, laws on policies and procedures governing transactions among companies and universities and government. And all that led to a unique and unprecedented set of policies, guidelines, procedures for joint ventures.

After two years of work, back in last fall [1989], we realized that there is just a tremendous potential. The country needs this kind of transaction mechanism for R&D joint ventures among government, industry, and university. And the idea we were bringing to it, which is the market-based transaction mechanism, became stronger and stronger. The more I talked with people, the more I got support for it: "Good idea. This country needs it."

So what you see here is a market mechanism for doing three-way joint ventures, sharing property rights and resources for common R&D goals but different end uses for the resulting technologies. Nothing is new, but collecting them all in one package is new.

DOWNEY: Wait. How does it work, if you're going to be a venture capitalist in a joint venture? A joint-venture capitalist, that's what you are?

SHARIQ: That's what we've become. Nonprofit, too. We realized nobody would know us at first, so what I did was I used my contacts in business, industry, and within NASA to pull together a joint R&D project. The first joint R&D project came into being because I knew a colleague at Ames. She was the one who really wanted to do it. I said, "Hey, why are you spending $200,000 a year on a grant? Why don't you multiply that through the joint venture process?" And so the whole thing got started. The second institute, the ACSYNT Institute, is pretty much the same way. I talked to a few selected people I know.

We said to these people, "We're not open for business right now, because we're still trying to work the bugs out the system.'' But we want to have some enterprising souls to work with. We can provide benefits for them, because AmTech offers a lot of free services: legal and financial research, deal-making negotiations, and so on. All of the contract negotiation and the entire deal package was put together by AmTech staff at no cost to Virginia Tech or other participants.

What we are doing now is marketing. I looked at the NASA budget for 1991 and made presentations to people in Washington at NASA headquarters whose money eventually winds up at the NASA research centers: Ames, Langley, Lewis, Goddard. We're not looking for lots of projects, three or four this year at the most. Each joint venture package is currently done in a customized way. We are still learning from these experiences.

People I talked to in the Council on Competitiveness, on the Quayle Council, or in the White House, they'd like to use this as a prototype.

DOWNEY: To keep it small, but make sure it's good, so you can show it as an example, a successful example?

This was the first time I had heard of the Council on Competitiveness, an advisory group legislated by Congress and with members from industry and universities appointed by the president. As I read its reports, I began to worry about building economic nationalism into AmTech, for the Quayle Council had already established a reputation for conceptualizing national issues in business terms. Its influence during the Bush administration became an issue during the 1992 presidential campaign. No formal connections between AmTech and the Council on Competitiveness were ever established.

SHARIQ: So the idea of establishing a nonprofit came because when I was talking to many, many people, it was very obvious that this organization must be a contractual party to the whole agreement. I found through a year's worth of legwork, many trips to Washington, that a group of trustees would help establish this organization only if it was a nonprofit. These people were very enthusiastic, and they would say, "Okay, if you are going to build a nonprofit, we'll be happy to serve as trustees.''

At the time I asked them I had known each of them for about two years, and had maybe five or six meetings and phone conversations. And I said, "Look, we're moving toward a problem here. We need to create a new organization so it can have its own destiny. Its destiny cannot be tied to NASA, nor can it be tied to the university or to industry. I asked them to work with me to create a fourth sector of the economy.

DOWNEY: Yeah, that's what's happening here. That's why I'm interested.

SHARIQ: It's not so different. There are lots of nonprofits. In fact, the independent sector in Washington, it's been there for a while. Probably you know, you were there. But what's not there is the market orientation.

DOWNEY: The market orientation, no. A nonprofit, market-oriented sector. [*Hesitatingly.*] A national concern that is realized through the maximization of private interests.

The implications of the concept are dramatic. Once the nation is reconceptualized in economic terms, AmTech and other similar organizations in the world of IUG actually become more representative of America than the federal government. While AmTech functions solely as a national, economic actor, nothing more, the government has many other dimensions: political, military, police, and so on.

SHARIQ: AmTech combines the public and private interests.

DOWNEY: You are trying to encourage private interests to pursue a public benefit, and you need to have a vehicle. [*Pondering.*] It has to be, this I hadn't thought of before, a nonprofit organization. If it were a profit-making organization, then there would be some distrust of its activities.

SHARIQ: Absolutely. I couldn't go into Washington. I couldn't go to the White House. My trustees wouldn't donate their time.

At the same time, however, I have difficulty attracting staff members to a nonprofit organization. Because there's no bonus pools, no personal wealth to be made. This is not venture capital. This is not shared ownership. This is a privately managed nonprofit, and there is no membership. It is very much of an anomaly. There's no precedent that I know of for it. I looked high and low at nonprofits and congressionally chartered corporations. I looked at the institutional arrangements that we have in our society, the corporation, the idea of agency, beginning from the late eighteenth century down to today. There is no place for something quite like this.

As we proceed down this path, I am trying to figure out What we are really trying to do and innovate. And what I'm following is an instinct, which says that there's a need for an organizational contract that can make the area of public-private R&D transaction efficient. And nothing makes it efficient unless it's market oriented.

DOWNEY: I'm fascinated by this ideology, by this institutional innovation whose justification is very much particular to our historical circumstances. If we didn't have the Japanese out there, this would not exist.

SHARIQ: Yes, I'd have no reason to do it. There would be no market for my idea, or for that matter, I would be told, "We don't need it. We are doing everything fine."

DOWNEY: Also, if the Soviet Union had been competing against us successfully, there wouldn't have been a need for you either, because we would have seen the battle as one between capitalism and communism. But Americans tend to perceive the Japanese as beating us at our own game: "If they're beating us at our own game, then we need to change our game."

SHARIQ: Or some way for us to do better.

DOWNEY: But there's still antitrust law. I mean there's still the whole American cultural tradition of market competition among separate property holders that you have to confront, as we saw in the ACSYNT Institute meeting earlier today. I mean, the industry guys didn't want to share their data with one another. One guy choked on naming the computer program his design engineers use. He couldn't say it because there are still proprietary issues. And that's why I was thinking about the problems that you must run into structuring legal contracts, because you've got a mess of cultural issues that have to be resolved in formal language.

SHARIQ: Absolutely.

Writing the New World into the Computer Program

The world of IUG also depended on integrating the ACSYNT code into industry practices. Remember that NASA engineers originally wrote it for research purposes only, and university researchers improved it. Semiannual meetings of the ACSYNT Institute brought the government and university people together with the industry people to review progress in rewriting the code, primarily through visual demonstrations on computer workstations, and to define objectives for future versions, or releases. Observers from other companies were also permitted to attend during the first day. At the third meeting, held at Ames Research Center, in the heart of Silicon Valley, Paul Gelhausen and Arvid Myklebust, codirectors of the institute, were hoping both to get industry members to commit themselves to use the code and to attract some new members.

GELHAUSEN [*opening the meeting*]: Let me first say, "Welcome." We're waiting for software to show up for our workstations so we can show the demos here this afternoon. We were supposed to have that all here yesterday and, you know, everything had to go bad, and that's where I was this morning. It will be here at noon, and I'll think, I'll skip lunch and we'll get it up and running for the afternoon demos.

The objectives of this meeting, my overall objective, is try to get industry participants to really come together and start giving us more feedback, I guess, on what we've got going. We spent the first year kinda like our grant years, where we just have been going off a little bit unguided, I guess. [When NASA gives a grant, as opposed to a contract, it is not permitted to direct the research.] I'd like to get more guidance in from the industry people. It's healthier if there's more industry participation, because it will help ACSYNT to become more useful to all of us. So one of our main objectives, I guess, is to get more dialogue going, I guess, with the industry folks.

This morning Arvid's going to talk about the technical accomplishments, and show off in viewgraphs and words what we have done with an overview of the upcoming release. Then this afternoon we'll get to see a demonstration of the upcoming release and some of the tools that are now available. Then we're going to have time for member input, which will include myself, and then we have a special guest, Boeing Commercial, I hope, has got a little bit to say, maybe, and then anything else from other members that are here.

Engineers at Boeing Military and Commercial working under David Grose had done a great deal of work improving the code in order to use it. Gelhausen was hoping that Grose would influence other members and prospective members to get involved and work together. However, Gelhausen raised it very tentatively because all the Boeing work on the code was proprietary. Would they share it?

GELHAUSEN: Tomorrow morning, at our meeting for members only, we will have a discussion of upcoming research and new ideas for research. I think we really need to spend some time together to generate some communication, to generate dialogue between the different members. So hopefully that will happen during the demos this afternoon, during the cocktail hour at dinner tonight, and then tomorrow afternoon. There's a lot of open space in this agenda that I think is mostly for kind of getting the input. With that, I'll introduce Arvid.

MYKLEBUST: All right, I guess the central concept here is to improve computer-aided conceptual design. There have been many objectives, but I'll just list three here that we've concentrated on for some time. First, we want to have a highly interactive graphics interface for conceptual design of aircraft. That should be obvious.

The significance of a highly interactive graphics interface for conceptual design is probably not obvious to most readers. Having a graphics interface means making it possible to generate visual images of proposed designs. Making this interface highly interactive means that engineers could make changes to the design and then view the results quickly and easily. The world of IUG thus offers industry a technical capability it did not have previously.

MYKLEBUST: The second objective is that we want to enhance the analysis and design capabilities of ACSYNT. So that the kinds of things we do on the screen, graphically and geometrically, can immediately be analyzed and, likewise, the kinds of things that come out of an analysis will be immediately represented as a geometric aircraft model without all the intermediate steps of putting in points, lines, curves, and surfaces, and all that activity that takes so long. Of course, the reason for that is, we want to be able to have very rapid design cycles, and to give the best possible design feedback we can have by doing that.

Representing shapes visually on the computer in most industries means drawing them one line at a time. The idea here is to bypass this step and have the computer draw the aircraft images automatically from data taken from the analysis process. NASA engineers originally wrote the ACSYNT code to do analysis, which involves figuring out combinations of characteristics that will produce an aircraft design that should perform properly with minimum weight. Industry members want additional forms of analysis because, unlike NASA, the companies also build airplanes. In addition, since ACSYNT was originally a research code used by a small group of people, its authors did not always follow rigorous standards for programming, much to the consternation of industry. Achieving the world of IUG means that every member must be able to see itself in the code.

MYKLEBUST: The third item, right here, is that you'd like device-independent computer graphics. As a result of that, we now have one version of code that runs on all our workstations, the only differences being a MAKE file at the beginning of it. The code is all the same.

The concept of device-independent computer graphics is a crucial one here. Industry engineers use many different kinds of computers. Writing a program that is device independent means that it will work on everybody's computers. This technical strategy helps construct IUG as a collaborative world by rendering insignificant one of the differences that separates the companies.

After giving his introduction, Myklebust launched into a half-hour lecture about how his university group is using the latest concepts in computer-aided design. These concepts included representing complex curves and surfaces, intersections between surfaces, various forms of color, shading, and lighting, and complex aircraft shapes. He also described changes that students had made so that the code might be easier for industry people to use. Except for Myklebust's colleagues and students, the audience of forty people knew very little about computer-aided design but loved being able to see visual images of aircraft designs. He then turned his attention to the half-dozen prospective members present.

MYKLEBUST: All right, for those of you that are new, let me review a few words about the ACSYNT Institute. The intent, of course, is collaboration between NASA, industry, and the academic world in improving aircraft conceptual design. The participants share the technology developed by the institute and the source code is distributed under license to the participants. That's a rare occurrence these days, and it's a bit of an experiment here to see how well this works out. We're very strongly committed to supporting this. Not only supporting what we're doing but supporting the source code itself. We need feedback, to know how we can help you, help the companies to be able to benefit by this.

This last paragraph reveals much about the technical strategy for producing a world of R&D that permits members to be both inside and outside at the same time. First, participants receive everything they need to modify the code for their own purposes because they get source code. Normally when one buys software, such as the WordPerfect I am using now, one can only interact with the code, not rewrite it. Giving the source code to all members brings everyone into the writing process, enabling them to write into the codes they possess both the collaborative world of the ACSYNT Institute and the local worlds of their companies.

Second, by providing maintenance support and assistance for the ACSYNT software, university faculty and graduate students blur the boundary around their activities and status as researchers. In exchange for accepting a new type of task, one that is normally an industry activity, they receive substantial financial support for their research. But, as Myklebust pointed out later in an interview, this arrangement is not without its costs, for AmTech takes ten percent off the top of everything.

MYKLEBUST: By forming this institute, we do get a great deal better leverage, a better return on your investment, than you would yourself in the companies. If you doubt that, you should talk to David Grose at Boeing Commercial Airplane now. He can talk to you about the advantages of doing it this way. We get support from the Commonwealth of Virginia, and we have much lower research and development rates at the university than you do in companies. All three of us working together can make significant strides at a much faster rate than companies can, I believe.

At present, it's open to U.S. businesses, universities, and government agencies only. We've had a lot of inquiries from people outside, but that's currently the policy.

Rolls Royce, for example, sought membership for some time but was denied on the grounds that it is based in Great Britain. I never heard any discussion about how to define the nationality of a multinational corporation.

Why Industry Engineers Participate

The industry engineers participating in ACSYNT are trying to enhance the influence of conceptual design, the earliest stage in the design process. Conceptual designers generally already have one foot inside the company and one foot outside it, in the sense that they are nonconformists, the "cowboys" lost in the world of aircraft design. Conceptual design groups are small, in sharp contrast with the more powerful groups in the subsequent stages of preliminary design and detailed design, where the groups number in the hundreds or even the thousands. Using an enhanced ACSYNT program that synthesizes considerations from different disciplines while also producing visual images may help conceptual designers better justify their choices of proposed designs to company managers. But integrating ACSYNT into aircraft companies also requires blurring some established internal boundaries.

GROSE: They asked me to head up a concept development group in advance product development and the biggest fundamental problem we had was what they wanted to do Boeing did not have tools to do it with in a timely manner. I was familiar with ACSYNT from when I was with NASA from the early seventies. We started to get into it and that's when we started to see how scrambled the code was. You know it was originally never really intended to be anything but an analysis tool at NASA and so there was really no configuration control or those type of things involved. The programming was representative of the era of fifteen years ago. So we bit the bullet and fixed it to the point that we could maintain it and work from there. I guess we started on it right about the first part of 1988.

I think it was around September of 1989 that Boeing made the decision to reduce the military part of the company. They had just finished about a ten-year, $2-billion expansion of the headquarters for military R&D in Wichita, facilities and everything, laboratories, the whole bit. It is not being utilized in the way it was intended. You hate to see that kind of an effort and then just kind of walk away from it, but the realities, I guess, for Boeing were that wasn't the place to invest their money—in the military side. They had about a fifty-fifty balance five or six years ago between military and commercial, and now it's ninety-three percent commercial, seven percent military.

DOWNEY: Could you give me a quick sense of the organizational structure that you're in?

GROSE: It's configuration development and engineering analysis for airplane design, and the conceptual design work is scattered all over the place. Each airplane project really does things totally different from every other project. When you have people spread out, you know, they are unfamiliar with how it gets done in different projects. So Boeing Commercial decided to create Configuration Development and Engineering Analysis as a group in its own right in the engineering technology staff.

DOWNEY: How big is the configuration development and engineering analysis group?

GROSE: It's got about 150 people in it.

DOWNEY: How big is the whole engineering technology staff?

GROSE: Oh my. Let's see, what's the population of Seattle? I would guess probably four to five thousand are involved in R&D work. Aerodynamics, propulsion, avionics [electronics and flight systems], and structures are the largest of the specialist R&D groups.

Now, they started a core group, which is supposed to focus on not what's going on now but what we anticipate in the future. We're trying to find a way of not only doing a better job of that initial definition of the design, which all the disciplines now work with and analyze, but also trying to find a way of integrating all those disciplines to where the work that's done is consistent.

A few minutes earlier, the head of a conceptual design group at Lockheed, Dalton Sherwood, had walked into the room and was listening to the interview. He jumped in to explain the major barrier that conceptual design faces in seeking a higher profile and greater authority in the design process.

SHERWOOD: This requires a culture change because each of these technology specialist groups are all used to keeping control of their numbers very closely, saying, You tell me what you want, I'll work on it and give you the numbers back. They did not want you to know how they did their study and what their assumptions were and they couldn't care less what the impact was on the overall study.

GROSE: I would say that very thing right there is the thing that hampers trying to do what we're doing more than anything. Because one thing computer-aided design or analysis has done is that it's integrating the data. The trouble is, some disciplines want to avoid that because they feel they're losing control of their own destiny. For example, in Wichita, the propulsion group was really opposed to what we were doing with ACSYNT, and yet in Seattle they're one of the biggest supporters of it.

Some want to get their software activity integrated into this thing, some don't; they are not going to give you their software. You know, you just tell them what you want and they'll give you the data type of thing, and that's not the way to design.

SHERWOOD: I've had guys curse me because I was taking their job.

GROSE: In reality, he was making it better, they just didn't realize it.

Getting the Code into the Company and the Company into the Code

GROSE [*in a presentation at the institute meeting*]: The work that was being done at Boeing Military was trying to look at restructuring ACSYNT so it could be integrated into a larger setting in our system. The primary issue there was to identify explicitly what was in it [*general laughter, since ACSYNT contained many mysteries*], and provide some kind of rigor and structure so we'll be able to add to the software in more of an automatic sense.

We have basically gone through the program line by line [*laughter again, because the original code had 60,000 lines*]. What we tried to is, in every case where there's equations, through conversations with Paul Gelhausen, we found out the reference and got a copy. Cases where an equation showed up, that it was not in a referenceable document, we then derived it to ensure its correctness. And we're embedding all these comments in the code so that people can look at a section of code and see the reference material. We'll have a set of reference materials that then they can go look at if they want more details on the issues there.

Again, we're trying to deal with the fact that, when engineers get involved in a design project, they may have a pet way of calculating a particular parameter, and they'll say, Well, how's this doing it? And it's not an issue that you're trying to promote the one over any other, but you can simply now readily show them what is in there.

An important feature of the ACSYNT code is that it is divided into large building blocks of subprograms, called modules, that are particular to each discipline. Each module is roughly independent of other modules and can be used or ignored in particular projects. The very structure of the code thus makes it possible for industry members to work within the world of IUG while remaining within the world of the company. That is, individual members can write, maintain, and use proprietary modules while the institute produces modules that all members share collectively, writing their collaboration into the code. Organizing all of this, however, will not be easy.

GROSE: Now the key to the whole thing is that we have to decompose the analysis modules to where you have fundamental building blocks of code and then pull the appropriate one in at the appropriate time. I think with the realm of problems we can envision with lots of people in the company trying to work with this, what we were trying to do is add as much flexibility as possible to it at the conceptual design stage.

And this will help us to try and integrate this into other software systems downstream in the design process, to start to pick up the information in them and couple better with their models. So this is the direction we're trying to go now, so we don't find ourselves spending all our time trying to get the program to work the way we want to get the answers, but we have the opportunity to do the fun part and that's actually to work the design problem.

Other conceptual design codes frequently do not distinguish so clearly among the disciplines. Without the modular design in the ACSYNT code, then, the ACSYNT Institute and its collaborative world of IUG could probably not have been possible.

GELHAUSEN: Well, I think that's one of the real key features of ACSYNT now, its modular design. Nobody else has done that.

DOWNEY: How do others do it?

GELHAUSEN: I think it's like one guy that sits down and kind of puts things together, pulling together different types of methods.

DOWNEY: Puts the modules together?

GELHAUSEN: No, there's no code that's as modular as ACSYNT overall, from what I've seen; my experience is in helicopter codes. You know, the guy is going along and he's thinking about aerodynamics and then he sizes his engine in the middle of his aerodynamics for some reason, you know, it's not as well organized.

DOWNEY: It's not subdivided by discipline?

GELHAUSEN: Those codes tend to be very inflexible or if they're flexible it's because you've got to have the guy who wrote the code with you to be able to modify it. It think my generation is always saddled with somebody else's code.

DOWNEY: Well, I watched the guy from Boeing, David Grose, paint kind of a nasty picture of the structure of the code and its documentations.

GELHAUSEN: Well, it's true. I mean, ACSYNT's documentation is pretty thin, I think because it's research code, but it's easier to justify ACSYNT than probably any of these other codes, except that the other codes were developed in-house in the companies. You've got, you know, people who say, "I know that the results of this code predict this and I trust that because we've done this before and it's been right." ACSYNT, on the other hand, was never in Boeing before so they're trying to justify it.

How Blurring Boundaries Transforms Government and University

By participating in the ACSYNT Institute, NASA Ames engineers gain the opportunity to direct university research, which they cannot do in a grant relationship, while university researchers accept the burden of delivering products.

DOWNEY: At what time did NASA Ames people say, hey, it would be nice to have a graphical interface. How did that come into the picture?

GELHAUSEN: We had always been playing around with graphics and doing graphical things and trying to make pictures of what our airplanes looked like. We've got some interesting pictures of airplanes with the tail flying alongside the airplane because something the optimization code did or some other bit of code that somebody put in, that kind of thing. So we, we'd been working on doing the graphics stuff.

At one time ACSYNT actually did have graphics in it. Sometime after 1975 they lost the punchcards that the graphics routines were on.

DOWNEY: Really?

GELHAUSEN: Yeah. So that was kind of depressing. I kept seeing all these great three-dimensional pictures of airplanes that came from ACSYNT but weren't there anymore.

Gelhausen described the initial encounter between his boss, Sam Wilson, and Arvid Myklebust, that led to a grant to write the graphical interface for ACSYNT. This technical work, including an initial collaboration between NASA and Virginia Tech, took place before AmTech and the ACSYNT Institute were founded.

GELHAUSEN: I was skeptical at first because I didn't know, but I'm generally skeptical because no one knows ACSYNT as well as I do. Or me or George, or Gregory down the hall. But anyway, I think I was really pleased with the professionalism of Arvid's students in putting together that first version of the graphics code.

We need now to get more interaction with Virginia Tech. Mark, my assistant, and I need to get in the middle of the graphics development somehow so that we can make the aerodynamics be fully interactive.

Linking graphics and analysis is a key point for the world of IUG. The graphics portion of the code is located in a separate module, thereby embodying the distinction between NASA and Virginia Tech. Gelhausen was fantasizing about blurring that distinction in the code, which would somehow involve blurring the distinction organizationally as well, through "more interaction."

DOWNEY: Boy, I see! See, this is what I was talking about. If you look at the actual development of the code over the past several years, you can see the various groups in it. In the newest versions, for example, there's Gelhausen in there, for your interests are making their way into the code.

GELHAUSEN: Yeah, because the earlier work was all grant.

DOWNEY: All grant?

GELHAUSEN: See, a grant is where I'm not supposed to interact too heavily. I went back to Virginia Tech and gave talks about what I'd like to see and where I'd like to go, but it wasn't until the ACSYNT Institute came along that I could go back and say this is what we have got to do.

In contrast with Gelhausen's technical vision of more interaction, Syed Shariq understands problems and formulates solutions through a legal discourse.

SHARIQ: We learned some things that we didn't know before we wrote the contract between these three parties. For instance, that technology rights issues come up when a federal employee spends time at a commercial company. What happens is a joint invention can have two authors, one a federal employee, the other a company. What happens to those rights? How do you negotiate *a priori* the disposition of those rights? There are certain federal laws covering federal inventions. But what about a joint invention? We are coming into some things that are fascinating, where we can make some contributions.

One thing that's happening is that there's more management on the projects we do, because joint ventures are managed differently than grants. In government, the idea of grants means you give the money and hopefully something comes out of it. You may not get anything.

DOWNEY: You cannot ask for deliverables.

SHARIQ: In a joint venture it's not the same thing. It is an obligation on the part of industry to join hands with government to comanage the project. And the university has to take direction and be managed somewhat more than is customary.

DOWNEY: Which can raise a lot of tension within a university. I think Arvid Myklebust regularly has to justify how his work on the ACSYNT code and in the institute make legitimate contributions to engineering research. There always seems to be a tension there. Of course, it's a tension that most of his colleagues face in their own way by taking industry money. So it's not an unusual thing. This tension has been around in a big way at least since World War II.

SHARIQ: The industry investors in the ACSYNT Institute do get to manage the research somewhat. Also, students get exposed to real company people. They will probably have better jobs/contacts because of the institute. Plus, Virginia Tech gets rights to royalties that come out of commercializing this software. So, for the university, the benefit from accepting some outside management is there. In addition to normal teaching, research, publications, promotion, and tenure, it can expect to place its students quickly and can look forward to royalty revenues down the road.

Conclusion: Success in an Economic Idiom

My fears about the world of industry-university-government reinforcing national chauvinism stem from conceptualizing entirely in an economic idiom. But if the organization were defined otherwise, would it ever get off the ground?

DOWNEY: Will you be able to solve all these problems?

SHARIQ: See, we have taken a very basic approach. We really will not solve the ultimate question of ideology. We cannot solve even the many questions of law and policy. I think if I was standing today in front of Congress, I'm sure we would be shut down by somebody or other, for some reason or other. So we really need to take a more pragmatic attitude: Let's just hide and do it under the legal banner of the national authority that exists already. Let's get it to a point where it's working well enough. Let people come and kick tires and look at it. And then they can decide whether they want to do it or not.

But rather than trying to publicize it prematurely, or "before its time," we're just doing it, in a real world, with real projects. On the other hand, it's an opportunity. In the many studies that have been put together on U.S. competitiveness, written by bright, competent people at MIT or Stanford Research Institute or wherever, the big question is always implementation. Who's implementing these great ideas? What's being done? Not enough ever gets into implementation. In 1980, legislation came along that gave universities and nonprofits the rights to retain federal technology developed under federal grants. Then the Technology Transfer Act came in 1986, which gave some impetus for government to transfer technology to the commercial sector. But the mandate for implementation and specifically through the market orientation, the incentives that could really bring these parties together, are not there.

DOWNEY: So how do you define success?

SHARIQ: As yet there is no clear indication of how you measure success in technology transfer. Where's the benefit? How do you quantify it? Meanwhile, in our system of doing things, we have located success in the sharing of resources. We have built indicators of success into the design of the joint venture: the commitment of resources by industry in return for technology is a fundamental requirement for successful commercialization, assuming that the joint project produces good technologies.

Consider how a project looks at the beginning. From the government's point of view, if a project is doubled in scope the industry commitment ensures that twice the resources have been invested. So just by having additional parties in the deal means that the accomplishments will be much bigger than any one party could do by itself. So there's efficiency built in here, in that sense that, why duplicate the same research separately in a corporate lab and a federal lab? Why waste the R&D resources of the country? It's actually a triplication here, because the university is involved, too. So this way R&D resources can be conserved and dedicated for something else that is important. This in a sense is a justification for joint R&D and at the same time a built-in or designed-in implementation of the philosophy that ensures success at the start. That is where we measure it, not at the end.

The confidence in this measurement comes from the fact that the companies are putting their own earned income on these joint projects. We don't allow them to put in federal money they might receive from work on government contracts. They cannot subcontract this, either. None of the government money can flow right back into these joint projects. This is one of the essential requirements. So, as long as industry is putting their own hard-earned money on the table, they clearly would not do these joint projects unless they see opportunities for profits. Each joint project, therefore, becomes successful by definition, through a commitment to a transaction mechanism that is based minimally on the needs of parties to fulfill their specific self-interests—thereby creating a market-based collaboration between government, industry, and university.

I am not satisfied with this account of success for the world of industry-university-government. I am not happy simply if the R&D projects are well-supported and the participating companies see the opportunity for profit. I understand that economic returns constitute a minimum condition of existence for these organizations. But if success is limited to an economic idiom, then the ACSYNT Institute becomes nothing more than a convenient mechanism for pursuing the sectarian interests of industry, university, and government simultaneously. The underlying sense of citizenship that also motivates participants and the boundary-blurring vision they build into both their organization and their code become short-term rationalizations for the longer-term maximization of self-interest.

Instead, why not let it get really big? Why not allow membership from other countries and make the world of industry-university-government a place where conceptual designers work together to improve the tools for reducing costs, reducing fuel use, and reducing emissions into the environment? The step to a global world of industry-university-government is not that much harder to conceive than the step already taken.

SHARIQ [*responding to the author after reading a previous draft and unhappy with using the above passage to conclude*]: You are right, that's where we need to get to, eventually. However, given the enormity of this challenge, it really comes down to the fact that we don't know how to make giant strides. All we know is how to take modest steps: a journey of a thousand miles must begin with a first step. Socioeconomic innovations and the creation of market mechanisms are complex problems, as the struggles of countries in the former Soviet Union and Soviet bloc demonstrate. Reimagining R&D in America's future is a problem of the same complexity. Nevertheless, while we think of better ways to resolve these complex problems, we need to proceed with and encourage responsible innovation and experimentation to foster solutions through experience. AmTech is one small effort in that direction.

Acknowledgements

Portions of this research were conducted with support from NSF Grant #8721941. Thanks to Joseph Dumit, George Marcus, Paul Rabinow, and other contributors to this volume for their helpful critical comments. I appreciate David Schabes's excellent copyediting, which significantly improved substantive content in addition to style and readability. Thanks also to Paul Gelhausen, David Grose, Arvid Myklebust, Syed Shariq, Dalton Sherwood, and other participants in the ACSYNT Institute and American Technology Initiative, Inc., for generously sharing with me their time, insights, and critical reflections.

ARMS AND THE SCIENTIST

These contributions are, of course, concerned with the classic question of the "conscience" of science in the nuclear age, but present a sense of it as the cold war tensions that shaped this age have eased. Hugh Gusterson focuses on the ambivalences of his own commitments in working with nuclear weapons scientists and on those of a particular subject in his research. Diana Hill's conversations with her father also reflect these ambivalences, but from another location in the community of such scientists and specifically in the context of the post–cold war moment. Gusterson's discussion of Hill's contribution and Kim Laughlin's commentary on Gusterson's piece, included in the extracts of the collective editorial meeting later in this volume, constitute a particularly cogent framework for a reconsideration of Gusterson's and Hill's presentations.

8

Trust but Verify: Science and Policy Negotiating Nuclear Testing Treaties—Interviews with Roger Eugene Hill

The Lake Trip

10 September 1988.

Diana: You were about to talk about a personal experience.

Roger: Oh, yes. One day, after our escorts had warmed up to us, they arranged a fishing trip so we could go catch carp with them at a lake at the test site. This lake, it turns out, had been created by a nuclear underground explosion back in the early sixties. When I was in Paris working on Project Plowshare, I gave a talk before UNESCO on "Management de Territoire dans L'Ans 2000." More or less the management of landscape on a large scale in the next century. They asked me to speak about what nuclear explosives could do in this realm. One of the things I showed was an International Atomic Energy Agency film, made by the Soviets, which showed the creation of this water reservoir. It was a flood control project. They created a dam and a reservoir. They did it with a 100 and some odd kiloton explosion all in one fell swoop. *Bommmm,* up goes all this dirt, suddenly there was this huge crater.

Diana: Where did all the earth go?

Roger: It went around the crater and made the rim of the dam. They did this next to the river and then diverted the river into it using bulldozers.

Diana: What about the radiation?

Roger: The film showed these men swimming in the water. One of the men was a man I knew. He was a very famous health physicist who had given talks about the radiological consequences of using nuclear power for peaceful purposes. So there he was in this film, Yuri Israel, swimming in this lake, and I thought, this guy is really putting his money where his mouth is. They claimed this was a safe thing to do and show this guy jumping off this boat and swimming on this water. Well, the lake they took us to on the test site was that lake. And I went swimming in there!

DIANA: You didn't eat the fish did you?

ROGER: It was not a problem.

DIANA: Did you take your radiation badge?

ROGER: I always have to have it with me and some of the field guys brought some equipment. Let me tell you, I wasn't sure about what we would find there. When I knew where we were going after wondering all these years and now twenty-three years later—when I got there I went off along the rim of the crater and there were reeds and birds and it was teeming with fish. There was absolutely no sign of environmental devastation. I went away from all the people because I wanted to sense this. When I lived on the mountain I tried to sense nature, tried to feel what she was telling me. It sounds strange coming from a scientist, but anyway, I had the sense that it was a vital living place: the mud stunk of algae and life. You know in a marshy environment you really sense life, it smells, there is decay, it was just full of that.

DIANA: Cute little frogs with three eyes?

ROGER: No, not that I noticed. Now, I didn't eat the fish. I wouldn't eat carp anyway, in any water. But a lot of guys went fishing and had a great time. I went swimming, others walked all the way around it. It was an interesting experience to end up there. I never in my wildest imagination would have thought when I gave this talk at UNESCO that seventeen years later I would be there swimming in it.

DIANA: I don't know how different types of radiation escape, but wouldn't a lot of it end up in the atmosphere?

ROGER: Actually, it is a fairly safe kind of situation because what they use for peaceful nuclear explosions is a fusion hydrogen bomb with a very low-yield fission trigger. So what you do is bury it deep enough that the cavity produced by the detonation itself, the fireball where everything is melted, is buried under a layer of rock that is several times the size of the cavity. The rock melts and these fission products which result from the fission trigger get captured in this molten rock which solidifies into what is like a glass, locking them in. You wouldn't want aquifers to flow through these things; it's not like these elements are perfectly trapped. There are gases which can work their way to the surface but with a low fission yield like that, those gases are gone in a few years. The tritium, which is the major by-product in the fusion reaction, gets dissipated in the atmosphere. Tritium has a biological half-life in the human body of about twelve hours. It's not like one of those dangerous radioactive nuclides that stays in your body forever, gets locked up in your bones or organs and emits particles incessantly. After twenty-three years it was obvious that this lake showed no signs of damage. I felt some vindication as somebody who had been involved in studying that kind of thing—not necessarily an advocate but, well, I was advocating it in a way, by claiming that these kind of things could be done safely, under the correct conditions if you used enough intelligence, and that it was possible to use this technology in a beneficial way. It was a sort of closure to find myself in a new era at that site I had talked about and wondered about for all those years.

DIANA: So there you were, the Soviet escorts and the U.S. team just having a nice day swimming and fishing!

ROGER: Oh, yeah. I have some pictures in my office of us all standing on the rim of the lake in Kazakhstan with the New Mexico Land of Enchantment flag, all these guys in shorts and 700 pounds of blubber between us but there it was.

In late January 1993 Roger Hill and I sat down in his living room for three days of interviewing and visiting. My father is a nuclear (particle) physicist who has worked in academics doing basic research, in industry, and is now working in arms control at the national laboratory at Los Alamos. On the first day we discussed his present involvement and responsibilities as the leader of the CORRTEX On-Site Nuclear Testing Treaty Verification group at Los Alamos. His area of responsibility in the group, as well as being its manager, is the design of data acquisition systems and data analysis. We talked about how his group and the laboratory is trying to envision its future and struggle to keep its budget alive during this transitional phase for arms control and the national laboratories. We discussed the possibilities for joint projects with international and Russian teams, and between government and industry.

On the second day we talked about the history of the Limitation of Underground Nuclear Weapon Tests (limiting underground tests to 150 kilotons, which was supposed to limit the development of new weapons) and Underground Nuclear Explosions for Peaceful Purposes Treaty and Protocols. We discussed all of the parties involved and the various relationships between them, as well as the process of meetings, talks, negotiations, visits and tests. Since Ronald Reagan agreed to move forward with the treaties which had been signed by Richard Nixon and Leonid Brezhnev in 1974 and Gerald Ford and Brezhnev in 1976 but never ratified, the negotiation process has included three Nuclear Experts Talks, familiarization visits to each other's test sites, the Joint Verification Experiments (JVE) at which "devices were detonated," one in the U.S., the other in the Soviet Union,[2] the Protocol Negotiations and the actual verification visits. The Soviets have verified two American tests and were present when a third was canceled. The American team is now in Moscow discussing plans to verify tests in Russia if and when the current testing moratorium is lifted.

On the third day we talked about his biographical background. My father's career as a physicist, his youth, and our subsequent nomadic family life around the U.S. and Europe is revealing of the social climate which nurtured a scientific imagination and how in turn that scientific imagination shaped his interpretation of the world. His social, religious, and physical displacements at once facilitated his move to a scientific world view but in turn further heightened these displacements at great personal cost. Global politics (through the media), family, and spirituality, as well as science serve as major reference points throughout the conversation. After a prolonged period of isolation and struggle to complete the most sophisticated computer simulation program for breeder reactor safety analysis then available, he quit his job in France and returned to the U.S. His involvement in arms control and his attempts to create for himself and his group "a balanced working environment where the process is as valued as the product," has been a conscious effort to redress some of the personal and social alienation that has run through his life and career.

Growing up, my father's work was as mysterious to me as the names of the companies were hard to pronounce. There were rooms full of colored wire, note books full of ticktacktoe games, and stacks of green and white paper filled with zeros and ones which he would spend hours pouring over. He could explain to me why the sky was blue and other mysteries of the universe but his work also brought up uncomfortable subjects like nuclear energy and bombs. Later the ambiguous nature of his work became more apparent. The purity of the science was matched by the complexity of its implications and my father always lamented the corruption—rational, moral, and social—that was involved in any practical implementation. Over the years, to my father's dismay, I insisted in trying to drag the family secrets out of the closet as well as spar with him over such subjects as the relativity of truth. I turned to cultural anthropology as I attempted to uncover the secrets of the social and cultural world. I went about celebrating the "corruption," "the human factor," that had interrupted my father's dreams. The corruption was, I suggested, everything cultural and everything which undermined the privilege of science.

These interviews might not have been possible a few years ago, but my father and I have never talked about his work and mine as much as we do now. One reason that we may be talking more about it now is that anthropology, and especially fieldwork, allows me to be interested in his work and be able to ask questions in a way which I could not previously. Previously, I had tried to talk science with my father only to be frustrated and confused and talking to him about the social consequences of scientific issues alienated us further, but my acting as an anthropologist to understand how scientists think and work in the world opened new arenas of discourse for my father and me.

Another important reason that we were able to do these interviews may be that my father has taken a concept he has found very exciting in his arms control work with the Soviet scientists and is now applying it to other relationships in his life. The concept is one that was introduced to replace the idea of "confidence building" that joint verification work implies. It was replaced with what the military thought to be a more politically neutral term but which the scientists adopted quite happily to express the nature of their scientific efforts: "increasing transparency" (albeit while protecting classified secrets). Over the days of our visit and interviews, my father thanked the editor of this volume several times for the opportunity for us to visit and for him to tell me about his life and work.[3] We are beginning to redress the silences and absences of our relationship.

The three days of interviews produced many more pages of transcript and areas of interest than can be presented here. I have concentrated here on some themes that were repeated throughout. I am relying on cut and paste editing to frame the themes against each other with the hope of evoking the particular relationships between them.

Scientific Culture

DIANA: If you could create your fantasy job, what would it be?

ROGER: Well, I've been so involved in doing what I've been given the responsibility to do that I haven't let myself develop a nice fantasy. If I really just let myself go, I guess it is what I need to do, at five o'clock I'll find out if they are cutting out our budget! My future has been pretty tied to the group but if individually I could do anything I wanted, I would probably enjoy staying at Los Alamos working on problems of the superconducting supercollider. Frontier research has always been my first love. It is exciting to deal with the frontier of knowledge, to deal with issues fundamental to our understanding of space and time. Once you experience the excitement of discovery it is hard to give it up.

In 1983 I began to work on a neutrino experiment at the Meson Physics Facility at Los Alamos as part of a team from the University of New Mexico. I was looking for a more permanent job. About the only jobs then available were in Star Wars-type work. They don't like you to call it that, of course. I had technical objections to that whole concept. I just thought it was a big con job and I didn't want to have any part of it.

Well, there is the universality of science that you recognize when you are dealing with common problems. It was wonderful to see—the notion that there is universality in science is commonplace—but it's really enlightening, uplifting, to see it in action. As part of the JVE we, the Soviet and

American teams, had to solve problems together. We also tried to do things independently. We would take a problem and go off to our own institutions and work on the issues and then come back together and bring them to bear and we found we had done almost exactly the same. We had used similar tools and there is no mystery to this: physics is physics and science is science. There is logic, there are the laws of nature we live under, and they have a certain consequence for ordering human affairs and activities. So here we were and we find, my God, they did things exactly the way we did. Or when we would solve a problem we would find we had the same responses. I'm talking clichés but with the universality of that we began to understand that we are colleagues, we are the same. We also recognized our differences, but they seemed less important.

We are continuously being bombarded by cosmic radiation—tens of thousands of particles are interacting with the cells in our bodies every minute. "The truth will set you free." It is a biblical saying, but it is so true and explains why science is important to human affairs. We can not organize human affairs on superstition or fear. The truth is the only way we can do it. We have to do everything we can to build the truth about our situation. Given that, we have every chance of surviving and prevailing. Without the truth, we are just going to be lost.

A scientist without his integrity has nothing; that is where his authority comes from. His integrity has to be based on a commitment to scientific truth and not some hidden agenda.

I've always had the impression that physics is a meritocracy: if you work hard and you know what you are talking about, it doesn't matter what race, sex, or nationality you are. Graduate students are the cannon fodder of research and the ones who do the lion's share of the work, but, at least it seems to me, speaking from my own experience, as long as they are good researchers they won't encounter problems of bias.

We have a certain culture. First of all, we are scientists and engineers, technicians. We have a particular culture, a particular way of doing things. It was sometimes awkward to work with the military On-Site Inspection Agency (OSIA) when they wanted us to be military. In fact, the problem with that lieutenant colonel, the marine who came to the Soviet Union with us, was that was his attitude: that we were temporaries. In fact, he called it temporary duty: we were TDY to the military. He expected us to have orders, to live in military barracks, to perform as military would. Now, we were willing to do that up to a point but there are certain aspects of that culture which are antithetical to the scientific. We need people who can think independently, who can challenge. The scientific mind is basically that of an open-minded skeptic, and that doesn't make for a very good foot soldier who just does what he is told without questioning. So there is a cultural problem and it was even called that, a cultural problem: military culture, scientific culture.

Military Culture

DIANA: So how do you train the new military people? What, as a team, do you decide they need to know?

ROGER: Well, I think our biggest challenge is to try to develop an atmosphere of trust. These, well, we're going to get back to this event in 1988 because it is the key which determined a lot of these things. It is called the Joint Verification Experiment. At that time, from the United States' point of view, it was a purely civilian operation run by the Department of Energy. One of the aftermaths of the experiment was the complaint on the part of some of the more conservative elements in the government, whom I would call cold warriors, primarily from the Pentagon, that the civilian scientists "gave away the store." That has never been defined and is, of course, not true. We were characterized as being too friendly with the enemy, too cooperative, and they wanted the word to be spread that verification was an adversarial activity. But we were involved in a *Joint* Verification Experiment and the rules of that game were based upon cooperation: we solved problems together. We and the Soviets had technical problems, and we solved them *together,* we solved them together in the spirit of "Let's get the job done. We are here to demonstrate the adequacy of this kind of verification system and to identify problems." Both sides had identified problems involved in using that kind of system and we wanted to resolve those problems and we *did.* As a result of solving those problems we got a treaty. Some of the people in the government didn't want the treaty. They don't want arms control, they don't want any of these things to really happen. And so one of the outcomes of that, these are all my personal opinions, was that the military was put in charge. They approached us with what I would have to call the prejudice that they were dealing with people that don't understand the difference between policy and science, that don't know the limits that anyone in this situation would face with regard to stepping on or beyond the bounds of technical issues into policy issues. They felt they were the only people able to identify that boundary and to have the discipline to observe it and so forth.

DIANA: But they are not scientists.

ROGER: No, they are not scientists.

DIANA: They said you couldn't do it because you weren't policy but they could do it without any scientific expertise.

ROGER: Exactly.

ROGER: The technical issues are very complicated. We have to explain to the military what the central issues are, including, why we have to protect certain treaty rights, what the security issues are.

DIANA: *You* have to tell *them* what the security issues are?

ROGER: Yes, we had to explain all those things to them. They don't usually understand how you cheat on a treaty like this, especially when they were rotating their officers through there every six months or so and most had no scientific background.

DIANA: So, it's funny that now after that the military steps in, the scientists have to go back and actually explain to them what it is and how it works.

After the JVE, the On-Site Inspection Agency was given managerial responsibility for this and other arms control treaties. OSIA represented the U.S. at the daily meetings between the Soviet and U.S. teams on the test sites and at the coordinating group meetings in the capitals. At these meetings important technical issues of intrusiveness were worked out. Since Roger's responsibility is data acquisition, the issue of intrusiveness in the field was also. He therefore was in a pivotal role and was the point of contact and negotiation between the groups. The military had its own advisors that I imagine were listening very carefully to what Roger and the Soviet counterparts were working out.

DIANA: Well, it sounds like through the treaty negotiations, the JVE, and all the work you had done together, you had developed a good rapport with the Soviet scientists. What happened when the military got into the picture? Did that muddy the water?

ROGER: That was very awkward. As I said, one of the things I was concerned about was telling the military, "You have to understand that the people that we will be sitting across the table from are our colleagues. Some of us even call them friends." They didn't like hearing that. We were to call ourselves inspectors and not acknowledge where we came from. There were strict rules on contact and familiarization. In describing those meetings I once asked them, "What are you going to do if one of us walks up and gives one of them a hug, how are you going to handle that?"

DIANA: What was their response?

ROGER: Awkward. They didn't let that really happen. I mean, it did happen, as humans we recognized each other, we smiled, winked, waved. But we had to do our best given the formality of how the military wanted the contact to happen: all through the colonel. There were some social occasions and we found time to talk and have contact but the bureaucrats did have a problem with it. They insisted that it be an adversarial process. Now, as scientists we had problems with that because the logic of it is simple: this is a type of verification which, by it's very nature, requires cooperation. To try to characterize it as adversarial is foolish and dogmatic. It is not facing reality.

DIANA: Was the idea of reciprocity and getting good data something the scientists had to push for?

ROGER: I think we, who have actually had to go over there, have had to keep the issue of reciprocity high on the agenda because it is in our interest that that is never forgotten. So wherever there is an arena where these things are being negotiated or discussed, I have always tried to ensure that there is someone there representing the issue of reciprocity.

DIANA: It seems like an issue, when it is represented as getting good data, that is a concern from academics and science spilling over into the policy arena.

ROGER: I can't deny it. That is there. For some of my colleagues it is very much there. Some who aren't in politics at all, for them it's just an experiment. "These guys are here to get good data. We know what that is and by golly they are going to get it." I'm sure there are people in our government who don't want to hear that kind of thing. The message we kept getting was, this is not a scientific experiment, this is verification, remember that.

DIANA: How were you supposed to remember that? What does that translate into?

The Policy Community

DIANA: Who spells out the issues?

ROGER: An issue like spoofing is clear. It was fairly clear from the beginning what the issues are: cost, intrusiveness.

DIANA: But those are Washington's issues. If this was "pure" science those wouldn't be the issues, so they are defining what the issues are for how you are looking at this problem. They set up the parameters, you fill in the blanks and feed it back to them.

ROGER: Well, explain it to them. You make your recommendations but they go through their own process. You don't think for them if that is what you are implying.

DIANA: No, I'm just trying to figure out what the process is, what you supply each other with.

ROGER: Ah. The kinds of things the policy people want from the technical people are options and a quantitative understanding of probabilities. If you say there is a possibility of something happening, they would like to know the extent of the probability and how confident you are in it. That is the kind of thing policy people need.

DIANA: So in your writing you have that in mind.

ROGER: Yes, you understand what they need and try to give it to them. Don't try to hide it. If you don't know them, you have a problem.

DIANA: You have to understand what they need—was something they explicitly told you or you implicitly understood in the process of dealing with them?

ROGER: It wasn't explicit but what they are there for is to serve as the body that makes a decision or makes a recommendation to the president. Any decision requires an assessment of consequences so you have to clearly spell out what are the consequences of the options available.

DIANA: That must have been very different for you from previous work where that was not a big consideration—basic research and publications.

ROGER: That is true. I found myself doing a lot of things I thought a lawyer would probably do; I found myself thinking very much like I would expect a lawyer to think—carefully wording things but I think that is from having been involved in the negotiations. In the final stages of the negotiations the last people to look at these things before they go out are lawyers, for good reason—that is their business. So, I felt I had to appreciate that.

DIANA: That must have been an interesting reinterpretation of the scientific process. Coming from "pure" science, how did this feel?

ROGER: I was aware that I was involved in a marriage between policy and science and that the child that that marriage produces is sometimes an ugly one. That is the nature of the world we live in and I do not work in pure science anymore.

DIANA: Did you feel they, the bureaucratic policy side, was as accommodating in realizing that they had to understand some of the science, or that it wasn't pure policy?

ROGER: There was a range of responses in that area. In the beginning I felt an openness and receptivity. My first contact was fairly positive. I felt like they wanted my opinion and they valued it so I was careful in what I said to them. A problem developed, because of the political agendas even among the scientists, the policy people were being given technical advice which was in conflict. One lab saying one thing and the other lab saying another. That was a very bad situation.

DIANA: So policy is saying, "Oh, look, science is like everything else."

ROGER: And they weren't too surprised, but the technical community lost credibility. A scientist's only asset is his integrity if he is going to maintain that credibility. The moment a scientist lets himself be guided by political concerns and is not functioning objectively—I'm not saying there isn't controversy among scientists, there always will be, some legitimate and some not. In the negotiations that went on in Geneva there eventually became a real breakdown between the policy and the technical community.

Negotiations and Translations

DIANA: We have pretty much covered the questions I had for today about the specific ways and the minute particulars of how these things work. I'm sure you've noticed I keep coming back to this idea of the different languages used by science, policy, the military, in your relations with the Soviet scientists and Russian itself and the translations that go into that process. Those translations seem to require a great deal of negotiation, mediation and even interpreters. I've been asking a lot of questions to try to understand on the absolute minute basis of daily interaction examples of this, like, when you are at the negotiating table how you negotiate over something that is to you a technical fact that has to stand as it is because it is an integral part of some system or equation but, to policy, may be something they do not want included because, "it's a secret," or they might say, "You can't say that because that isn't clear" which gets you into a discussion about language.

ROGER: Well, those layers of language often came up in the interaction among the whole delegation. Typically, how this would happen, as I recall, was that we would have meetings every morning, staff meetings to discuss what was going to take place at the next session with the Soviets, to discuss guidance we had received during the night from Washington, and to develop a consensus between the policy people about the position the ambassador would take. He had to be sensitive to input by the policy people as well as technical people. It was usually done in a very classified meeting so they could discuss freely what the negotiations would be like and not worry about security. These things were hammered out and this is often where the technical people would be exposed to policy concerns. The technical people would be asked what the technical issues were and what we could do. We would describe it and they might say that they didn't want it subject to negotiation. Washington would say if we couldn't discuss one subject or another in more detail. The ambassadors did have some latitude but there were times they were under pretty strict guidance from Washington. And these people at the meetings were also in touch with their own agencies back home, each lobbying for the agenda of their agency within the interagency group, so it was kind of an awkward thing for the ambassador that he was getting guidance, supposedly from outside, back in Washington, but some of that guidance reflected differences of opinion that he was trying to solve right in his own delegation.

DIANA: And all of that rubbing up against technical, scientific objectives as well.

These first sections show the tensions between the objectives and workings of the scientific, military, and policy communities. The next sections contrast these antagonisms to the working relations developed with the Soviet scientists around a common goal, task, language, and human contact.

Confidence

ROGER: It became clear in these interactions with the Soviets that they had a technology similar to CORRTEX—they were in the process of developing a technology almost identical to ours. The concept is that if you can measure the velocity of the shock wave underground then you can get a measure of the energy that is driving the shock. So, to have the position of the shock as a function of time you look at it at various times. What the Soviets had always done is put contact switches down on wires underground, when the shock wave passed it closed the switches so there would be a voltage pulse created and they would measure the time of each pulse. The contact switches are located at discrete points in space. Our technology is to use a cable and to measure almost continuously the length of the cable as it is being crushed or shortened by the shock wave. CORRTEX stands for the Continuous Reflectometry for Radius versus Time Experiment. It is almost continuous, every twenty millionths of a second or so. Our issue is to locate the measurement in space. They had theirs located in space and their problem was to locate it correctly in time. They were kind of complimentary technologies. What happened after our first discussions with them was that they developed a continuous, exact compliment to CORRTEX called MIZ. They were able to prepare a version of MIZ which was fielded on the American test during the JVE.

DIANA: What is the nitty gritty of how we establish that confidence? How do we do that?

ROGER: We haven't yet. That is established over time. I think it is a process—

DIANA: There have already been some building blocks. So far we have talked about the confidence that came about just from being in the same place at the same time with these human beings and working on a common project. The project could have been knitting a sweater, it could have been, well, any kind of directive, "We have to do this thing together," it could be anything.

ROGER: Absolutely correct.

DIANA: Just that accomplishment of a common goal seems to be—

ROGER: I think you are right. The person-to-person exchanges are the little bridges that build the confidence that will lead to these larger objectives. But the JVE was particularly important because it was dealing with this "priesthood" associated with nuclear energy, it was dealing with the power structure. These are the guys who are wielding the levers of the most important aspects of superpower status. On the Soviet side, their nuclear capability is the essence of their superpower status, and their space program to a certain degree, but by definition today, well, let's say two or three years ago, a superpower was someone who could annihilate the world. They had that power and those people who had their hands on the levers of the technology that could be used to annihilate the world were the people involved in this particular exercise in building confidence and that was the unique characteristic of this activity. I think it had a real impact on history.

DIANA: I think you are right and that having the common scientific language really helps get around some sticky cultural and translation problems. That you did trust that language, that both sides trust that language, I think that helps you out of a big quagmire that other kinds of diplomatic missions have fallen in.

ROGER: Yeah, the solutions to some of these problems were very technical and we recognized the validity of the solutions and developed respect for each other's technical capabilities and commitment to the resolution of these problems. I think after the last verification exercise in the United States, the only one that has really taken place so far, the Russian scientists, in my opinion, went home with a conviction that verification works: that we are going to abide by not only the agreements but the spirit of the agreements. Which says that we want you to walk away from this with good data and the understanding that we are not cheating on the treaty and we want you to have everything you need to be convinced that we are not nor have any intention of cheating.

Transparency

ROGER: One of the most important aspects of arms control, in my opinion and of the colleagues that worked with me on the Joint Verification Experiment, was the concept that arms control has a confidence-building component. These old cold warrior types don't recognize that, they don't want to even hear that. There is a new buzz word that has come along recently to replace *confidence building,* it is now *transparency.* I like that word; it's a very good description of the outcome of the activities we have been involved in with the Russian and Soviet scientists. It has created transparency. The iron curtain created for many people a barrier. While it was a symbol of a position it was also a barrier to understanding. We had an awful lot of misconceptions on both sides about our motivations, our technologies, our capabilities, our intentions. All those kinds of things began to clarify with really outstanding consequences. I really feel that increase in transparency with this particular group of people—these people were the influential policy makers in the nuclear community—that increase in transparency, I think, had a big impact on the events of 1989. That's just my opinion, but the fact is, the people we were dealing with, the guy I sat across from at the table at our daily meetings arguing back and forth about these various issues that came up in the JVE, he went on to become the cabinet minister in charge of all the nuclear projects both military and civilian and this is a man for whom transparency was greatly increased.

There were some tough moments in the daily meetings at the test site, because their team chief, the guy who went on to be minister, he was a very strong individual with a very determined attitude about why he was there and what he wanted to accomplish and had very strong ideas about how things should be done and they wanted to push things a little bit to see how we would react to certain situations. They brought some equipment, for example, which was indicative of the state of the art of their technology and impressed a lot of our folks. It was an oscilloscope which was at that time one of the fastest in the world. They were way ahead of us. Now some people in our country deny that that is the truth but I've asked other experts and they said that the scope did in fact represent something better than what was available to the U.S. at that time. The Soviets were proud of it, and they brought it with them. However, it was potentially a very intrusive piece of equipment. It had the potential for causing some trouble: it was capable of recording the type of information we did not want recorded. So I had some very interesting interactions with this man regarding the use of this scope questioning whether it could be used at all, and, if it was, how would we control it and make it benign and what kind of procedures would we adopt to screen the accumulated data before giving it to them. So we had to work out things like that and it got a little tricky. At the same time we had a team in Russia getting ready for the test over there. Half of our experts were already there. Every thing I did over here had direct consequences for them over there so it was important.

On a personal level we began to understand each other's motivations of patriotism and love of family. I'm still moved when I think of some of the revelations that happened to me on the JVE. Walking into their dining hall on the day of the test, they took us into this banquet, a celebration of what we had done and it was the first time I had been in their dining hall (we were confined to a particular hotel and we had all our meals in that hotel while they were across the street). This is a military base in the middle of the Soviet nuclear test site at Semipalatinsk, and there is only one piece of artwork on the walls of this dining room and it's a huge photograph of a playground, a child on a slide, a mother pushing a baby in a pram. The symbolism of that was really powerful to me because it was obviously one of their central values, and, when I visited the city nearby, one of the things they wanted me to see was the playground and how happy their children were, how much importance they put on their families. That affects all of us. Working in this area of arms control to make the world a safer place for your children, it comes home to you, peace is what we all want. They thought of us as warmongers, they admitted that. When they first met us they thought of us as, in fact, they used the word, *devils*. By the time we finished, I can't give this man's name, because it could give him problems with security, he said, "One of my jobs is to target U.S. cities with nuclear weapons and after this there is a human face on some of those targets. Los Alamos—I will see your faces. I don't want to do it anymore.'' So, I feel that we really did make a contribution. We have been vilified by the military. Like I said, they characterize us as giving away the store. We have had absolutely no recognition by the government for what we did. I did get one letter from one of the DOE officers and that was it. Oh, and the ambassador to the talks sent me a very nice letter recognizing the personal sacrifice involved for a lot of us. And there were some personal sacrifices—a couple of divorces—those were very hard times, demanding times. I think someday the history of that will be written and transparency will be recognized as a contribution and not as a detriment. On the Russian side they recognize that. All those guys that we worked with have been promoted, some to very high positions. As I said, the team leader, whom I worked with, is now a cabinet minister. All the second team that we dealt with replaced the first, and we have been put lower in the Russian eyes. They are very hierarchical. When I went to Moscow, the second team was now sitting across the table and I had to go through this lieutenant colonel in the marine corps who didn't understand at all the background or history of the team he was now technically in charge of. It was a very awkward situation personally.

DIANA: So, you think that science was a major factor at getting at transparency?

ROGER: Oh, I didn't finish that thought. I think the universality of science contributed to the transparency.

DIANA: It sounds like that experience helped you to understand them as people, that that translated into some kind of human relationships.

ROGER: Yeah, commonality is the beginning of understanding. When you have something in common it's a place to start. It wasn't hard to understand the differences either when they would talk about them. This man who is now a cabinet minister would talk about his life, he was born the same year I was, 1936, and he grew up in the midst of World War II, what they call the

Great Patriotic War. I was in Los Angeles, there were air raids and there were balloons and reports of submarine shellings on the beaches and things like that, but he really lived it. One of the other guys we worked with was out killing Nazis in the forest when he was ten years old. These are people that went through hell, and when you start to understand, as a colleague, what that would to do to anyone and then you understand the terror of Stalinism and what they went through and—you really began to develop compassion for them as people. Some of the people survived through the Stalinist system and came out with a position of prestige and authority so they clearly participated at some level. But I don't think they did so in the worst of the Stalinist times, these people were still pretty young.

DIANA: I think it was interesting the use of science for this idea of transparency, science as the idea of a common theme. On one hand there are these stereotypes on each side that we are warmongers, and they, as you said, ideologues, and both sides have employed science and technology against each other. It comes back around on the other hand, here, that science is the arena for cooperation, for this new found transparency. It strikes me also that while you talk of transparency, it is trust and cooperation that underlie this whole endeavor. It was a value judgment made, a human decision that you were going to trust each other, at least enough to negotiate the treaty, that has led to this and had nothing to do with science.

ROGER: The use of the word *trust,* hmmmm. The theme of the JVE and they passed it out on souvenirs, I can show you some of the paraphernalia, the theme was "Trust but Verify." It was a Russian phrase that Reagan had repeated in front of Gorbachev. Trust but Verify—so I don't know if *trust* is the right word. I think transparency is a good description of what developed.

DIANA: Verifying gets you back to science rather than trust? I'm just coming back to this thing about science because there have been other kinds of groups that have tried to make diplomatic missions to deal and cooperate with the Soviets. It's interesting that the fact that you had this language of science allowed you to get a lot further a lot faster. I was also wondering about you having to protect from the intrusiveness of it and certain kinds of information but it seems that working on the same things, and both knowing how they work, those secrets are kind of a myth. There is a logic to it so they could figure it out, too. What can you hide from them that they couldn't apply their same scientific imagination and logic to figure out for themselves?

ROGER: Exactly. It's absolutely true. In fact, in some of first contacts we had to display our concern about these aspects of verification and what we would do.

DIANA: So were you winking at each other?

ROGER: Well, it's a law. We are required to protect restricted data. To divulge it knowingly is subject to criminal prosecution. It's a crime, so we are bound by our position. In fact, when we accepted our security clearance we accepted the equivalent of an oath to protect this information. So, it's a law but in our first contact with them on this issue they (a Soviet general) said, "Oh, it will be between this and this." And of course they know that. Other people have tried to point out that the type of information we are trying to protect we used to not protect when we were doing atmospheric testing. They could have gotten all the information they wanted on those matters from airplanes observing U.S. atmospheric tests. So a lot of the scientific community has challenged the validity of classifying that information.

DIANA: I would think that would make for another level of transparency, another bond between you that through science a lot of this stuff is going to be obvious. The secretness of it seems kind of arbitrary.

ROGER: In some respects I think you're right. The particular number I'm thinking about is controversial. Like some of our people have said, it is restricted and we have and will keep it from them.

DIANA: The people you have actually worked with over there, do you think they care about that particular number?

ROGER: We have to play the game. We have to assume that if they had the opportunity to find that number out that they would. We, as scientists, fortunately we have the support of the Department of Energy on this, we have said that we will not engage in espionage as part of verification. We can not do our job if there is any suspicion that we are over there for any other purpose. When we are over there it is to measure the yield of a nuclear explosion, not to find any secrets or observe their military system.

DIANA: You don't have to be debriefed?

ROGER: Well, they do it any way. We said we would talk about anything we saw in the course of our work, but don't ask us to do anything covert because we won't. That is against the law. I'm not going to send people over to Russia to do things they could be put in prison for. They are there on a scientific mission.

Humanizing

ROGER: I want to comment on something you said earlier that I didn't really answer but I thought you might be interested. You said that when we returned from the JVE in the Soviet Union that I reported feeling moved by the experiences and I was recalling what some of those feelings were like. If you put it in context, just imagine that I was a kid born just at the beginning of World War II. When the first atomic bomb was dropped I was nine years old. During the early fifties and my school years, the fear of nuclear war with the Soviet Union was a very real thing. We practiced drills to fallout shelters, all that kind of thing was part of my experience as a kid. So the threat of Armageddon, the threat of Soviet nuclear power was very real to us. The Cuban missile crisis was something that I went through at twenty-five, it was very real to me. When I found myself sitting in a Soviet trailer next to the monitoring station where the firing party was about to detonate a Soviet nuclear device only four kilometers away from where I was standing—as a matter of fact, it was a rare experience for an American to be so close to an explosion. Our practice at our test site is that we are twenty or thirty miles away from these things, and here we were not two and a half miles away. So there was that but I was conscious of the fact that it was a very historic occasion. That here I was, an American, so far from home, in the middle of Kazakhstan, so far from everything I knew and found familiar, standing with all these Russian people about to observe the explosion of a Soviet nuclear device and become part of the process to make that threat to my children and the world less real. It was such a tangible sign of it, that we were there together. Sitting in the control room that morning I was listening to Eliza Gilkynson on my walkman and began to write a letter to her about my feelings on hearing her music at such a time and I noticed my colleagues were engaged in similar things. We were all writing our feelings down and reflecting—trying to use that moment before the explosion. We were perfectly ready, we were just waiting for the moment and all of us were writing our feelings and what we were sensing at that time.

DIANA: Was that on both sides?

ROGER: I think the fact that we were there was something, yes, to them it was a very significant fact. They were especially busy, but their press was there. I was interviewed by their press and shown on national television that night. I don't have a copy of that, but I'd like to. Someone in the government has it. I'm wearing a badge with my name in Cyrillic. The TV crew came into where we were and asked what we were doing, we told them and they translated.

DIANA: When they were here there was none of that.

ROGER: None, not by the media. There was a movie made about the JVE that has not been declassified. It was narrated by Charlton Heston. A great effort was made to show the real tone and tenor of the interaction between the Soviets and Americans at the Nevada test site. They filmed us playing softball. It was really fun, we had this game of softball going and they put their hearts in it, believe you me. We played volleyball with them all the time and it showed all that. It showed our classrooms where the Soviets taught us about their equipment. There were a lot of interchanges and it is all recorded on the film. The film will probably never see the light of day.

DIANA: Why was it classified?

ROGER: I think all that would be needed to declassify it would be to get permission from the Soviets to do that, but I doubt that they've ever been asked. I think the JVE was not something that was found in favor by the people who were—as I said, this is my personal opinion—not terribly interested in building bridges between our two cultures, who are not interested in confidence building, and transparency is not something that they really wanted. The events of the day brought it about but they weren't very happy about it.

Familiarization Visit Prior to JVE

ROGER: It was a very interesting experience. They came to Nevada. It was the first time a Soviet had ever been on our test site. They brought with them the interpreter who was Gorbachev's interpreter, so you had the feeling this had a pretty high Soviet blessing. There were several Soviet generals in it and some deputy ministers. It was a high-level visit.

DIANA: But they weren't scientific people.

ROGER: They had their scientific people with them as well. Simonenko and Voloshin were there, who had been dealing with Eilers and the other scientists at the earlier talks. We presented an agenda of issues to them. My role was sort of a foreshadowing of my role through the history of the project. I was going to deal with the issue of intrusiveness right up front. I was going to present to them the concerns we had about the ability to detect sensitive information at the same time they were doing measurements of the yield. So I was going to describe as best I could without stepping outside of classified limits, what our feeling was about the intrusiveness of this technology with regard to picking up sensitive information. I think I really sparked their interest because it is a subject, well, talking about U.S. secrets and recognizing that we had secrets, just admitting that we had them, was delicate but obvious. What those secrets were, I had to imply that, and not say it very clearly, but they knew immediately, of course. This is where I told you earlier, one of the generals said, "It will be between this and this." I couldn't respond but they made it very clear they understood and had the same concerns. But it was interesting to bring up that kind of issue. I had to be very careful about classification and it was my first contact with the Soviet nuclear testing people, so it was a delicate presentation. So that began the interaction between the two sides on the issue of intrusiveness. My presentation was the first.

ROGER: The visit went on for a number of days. They were given tours. We went and showed them our equipment and I went with them. We showed them the typical ground-zero situation, what the site of a nuclear test looked like. We took them out to a site and they saw one of the canisters in which we put our bombs. The man who became the Minister of Atomic Energy was photographed inside one of these things. They had a great time.

DIANA: Really? What was the mood?

ROGER: During the briefings very professional, very technically oriented.

DIANA: Did they have questions?

ROGER: Lots of good questions and they wanted us to know that they knew what they were talking about. I think one of the things they were most concerned about was this sense of equality. When we knew them better later and saw what their society was like, I could understand this somewhat defensive position that they might have been coming from. I had no reason to expect them to be in that position. What I am saying is that they come from essentially a backward economy and they come from a third-world country with nuclear weapons and a space program. We didn't know that. I didn't realize how much of a third-world country, in a sense, some parts of the Soviet Union were in the sense of the struggle for local economy.[4] But this establishment that these people represented was thoroughly professional and these were first rate people and they just wanted that recognized. They wanted us to understand that we were professional equals and that they were not in any way inferior to us technologically. I think that was part of their mind-set. Some of them seemed a little bit arrogant to me, but considering where they were coming from, I understood it, later especially. The mood was excited, having "the enemy" on our test site, standing inside our bomb canisters and looking at everything.

ROGER: After the familiarization visits the Soviets were going to be there in June for the JVE. We began to make the plans for the experiment. It was a busy, exciting summer. Having them there was really an exciting experience.

DIANA: So the next thing that happened was their arrival.

ROGER: They began to arrive but I had stayed on in Geneva with the protocols negotiations. We finished the JVE agreement in May, it was signed on the thirty-first but I stayed till the middle of June. The object of these talks was to produce new protocols allowing for this kind of verification to take place. In other words, the treaty itself limits the size of underground nuclear explosions to 150 kilotons. The protocols are, in a sense, an annex to the treaty which outline the rights and responsibilities of each side in reference to verification of compliance with that treaty. So we began to negotiate the protocols to the treaty. I had to get home because I was going to be involved in the JVE. In fact, by the time I got home the Soviets were already there. The first teams had done the geophysical logging of the satellite hole. I got there about the time their hydrodynamic team arrived. Some of key people from the negotiations were there and a lot of new people I hadn't met before. It was a very interesting experience. I met their younger people. I had no idea what their computer literacy was like. Some of the first people I met worked in my specialty, if you will, data analysis and the intrusiveness issue, which meant I was going to be concerned with the data they obtained; so I met almost immediately with their data analysts. They turned out to be young computer hackers wearing T-shirts that said Nike or something and, like the interpreters, were really nice young men who were very typical of their generation. Maybe it was a studied thing, I don't know, but they certainly didn't feel foreign to me. Some did, but others didn't. There was one in particular who reminded me of graduate students I had had. They were people I really liked in the sense that they were very hard working, very interested in what they were doing and tried to inject into the process some human warmth. So I met these young hackers and they spoke English.

DIANA: At least computerese.

ROGER: Yes, and they spoke really good English. They really wanted to show me what they knew. Not in a bragging way, they were just proud of what they knew and proud of their country. They were all patriots. They did things like program their computers to draw an American flag in color, and they brought me in to show me that. And I was able to talk to them a little bit about their personal ambitions, their schooling. Two of them wanted to set up a company. They had a very entrepreneurial attitude.

DIANA: Sounds like you worked by yourself and with these young guys.

ROGER: Yes, in my particular area of responsibility for the experiment, which was taking care of their data, making sure they got their data, I was just there to work with them. I also had the daily meetings to deal with and they covered all aspects of the experiment. Everything came up in these meetings.

DIANA: When they were here where did they stay?

ROGER: They took over a dormitory. Each of the laboratories has a dormitory at the test site. They took over part of Sandia's dormitory.

DIANA: Did you eat together?

ROGER: Not really. They had a room at the Steak House, which is the restaurant available to the people at the test site. They had a room of their own, however; they could, if they wished, invite us in. It was an attempt to let them have some privacy so they could be among themselves and not always surrounded by other people but they could invite you in. And they did. There were some joint dinners. Some of them preferred not to have this larger dinner and they could go through the cafeteria. Sometimes I found myself in the cafeteria with one or two of them talking. That was a pretty relaxed atmosphere. During the actual verifications as it developed later those kinds of things would not be allowed. This was under the DOE and there had not been restrictions on contact established. Later, when the military took over, there were. There were people there even then who were concerned about the interactions and observing them—I can't be more specific than that.

DIANA: The tourists [the name the scientists gave the intelligence people who accompanied the American team] were there.

ROGER: Yes, they were there looking. The interactions were being examined but there were no specific regulations saying I couldn't sit down with one of the guys and have a meal in the cafeteria. Now, with actual verifications, and host protocol was in the hands of the military, Soviets have to be "accompanied" at all times so this guy going through the line on his own would not happen.

DIANA: It is so strange that they would draw the line after the JVE was done, which included a real test, to suddenly impose these kinds of regulations.

ROGER: The key word was that this was a *joint* experiment; we shared data. Subsequent verification was not to be a joint effort.

DIANA: By the end had you established some nice social relationships?

ROGER: Yeah, after spending a couple of months working with these guys on a joint effort, solving these practical problems as they came up, you develop respect for each other's capabilities. The man who was the team leader, for example, I developed a lot of respect for his leadership and technical ability. I developed a sense for all of the people I worked with—the interpreters, the young hackers, these postdocs, were a pleasure to work with. I liked them a lot. I would have given them a job in a minute, if I could have.

I have wonderful pictures of some of our people visiting a market and the children just flocking around, "Amerikanski, Amerkanski." This fellow Mike O'Conner, in our group, was truly an ambassador of good will. He is a technician, nonpolitical, nothing to do with any of this stuff. He just really liked the kids and they loved him. The photographs are fairly touching, and when I give a talk about the JVE, I usually wind up my talk with a picture of him with all of these children around him and say something like, "There is another side of this activity which should not be forgotten."

Cultural Differences: Soviet and American

ROGER: Now, it turned out to be quite different working with them there—cultural differences. They were back on their own test site under very different constraints. I was able to work, to visit their area, for example, when they were in Nevada. We had set up an area that was, strictly speaking, theirs, but we could come in at their request which we did. When we were in Kazakhstan, they were kept some thirty or forty miles from where we were. We had no direct contact. I couldn't call them on the phone but I could see them in meetings. They had a different set of constraints over there and I didn't have so much contact with them working there. But we had already worked out all the precedents so it was okay. And I had a couple of weeks to put it all together. We had the daily meetings again, and again intrusiveness was the stickiest issue. Typically what happens when you have a difficult issue like this, the resolution of that issue gets put off and put off and put off until, in the end, it is almost a crisis. It happened in Nevada and it happened in Kazakhstan in almost the same way. The last day or so is just a panic, suddenly paranoia flares—What are they doing? Are they really out to hurt us? You have to be very careful and very clear and firm cause you have to get a job done and you couldn't let their paranoia stop what you were there to do.

DIANA: Was that the rhetoric that was used to get through the crisis?

ROGER: No, it got to points like, "If you don't do it we'll leave." It got to points where you had to be tough. It was hard. What you are dealing with usually is paranoia of some kind or another. It flares up under circumstances like this. It was very revealing to me to see what the culture over there was like in the sense that the people I ran into over there were those that had not been in Geneva or Nevada. These were people at their test site who had been there for a long time. I felt like I was dealing with people that had lived in an atmosphere of abuse and fear. They had a hard time looking you in the eye. They never smiled. There was just a feeling of—I found out later that that test site was run by Baria, Stalin's hatchet man. He ruled by terror and if you did something wrong in the old days they'd probably just hang you or shoot you. So some of that atmosphere lingered and people would make remarks like, "This place is a lot closer to Siberia than Moscow." They would say things like that to us. There was a sense that if they screwed up there was serious consequences for them. So this question of paranoia—you know it's paranoia—it's not excusable, but nevertheless, it had an intensity about it that we didn't have over here.

DIANA: You noted some cultural differences when you arrived there, people not smiling or looking at you, which could just be standard behavior toward strangers. Were there other things?

ROGER: Well, that behavior might also have been because of their stereotypes of us as annihilating warmongers, because what I noticed most was the change. There were some people there who were assigned to escort us. They really warmed up to us. They gave us gifts, the shirt off their backs, practically. They wanted us to be aware of their good will toward us. Some of the people who escorted us around the test site were very solicitous, they took us to—an experience that was to me so interesting, on one of our days off, the escorts took us fishing at a very special lake on the Test Site.

The cooperation and interactions of Soviet and American nuclear scientists is either a historic occasion with far-reaching implications or can be read as yet another self-serving maneuver on the part of the scientific-military complex to dictate the terms of international relations. Dependent on our ability to verify each other's destructive force, we engage in an absurd measurement of peace—below 150 kilotons. Trust but verify, which was the motto of the JVE, the scientists seemed to have "naturally" assumed it to mean verify the truth. This is something which is well within their parameters of reasonable behavior. What of *trust?* A paper presented by the Soviet team of experts had this to say: "The main thing that the negotiations provided was the openness of professional discussions of many scientific problems and the mutual understanding of the complexity of the scientific and technical aspects of monitoring in regards to limiting nuclear weapons tests" (Novaya Zemlya 1991: 1). The trust which was required in order to verify effectively was worked out by the scientists through, on the one hand, the common language of scientific systems designed to cut out the ability to measure intrusive material and equally on the other hand, by discovering their common humanness through their joint efforts in a common task at a historic juncture.

What becomes clear through our discussion is the extent to which policy, and especially the military, are excluded from this. Though they are present for the daily on-site meetings and the negotiations leading to them, they don't have the universal language of science to assist them, and the human contact eludes them in the logic of war as anything other than a series of intelligence-counterintelligence maneuvers.

In the very complex web of relations between these scientific teams and their government, society and each other, the contact between the Soviet and U.S. scientific teams may have "increased the transparency" which may turn out to have valuable implications for the relations between the nations. However, before "the universality of science" is heralded as the savior of humanity, the question has to be asked, Was it not these same institutions which created the nuclear threat, the cold war? Or was it? Fifty years later, the workings of Washington, the ties between the military and the national labs, the structure of the labs and the creation of such things as the interagency group, has created processes of/for interactions between these institutions which have heavily mediated possible alliances and made them more volubly and visibly negotiated. Could it be that the tenure of those relationships may have a greater impact for the future of arms control rather than other outside circumstances or the ability of science to transcend national borders? Then again, the man who is now the minister of Russian atomic energy (responsible for nuclear power, testing, and weapons) keeps the photographs and mementos of the Joint Verification Experiment on his desk; Soviet and American scientists sitting around the negotiating table, at a ceremony exchanging commemorative plaques, standing smiling in each other's bomb canisters, swimming in the lake.

Notes

1. This fact was challenged by one of the anthropologists who suggested that tritium had a half-life of 12 years. Roger responded that tritium has a radiation half-life of 12.3 years in the atmosphere but a biological half-life of 12 hours in the body.

2. Because most of the activity we discuss happened prior to the breaking up of the Soviet Union and because not all of the scientists are Russian, I use the term Soviet and Soviet Union.

3. In deference to my mother and my sisters, I refrain from reviewing only my father's version of his life, which includes the family's history.

4. In about November 1992 the team learned that while the Russian labs were still functioning, one of the element physicists told his American visitor that he and his family would make it through the winter because their backyard potato crop had done well. This was about the time there were also reports of missile engineers being stopped on their way to North Korea where they had been offered work with salaries about forty times what they were making.

Reference

Novaya Zemlya: Ecological Safety of Underground Nuclear Tests. Working group of Soviet experts at the Soviet-Finnish meeting of experts, 28 February 1991, and at the International Symposium on Underground Nuclear Testing, Ottawa, 21–26 April 1991. United States Department of Energy, Nevada Field Office, Las Vegas, Nevada.

Hugh Gusterson

9

Becoming a Weapons Scientist

This paper is about the charged intersection between my own life and that of Sylvia,[1] a key subject in my fieldwork, at a moment when we were both undergoing important processes of transition and apprenticeship in our professional lives. At the time I was carrying out an ethnographic study of the Lawrence Livermore National Laboratory, a federal laboratory near San Francisco which has, for the last forty years, devoted about two-thirds of its resources to nuclear weapons design. It was Livermore scientists who designed, for example, the neutron bomb and the warheads for the MX and ground-launched cruise missiles.

Sylvia, who is my age, had earned a Ph.D. in physics from MIT. Although she had not yet designed a nuclear device when I first met her in January 1989, she was in the process of mastering her trade and joining the tiny and mysterious elite within the Lawrence Livermore Laboratory by becoming a nuclear warhead designer. I had, in the mid 1980s, been an antinuclear activist with the Nuclear Weapons Freeze Campaign in San Francisco, an organization dedicated to the termination of the Livermore Laboratory's principal mission. By the time I met Sylvia I had completed coursework in graduate school, ceased my work on behalf of the nuclear freeze, and was attempting to master my new trade as an anthropologist. Making a highly unconventional choice of fieldwork site, I was attempting to earn my own Ph.D. by studying Sylvia and her colleagues rather than the peasants, nomads, and slum-dwellers who are more customary targets of the anthropologist's exoticizing gaze. Although I did not know it at the time, my writing about Sylvia and her colleagues would eventually lead me, in another professional transition, away from Livermore to MIT, the university where Sylvia was trained as a physicist, and where I now teach the anthropology of science and train students to study scientists like Sylvia.

Thus ran our lives, then, both parallel and opposed. At a moment when anthropologists are questioning the grounds of cultural knowledge and the established canons of ethnographic writing, I would like to speak here about Sylvia's world in conventional ethnographic terms but also, more reflexively, as an entity that has been refracted in my writing through the prism of symmetries and contrasts between our two lives. My account, then, will seek to shed light on Sylvia's world, but in doing so it will also accent the ways in which my knowledge of Sylvia and her world is what Donna Haraway (1988) has called situated knowledge—knowledge which is, without being untrue, never finally true; knowledge which is partial; knowledge, the shape of which is determined in part by the particular questions, angles of insight, and blind spots that I brought to my research as someone who is, in spite of my best efforts to be "detached," an actor with my own social history and my own positioning in a web of social and intellectual relationships. I say "in part" because, while my own history and identity affect the way I tell Sylvia's story, it should also become apparent that her story often ruptured the schemas with which I intended to contain it. This is what it means to say that doing fieldwork and writing ethnography are dialogic processes.

But before we go any further, let me tell you about Sylvia.

A Typical Weapons Scientist?

Looking back through my fieldnotes, I find these scattered remarks about Sylvia, written after our first encounter:

> Sylvia has worked at the lab for about two years. She is 30 and lives in a suburban house on her own except for a huge golden Labrador. On the phone I was struck by her openness and friendliness. The interview was fairly short because she gets self-conscious talking at great length and soon trails off in embarrassed laughter. Before the interview she asked me to give her some suggestions for reading in the history of sociology. At the end of the interview she peppered me with questions about the protestors. Why do they do what they do? What do they think about the Russians? Do they know anything about science? At one point in the interview we're interrupted by a phone call from a computer programmer who's running one of her programs. (It's 8 P.M., but work never ends.) He can't get into one of her files. To my amazement she rattles off the file name by memory, and it goes something like this: AB666589:!!$#:lllbm58745.

After Sylvia and I had gotten to know one another better she once said jokingly to me, "I think of myself as a six-foot-tall, blond, blue-eyed male." Since Sylvia is rather less than six feet tall, dark-haired, dark-eyed, and Japanese-American, this is a striking remark. It is also an ambiguous remark. It hints at the possible psychological costs inherent in working as a minority woman among Livermore's predominantly white male physicists, but it also bespeaks, metaphorically, a self-confidence which, over time, I came to see as an integral part of Sylvia's persona. She is, to my knowledge, the only nonwhite member of the laboratory's elite cadre of warhead designers. Having lived for a year in Japan, Sylvia speaks Japanese and says she identifies quite strongly with her Japanese ancestry. During the Second World War her mother lived in Japan and her father, who is now a successful engineer, was put in an internment camp by the U.S. Government. On the day the first atomic bomb was dropped, Sylvia's aunt was taking her grandfather to Hiroshima. She sustained a dose of radiation high enough to induce radiation sickness, though she did survive and now lives in California. During one of our conversations Sylvia recounted for me her aunt's experience that day in 1945, as it had been told to her by her aunt herself, complete with a horrifyingly powerful image of the aunt standing on a bridge on the edge of Hiroshima looking down into a river choked with corpses and red with blood. How does your family feel about your nuclear weapons work? I ask. "They have this unspoken way of expressing themselves," she says quietly. "I think they're against it. They'd rather I was a doctor."

Sylvia's account of her aunt's experience in Hiroshima affected me deeply. When I was an antinuclear activist I had, like many other antinuclear activists I knew in the 1980s, had nightmares about dying in a nuclear war. Once, while sitting on a hill overlooking San Francisco, I had even had a strange waking vision of a mushroom cloud over the city. During my fieldwork, as I explain in more detail below, these nightmares and fears receded. But the night of my conversation with Sylvia about her aunt, I had a nightmare that there was a bloodstained body in my bedroom. It was the only nightmare I had during my final year of fieldwork.

When I first met Sylvia I was bewildered that anyone whose aunt had been bombed at Hiroshima could now earn a living designing nuclear weapons. Surely she of all people should know better, I thought. However, such an attitude on my part preempted the fundamental question in regard to nuclear weapons—the basic question at stake in my fieldwork—namely, whether nuclear weapons protect their bearers from danger or are themselves the danger from which we need to be protected. My own evolution into an antinuclear activist in my early twenties was associated with a sense of awakening, of pulling off layers of numbness and denial about the terrible things nuclear weapons could do to nations and to bodies. For me, any sense as intimate as Sylvia's of what nuclear weapons actually do to people could only be linked to an antinuclear sensibility. Sylvia confronted me with a different reality: for those who are persuaded by the arguments in favor of nuclear weapons, a stark knowledge of what happened at Hiroshima may simply reinforce the notion that it is important for one's own country to have such weapons. Just as those whose relatives have been shot may decide to buy guns, so those whose relatives have been atom-bombed may decide to work on atom bombs. The antinuclear movement of the 1980s sought in its literature and iconography to appropriate the shattered bodies of Hiroshimans as incontestable signs of our own imminent extinction should the arms race continue, but this is not the only way these bodies can be read.

At least as puzzling for me as Sylvia's relationship to Hiroshima was the fact that she was a woman. The antinuclear movement in which I had participated in the early 1980s was deeply shaped by an array of feminisms from mainstream equal rights feminism to ecofeminism and the separatist feminism that informed the vision of the large, women-only peace camps at Greenham Common in Britain and Seneca Falls in New York.[2] I had learned from the movement that, to put the message in caricature, men make war and women make peace. And indeed many of the most influential and charismatic national leaders in America's antinuclear movement were women: Randall Forsberg, Helen Caldicott, Pam Solo, Jane Gruenebaum, and Jesse Cocks, for example. Meanwhile the disparate but connected ideas of such feminist writers as Starhawk, Carol Gilligan, Susan Griffin, Nancy Chodorow, Gloria Steinem, and Ursula LeGuinn pulsed in pristine and mediated forms through the movement which, however imperfect its actual record in regard to women, was permeated by women's symbolism and women's consciousness. Many antinuclear affinity groups, for example, took women-oriented names such as Ovary Action, Princesses against Plutonium, Spiderwomyn, and Moms Against Bombs. There were also such groups as Women's Strike for Peace, Women's Action for Nuclear Disarmament, and the Women's International League for Peace and Freedom, the oldest peace group in the United States. At antinuclear demonstrations it was commonplace to see signs such as Pay Mothers, Not the Pentagon, Take the Toys Away from the Boys, and (accompanied by a picture of the earth) Love Your Mother. In this movement, especially in the more radical direct-action wing of the movement, it was in varying degrees taken for granted, by many men as well as women, that the militarist project was in some sense a masculine one, that women (especially mothers) represented a "different voice," a more peaceful voice, and that—as Carol Cohn, Jean Bethke Elshtain, Cynthia Enloe, Sara Ruddick, and others have argued—the project of making peace was deeply connected with the transformation of America's gender system. As the handbook for the women's peace camp in Seneca, New York, put it:

> Feminism is a value system which affirms qualities that have traditionally been considered female: nurturance of life, putting other's well-being before one's own, cooperation, emotional and intuitive sensitivity. . . . These traits have been discounted by societies which teach competition, violent conflict resolution, and materialism.

> Feminism insists that the qualities traditionally considered female be recognized as deserving respect and manifesting great power.
>
> (Krasniewicz 1992:48)

The general sentiments of this passage, if not necessarily its essentialist phrasing, were widespread in the movement, and it was within the movement that I myself received my earliest and most thoroughgoing exposure to feminist thinking and practice.

In terms of the framework, I had internalized in the antinuclear movement, then, a woman warhead designer was an anomaly indeed. Sylvia, however, does not see things this way. Although she works on weapons, she describes herself as a feminist, and I found her acutely aware of the problems women face in science. She was well aware, for example, that women accounted for only twenty-six percent of the employees at the laboratory, that most of these women were secretaries or were concentrated in the fields of biomedicine, computer science, and environmental science, and that there was evidence suggesting women at the laboratory tend, statistically speaking, to be paid less than men for comparable work (Rogers 1989; LLNL Women's Association 1988). Just under six percent of the laboratory's physicists are women—a number that is in line with the number of Ph.D.s in physics given to women nationally. Among the elite caste of warhead designers at the laboratory I knew of only three women, including Sylvia. "The only other women that I see regularly [at the laboratory] are computer analysts and secretaries," she says. "There aren't that many engineers or scientists who are women and, in general in science, as soon as women move into a field, the average salary drops."

Sylvia has strong feelings about the male-dominated culture of university science departments, and is particularly critical of MIT. "My observation was that it was all right for a graduate student to be a woman because the professors had control over the students, but the competition from male faculty if you were a woman was something else. I know women faculty there, actually, who have had the chairman of their department saying that they will never get tenure as long as they were the chairman there."

At the laboratory Sylvia has had no problem finding mentors among the senior male warhead designers, two of whom have worked closely with her and have passed along a substantial body of cultural as well as technical knowledge to her. On the whole Sylvia finds the laboratory's treatment of women more favorable than MIT's. Still, she has run into a few problems—technicians who, perhaps deliberately, open drawers containing centerfolds, even though such pictures are officially banned at the laboratory, and invitation lists to design review meetings from which her name is omitted. Attendance at such meetings is important because this is how designers maintain a profile in their community while assuring access to the latest ideas and data. She said she was not finally sure whether her name was, in her early years at the lab, sometimes omitted because she was a woman or simply because she was a new designer.

I once asked Sylvia what it meant to her to call herself a feminist. Arguing firmly against versions of feminism that essentialize or sentimentalize women by attributing to them a different voice, a separatist agenda, or a unique consciousness grounded in the particular experience of motherhood, she replied, "I'm fighting for everybody's rights, not just women. I just wish that people would let people be the way that they are, and not put them in little boxes." I asked if she saw a connection between masculinity and the arms race on the one hand and mothering and participation in the peace movement on the other. She replied, "I think I'm protecting children. I feel as if I'm protecting, helping to protect the country. Anyway, I don't think that looking after children is inherently men's or women's work. I think that we get socialized to take it that way." She added mischievously, "I'm looking forward to the day scientists make it possible for a man to carry a baby. Then I can get on with my work." Her response, eschewing the essentialist grounding of womanhood in motherhood so popular in the peace movement,[3] appropriated Nancy Chodorow's vision of shared parenthood as a basis for the mutual liberation of men and women but on behalf of a militarist-feminist program of liberation. When I left the field, Sylvia was busy raising money in her spare time to fund a special summer workshop to prepare local high school girls for careers in science.

Sylvia says she is not very political, but as an undergraduate she did protest her university's investments in South Africa. "If I had to label myself," she says, "I'd say I was a Humphrey Democrat." The reader may be as surprised by this as I was. When I arrived at Livermore I expected its weapons scientists to all be communist-hating, Reagan-Bush-loving conservatives, but it turned out that I was wrong and that Sylvia was far from unusual. Take Sylvia's colleague Clark, for example. He used to be a warhead designer and is now making a name for himself in weapons policy studies at the laboratory. He has also been a member of the Sierra Club, an active supporter of women's rights, an opponent of U.S. intervention in Central America, a supporter of gun control and, in the 1970s just before he came to the Livermore Laboratory, an active protestor against the Vietnam War.

A number of the scientists I interviewed said they had been opposed to the Vietnam War. Others were environmentalists who had been active members of the local Sierra Club before it took a position in favor of a nuclear freeze in the early 1980s, at which point its Livermore membership plummeted. One young scientist at the laboratory told me she was so enraged by the Exxon Valdez oil spill in Alaska that she cut her Exxon card in two, soaked it in oil, and mailed it back to Exxon. And Mark, a weapons designer on whose living room wall I noticed a Gandhi poster, told me about his occasional daydreams of saving whales—a cause as classically liberal as his methods were not.

> "I had fantasies of being Captain Nemo in *Twenty thousand Leagues under the Sea,* torpedoing the whaling ships. What's wrong with that? They're willing to kill whales, so why not blow up their ship and leave them to figure out where to go from there?"

Many laboratory scientists had also been active in the civil rights movement. One warhead designer had helped organize a campaign in Livermore in the 1960s to prohibit racial discrimination in jobs and housing. Jeremy, a deeply religious man who now worked as a weapons physicist, had spent part of the 1950s working with a worker-priest for racial integration in the American South before coming to the laboratory. And Phil, a warhead designer who had some union-organizing experience and who liked to complain to me about the domination of American politics by corporate interests, told me he left his church because the minister opposed a social action program on behalf of minority inner-city residents. He also got into an argument with some high-ranking military officials when they visited the laboratory.

> "We had some colonels or lieutenant colonels over, and they were talking about something, and it was, they were here to defend capitalism versus communism. And I called them on it. I said, 'You've got things all screwed up. I'm not supporting this country because it's a capitalistic country. I'm supporting it because of its form of government.'"

Just before the 1988 election, one weapons designer took a straw poll of his colleagues to see for whom they planned to vote. He found they were split down the middle between George Bush and Michael Dukakis—roughly the same spread I found in my own interview sample (two of whom told me they had voted for Jesse Jackson in the primary).

So, the laboratory was, to my surprise, a place of political diversity. It was a place where Reagan-Bush supporters, those with no great interest in politics at all, and liberals who had struggled for civil rights and against the Vietnam War could all work together in the development of nuclear weapons.

Thinking through Identity

As my fieldwork proceeded, I began to wonder about the grounds of this collaboration. How were conservative and liberal scientists able to work together on nuclear weapons in the context of a society that, in the decade of Reagan's defense build-up and the nuclear freeze movement, was contentiously divided, largely along liberal-conservative lines, about the need to keep building nuclear weapons? How were liberals—some of whom had participated in the peace movement of the sixties and seventies and were critical of many Reagan administration policies in the 1980s—able to feel committed to their work developing nuclear weapons for Ronald Reagan's arms build-up? And if the kinds of overt political ideologies—liberalism and conservatism, Democratic and Republican Party political affiliation—celebrated by the American media as the schismatic indices of political identity were not the ideological glue holding the laboratory together, then what was? Evidently I had to think about political identity and ideology in new ways, not simply in terms of America's conventional political labels, if I was to understand the political integration of the laboratory that enabled its mission to proceed.

Thus, instead of thinking about the laboratory's integration solely in terms of its ability to recruit a particular type, I began to think about the practices through which the laboratory resocializes recruits and actively produces new thinking, feeling, believing, acting selves among its scientists. I also began to think about science itself—the ideology that claims not to be one—as a source of binding energy capable of holding the scientists together despite their apparent political differences. I began to see Livermore's scientists as united not by a commitment to any overtly political program (for example, "push the Soviets out of Eastern Europe") but by a shared dedication to a technocratic program ("make sure the nuclear arsenals are safe and stable") that was avowedly transpolitical. I began to think of the socialization of scientists into the laboratory as, in part, a process whereby political questions were transformed into technocratic questions, and I became interested in how this process was accomplished and experienced.

Insofar as other writers about weapons professionals have adopted a process perspective, and I am thinking of Robert Jay Lifton and John Mack in particular here, they have tended to present the processes involved as largely repressive or subtractive: ethical questions are avoided, feelings are denied, and fears are repressed. In such analyses, weapons professionals are defined as much in terms of what they lack as what they are. While part of the work of becoming a weapons scientist does indeed involve learning not to attend to particular fears, feelings, and questions—just as part of the work of becoming an antinuclear activist also involves a selective learned inattention—it also involves the active learning of discourses, feelings, and practices. To take the example of ethics, as I have argued elsewhere (Gusterson 1992; forthcoming), rather than ignoring the ethical dilemmas of their work, weapons scientists learn by listening to their colleagues and other members of the local community to resolve these dilemmas in particular socially patterned ways. They learn, for example, to think about the ethical problems of weapons work in private, not to raise them in public or over lunch with their colleagues. They also learn to think of these issues in instrumentalist, technocratic terms, and to reject discourses that do not recast these moral problems in technocratic terms. In other words, becoming a weapons scientist involves much more complex and creative social and psychological processes than repression and avoidance. As Michel Foucault says, "power would be a fragile thing if its only function were to repress" (1980:59). The power of the social processes sustaining the laboratory's work lies in their ability to teach scientists new ways of thinking, to positively reshape the identities and discourses of its employees as they are transformed from neophytes into mature weapons scientists.

The process of social and psychological engineering involved here is ideological, but in a more fundamental way than we often mean when we use the term *ideology.* Raymond Williams (1977) argues that we must think of ideologies not only in terms of discourses and ideas, but also as "structures of feeling"—ways of experiencing and living in the world that profoundly reshape our emotions, bodily reflexes, and fantasies, as well as our ideas and beliefs. As Sharon Traweek has put it, taking up some of Williams's ideas in her own ethnographic study of physicists at the Stanford Linear Accelerator in California and the KEK facility in Japan, "Novices . . . must become unself-conscious practitioners of the culture, feeling the appropriate desires and anxieties, thinking about the world in a characteristic way" (Traweek 1988:x–xi).

I should add here that my awareness of the culturally transformative power of particular milieus and social practices derived partly from a process of reflection on changes in my own identity as my fieldwork progressed. When I arrived in Livermore in 1987 I was no longer a practicing peace activist, but in many ways I still had the consciousness of a peace activist. I had been very concerned about the potentially catastrophic consequences of the arms race, a concern that had manifested itself in, among other things, insurgent and graphic mental images of nuclear destruction; and, during my first interviews with Livermore scientists, I felt deeply conflicted about my role as polite ethnographer, which felt a little like a pose, and I wondered if I should stop even trying to find out how these people had become deranged enough to work on nuclear weapons. However, over the two years that I spent talking, eating, hiking, worshiping, and celebrating with nuclear weapons scientists, through a process I still find mysterious, my fear of nuclear war receded, my nightmares about nuclear war disappeared, and I began to find the dark apocalyptic rhetoric of antinuclear activists increasingly quaint, even tiresome. Moreover, after months of disciplined interviewing, note taking, hypothesis testing, and academic writing, the notion that I was an anthropologist, which had at first seemed so hollow, took on increasing substance and became the anchoring principle of my existence. By behaving like an anthropologist, I had become one.

But enough about my voyage of initiation. Let us return, instead, to Sylvia's.

Becoming a Weapons Scientist

Moral philosophers have distinguished two basic positions in the debate on nuclear ethics: the deontological and the consequentialist. Deontologists believe that, if it is wrong to destroy an entire city, then it is wrong to threaten to do so. Consequentialists believe that actions should be judged not by their intrinsic purity but by their consequences; hence, if threatening to destroy an entire city helps save the city, then it is moral to make the threat.

Unsurprisingly, Livermore scientists are consequentialists, and part of the process of becoming a nuclear weapons scientist involves internalizing a commitment to what we might call the central ideological axiom of laboratory life: that scientists at the laboratory design nuclear weapons to ensure, in a world stabilized by nuclear deterrence, that nuclear weapons will never be used. The pragmatic spirit of this central axiom enables it to transcend conventional political divisions of left and right and to unite weapons scientists, in theory at least, around the technocratic project of figuring out what works best. To antinuclear activists and laboratory critics this central axiom of laboratory life seems like a hollow and dangerous cliché, but then it is in the nature of deeply held ideological beliefs that they appear to those who hold them so self-evident that they require no elaboration, whereas to others they may seem so bafflingly wrong that they defy explanation—one group's common sense is another group's nonsense. Part of the process of maturing as a weapons scientist, quite apart from learning the physics and engineering, is coming to see the laboratory's central ideological axiom not as empty cliché but as comfortable truth, of coming to feel in one's bones that theories of nuclear deterrence describe the world as it really, inevitably is.

From the perspective of laboratory critics, antinuclear psychologists, and peace activists, this process is one of denial, avoidance, and repression. Critics and antinuclear activists often accuse weapons scientists of spiritual deadness and a lack of imagination, and criticize them for not thinking about the moral issues raised by their work. For example Hugh DeWitt (1989), an internal critic who has called for an end to weapons research at the laboratory, calls his colleagues Ph.D. peons who cannot see beyond their "high pay, job security, good benefits, excellent physical facilities, travel to scientific meetings, and good retirement programs" (3). He says they do not think about the ethics of their work.

I will leave it to professional psychologists to decide whether Sylvia is "in denial," but we certainly cannot argue that Sylvia has not thought about her work, and it became increasingly clear to me that Sylvia deliberately sought out highly conflictual situations in order to incite such thinking. She approached these conflictual situations as if they were experiments.[4] For example, the summer after she accepted the job at Livermore, but before she arrived at the laboratory, Sylvia made a point of traveling to the Soviet Union, and even telling some Russians what kind of job she had been offered, to see if she would still feel comfortable working on nuclear weapons by the end of the trip. She did. Later, she made a point of traveling to Hiroshima and seeing the Peace Museum there, a museum which memorializes in the most graphic ways the damage inflicted on the people of Hiroshima by the first atomic bomb. She also, at one point, asked if I would take her to an antinuclear meeting and, on another occasion, asked if she could borrow an antinuclear video I had so that she could see for herself what antinuclear activists were saying.

My discussions about nuclear ethics with Sylvia and her colleagues often surprised me. I was surprised, for example, when she told me of her disgust at the effects of American nuclear testing in the 1950s on Pacific Islanders. "It was very disturbing to me to see that these people had been screwed. Their land was taken away. If I had been around then, I hope that they wouldn't have done that. Either that or I would have quit." And I was surprised that, although she felt comfortable working on nuclear weapons, she said she would have difficulty working as a defense lawyer trying to secure the freedom of criminals. I was also surprised by her colleague Clark who told me he felt more comfortable working on nuclear weapons than he would working on conventional weapons. From my perspective as a lapsed antinuclear activist, nuclear weapons were more immoral than conventional weapons because they could kill so many people and kill them so indiscriminately: they were genocidal weapons. For Clark, however, it was precisely because conventional weapons were less destructive that they were routinely used to kill people and, for this reason, he would have had difficulty working on them. As one of Sylvia's colleagues put it, "The moral questions aren't simple. Your conscience should trouble you either way. If you do work on nuclear weapons, think of all the people those weapons might kill. If you don't work on nuclear weapons, think of all the people you may be endangering by leaving them undefended."

Seen in this light, then, in a situation of grave ethical and practical ambiguity, the process of becoming a weapons scientist is one of becoming increasingly certain that nuclear weapons are reasonably safe, and this has been one of Sylvia's central achievements in her first seven years at the laboratory. Sylvia had largely completed this process of self-assurance by the time the cold war ended and she now believes that nuclear weapons work should continue in a post–cold war world in order to improve the safety of the nuclear stockpile.

When I asked Sylvia why she took the job at Livermore, her response was, compared to those of other weapons scientists answering the same question, unique. Most scientists at Livermore told me they accepted job offers at the laboratory because the work seemed interesting and challenging, and they liked the laid-back, collaborative ambience of the laboratory (as opposed to the cut-throat competitiveness of university physics departments). One of Sylvia's senior colleagues, for example, explained his decision to come to the laboratory many years earlier in these words:

> I had seen enough at MIT of young professors coming in and the scramble to achieve tenure and the tactics that were required or seemed to be required for them to receive good standing in the eyes of the department head and so forth, and I thought that was really disgusting. At the same time, when I talked to people at Lockheed, I really felt uncomfortable about any sense of freedom; I really felt uncomfortable about the push based on the profit motive. I had seen enough of Livermore to know that it seemed to offer quite a bit of academic freedom, quite a bit of time for self-improvement and for satisfying things that were interesting specifically for you—a characteristic of the university without that "after three years you will be put on the auction block and either accepted or rejected from this particular institution." And there was a great deal of adventure in those days. With atmospheric testing, the way you collected samples was by climbing in the back seat of an air force airplane and taking off and having filter papers on either wing, and after the bomb goes off and you have a *lovely* mushroom cloud, then in an hour or so you make a quick pass through the cloud and expose the filter papers and collect samples and bring these back to the laboratory. I thought that sounded *absolutely* wonderful.

The explanation Sylvia gave me of her decision to work at Livermore was decidedly different. Unlike the account any other weapons scientist gave me, Sylvia's explanation foregrounded her own initial doubts and fears about nuclear weapons themselves. "The work is quite interesting, and that was definitely a consideration. But I decided to work at the lab, I think, because I had a fear of big weapons. I really wanted to see what was happening for myself. I wanted to see what was going on, rather than take other people's word."

My own fear of nuclear calamity, a fear that was socialized by the antinuclear movement and abetted by the extravagant military rhetoric about winnable nuclear wars during the first Reagan administration, was organized around scenarios of nuclear wars blundered into or started deliberately by leaders who were either mad or so hyperrational about the calculus of terror that they had lost touch with the human consequences of war. Projecting my concerns onto Sylvia, I presumed that she too feared an all-out nuclear war, and was perplexed that she could hope to allay such a fear by learning more about the processes of nuclear weapons design rather than the processes of international relations and psychology. It took me a while to realize that she presumed our leaders too rational to start a nuclear war. "I'd say the odds of a nuclear war in my lifetime are very small and very low. I don't think any rational person would use them," she said. What mainly concerned her was not a deliberate nuclear war but the possibility of an accidental explosion of a nuclear weapon. We had each turned nuclear safety into the kind of problem we felt equipped to solve—in my case, a social and political problem, in her case, a technical one.

I have argued elsewhere (Gusterson 1992; forthcoming) that nuclear weapons tests are ritual simulations of human control over the awesome power of life and death embodied in nuclear weaponry, and that it is by participating in nuclear tests that nuclear scientists become confident in their mastery over these weapons. Drawing on Bronislaw Malinowski's notion that rituals alleviate anxiety about human vulnerability in the face of cosmic mysteries, I have suggested that nuclear tests matter to warhead designers not only as scientific experiments but also as rituals that experientially validate the designers' claims that nuclear weapons are indeed subject to human control. I have also suggested that nuclear tests might be seen as rites of passage through which young scientists are transformed into established designers as they display their mastery over nuclear weapons to the community of weapons scientists at large—an argument, I might add, that has not, on the whole, been well received at the laboratory.

In the course of my research, Sylvia worked her way through the nuclear cycle. While I was doing my fieldwork, she participated as a neophyte mastering her trade by assisting other designers in their tests. Throughout, at the same time as she was on her personal odyssey to discover whether nuclear weapons were indeed safe, she was learning, both culturally and technically, how nuclear tests are done. Finally, just as I was finishing my dissertation, completing my own professional rite of passage, she oversaw her first test as lead physicist—a test that, so far as I have been able to ascertain, went well.

The last few times I spoke to her, Sylvia seemed changed in subtle but important ways. When I first met her, she was fascinated by the counterculture of Berkeley, whose streets she liked to roam while thinking through difficult calculations, and she sometimes talked about moving there from her suburban home. The last time I spoke to her, however, she had given up thoughts of moving to Berkeley. "I'm working much harder now, and the commute would take up too much time. It's better to live close to the lab." Sylvia seemed preoccupied. Although she had been publishing in unclassified areas so that she could get a job outside the laboratory if she needed to, she was concerned that the U.S. government's restrictions on nuclear testing were creating a logjam of experiments, thus slowing down her and her colleagues' work. She opposed an indefinite suspension of nuclear testing, which the government subsequently implemented, much more confidently than she had when I first met her.

> I would like to see testing of things that have already been built, just to make sure nothing has happened, like quality of the sample, because things change. So knowing nothing, I mean, knowing incomplete information and then having that information change on you is very risky. Of course this would be my personal bias. I like to poke things and tear them apart. I would like these systems to be as predictable as possible.

Sylvia had become a weapons scientist.

Dialogue

As the reader will probably have gathered by now, although she has been subjected to the same social processes as other weapons scientists, Sylvia has worked her way through them in her own unique way. For me, Sylvia was fascinating for her willingness to place herself in conflictual situations and for the ebullient uniqueness of her life—a life that fulfilled the career ambitions of an upwardly mobile Japanese-American family and of a generation of feminists but in the most improbable way either could have imagined. I also found myself drawn to Sylvia because of her energetic interest in reflecting on the laboratory and her life within it, and in understanding something about the antinuclear activists who stood opposed to the direction her life had taken. Traditionally anthropologists call people such as Sylvia—"natives" with an unusual interest in reflecting upon and explaining their own culture—key informants. I dislike this term not only because of the connotations of surveillance and betrayal inherent in the word *informant,* but also because the hierarchy built into the word fails to convey the discursive, mutually challenging nature of my conversations with Sylvia—conversations which, despite the roles of native and ethnographer that formally underpinned them, were informing, challenging, and rewarding for both of us. Thanks to these interactions not only did I learn about the laboratory, but Sylvia learned about the antinuclear movement and, every time she violated my expectations about warhead designers, I learned about myself and my own unexamined assumptions.

These conversations with Sylvia became conduits for all kinds of communications that would not otherwise have taken place. For example, Sylvia used her relationship with me to find out more about the protestors, about whom she was greatly curious, and once even asked me, only half playfully, I think, to let the protestors know that her division had enjoyed the theatrical nature of the latest protest at the laboratory gates and that she had admired the giant puppets the protestors made especially for the occasion. As for me, thanks to my friendship with Sylvia, I was able to find out not only what it meant to her to be a weapons designer, but also what her colleagues thought of particular hypotheses I was developing or papers I was writing, and how I was perceived among the community of warhead designers. Among other things I learned, for example, that the warhead designers sometimes joked about collaborating in feeding me plausible but spurious accounts of their secret lives behind the barbed wire fence.

Above all, the reflexive nature of my conversations with Sylvia helped me realize the extent to which, despite my best efforts, I often read into the lives of Livermore scientists a pervasive sense of guilt and conflict about their work which, I now believe, existed far more in my mind than in theirs. A brief anecdote will illustrate the point. Sylvia once asked me if I would take her to an antinuclear meeting to rally supporters for an upcoming protest at the Nevada nuclear test site. As we drove back from the meeting it was clear to me that Sylvia was upset about something. I presumed that the meeting, where speakers held forth at length about the evils of the arms race and the moral callousness of weapons scientists, had brought to the surface her deep psychic conflicts about her work. No wonder she was upset. No, she said, she was angry at me because, on the way to the meeting, I had told her exactly what to expect in such a way that I had, as she put it, "ruined the experiment."

Since I started to write about my fieldwork, Sylvia's response to my ideas has been ambivalent. When I wrote a series of articles about the laboratory's reaction to the end of the cold war for a local newspaper in Livermore, Sylvia wrote to tell me, "people on my corridor brought me zillions of copies of all three articles—they all thought, as do I, that the articles accurately reflected the mood around the lab." She also felt, unlike many of her colleagues, that there was something to my notion of nuclear tests as rituals, even if she did not completely agree with it. Just after her own nuclear test she wrote:

> I wanted to convey my congratulations to you on the completion of your Ph.D. At about the time you were in Palo Alto [submitting my dissertation], I was probably in Nevada, going through a ritual you may have described in your dissertation. The note that Harry [her mentor] sent a couple of months ago was shown to me before going out—I regret not adding my own comments to balance and temper his, but I was in the midst of the process. Let me just say I enjoyed reading about the rituals A-Division designers go through (I actually saw them do some of them) during a test. Harry believes you were misled because he's never seen this stuff, but some of it happens.

She was much less happy, however, about a passage in my dissertation where I analyzed one of our conversations. In that passage I mentioned that she had watched a television docudrama about the bombing of Hiroshima and had answered my question about the feelings the program aroused in her by saying "it was poorly executed. They were quite accurate about the physical effects of the explosion, but the accents were all wrong." I used this response to illustrate my observation that a scientist's learned attentiveness to detail and objectivity could distance him or her from the suffering of human bodies. Sylvia responded quite angrily to this:

> It is very natural and automatic for me to be able to distinguish between native and nonnative Japanese speakers. I was not able to suspend my belief that this was not "real," just as I am not able to do so at a badly done film of any kind, and so it was difficult for me to empathize with the pain that those actors were trying to portray because, to me, they couldn't be real bomb survivors. But that does not mean I do not or cannot feel pain for the bomb victims. I remember your expression of disbelief when I described how some physicists have cringed when driving by a roadkill. But why is *this* so hard for you to believe? Why didn't you use *that* as an example of something?

For better or worse, Sylvia and I have become part of one another's lives. I must live with her criticisms of my work, and she must live with my interpretations of her life. The intrusiveness of this situation is encapsulated nicely in a card I once received from her when she was traveling in Japan. It said, "I had a very strange dream about going through the Peace Museum in Hiroshima with a tape-recorded tour piped into my ear, except that the headphones were attached to a microphone into which *you* were speaking. And there was no stop button."

Conclusion: There Is None

When I applied for funding for my research from the National Science Foundation, one anonymous reviewer recommended that I be denied funding because, as a former antinuclear activist, I would not be objective about nuclear weapons scientists. Just as I have left it to others to decide whether Sylvia is "in denial," I will leave it to the reader to decide whether I have been "objective" about weapons scientists. Returning to Donna Haraway's notion of "situated knowledge" I will say, however, that no knowledge is disinterested, no ethnographer capable of what Haraway calls the "God's-eye view from nowhere." From my perspective, it is my very identity as a former antinuclear activist that gave me an angle, a set of questions, a terrain of engagement in my encounter with the weapons scientists. Many of my earlier presumptions about nuclear weapons scientists perished on that terrain of engagement, as I have tried to show here by giving a processual rather than a finished account of my fieldwork—by, to adapt a phrase of Bruno Latour's, giving the reader ethnography-in-the-making rather than ethnography-ready-made. Following Latour, I have taken it as axiomatic here that a postmodern stratagem of objectivity would be a reflexive one that interrogated itself while focusing on the making as well as the presentation of interpretation. This is one reason for the dialogic mold of this article. Let me now turn to some others.

Sylvia's allegorical observation that our encounter came without a stop button evokes a characteristic feature of anthropology as it is practiced in the late twentieth century that is particularly marked in the case of my own project. Where fieldwork used to be a bounded and hierarchical encounter between the knower and the known aiming to produce an authoritative summation of the observed culture, today it often has more of the qualities of an ongoing, albeit asymmetrical, dialogue.

To begin with, the ivory tower from which I write no longer affords that "God's-eye view from nowhere," in part because my location has been situated and problematized by Sylvia and her colleagues. If in my writing I raise questions about the way they make their living and the institution for which they work, I cannot help but notice that, in their answers to my questions about their choices, they raise questions about perceived sexual discrimination and abuse of junior faculty at the institution where I work, saying they would rather work at a nuclear weapons laboratory than at a university like mine. The process of cultural critique here, while not equal, is mutual, and it leaves me with a lot more to think about than just the socialization of weapons scientists and the contradictions of other people's lives.

Moreover, unlike a former generation of ethnographers whose works were not read by their subjects, anthropologists today must live with our subjects' responses to our work—especially if we practice "repatriated anthropology." The days when our subjects had culture and we had culture theory are, if they ever existed, now gone. In my case, before I interviewed Sylvia, she quizzed me about the interpretive turn in anthropology and, after I wrote my dissertation, she disputed the terms of some of my interpretations. Other "informants" disputed my reading of Foucault, and so on. Addressing this kind of situation, Renato Rosaldo has written that

> we should take the criticisms of our subjects in much the same way that we take those of our colleagues. Not unlike other ethnographers, so-called natives can be insightful, sociologically correct, axe-grinding, self-interested, or mistaken. They do know their own cultures, and rather than being ruled out of court, their criticisms should be listened to and taken into account, to be accepted, rejected, or modified, as we reformulate our analyses (Rosaldo 1989:50).

But the dialogue goes in the other direction as well. Not only must I deal with Sylvia's interruptions of my professional conversations, but she and her colleagues must now deal with my presence in their conversations, too. Whether they like it or not, my notion that nuclear tests are like "primitive" rituals and my newspaper articles delineating the laboratory's response to the end of the cold war have now been absorbed into their discursive economy. And, even though I am no longer in the field, I sometimes get calls from journalists in Livermore hoping for a gnomic pronouncement on the laboratory's latest crisis—pronouncements which, if quoted, sometimes provoke correspondence from the laboratory.

If my fieldwork was a sort of dialogue, then, that dialogue is still continuing, albeit in a new register. There is no stop button. And surely this is what we should ask of anthropology as we enter the twenty-first century—not that it settle conversations, but that it start them.

Acknowledgements

This article has benefited from the comments of a number of colleagues. I am particularly grateful to Leslie Eliason, Michael Fischer, Jean Jackson, Evelyn Fox Keller, George Marcus, Kim Laughlin, Sherry Turkle, and Sylvia Yanagisako for their comments. I would also like to thank the NYU anthropology department, the Williams College anthropology department, the cultural studies group at the University of Washington, and the Program in Science, Technology, and Society at MIT for inviting me to present earlier versions of this piece, and for the helpful advice I was given at those presentations. Finally, my thanks to everyone at the Late Editions editorial workshop at Rice University.

Notes

1. As with all the names in this article, *Sylvia* is a pseudonym.

2. For ethnographic commentaries on these women-only peace camps, see Krasniewicz 1992 and Wilson 1988. For accounts of the San Francisco Bay Area antinuclear movement that emphasize the importance of women's leadership and feminist consciousness in that movement, see Epstein 1985, 1988, and Starhawk 1982.

3. Caldicott 1986 and Ruddick 1989 offer two influential examples of essentialist texts that argue for a triangular correlation between womanhood, motherhood, and peace consciousness.

4. My thanks to Rich Doyle for this suggestion.

References

Caldicott, Helen. 1986. *Missile Envy: The Arms Race and Nuclear War.* New York: Bantam Books.

Cohn Carol. 1987. "Sex and Death in the Rational World of Defense Intellectuals." *Signs* 12:687–718.

DeWitt, Hugh. 1989. "The Nuclear Arms Race as Seen by a Nuclear Weapons Lab Staff Member." *SANA Update: Scientists Against Nuclear Arms Newsletter* 74:2–4.

Elshtain, Jean Bethke. 1987. *Women and War.* New York: Basic Books.

Enloe, Cynthia. 1983. *Does Khaki Become You? The Militarization of Women's Lives.* Boston: South End Press.

———. 1990. *Bananas, Beaches, and Bases: Making Feminist Sense of International Politics.* Berkeley: University of California Press.

Epstein, Barbara. 1985. "The Culture of Direct Action." *Socialist Review* 82/83: 31–61.

———. 1988. "The Politics of Prefigurative Community: The Non-Violent Direct Action Movement." In Mike David and Michael Spriker, eds. *Reshaping the U.S. Left: Popular Struggles in the 1980s.* New York: Verso Books.

Foucault, Michel. 1980. *Power/Knowledge.* New York: Pantheon Books.

Gusterson, Hugh. 1992. "Coming of Age in a Weapons Lab: Culture, Tradition and Change in the House of the Bomb. *The Sciences* May/June.

———. Forthcoming. *Testing Times: A Nuclear Weapons Laboratory at the End of the Cold War.* Berkeley: University of California Press.

Haraway, Donna. 1988. "Situated Knowledges: The Science Question in Feminism and the Privilege of Partial Perspective." *Feminist Studies* 14:575–99.

Krasniewicz, Louise. 1992. *Nuclear Summer: The Clash of Communities at the Seneca Women's Peace Encampment.* Ithaca, N.Y.: Cornell University Press.

Latour, Bruno. 1987. *Science in Action: How to Follow Scientists and Engineers through Society.* Cambridge, Mass.: Harvard University Press.

Lifton, Robert Jay. 1982. "Imagining the Real." In Robert Jay Lifton and Richard Falk, *Indefensible Weapons: The Political and Psychological Case Against Nuclearism.* New York: Basic Books.

LLNL Women's Association. 1988. *LLWA Salary Study Committee Report.* 26 July.

Mack, John. 1985. "Toward a Collective Psychopathology of the Nuclear Arms Competition." *Political Psychology* 6:291–321.

———. 1986. "Nuclear Weapons and the Dark Side of Humankind." *Political Psychology* 7:223–33.

Malinowski, Bronislaw. 1948. *Magic, Science, and Religion and Other Essays.* Boston: Beacon Press.

Rogers, Keith. 1989. "Lab to Improve Minority Hiring Policies." *Livermore Valley Times.* 25 May.

Rosaldo, Renato. 1989. *Culture and Truth: The Remaking of Social Analysis.* Boston: Beacon Press.

Ruddick, Sara. 1989. *Maternal Thinking: Toward a Politics of Peace.* New York: Ballantine.

Starhawk. 1982. *Dreaming the Dark: Magic, Sex, and Politics.* Boston: Beacon Press.

Traweek, Sharon. 1988. *Beamtimes and Lifetimes: The World of High-Energy Physics.* Cambridge, Mass.: Harvard University Press.

Williams, Raymond. 1977. *Marxism and Literature*. New York: Oxford University Press.

Wilson, Lynn. 1988. "Epistemology and Power: Rethinking Ethnography at Greenham." In Johnetta Cole, ed. *Anthropology for the Nineties: Introductory Readings*. New York: The Free Press.

SCIENCE AND THE HOPE OF NATIONS

In diverse situations, these contributions expose how scientists are embedded in various projects of nationalism, often in the vortex of momentous contemporary events. Kim Laughlin provides an intimate view of Indian leftist scientists-turned-activists, in relation to the Bhopal toxic disaster, who attempt to keep alive public and official attention to the far-reaching consequences and lessons of this event. Kathryn Milun and Leszek Koczanowicz provide access in very different sites of scientific practice and possibility to the dramatic effects of the fall of the Soviet empire on the kinds of science that had developed within it. Through a sensitive memoir of her reactions to her life within a contemporary science city, Sharon Traweek explores how science is literally constructed on a landscape that reflects the ambition and confidence of both Japanese state and society as well as a kind of distinctive internationalism alive in contemporary physics, among other sciences.

Kim Laughlin

10

Rehabilitating Science, Imagining "Bhopal"

The research for these interviews was done during fieldwork in India, October 1989 through June 1992. I was based in Bhopal, living and working with a group of political activists responding to the continuing problems emergent from the Union Carbide gas-leak disaster. A primary source of field data for me was the casual conversation which filtered through our work and living. Dialogue was among members of our own household and with other activists working in Bhopal, particularly those working at Eklavya, an organization which provides science curriculum to secondary schools based on John Dewey's pedagogical commitment to experiment and participation. We also talked with the steady stream of visitors who came from outside Bhopal, both those coming for short-term projects with gas victims and those working in nearby rural areas for which Bhopal was the state capital.

Because I choose to conduct my fieldwork in a deeply participatory manner, transcription of the discussions among activists was impossible. My goal was to share in the process of understanding and work rather than observe from a distant and neutral position. Thus, the interview recorded here is constructed. I have tried to mime both the form and content of actual conversation but have also taken editorial license in the focusing of issues. Such license is not only inevitable but part of the project to "repatriate" anthropology so that cross-cultural perspectives can become part of social critique at home.

It takes little constructionism, however, to imagine discussions among activists on the scientific imaginary. It is widely acknowledged that scientific progress is the overt reason of state in India. It is thus imperative to question the ways that scientific expertise operates both as a resource, and as a mechanism of exploitation. A forceful lesson I learned from the activists with whom I worked was that oppositional political agendas must both expose the disaster of science, and capture the rhetorical and material artifacts of the scientific tradition.

The process of capturing science is both enabled and complicated by the fact that many activists are trained as scientists, engineers and medical doctors. Their social position provides the resources of critique through complicities that must be continuously acknowledged and undermined. Thus, these interviews give attention to the caste and class backgrounds of activists, to the emergence of their critique of dominant uses of science and to the ways these critiques are carried out in political work. The activists identified do not share a unified, antiscience position. The ethics and strategies of alternative science are vehemently debated and constantly modified by practice. This interview will describe a range of alternative approaches as they have emerged in Bhopal and elsewhere in the Indian environmental movement.

The two activists interviewed represent two ideal types, loosely defined. Rita is a Delhi-based feminist politicized as a Marxist during her years at Jawaharlal Nehru University. She is trained as an engineer but has chosen to work full-time in a women's group. Her family is Tamil Brahmin but she has resided in Delhi since she was young because of her father's military appointment. The language in which she most comfortably talks politics is English.

Pradeep is from Jabalpur, a small city in Madhya Pradesh. He trained as a medical doctor in Jabalpur and was politicized through work with local socialists. He tried to mobilize other students to join the work but the university never became a locus of political activity. Pradeep's family is middle-class, middle-caste. Hindi is the only language spoken at home and is the language through which Pradeep thinks politics. Like many other socialists, Pradeep believes that use of the English language is complicit with the cultural structures which his politics challenge. Since university, Pradeep has been involved in a series of alternative healthcare projects and in the continual work against communalist hostilities.

Rita and Pradeep represent configurations around which the identities of activists cohere. Of particular note is the difference between activists trained and politicized in the cosmopolitan universities and those trained in regional colleges. This difference often marks a separation between Marxist and Gandhian perspectives. However, as the discussion should indicate, these identities are in a continual process of revision. My representation does not focus on the often vehement tension between these and other perspectives. Instead, I have highlighted, through both form and content, the equally vehement attempt to forge a shared outlook and cooperative practice.

Redefining Progress

KIM: In their annual meetings, Union Carbide always tells their shareholders that the Bhopal tragedy is behind them, that a future of growth and profits lies ahead. Do environmental activists in India consent to this version of history, and its aftermath?

RITA: In Bhopal, there can never be any end to the tragedy; rehabilitation is not a possibility. Even if there were successful tort action against Carbide, real redress would not be possible. How can money replace dysfunctional lungs? How can money remedy problems of social structure that make victimization a continual inevitability?

PRADEEP: A slogan often repeated by gas victims is that "Bhopal is no isolated misery." Historically, this slogan exposes how the Green Revolution, which the Carbide plant purported to serve, was a major cause of agricultural labor displacement and migration to the cities. Thus, many of those living near the Carbide plant had already been victimized by the same processes which culminated in their 1984 exposure.

The slogan also points to the victimization process which led to certain communities living in such close proximity to a hazardous industrial facility. The most affected colonies are among the poorest in Bhopal. They are situated just adjacent to the plant so residents were aware that "poison" was being produced because of the frequency of small leaks and routine emissions that caused nausea and other effects. They were never in a position to protest because they had illegally set up their houses, without land ownership or government permission. They continually feared that the slums, not the plant, would be relocated.

RITA: Even when union workers tried to mobilize their support in pressing for greater safety standards, residents felt the security of their homes as the greatest risk. They saw the workers as an elite group trying to protect itself. Ironically, in the elections just preceding the disaster, these residents were given *pattas* [land rights] by a politician trying to pull their votes. Later, trying to justify these grants, the politician insisted that he never saw them as a "final solution."

PRADEEP: Focus on the need to establish community control is a way to face up to the deepest critique that environmentalists must make. At independence, the government of India promised that scientific socialism could provide democratic distribution of resources and rights. The failure of this promise is particularly clear in Bhopal. The Union Carbide plant was set up in 1969 as part of the effort to bring Green Revolution prosperity through high yield agriculture that is dependent on heavy inputs of chemical fertilizers and pesticides. The plant was located in an already densely populated area despite city planning codes which require facilities handling hazardous substances to be located away from human settlements. Nonetheless, the plant was not designed to fully accommodate safety precautions; it was poorly maintained and negligently operated. Meanwhile, government officials were entertained in Union Carbide's posh guest house.

RITA: The most benign interpretation suggests that the government failed to enforce regulation of safety standards at the Carbide plant because they remained convinced that the benefits of foreign investment and technology would bring "development" to India. The truth is probably better told by the hard facts of the government's response since 1984.

Since the disaster, the government has worked for control, not remedy. Individual medical files were only released from the Official Secrets Act in 1991. Their contents are not structured for diagnostic response but for easy categorization of victims. Despite mass indicators of widespread and increasing morbidity, official categorization only places thirty thousand individuals within compensatable categories. "Health" is understood as restored once an individual is labeled as "permanently injured, but not disabled."

PRADEEP: The sabotaging of history by Union Carbide and the government of India is perhaps most blatantly seen in the medical categorization of gas victims. In October 1991, the Indian Supreme Court upheld an out-of-court settlement of the Bhopal case for $470 million. The judgment did not specify how this money was to be distributed. It was indirectly indicated that disbursement would be according to the medical categorization data provided by the Madhya Pradesh government. The quality of this data has been severely criticized by a broad spectrum of medical professionals. The data is not based on sufficient diagnosis or laboratory tests. As consequence, the results are unrepresentative of the severity and distribution of injury among gas victims. Out of a gas affected population of over six hundred thousand, only forty individuals have been categorized as permanently disabled. In J. P. Nagar, a severely affected area, only forty-nine people are categorized as C—permanently injured. Meanwhile, Kotara Sultana Bad, an unaffected area, has fifty-six people categorized as C. Clearly, if the medical categorization data is used as the basis for disbursement of compensation, there will be few beneficiaries and they will not be those most in need.

RITA: The voluntary sector in Bhopal has been unequivocal in insisting that the medical categorization not be used as the basis of compensation disbursement to gas victims. The alternative must be immediate, equal distribution to all those living in the thirty-six wards designated as gas affected by the Indian Council of Medical Research. Only through equal disbursement can compensation reach victims with minimum delay and minimum disruption of their lives. This would also offset bureaucratic control over the history of the disaster, undermining government license to label who counts and who doesn't.

PRADEEP: The debates over medical categorization of victims have involved a strange tension between use and critique of scientific rhetoric and practice. We have tried to capture the legitimacy of science in our attempts to document the cause and continuing effects of the disaster. Simultaneously, however, we must argue against the claim that scientific certainty must precede all efforts toward change. Partly due to the politicization of official research, there is insufficient data to substantiate a causal link between gas exposure and many of the disabilities suffered by victims. Thus, we must take care in arguing that science provide the basis of rehabilitation.

KIM: The tension you speak of comes up often in environmentalist debates in the United States. The precedent established by the Montreal Protocol to reduce CFCs [chlorofluorocarbons] in the atmosphere is cited as particularly important because it mandated change before there was scientific consensus on the cause or consequences of global warming. Toxic tort cases are also struggling with the lack of precedent-legitimating evidence based on statistical rather than causal correlations. The courts haven't yet taken responsibility for questioning the way appropriate methodology is context specific.

PRADEEP: Such questioning could also be of use within the medical establishment. As healthcare professionals we should not consider ourselves exempt from the responsibility for defining what constitutes "health." In Bhopal, rehabilitation cannot be identified as individual freedom from disease to be brought about through disbursement of cash and antacids. The situation in Bhopal makes it clear that health is not only a medical problem. True rehabilitation will require major reorganizations of the social structure so that communities can gain control over a range of life support systems. Thus, we have also critiqued a scientistic approach to the organization of rehabilitation, insisting that the positioning of expertise in conventional hospitals and work places structures in victimization by keeping people uninformed and out of control.

RITA: It must be remembered that lack of information not only haunts the process of responding to disaster, but also the experience of disaster itself. The Bhopal case is a tragic example of this. From long before the disaster, there was a manufactured silence that barred any preventative action. Since, it has been a long nightmare of ignorance and fear.

The events of 3 December 1984 left Bhopalis shocked and keenly aware of their vulnerability. No alarm announced the presence of deadly gases in their homes. They woke, thinking neighbors were burning chili peppers. When it became impossible to breathe, they fled into dark streets. With no information on what was happening, they simply ran with the crowd, catching rides on passing vehicles whenever possible. Families were separated; old people and children abandoned. Many ended up on the outskirts of the city or beyond, without money and only the company of strangers.

But it was only when the sun rose that the magnitude of the disaster became apparent. Dead bodies of people and animals blocked the roads; vegetation was yellowed and shriveled, the smell of burning chili lingered in the air. Meanwhile, Carbide officials had informed hospitals that methylisocyanate (MIC) was like tear gas and could be treated with oxygen, antacids, and water-washing of eyes. Police were making rounds of the city, informing that the danger was past and asking people to return home. But no one believed them. Already there were rumors that the police had started dumping bodies in the river in a conspiratorial effort to hide the devastation.

PRADEEP: Rajiv Gandhi announced that it was safe for victims to return home even before an official assessment was made. Meanwhile, it became known that food was being brought in from outside for officials and other elites. A hard, green crust had appeared on the surface of stored food but people were told that there was no "scientific reason" for concern that it was contaminated. A helpless dread spread throughout the city, with people very conscious of their lack of control over the continued risks.

RITA: When victims approached the Carbide factory for medical aid and information, desperate managers afraid of rioting suggested that another leak may soon occur. This set in motion a second mass flight and intensification of the uncertainty. People continued to die but there was no information on possible antidotes or remedies. Clearly the effects of the gas were not only temporary and there was increasing fear of long-term consequences. If they lived, would their sight be restored? Would they be able to breathe easily? How were they to earn a living if disabled? What about the effects on unborn children?

PRADEEP: Immediately following the disaster there was a real institutionalization of silence. Voluntary workers had flocked to the city to render any aid possible, but they were completely barred from official information sources. The clamp-down reached from the highest city authorities to the ward nurses at government hospitals. One doctor, when queried about the extraordinary rate of spontaneous abortion, told us that, officially, he couldn't even tell us his own name.

RITA: Two thousand died during the first few days after the leak. Approximately one person has died every day since—raising the death toll to over ten thousand. And Carbide claimed that MIC was only a mild irritant, despite their own safety manual instructing that it is deadly even in very small doses. And the government has concurred in both denial of death figures and in refusing to acknowledge long term effects. The government must limit Carbide's liability so to hide it's own. There were gross regulatory failures that allowed the conditions for the disaster to arise.

PRADEEP: The most obvious show of government complicity involved the controversy over administration of sodium thiosophate, an antidote to cyanide poisoning. Autopsies showed signs of cyanide poisoning; morticians themselves were overwhelmed by cyanide fumes when cadavers were opened. Physicians first administered sodium thiosophate to themselves; noting its effectiveness, they administered available supplies to victims. The result was immediate relief. Sodium thiosophate is not known to have negative side effects even when identification of cyanide poisoning is mistaken.

Carbide denied that cyanide could have been a by-product of the MIC released, despite their own doctor having recommended sodium thiosophate immediately following the disaster. The doctor retracted his recommendation, claiming that he was sleepy and confused when it was made. The government of India conceded. A ceiling was set for the administration of sodium thiosophate; extra supplies were not made available. One of the most brutal attacks on volunteers was a police raid on a clinic set up to administer sodium thiosophate. Police beat up health workers, smashed the premises, and, most importantly, destroyed all documentation showing the antidote's effectiveness.

RITA: Strangely, the brutal destruction of documentation during the sodium thiosophate controversy reminded us of means of opposition on establishment terms. Rigorous empiricism has both the legitimacy of the scientific tradition and the mechanisms needed to expose grossly illegitimate abuses of power. The process of "advocating Bhopal" can thus be strategized within conventional forums such as the courts and the mainstream press without recourse to ideological denouncements. Close description carries its own critique, particularly if there is awareness that there is politics to the definition process. Carbide defines the Bhopal disaster as a two-hour interval in 1984. We must push the parameters of this definition, extending the story to include the history of negligence in the plant, the continued failure of rehabilitation and the ways liability is tangled with international processes such as the IMF [International Monetary Fund].

The reductive tendency of official descriptions has been a political process with deadly implications. False or simply absent information has directly exacerbated the crisis. To protect themselves, both Carbide and the government have represented the Bhopal case as an isolated event, without source and finalized by distribution of cash compensation. We have to resist the reductivist tendency, the urge to encapsulate, all within the logic of the market. Opposition to corporate and official handling of the Bhopal case can be understood as a multifaceted challenge to this encapsulation.

PRADEEP: One thing we have realized within the legal work on Bhopal is that advocacy is not just a matter of alternative proposals for the future, but also a matter of conflicting constructions of the past. In response to this, our group has consistently tried to publish the stories of victims themselves, recuperating the "native point of view." Hopefully, we haven't been complicit with that tradition of anthropology which used such documentation to fine-tune the control of colonial administrators.

KIM: Can you describe the broad agenda of the group you work with?

PRADEEP: The Bhopal Group for Information and Action (BGIA) was established in 1985 by a group of activists who had worked in the coalition which formed immediately following the disaster. Since then, the size of the group has ranged from two to seven persons. The agenda of BGIA is to attend to the whole range of problems emergent from the disaster. Local issues of healthcare, vocational rehabilitation, and housing are necessarily connected to broader issues of political corruption, development policy, and cultural imperialism. Response strategies include attempts to mobilize victims, attempts to document and disseminate information on health and economic conditions, and attempts to build alternative rehabilitation structures. Obviously, we are very constrained by lack of resources since we won't accept money from the government or international agencies. All our funds come from individuals or local trade unions.

In 1992 we tried to set up a small clinic in which victims would be trained to participate in the healthcare process. Funds never came through so we had to continue an ad hoc approach of helping people get admitted to government hospitals and then trying to monitor the care they received. This process is beset with contradictions. In 1991 we did a survey which documented extraordinary incidence of corruption and malpractice in the government hospitals. Patients had to pay for drugs which should have been free; many of the drugs distributed are considered "irrational" by international standards; there is an almost total reliance on broad-spectrum medicines that only offer symptomatic relief and allow an evasion of diagnostic specificity. The latter has particularly insidious implications since gas victims are obliged to prove that their disabilities are gas related to qualify for compensation.

RITA: Attempts to produce information reveal the essential structure of the grass roots. By definition, we do not have access to official sources, nor the funds required to produce "solid" alternatives. Thus, much of the work is through circuitous logics that bring together pieces of the picture that would otherwise remain disparate. An example of the way this works is particularly evident in the ways women's health issues became the basis of wider critique. In the months following the disaster, the rate of spontaneous abortion and high incidence of menstrual disorder became a matter of great concern among volunteers. It indicated that toxins were residing in the body and could pass the placental barrier. This substantiated the fear of long term and multisystemic effects. When the abortion rate remained extraordinarily high even one year after exposure, it came to be understood as a marker of chromosomal aberration and the possibility of long-term effects like cancer. This understanding was verified by a 1990 article in the *Journal of the American Medical Association.* Our reports were based on interviews with women, without laboratory evidence. They nonetheless anticipated similar conclusions.

KIM: I know that BGIA has regularly published a news sheet for victims and also a range of publications for a broader audience. Tell me about how you approach this task.

PRADEEP: The publication we are most proud of is a pamphlet titled *Voices of Bhopal,* which we produced for the anniversary commemoration rally in 1990. Focus on the stories told by gas victims themselves is a strategic move that counters official attempts to control information and action. Such control was blatantly visible when the government of India accepted an out-of-court settlement with Union Carbide without plaintiff consent. Later, the Supreme Court of India acknowledged a "denial of natural justice" but justified the denial by appeal to the "situational particularities" of the Bhopal case. As these interviews tell, the situational particularities of Bhopal demonstrate a failure of rehabilitation that the government of India is not prepared to acknowledge.

RITA: Victims of the Bhopal disaster are mostly poor, illiterate, and otherwise in need of interlocutors between themselves and the official rehabilitation apparatus. BGIA thus provides crucial, daily support, particularly to BGPMUS, the largest organization of gas victims in Bhopal.

PRADEEP: After the disaster, victims participated in a range of protests, coordinated by a range of groups, including the major political parties. The most sustained organizational outcome has been the Bhopal Gas Peedit Mahila Udyog Sangathan [Bhopal Gas-Affected Working Women's Union]. Today, the *sangathan* is the largest organization of gas victims and the only one to regularly articulate victims' demands through public protest, press coverage, and litigation. With a formal membership of fourteen thousand women, the *sangathan* has been at the forefront of the struggle to establish Carbide's liability and to force payment of sufficient funds for rehabilitation. Over time, the *sangathan* has developed an ideological and strategic program which demonstrates that sustainable response to the problems in Bhopal challenges not only the adequacy of government initiatives but also conventional understandings of "rehabilitation," and the progress toward that goal.

The primary fight of the Sangathan has been for restoration of local control through extended participation of community members. This has led them to challenge rehabilitation programs that require dependence on the government, hospitals that won't allow patient participation in the healthcare decisions, local solidarities based on communalism and other forms of political manipulation, and family structures that curtail public involvement of women. The continuing challenge of the *sangathan* is to evolve an organizational structure which can carry these commitments.

RITA: The initial mobilization of BGPMUS was to protest the closure of a government-sponsored sewing center which provided gas-affected women with stitching orders. Many of the six hundred women employed were widowed or had husbands incapacitated by the gas. Their average wage of 320 rupees [$7] per month was often the only income for an entire family. Hence, when the center was shut down in December 1986, and job orders terminated, the survival of many families was at stake. Within four days the women formed an organization and staged a demonstration in front of the Chief Minister's residence. Various collective actions occurred over the following months and, ultimately, the center was reopened and the demand was met for an increase in the number of women employed.

Since 1986 it has taken continuous pressure to keep the sewing center open. In the spring of 1993 the sewing center was closed and hasn't yet been reopened. The government's logic is that the end of the legal proceedings ends their responsibility for rehabilitation. All of the job provision and housing development programs have stopped. The government assumes that cash compensation is the final solution, although only promised to about five percent of the gas-affected community. Less than thirty thousand victims even have access to the compensation courts, although six hundred thousand have been geographically identified as exposed. So far, the courts have only dealt with death claims and in over nine months of hearings have only made awards to two families. When claimants appear before the court, they are expected to "prove" cause of death with official documentation like hospital records and pharmacy prescriptions. Obviously, these documents are hard to come by, particularly since government doctors have persistently refused to identify many illnesses as gas related—the gas-damaged respiratory and immune systems, among other things. Yet tuberculosis patients are not considered eligible for admission to the MIC wards of the hospitals.

PRADEEP: Because of the interlocking set of issues that affect victims, BGPMUS has extended their commitment beyond the sewing center to a range of issues shared by gas victims. During the first few months of spontaneous organization, the most involved women formed a steering committee to coordinate activities. Soon after, BGPMUS began having weekly public meetings. These meeting were and remain a significant source of information for the gas-affected community. Updates are given on the state of the legal proceedings, on healthcare options, and on possibilities for income generation. The meetings also provide practical help for dealing with the relief bureaucracy, particularly the problem of claims processing for those who are illiterate.

PRADEEP: Experience has shown the *sangathan* that whatever justice they get from the courts will not be because of the intrinsic fairness of the law but because of vigilant pressure from those affected. Many observers argue that the successes finally received from the courts have been a direct result of the sustained protest organized by BGPMUS. Particularly, the granting of interim relief and the reinitiation of criminal proceedings has been attributed to extralegal pressure. However, all acknowledge that the successes of the legal battle have been limited. One outcome of the frustration was an incident of property damage at the time the settlement of the Bhopal case was announced. During a demonstration protesting the adequacy of the settlement, women participated in the bashing of Carbide offices. Most still see this as an appropriate response to their victimization and as an effectual use of their limited means of protest.

RITA: In an attempt to offset the limited resources of victims, BGPMUS is provided support by a range of middle-class groups. Groups from all over India have used petitions and other mechanisms to show their solidarity. These groups regularly publicize support for the demands made by BGPMUS and also contribute to the ongoing process of keeping Bhopal in public memory, as a marker of environmental crisis linked to development policy dependent on foreign investment. These groups recognize that "Bhopal" must become a symbol and lesson for the future. Their attempts are akin to those of Japanese writers who write about Hiroshima, stressing the importance of recognizing not only the awesome scientific achievement of technology but also its capacity for devastation.

The group I am with provides more specific support. BGPSSS [Bhopal Gas Victims Solidarity Organization] is a Delhi-based group comprised of lawyers, doctors, and journalists, who are able to formulate legal arguments, supporting evidence, and demands for alternative rehabilitation programs. We also work on the international issues, in coordination with the International Coalition for Justice in Bhopal [ICJB], a New York–based group which has waged a campaign to expose the connection between industrial hazard and human-rights violations. ICJB has also supported the efforts of workers to document the negligence in the Carbide plant in the years preceding the disaster as part of a broad attempt to undermine Carbide's "sabotage theory" wherein total blame for the disaster is placed on a "disgruntled worker" who sought revenge against the company through spoiling the contents of the MIC storage tank.

PRADEEP: Carbide's sabotage theory was first fully presented at an independent conference of chemical engineers in London during the spring of 1988. The paper "Investigation of Large-Magnitude Incidents: Bhopal as a Case Study" was presented by Mr. Ashok Kalelkar, a representative of the public relations firm of Arthur D. Little. Mr Kalelkar's primary goal was to challenge the water-washing theory put forth by journalists and by India's Central Bureau of Investigation, a theory constructed through the testimony of plant workers.

The water-washing theory implicates Carbide management both for decisions immediately prior to the disaster and for long-term processes of plant design, maintenance, and personnel training. In his refutation, Kalelkar insists that "salient, nontechnical features" be investigated, requiring "an understanding of human nature in addition to the necessary technical and engineering skills." Kalelkar then refutes the water-washing with a technical analysis grounded on the claim that "there is a reflexive tendency among plant workers everywhere to attempt to divorce themselves from the events surrounding any incident and to distort or omit facts to serve their own purposes."

Because of this "reflexive tendency" among workers, Kalelkar bases his investigation on the accounts of peripheral witnesses. A primary witness was an instruments engineer who claimed that, on the morning following the leak, he noticed that the pressure gauge on tank 610 had been removed, leaving an opening through which water could have been injected from a nearby hose. It took Mr. Rajan over one year to remember this detail, after which he was comfortably relocated to Bombay. The other primary witness was a twelve-year-old tea boy, retrieved after much effort from his native Nepal. The tea boy is said to have been on duty the night of the disaster and describes a tense atmosphere just preceding the leak, thus "verifying" that all workers on-site were involved in a conspiratorial cover-up.

RITA: Carbide's sabotage theory pins the cause of the disaster on a five-minute interval when a "disgruntled" worker attached a water hose directly to a storage tank filled with methylisocyanate. No mention is made of the reason why there were no safety mechanisms to prevent unauthorized inputs. No mention is made that, according to Carbide's own regulations, the tank should only have been three-quarters full and an adjacent tank empty to allow for transfers in case of emergency. No mention is made that there was no early warning system that would have tracked subsequent rises in temperature, indicating that disaster was imminent. No mention was made that four of five major safety systems failed to work due to indifferent maintenance. No mention was made that these systems were all underdesigned to accommodate mass escape of gas at pressures up to 720 pounds per minute.

PRADEEP: Workers challenge the sabotage theory because it shifts liability away from management and diverts attention from the total context. T. R. Chouhan, who is writing a book on the subject, challenges the sabotage theory with detailed documentation of the safety lapses, maintenance failure, and general negligence in the plant since he was hired in 1979. These details attempt to breach the claim that the cause of the disaster was one nonmanagement person's activity in an interval of less than five minutes. Chouhan insists that the whole story has more of a history, both within the plant itself and within the broader processes of transnational capitalism.

RITA: Generally, Chouhan situates Carbide's activity within a changing market for chemical agricultural inputs. When the MIC unit was set up in Bhopal in 1979, a plant design was adopted which allowed bulk storage. This was done in anticipation of selling MIC produced in excess of their own formulation requirements to other suppliers of MIC-based pesticides. Clearly, this indicated expectation of a huge growth in demand, as no doubt promised by Food and Agriculture Organization (FAO) officials, especially since MIC-based pesticides were being promoted as environmentally sound. It was not anticipated that farmers would be disillusioned by the cost or by increased crop vulnerability to new pests. It was not anticipated that there would be famine years in which farmers were simply unable to produce the cash necessary for purchase of chemical treatments. Also, no one expected the spread of small pesticide producers who relied on cheaper components.

The Bhopal plant was licensed to manufacture 5,250 tons of MIC-based pesticides. However, peak production was only 2,704 tons, falling to 1,657 tons in 1983. In 1984, the Bhopal plant manufactured less than one-quarter of its licensed capacity. Union Carbide India, Limited, was thus directed by Union Carbide, Hong Kong, to close the plant and prepare it for sale. When no buyer was available in India, plans were made to dismantle the factory and ship it to another country. Negotiations toward this shutdown were completed in November 1984.

Chouhan's book includes graphs showing the connection between the changing market context and the management decisions leading to the disaster, retelling the oft-told story of how corporate loss is externalized and the risk burden borne by workers and communities. Chouhan's documentation of intensified personnel reduction runs a parallel chronology to Carbide's plans to relocate.

PRADEEP: It was not, however, only after dramatic losses that conditions at the Bhopal plant were unacceptable. Chouhan reports that as early as 1978 workers joked that "anything could happen in this plant." The event which provoked this remark was a fire in the plant during which the company fire truck was sitting up on a jack with all tires removed. The company "managed" the fire by removing outspoken union leaders and entertaining government officials at posh hotels. Periodic safety audits were managed in a similar way. Years before the disaster these audits pointed out major safety concerns and explicitly warned of the possibility of a runaway reaction in the MIC storage tanks. They also recognized that there was no evacuation plan for the surrounding community. Bhopal managers told Carbide headquarters in Danbury, Connecticut, that to coordinate such a plan would overly publicize the dangers posed by the plant, which had to be avoided since they had recently avoided relocation away from the city by having their production labeled nonnoxious by zoning authorities.

RITA: Despite this history of negligence, the Bhopal litigation was settled out of court. Carbide shares rose two dollars, or seven percent. That same year (1989), Carbide doubled its profits from chemicals and plastics business. According to business analysts, these improvements were not due to improvements in management. When shares plummeted just following the disaster, Moody's lowered Carbide's rating to the lowest investment grade.

Their reasoning stated "fundamental weakness" in the firm's day-to-day business operation. Carbide's rise after the settlement wasn't seen as a reversal of this weakness but merely as a market correction, an effect rather than a cause.

KIM: There was an article in the *New York Times* discussing this. An analyst from Solomon Brothers was quoted as insisting that "the so-called turnaround at Carbide is a result of the improvement in petrochemical markets and not much more." The same analysts nonetheless described the settlement in euphoric terms, saying that "psychologically, it's terrific. Financially, it's reasonable. This relieves the pressure on Carbide, and the stigma."

RITA: Bud Holman, Carbide's chief legal counsel in the United States, was interviewed by the *New York Times* and he described the settlement as almost revelatory, opening a new world for its beneficiaries. He described work toward the settlement as "like walking up a pitch black, winding staircase, and you never know how much farther you have to go in darkness. Then, all of a sudden, it's light and it's all over."

PRADEEP: Holman's client got off for $470 million and with no liability legally established. All but a small portion of the settlement money was covered by Carbide's insurance. Estimates of projected rehabilitation costs are over $3 billion. And Holman's only regret? That the case was not settled earlier.

RITA: One way to understand the whole of our effort in response to the Bhopal disaster is as a challenge to Bud Holman's interpretation of history. Yet, we are the first to admit that suffering can never be recuperated in historical terms, rendered real through words or images.

PRADEEP: A tension between the treason of telling and the treason of forgetting has haunted efforts to keep "Bhopal" alive. What we try to remember is that "Bhopal" is a symbol. Writing Bhopal brings things together but there can't be a complete telling, a closure of the account.

RITA: Accounting for Bhopal could easily get bogged down in a morass of complicities and contradictions. Daily work on specific problems is what grounds us. We have a very paradoxical relation to data. We spend much of our time arguing that the problems are beyond technical description, that they can not be accounted for within market analysis or through any documentary proof offered by victims. Yet we also, always, offer our own accounting, fighting over what counts as "the truth," the "whole story."

PRADEEP: As we mourn the treason of telling, we also know that over 600,000 claims have been filed; 521,262 people are geographically acknowledged as exposed and, according to the Indian Council of Medical Research (a relatively autonomous but still government-guided and thus conservative institution), 32.52% of those exposed (169,514 people) are overtly symptomatic, with that number increasing year after year. Lung function tests show 97.5% of the exposed population (508,230 people) as having small airway obstruction, which could lead to future morbidity and now impairs the ability to work. No information is available to document or predict the onset of long-term effects.

Activism and Expertise

RITA: What is it to be an environmentalist when the label is shared by George Bush and a tribal from the hills of India? Is the mine worker who cooks his rice with firewood from ancestral lands a "forest thief," as conservationists contend? Are gas victims of Bhopal part of the problem or part of the answer?

These are questions we ask in Bhopal, and elsewhere in India, as part of a broad effort toward indigenous environmentalism. The difficulty of answering these questions reveals how environmentalism itself has become a contested terrain that provokes radical questioning of traditional leftist strategies and their vocabularies of opposition.

PRADEEP: *Environmentalism* entered mainstream Indian vocabulary after the United Nations conference on the environment at Stockholm in 1972. At this conference, Indira Gandhi gave a major speech insisting that "poverty is the greatest evil." Her rhetoric was part of a general argument that the first world was using the "pollution problem" to inhibit third-world development. This accusation stimulated awareness of the conflicting agendas called for by simultaneous plans for rapid industrial development and increased regard for environmental concerns.

In India, this conflict was responded to with the 1976 constitutional amendment which made protection of the environment a constitutional mandate. By 1980, India had created a Department of the Environment to carry out the mandate through research, public education, and legislation. Simultaneous with the development of an environmental regulatory structure, there developed an organized response in the voluntary sector. There was also a slow but steady response within the commercial sector. Some management strategies began to acknowledge that the total cost of production involves environmental degradation. More commonly, like their business partners in the West, Indian business learned that "going green" is an important public relations strategy.

RITA: Indian environmentalism was quickly mainstreamed. The state and its beneficiaries adopted the language to modernize independence-era rhetorics of socialism and democracy. The first environmental law in India was passed during the "emergency," a time of terror when Indira Gandhi suspended the constitution after the courts declared her election a fraud. Legislation to save trees and tigers occurred alongside censorship, imprisonment without trial, and forced vasectomies. Elite environmentalists received world renown for saving tigers. The displacement of tribal peoples from ancestral lands was not mentioned at award ceremonies.

PRADEEP: Response to the mainstreaming of environmentalism has occurred within a complicated context. Historically, the Indian Left has been divided by the seemingly oppositional agendas of Gandhianism and Marxism. Both the means and the content of socialism were disputed. Perhaps most centrally there were radically different perceptions of the appropriate role of science and technology in the development of Indian society.

Traditional Gandhians argued that industrialization would require centralized and authoritarian governance. Small-scale, village-based production was seen as the institutional basis for democratic social organization. The means to socialism was to be a covenant of trusteeship wherein the rich would assume responsibility for their less-fortunate neighbors. The logic was of harmonious interrelation through which differences of interests could disappear.

Traditional Marxists based their agenda on a conflict model of change through which the oppressed classes would appropriate the state apparatus to develop a national, industrial economy. Industry was conceived both as the means to provide for mass needs and the institutional basis for democracy, through factory organizing.

By the 1970s both these agendas were undergoing critique from within. Gandhians were confronted with the impossibility of localized economy and politics with the spread of Green Revolution agriculture. Dependence on off-farm inputs of high-yield seeds, fertilizer, and pesticides made it impossible to retain village autonomy. Environmental degradation contributed to the problem by destroying local means of sustenance. Environmentalist legislation itself only exacerbated the problem by taking away access to communal lands and by bringing in outside authorities for environmental control.

Marxists were confronted by the continued failure of factory-floor bargaining, ensured both by state control of most industry and by the numerical fact that eighty-five percent of the Indian population was still rural. These confrontations led to splits in the Communist Party due to different perspectives on the proper basis of organizing, whether urban or rural. Urban-based organizing, following Soviet focus on the industrial workplace, was criticized for privileging workers in ways which elided the interests of those still lower in the social structure, particularly the rural masses. Rural-based organizing was criticized for cohering around the middle interests of small landholders instead of around the needs of agricultural laborers. All sites of organizing were critiqued for compromise connected to constituency building and the wager for state power.

Both Marxists and Gandhians confronted the overall rhetorical impasse created by a nationalistic, independence-era commitment to socialism. Whatever their policies, governments at the center spoke a language of social and economic uplift that was to reach the poorest of the poor in the farthest village. Opposition of all forms was thus caught by language, held by a rhetoric that progress was on and disaster behind them—caught by a history which marked development with aggregated accounting like the GNP and with the number of trained scientists rather than by female mortality rates.

Thus, shared challenges emerged. The initial move of opposition, whatever the ideological agenda, was to challenge the Nehruvian vision and rhetoric which had dominated political discussion since independence. The Nehruvian vision promised an equitable society following in the wake of massive technological advance. Logically, it elides issues of social structure through faith that increased material prosperity will naturally cure all social ills. Social stratification based on caste and class are not addressed. Attention not given to the inequitable distribution of both risks and benefits that accrue through industrial development.

RITA: The shared challenge of contesting Nehruvian politics was thrown into high relief by the Bhopal disaster. Union Carbide's pesticide plant in Bhopal was to bring science to India. Methylisocyanate, the chemical released over the city, was a component of the pesticide Sevin. Sevin was an essential requirement of Green Revolution agriculture, particularly since it was being promoted as "environmentally sound." According to Nehru, the Green Revolution was to provide national food security and a surplus labor force to power industry that would provide for all needs, whatever caste or class.

Instead, Union Carbide brought the Bhopal disaster. The communities exposed to the leak of methylisocyanate were among the poorest in Bhopal, slum-dwellers who had no legal voice after migrating from villages where Green Revolution agriculture had taken away their land. The gas leak was caused by negligent management immune from regulatory intervention because of the need to coddle foreign investment and technology transfer.

KIM: Immediately following the disaster, how did the Left respond?

RITA: Voluntary efforts to combat the suffering and uncertainty in Bhopal came quickly. Initially, organization was completely ad hoc. Despite their exhaustion and sickness, victims themselves showed extraordinary resourcefulness and compassion. They were joined by outsiders who flocked to the city to render any assistance possible. Within days, a *morcha* (organization) was formed to coordinate these efforts.

The *morcha* brought together activists from diverse backgrounds. Local people worked alongside professionals, academics, trade unionists, and others in a mammoth effort to respond to the urgency of the situation. In so doing, however, the *morcha* did not simply prioritize relief and relegate critical perspectives to secondary importance. From the start, the *morcha* was as concerned with means as with ends. There was continual attempt to evolve a democratic organizational structure, with particular attention to the role of local people and women.

Further, the *morcha* took a very critical stand on Bhopal as an issue in the ongoing debate about the role of technology in the development of Indian society. The *morcha* represented the disaster as the outcome of economic planning which prioritized high-speed growth and imagined bureaucracy as the appropriate response to unequal distribution of its benefits. Perhaps most significantly, they questioned the ability of technology to solve problems of its own creation. The *morcha* worked on the line that there can be no true rehabilitation or justice in Bhopal, that there are "no technical solutions to social problems."

PRADEEP: The *morcha*'s exposure of government culpability was not, however, limited to a critique of planning orientations. The *morcha* also exposed the very direct blame of local officials in allowing Union Carbide to be operated so negligently and so near residential colonies. They also exposed the inefficiency and corruption of official relief efforts, publicizing the government's attempt to avoid welfare responsibilities through overall denial of the magnitude of the problems. However, the *morcha* recognized that the voluntary sector could never provide more than a modicum of the required services and thus engaged in a demand campaign to push toward the greatest possible government accountability.

RITA: Despite the frustrations, many activists have remained committed to work on the issues of Bhopal. The orientation of this commitment was seen at a national convention, held in Delhi 8–9 April 1991. The "National Convention on the Bhopal Gas Leak Disaster and Its Aftermath" was organized to revitalize awareness and support for voluntary-sector work. Participants included journalists, trade unionists, representatives of political parties, and activists from a range of people's struggles.

The conference concluded with a powerful statement by Praful Bidwai, a noted Delhi journalist. Commenting on his reporting experience, Bidwai stressed that, "Given the limitations of state intervention and technology based solutions, we must ask if there is a different way of organizing society. Within the existing structure of state, there is an accepted notion that certain processes are excessively toxic in relation to conceivable benefits. We must extend the logic. Nuclear power as a means of electricity. Cyanide as a means to electroplating. Sevin, Temik as pesticides. Technology must be differentiated. We must disallow capital's domination of social considerations."

PRADEEP: If you are going to cite "capital's domination of social considerations," mention must be made of the way "Bhopal" creates the very structure of our own privilege, albeit ironically. Many of us contesting the technological fiasco of Bhopal are trained as technics, often at the most elite institutions in the country. As professionals, we are part of the surplus of the Green Revolution. An extraordinary number of activists in the Indian Left are trained as scientists, engineers, or medical doctors. Yet a primary leftist critique is of the role science and technology has played in the development of Indian society. This obliges us to continually ask if there is any possibility of an "appropriate technology of expertise."

KIM: Questions about the connection between scientific expertise and environmentalism are also debated in the United States. Many people question whether institutions like NASA "found" the hole in the ozone just when they were threatened with major funding cuts due to the end of the cold war. Similar questions are directed toward advocates of the Endangered Species Act, drawing a connection between the "fragile environment" of biological research and the evolution of careers in species protection. Even in social science there are questions to ask. The emergence of "science and technology studies" as a field of enquiry was part of an overt attempt to reposition the relation of the university to broader contexts, taking responsibility for the ways research interacts with wider social processes. Yet, as part of this effort, it is possible that we participate in the fetishization of science and its mechanisms of legitimacy.

RITA: In India, the irony is also structural, but with deeper roots. Since independence, government allocation for education has been skewed to privilege the development of a technocratic elite. As a result, India has both severe underdevelopment, partially marked by an extremely high rate of illiteracy, and the world's third largest professional class. In 1981, there were more than three million Indian scientists, half of these living in the country and half living abroad.

PRADEEP: But of those at home, about twenty percent are unemployed and the majority of others are in nontechnical, administrative jobs. In 1981, only about fifteen thousand scientists were doing creative research. And this number probably includes the few of us officially unemployed but doing "creative research" against the scientific establishment itself.

In India, as elsewhere, science is a mechanism of state legitimacy. Its purpose is show and control, not public welfare. We are trained as doctors to carry the authority of science, as emblems of India's progress. No matter that we have done little to improve the overall health of the nation.

RITA: The problem is not expertise itself but in the way it circulates. The problem is with the social and economic system in which expertise is embedded. Thus, by starting with acknowledgement of our own privilege, we break open a much larger critique. We come to admit that health is not merely a matter of physical care through expert intervention but a matter of structure. The appropriate response cannot only be pharmaceutical.

Real response to the health crisis in India will require major shifts of both political economy and culture. We must both allocate resources differently and coin a definition of health that is appropriate for the Indian context. By default, we have always relied on Western models. The efficacy of these models is unquestioned so even the most progressive political programs advocate simple extension. The contextual and institutional requirements of Western medicine are never addressed. Nor are the side effects.

PRADEEP: The ignorance is feigned. The problem is not that we are bound by Western logic. The logic is a veiling mechanism that hides failures of political will. Advertising the authority of science is a Machiavellian move which allows both scientists and politicians to protect their social position.

The first health plan after independence was based on the report of the Bhore committee. While the recommendations did recognize the need for decentralization through district-level programs, it retained commitment to a universal schema. This was evident both in its recommendations for standardization of medical care between urban and rural areas and through assumptions that public health was dependent on the extension of allopathy. Implicit to this was a bias against indigenous systems of medicine and all other context-specific programs.

Attempts to standardize healthcare provision on the Western model were recognized as a failure by the 1970s. Blame was primarily put on the insufficiency of trained health workers, which was exacerbated by the declining legitimacy of traditional health practitioners. So in 1977 the government introduced a program for training village-based health auxiliaries who were to implement a combination of allopathic and indigenous skills. But the purported recognition of indigenous skills was really only an effort to relegitimate traditional workers and accrue their efforts as part of the government's success.

The continued bias against indigenous medicine was seen in the training programs for health auxiliaries. The workers were not seen as holders of a knowledge system that could be tapped for greater provisions of health care, but as clean slates upon which new theories could be written. Knowledge about indigenous medicine was purely anthropologic, for the purpose of control, not integration. This attitude toward traditional workers was even made policy by the World Health Organization (WHO). In a 1966 report, the WHO recommended programs to increase the skills of traditional birth attendants because of the "repeated failure of legislation to prevent them from practicing." In 1985, the WHO had progressed to the point of recognizing traditional practitioners, "given the scarcity of modern healthcare personnel."

RITA: In failing to recognize indigenous medicine, India is not only blindly acquiescing to Western models but also neglecting the example of countries like China. China has benefited from allopathic medicine but in conjunction with traditional practices. Toward egalitarian distribution of healthcare resources, China now has over one and a half million "barefoot doctors" who offer village-based services which combine the medical knowledge of allopathy, Confucius, and that of minorities such as the Tibetans and Mongolians. Chinese hospitals also combine different traditions of medicine, giving equal status to all practitioners.

Witness the comparison. According to 1989 figures in the World Development Report, India had 2520 persons per physician; China had 1010 persons per physician. Yet female life expectancy in India was only fifty-nine years, as compared to seventy-one years in China. The maternal mortality rate in India was 500 per 10,000 births; in China, it was only 44 in 10,000. In India, the mortality rate for females under five years of age was 134 per 1000; in China, only 31 per 1000.

PRADEEP: Chinese attempts to integrate different traditions of medicine has many lessons for India. It is clear that allopathy has not offered sufficient remedy. There are problems of both resources and organization. Yet we ignore possibilities within our own tradition. In contrast to the Bhore committee's recommendations for standardization, *ayurveda* insists on the importance of context. The books of recipes for herbal remedies are written for specific areas so that local material can be used for formulation. This follows the dictum that for a person belonging to a particular country or region, *aushadhis* from the same region are the most wholesome.

The emphasis on regional integrity is broadened by the insistence that *shaastric* principles are not meant for mechanistic application. An important *shokla* insists that "a *vaiddya* who comprehends the principles of Rassa would discard treatment if not wholesome to the patient in a given situation, even if it is prescribed in the texts; on the contrary, he would adopt treatments that are helpful, even if they do not find mention in the texts."

RITA: The Indian policy of developing human assets stands in marked contrast to that adopted by Maoist China, where education policy reflected the fear that merit-based advanced schooling would nurture capitalist rationalizations of hierarchy. It also ignores Gandhi's attempt to disrupt the elite's hold on Indian culture by emphasizing anti-intellectualism and the ethical superiority of folk Hinduism over the Brahminic rites based on Sanskritic texts. Prioritizing elite, scientific control over nature is a logical extension of Brahminic obsession with control over personal pollution. Thus the religious fervor with which our parents push us through the technical education process.

The Indian fetish for science and technology has roots in both traditional, caste society and in the American faith that material abundance will automatically cure all social ills. Nehru followed this faith in his assumption that industrialization would alleviate the need for land reform and other challenges to Brahminic control.

PRADEEP: Nehru's commitment to technological self-reliance was tragically ironic. Instead of promoting an indigenous technology which accommodated local conditions, his policies sought to indigenize foreign technology by pouring resources into research toward import substitutions, efforts to copy technologies developed outside. We thus play a tragicomic game of chase, never getting the work done before it becomes obsolete.

RITA: Nehru may have honestly imagined capital intensive technology as the means to socialism. His failure may not of been in intent but in not conceptually connecting means and ends. This would have required a move beyond his famous scientism, which he was incapable of. Scientism posits necessary development to a certain conclusion once conditions and variable are defined. What it forgets is the impossibility of such definition, that a new society is not built on a tabula rasa on which all parties can be equally contracted, once the law designates as such. Nehru did not recognize any need to be specific to the Indian situation, except in retaining Brahminic control.

KIM: Do the critiques of Nehru's scientism lead to antiscience positions?

RITA: There are many different critiques, often coming from different ideological traditions. One task of the environmental movement in India is to collaborate the different critiques into a shared political platform.

PRADEEP: Questions about the role of science and technology in the development of Indian society have been a part of Indian political discourse since independence. Popular discussion often opposes a Gandhian, antitechnology position to a Marxist faith in industrialization. Environmentalism has brought these positions into collaboration. A primary tension within environmentalism is caused by the need to recognize science and technology as resources, both material and methodological, and as mechanisms which sustain environmental devastation and social inequality.

Negotiation of this tension has created vehement argument. There are antiscience positions which insist that science is authoritarian, both institutionally and in the way it organizes knowledge. There are arguments for an indigenous science grounded in Indian cultural and religious traditions. There is advocacy for the validity and scienticity of the knowledge systems of tribal peoples. There are arguments for a "people's science" which would use mass educational campaigns to "de-expertise" scientific method as a basis for disruption of traditional structures of exploitation.

RITA: The People's Science Movement bases its work on the argument that teaching scientific method to the exploited classes will allow development from the ground up. They argue that this will both undermine traditional exploitative institutions like caste and "de-expertise" technology, taking control away from elite interests.

Kerala Sastra Saitya Parishad (KSSP) was formed in 1962 under the slogan, Science for Social Revolution. In 1978, KSSP organized the first convention of the People's Science Movement, hoping to build a network of activism committed to the democratization of scientific knowledge and resources. The success of the efforts was particularly visible in 1987 when the Bharat Jan Vigyan Joshan brought twenty-six organizations to Bhopal to express solidarity with the demands of gas victims.

The initiatives of KSSP have included appropriate technology projects, activism against environmental destruction, and educational projects, including innovative use of traditional folk forms to popularize scientific analysis. KSSP initiatives are undertaken outside party politics, principled on inclusion of all those who share the view that science can be used to advance both the separation of politics from the people through the party process and the separation of legitimate knowledge from social processes.

PRADEEP: But KSSP is not without problems. Because of its commitment to science, there has been slippage in their commitment to other issues. The most heated debates have revolved around their reluctance to support campaigns against nuclear power. KSSP activists seem to understand science only as a knowledge system, and not as a way of organizing people. The antinuclear slogan insisting that "nuclear power mandates authoritarian governance" does not make sense to them. Nor does it make sense that, sometimes, science itself fails.

RITA: Critics of KSSP are often those who argue for more indigenous models of change, arguing that science is a legacy of colonialism and that "people's development" can only come through revalidation of local knowledge systems. These activists point out that the sustainable life-styles of tribals have been encroached on both by material appropriation of their resources by "scientific" development schemes and by the cultural legitimacy of scientific thinking, which labels their knowledge as superstition.

PRADEEP: Difference of opinion on the proper role of science in society has been mirrored within medical care initiatives in Bhopal. There has been enormous conflict over appropriate means of intervention. One set of doctors insists that the push must be for state-of-the-art care on the Western model. Their emphasis is on making resources now only available to elites available to marginal and victimized sectors. They want better government hospitals with comprehensive laboratory facilities and full availability of medicines, whatever the costs. These doctors are often those with previous or ongoing ties to Marxism. To speak with them, you must join the long queue at some government clinic where they are trying to be functionaries in the slow building of a progressive state apparatus. These doctors are often very hostile to voluntary-sector efforts as a diversion from this traditionally Left agenda. They also see praise of indigenous medicine as a fashionable excuse for continuing the deprivation of the lower classes.

The other set of doctors consider government hospital care continued victimization. They insist that allopathic medicine harms as much as it helps, particularly since its promoters are pharmaceutical corporations. They point out that even the diagnostic basis, particularly X rays, are invasive. They also insist that the social relations maintained under allopathic care depend on expertization and thus sustain dependency. These doctors push for decentralized, community-based care using local health workers and depending as much as possible on nondrug therapies such as nutritional supplements and exercise programs. These doctors try to avoid allopathy as much as possible and, instead, rely on the knowledge of both local traditions and indigenous systems of medicine such as *ayurveda* and *urani*.

RITA: The critical tradition followed by the latter set of doctors is often that of Ramanohar Lohia, a prominent Indian socialist who worked to reconcile the insights of Gandhi and Marx. By the 1950s Lohia was already pointing out that Eastern Europe had put their technics to the same work as the West, prioritizing capital intensive industry, particularly the military. He argued that it was inevitable that their "development" would come to crisis; but he saw the failure not in socialism but in the organizational structure of technology. As one activist interprets, "you can't accrue wealth to elite institutions and still expect democratic rewards. The process is the same, East or West."

PRADEEP: Lohia's arguments are seen as particularly relevant today because they remind that technology carries politics, that the adoption of any technological form requires a certain structure of society. Activists explain how the determining nature of technology is blatantly seen in an artifact like the atom bomb, or even in nuclear power generators. Due to the hazards associated with these technologies, there must be a control structure commanded by a technical elite which is insulated from all influences that might distract from performative excellence. Democratic involvement would disallow necessary control. Antinuclear activists thus argue that "to accept nuclear technology is to accept authoritarianism."

RITA: Less obvious are the ways capital and energy-intensive technologies of all kinds beget elite control over all aspects of society. Activists thus insist that it is not only the technology itself which is controlled, but all it's produce, including new ways of thinking about the purpose of society and the role of individuals within it.

A pesticide plant like that in Bhopal not only had a rigidly hierarchical chain of command within its premises but throughout the multinational corporate structure of Union Carbide. Thus, accountability for the Bhopal disaster can be traced from the managers in Bhopal, to Indian headquarters in Bombay, to regional headquarters in Hong Kong, to general headquarters in Danbury, Connecticut. And the control ultimately exercised by executives in Danbury is not limited to control over manufacturing facilities.

A local economy dependent on rapid growth gave Carbide influence over the functioning of the city of Bhopal and more generally over the state of Madhya Pradesh. A very direct indication of this was the site choice for the Bhopal plant. The initial site choice situated a hazardous production facility adjacent to already existing residential communities and barely half a kilometer from the main railway station. The plant fell out of compliance with city ordinances when, in 1974, it began manufacturing rather than simply formulating pesticides. As the twenty-first largest company in India and employer of over ten thousand people, Carbide was sufficiently influential to override the city's objection. Even politicians argued on Carbide's behalf, saying that a factory "was not a stone that could be picked up and moved at will."

PRADEEP: But the real grip of Carbide was broader still. An agricultural sector dependent on chemical inputs gave Carbide influence over farmers across the country. In turn, rich farmers increased their power as small growers lost access to surpluses necessary to maintain the reproductive cycle.

Dependence on industrial agriculture creates dependence on imports of both technology and petroleum and thus requires a local economy geared for export to generate foreign exchange. Cash cropping thus displaces subsistence-oriented farming and transforms the very logic of traditional agricultural production. "Value" is no longer tied to utility but also based on markers defined far from home.

RITA: Responding to questions about the appropriate role of technology in Indian society has brought the fissures in the traditional Left into sharp relief. Yet there is also awareness that none of the ideologies relied on in the past are sufficient. As one activist explained, "We have often called treason on each other failing to realize that oppositional politics now requires a crossing of allegiance that was once abhorred. We are operating in a time when the integrity of the Left must itself be radically questioned. This requires redefinition of what constitutes allegiance and loyalty."

PRADEEP: A friend of mind described the task as that of a bard, insisting that "we must narrate a new story that admits to the exhaustion of old blueprints, a story that carries traditional leftist aspirations but with acute sensitivity to our placement in the final years of the twentieth century. Old aspirations must be upheld without nostalgia or unwillingness to admit the failures of prior efforts. Tiananmen Square, the 'fall' of Eastern Europe, even the social distress that has accompanied the apparently successful welfare states of Scandinavia. All this must find a place in the narrative.''

Subverting tradition from within is a primary strategy. Activists insist that when they rewrite the Left they must make it an Indian story, but that does not mean a return to national chauvinism or even simple inclusion of dominant cultural motifs. It can not be forgotten that efforts occur in a context of rising religious fundamentalism, institutionalized by the political parties.

We must take advantage of the fact that definitions are politically constructed, especially since the Hindu fundamentalists know this also. They can't seem to decide if they should seize the term *secularism* for themselves, insisting that Hinduism is the only truly tolerant religion, or if they should damn *secularism* as an imported term, making an overt call for a theocratic state on moral terms. Meanwhile, they shout about "the perversion of Indian political parlance'' wherein certain words have "defied dictionaries.'' A Communist Party pamphlet quotes them bemoaning the fact that "contradistinctive labels—leftist vs. rightist, progressive vs. reactionary,'' haven't been apportioned by an "impartial'' tribunal like the Election Commission. This exposes the fascists they are. Control language and control the world, no? Or at least one of our most effective means of resistance.

RITA: Activists must admit to a compromised position, often reminding themselves of the irony of how much they are like those they fight. We speak their language; we are experts in the very forums which are the basis of contemporary exploitation. The question is whether the resemblance can become a means of resistance, subverting from within. Can we return the look of surveillance? Can we forge an appropriate technology of expertise? It's a dubious battle.

PRADEEP: Our resemblance to our enemies could be our greatest weapon; it is what they fear most. It was this fear of hybridization that made the Eurasian such a menacing figure in Rudyard Kipling's work. The child of an English official and an Indian woman was never allowed to live; it posed a real threat to the distinction between ruler and ruled that justified the Raj. And the lovers themselves always suffered terribly for their transgression. In one story, the man was stabbed in the groin and the woman's hands were cut off at the wrists.

RITA: The adult Eurasians that filtered through Kipling's work always epitomized evil, symbols of the greatest moral threat to Mother England. The culturally Europeanized Indian man was not so much of a threat because he was never quite there, always a child in his mimicking of gentlemanly gestures. As activists we must reappropriate the moral threat which the Eurasians represented.

Kathryn Milun(aitis)

Of Beets and Radishes: Desovietizing Lithuanian Science

The Ignalina nuclear power plant built in the eastern part of Lithuania during the Brezhnev administration was meant to be the largest nuclear facility in the world. A product of the same gigantomania that swept the United States during the cold war, Ignalina now stands as a monumental albatross to the four million inhabitants of the tiny republic which is now struggling to disentangle itself from the wreckage of the former Soviet Union. While it is the main source of energy for the country at the present time, it is also a source of great anxiety: the reactors are built on the same model as those at Chernobyl. Lithuanians were able to stop the extension of the plant in 1980 but the oldest reactor is currently running at full capacity while a second runs at fifty percent. Ignalina reveals the ecological, political, and economic problems that follow in the wake of the Soviet propensity for centrally managed large-scale technologies with an ethnic division of scientific labor.

"As far as I know there are no Lithuanians running the plant. Most of the technicians and physicists are Russian." Rimantis reminds me that, as in the U.S., nuclear plants were obviously the domain of the military and therefore surrounded by secrecy. In the last two years Lithuanians have had their neighbors the Swedes come in to assess and monitor the situation. "In Lithuania we didn't develop scientific expertise in the area of nuclear physics, but we'll have to do this as soon as possible." Rimantis Slizis is now the director of the Lithuanian Science Council, a group of thirty-six scientists who advise the president on issues of science policy in research and education. The Swedish team has told them that while the immediate addition of certain costly safety features is essential, the plant must nevertheless be closed down as soon as possible. It's a double-bind. "We desperately need the reactor to solve the energy problems of Lithuania, but if we don't follow the experts safety advise, there may not be a Lithuania at all."

Does the breakup of the Soviet empire leave Lithuania the ominous distinction of being the smallest country with a nuclear reactor to manage? Armenia, another former republic of the Soviet empire, opted to shut down their reactor in 1989 (although there is talk of reopening it again soon). For the small Baltic country issues of national security appear paramount: as Rimantis reminded me in the winter of 1992, one hundred thousand Soviet soldiers who take their orders from Moscow are still stationed within the borders. And despite their success against Moscow's attempt to crack down on national independence in January of 1991, Lithuanians still feel that their fate follows upon the political stability of Moscow. Echoing that sentiment, two years after that bloody January coup attempt, Lithuanians have still not removed the barbed wire and concrete barricades which surround their parliament and block major thoroughfares in the capital.

What will be the legacy of the large-scale technological and scientific organizations which were set up for the centralized rule of the Communist Party over its colossal holdings? "Science in Lithuania," Rimantis tells me, "was completely tied to the Soviet Union for the last fifty years." Scientific specialization by region was not uncommon. Rimantis, for example, is an electrochemist. He tells me that Lithuania "has the largest institute in the world for a very narrow subfield of electrochemistry. More than two hundred Lithuanian scientists work at this institute which has over five hundred support staff. This particular branch of electrochemistry was even encouraged in the Lithuanian school system because it fed the needs of the Soviet Union." But the country no longer has a need for such specialization, and, in fact, Moscow stopped funding the institute in 1990. "It's very painful now, our economic situation. It's not that we're doing poor work. Our scientific standards are excellent. Rather the problems lie in the fact that some fields are disproportionately large for our population."

In February 1992, Rimantis and I were discussing the tribulations of Lithuanian science in the living room of my relative Voldis' apartment in Vilnius, the capital city. Rimantis lived just upstairs in what turned out to be an apartment building filled with Lithuanian scientists. Voldis is my father's first cousin. He is a veterinarian who directs Lithuania's main laboratory where animals were kept and bred for medical experiments. Since the collapse of the Soviet Union all such institutes were left to earn their income by their own wits, and like most Lithuanians, Voldis looks to the West when thinking of a way to support his lab. We joked about setting up a Western market for the well-documented laboratory mice he raises. "In the United States these animals go for about thirty-six dollars apiece. I think I could do it for ten," he said. I imagined myself picking up the first cratefuls of these genetically transparent critters at the Intercontinental Airport back home in Houston and trying to help him make his first connection in the U.S.

Actually, it's an interesting question, the role of intercontinental family ties being called on after two or three generations in order to help set up joint market ventures. This was my first trip to Lithuania. I had very faint memories of my father's mother who had left her family's prosperous farm in Lithuania in 1913 to follow her husband to America. My father grew up in Kearny, New Jersey, where there was a large Lithuanian and Polish population. His own father, despite being a closet socialist and student of Esperanto, was a strong nationalist and named all his children after Lithuania's epic heroes and heroines. Although the children attended language schools run by the local Lithuanian Catholic church, only my father's sister is still fluent in Lithuanian. She helped me set up this trip.

Voldis is in his midfifties and Rimantis perhaps ten years his senior. They appear to be good friends and sneak off into the kitchen for a quick smoke. In the living room Voldis' wife, Vanda, and I smile at each other. We are almost desperate to communicate but don't speak each others' languages. We've been relying on her son, Audris, who is twenty-six and speaks English remarkably well. Audris has finished his studies in biology and now works for a company that produces color film, a job he finds quite boring and not sufficient to support his passion for video equipment. Voldis' mother, my grandmother's sister, is patiently listening to our cross-fired conversation via Audris. All three generations share this apartment and I am humbled by the respect they show for one another. She is ninety years old and looks like a heavier and happier version of my grandmother. Both she and Vanda have been feeding me constantly since I came for this four day visit. I know how bleak the food situation is and ask them how they've managed to come up with the meat and vegetables they've so kindly been serving me. "We have a small garden outside of Vilnius and every year we can grow plenty of beets, peppers, and the like. But we never thought we'd actually be living off of this the way we are now." Vanda works during the day in a government office that fixes the prices of goods—a most confusing job these days. A few weeks before I arrived you could buy 175 rubles for one dollar. Now it has gone down to 80, still a frightening exchange when you consider that in real values one dollar is worth about 10 rubles. Grandmother has been taking care of me while the others are at work, and despite the language barrier we've managed to get along.

Also in the room with us is Noemi Vanagiene, another neighbor whom Vanda and Voldis have asked to come over to help with translating. (All the Lithuanians present also speak Russian except my grandmother's sister, whose conscious life actually predates the period of Soviet domination.) Rimantas speaks a very halting English and Noemi is a professional translator whose husband, Vlades Vanagas, was one of Lithuania's most famous theoretical physicists. He died of a heart attack in 1990 at Yale University where he was a visiting professor. Noemi's strong, deep voice casts a haunting accent over the English words she commands with such skill. Her voice bespeaks the determination of her character. Six years ago her nine-year-old son came home from school talking excitedly about some brave Young Pioneer Communists he had read about in class who had crossed over an icy river. The next day he went out on the Neris which runs through Vilnius to try this himself. He fell through the ice and drowned. Noemi's other child, a daughter, is studying medicine in Israel. "Most of my Jewish friends have left for Israel in the last few years," she tells me. "Anti-Semitism is on the rise once again, but I love Vilnius. Lithuanian is my first language and I plan to stay here." Noemi's mother, who had traveled to Strasbourg, France, to study pharmacology in the 1920s is still living in Vilnius and Noemi visits her every day. Before the Second World War, Noemi tells me, there were nearly two hundred thousand Jews living in Lithuania. Now there are fewer than five thousand. As a Jewish doctor in the Soviet military, Noemi's father was in constant fear of being sent to Siberia. Especially in 1953 when Stalin, a few months before his death, announced that there was a secret group of doctors linked to an international Jewish bourgeois organization conspiring to kill top leaders of the Soviet Union. Fortunately, Stalin died before carrying out his paranoid punishments. Noemi's husband, a gentile, had a different set of problems with the Soviet state. He gained honors for his work in physics, but he refused bureaucratic pressure to become a member of the Communist Party. "I hated their hypocrisy," Noemi told me, "but my husband always told me not to speak so openly or we would soon be in Siberia. I always said, 'Why don't we debate openly like Sakharov?' But my husband would reply, 'There is only one Sakharov and I am not him. I am not brave enough and I want to work.' "

When Voldis and Rimantis returned we resumed our discussion of the issues facing scientific work in Lithuania today. "One problem we have is with the system of peer review. Our scientific expertise is strong, but we all know each other. How do you review the work of a colleague you know personally. We're a small country. Imagine, even the United Kingdom with a population of fifty million has twenty percent of their expert advice coming from abroad. So we need to get scientists from abroad in all our research institutes and in all our universities. This will cost a lot, but we absolutely need to have a complete reevaluation of our science programs right down to the high-school level. We're trying to have this done in the next year. Practically speaking, we've stopped all reconstruction of our science programs until things have been reviewed by outside experts. Bad expertise is worse than no expertise at all."

I can't help but think of the effects of the cold war on science education in the United States where in many disciplines it was the military that set the paradigm for post–World War II science. Military money flooded both industrial and academic laboratories, as well as departments of social science with contracts for both applied and classified research, redefining the relevant academic research programs. Thus in the U.S. as well, it was the state which had a direct hand in shaping what it meant to be a scientist or engineer, in deciding what should be studied as well as what would be built.[1]

Did the secrecy demands of the national security state that emerged in this era for both superpowers create similar problems in terms of peer review, what some scientists believe to be the scientific parallel to the political structure of checks and balances? President Eisenhower warned of this when in his famous farewell address he identified the threat of the "military-industrial complex" to both political and intellectual freedom. Even before this, in 1947, chief bomb-development physicist J. Robert Oppenheimer observed the value of peer review to the scientific enterprise when he expressed his growing concern over the way in which military secrecy was influencing scientific work, in this case at the Hanford, Washington, site. "To keep secret the fact that we don't know how to do some things," he said, "may be controversial because it may be that we really do need some ideas, and the classification, or keeping secret, of our ignorance of an area in which we haven't been able to make any progress may in some cases be a very serious hindrance to getting the insight, the bright ideas and the progress which would come if a much wider group of people could be interested [allowed access to data]" (Gerber 1992: 215). This complaint over how state secrecy hindered free communication among American scientists and helped structure science education is heard throughout the cold war period. But while these issues of state secrecy—as well as other problems which appear to plague the current system of refereeing in scientific journals (Armstrong 1993)—may have produced similar structural problems for scientific work in both the former and remaining superpowers, there are also some striking structural differences.

"One of the 'achievements' of real socialism," Rimantis continued,

> was the separation of research from teaching. In the United States scientific research is concentrated in universities, but that is not the case here. This separation of research [which is under the auspices of the Academy of Sciences] and [university] teaching is much deeper in the Soviet Union than it was in Poland, Hungary, or Czechoslovakia, but they have similar problems. The situation it creates for science is actually quite different than what happens with the social sciences and the humanities. In the hard sciences, textbooks need to be kept current. Our teachers were using outdated textbooks and were themselves far removed from new developments in research. This produced the awkward situation where proud university professors are still voting to keep the systems separate so they do not

> have to face the changes in their fields. And then there is the problem of the Academy of Science, the research branch, being a division of the Ministry of Science, you know, a branch of the government.

During the cold war Soviet scientists, unlike their colleagues in the humanities and social sciences, were never cut off from reading foreign journals. The problem was rather that in order to publish scientific results in foreign journals, they first had to be published in Soviet journals. "But who reads Soviet journals?" Rimantis tried to explain the dilemma. "We'd get money for our research, but nobody seemed to care about the results. You'd write up the results of your work and then nobody, not even Soviet scientists, would read them." It's a vicious circle. "Expertise, of course, has to do with the citation index. Who are the experts? The ones who get cited. But we would never be cited in the foreign journals and, of course, this did not help our situation." Now Rimantis is lobbying for Lithuania to have a system of grants from both private and government sources which would be competitive and of a limited duration. After the grant period is up, the work would be evaluated to see whether it should be continued. It sounds logical enough, and in fact one reads of similar strategies being mounted throughout Eastern Europe (Lepkowski 1992).

I asked Rimantis to tell me what kinds of problems were arising now with the shift away from Soviet-style science. One of the biggest problems was with teaching. "The professors who taught Marxist-Leninist economy are the very ones who are now teaching modern marketing. You can't reconstruct things overnight, but in fact at the NATO conference which I attended last month [January 1992] we were promised help in the form of foreign expertise on this matter. In Lithuania," Rimantis added, "there have been no history textbooks for two years. The teachers are thankfully no longer teaching the lies of Soviet history, but they are the same teachers who were trained in this history."

Relating his surprise at the lack of historical and geographical knowledge on the part of his Russian colleagues in science, Rimantis asked whether I recalled how in 1991, when Lithuania had declared its independence from the Soviet Union, Gorbachev had claimed that Lithuania had a strong historical tie to Russia. "He claimed that two hundred years ago Lithuania voluntarily saw itself as a part of Russia and that these were very good times. I was astonished. I saw that even many of my scientific colleagues who were Russian did not know history, did not know how Lithuania had in fact been incorporated into Russia. But you know, now many of them have sent us congratulations on our independence. Still others cannot understand why we want to be independent. They think of Lithuania as a part of Russia. If you had asked someone ten years ago whether they remembered how Lithuania had been incorporated into the Soviet Union in 1938, they might have remembered. Even in eastern Germany, if you used the German names for these Lithuanian towns, they too would probably remember. But, come to think of it, I often mix up the names of the capitals in the Soviet republics in Asia. Perhaps it's quite natural that this should happen with Lithuania, Estonia, and Latvia.'' Recalling the notoriety of Lithuania gained throughout the former Soviet Union for holding off Soviet forces in 1991, however, he laughed. "They probably won't be mixing up the names of our capitals anymore.''

Neither Rimantis nor my father's cousin Voldis were ever members of the Communist Party. The same was true of Noemi's husband. I asked them whether being a member of the Party compromised one's scientific work. This question seemed to set eyes rolling and lips curling. "How do you answer such a question?'' Rimantis pleaded. "On the personal level I know what Voldis would say. If you're a member of the party, you're a bad person. My personal opinion is a bit softer than that, but, in fact, I think he's quite right.'' Laughter breaks out. But I persisted. Was this a scientific issue? Were communists bad scientists? I was driving them to culinary metaphors. Voldis: "I would say that a communist scientist is like a cracked egg. Now, a cracked egg may very well be edible, but since its cracked you can't be sure. You'll have to check it carefully to know.'' Rimantis: "We have two sorts of communist scientists and they are like two sorts of red vegetables: beets and radishes. If you scratch a beet, it stays red. But if you scratch a radish, you'll see that it is in fact mostly white. Most of our scientists now are radishes. You scratch them a bit and you'll see. Those radishes must be accepted now as the good ones. As good, cracked eggs in Voldis's terms. You can still use those eggs. As for those who really believed in the ideas of Marxist-Leninism, there was probably something wrong in their brains. Really. I never met a Communist who was a beet and also smart. Never, never, never.'' Audris and Voldis chime in, "Never never.''

Before we broke up for the night, I asked them about Chernobyl. Was Lithuania affected? Rimantis had heard that the southern most parts of Lithuania were contaminated. "Physicists were not able to do or publish research on this issue until very recently, so we will have to see.'' Noemi, who works as a medical translator at a kidney center, tells us that the head of her department said that kidney diseases have steadily increased since Chernobyl. "And there is more cancer, too. The doctors are connecting the rise in slow developing diseases to Chernobyl. They speak quite openly about it now. Perhaps someone will do a study, but right now we are so overwhelmed by the many patients we have and the lack of medicine and equipment. Doctors are doing very dangerous procedures with only minimal means.''

To follow up on how Lithuanian scientists are responding to the ecological devastation of their country, I took Noemi's suggestion and tried to contact

Voldis Adamkus, the Lithuanian-American who has been instrumental in setting up the ecology programs at Lithuania's two major universities these past years. In Lithuania my relatives called me Kathryn Milunaitis, readding that telling suffix that my father had dropped in the 1950s after he became a scientist himself and moved from the ethnically diverse East Coast to the more homogeneous Midwest. So when, in the spring of 1993, I called Adamkus at his office, I used my Lithuanian name. I believe it worked to get through the bureaucracy. Although I assume he is extremely busy as the head of the region, Adamkus was most gracious in responding to my questions from his office in Chicago:

> I've been going back to Lithuania lecturing on the environment as such since 1972 when I joined the EPA. For a long while nothing quite seemed to sink in, then in 1974 I was asked to help create an environmental agency following the model of the EPA. We began with a center for genetic testing and are now focusing on water quality. I can tell you that the situation in Lithuania is tragic, it has come close to the point of no return. The toxicity of the water—well, in all the major cities, Vilnius which has eight hundred thousand inhabitants, Kaunus with six hundred thousand, and Siauliai with three hundred thousand, there is no biological treatment of waste water. Everything is dumped directly into the rivers. I can safely say that there are no unpolluted rivers or streams in all of Lithuania and all this just flows out to the Baltic Sea. Now there is an ethical understanding of the situation and laws actually on the books but implementation is close to zero. It's amazing how the scientific community is attempting to work on these programs when basically the economy is in such a state of chaos and people literally do not know where their next bite of bread will come from.''

Adamkus had just returned in March 1993 from his third trip to Lithuania in the past year. I asked him if things had gotten worse since I was there in February of 1992. "Much worse, much worse,'' he said, confirming what I had been hearing through my relatives. "You know, Lithuania's largest export product was always agricultural. But between 1990 and 1991 they allowed their collectivized agricultural base to collapse. People stole the equipment and sold it, forty percent of the animals were slaughtered for food—now what do you think will happen next year for food? It takes more than a year to raise livestock—only forty percent of the available farm land will be cultivated this year, farmers are desperate for seed.'' Instead of trying to use the infrastructure that was in place after fifty years of collective farming, Adamkus reported how individual greed had run rampant and dismantled the very means of food production and distribution. In March 1993 the country elected a former communist as its new president and it will be interesting to see if this means that the society is calling for the reinstitution of some collectivized forms of social organization and a retreat from the fast track to capitalism.

I was very curious about the ecology programs Adamkus had helped set up in Lithuania's two major universities, in Vilnius and Kaunas. When I talked to him, the team of Lithuanian scientists he had arranged to visit ecology departments in the U.S. had just left. He informed me that

> 1991 was the first year the ecology program began accepting students. There was a great deal of competition and only sixty spots available. We'll have the first Bachelor of Science graduates in three years. Ten years ago we started planning for this. Basically I just gathered up the course catalogs from a number of ecology programs in the United States and took them over there. And we've been arranging for Lithuanian scientists to come over and see what is being done on our campuses. Of course, the emphasis in the two curricula will be different. For example, here in the U.S. we're monitoring and cleaning up after the petroleum industry but in Lithuania the main problems come from controlling toxic runoff from the highly chemicalized agricultural industry.

I cannot help but think of my husband's parents, farmers in the Midwest who both died this past year from cancer. For years now, everyone in their farming community has been buying bottled water, as it is well known that due to the chemicalization of agriculture since the sixties, the tap water is undrinkable. Since the collapse of the Soviet state, the U.S. press has been diligent in reporting on the devastation of the environment in the former Soviet Union, showing an almost voyeuristic need for a foreign screen on which to project the repressed fears of American citizens over the toxicity of their own soil. It must be said that, despite the integrity of individuals who work there, the EPA itself is monitored by many nongovernmental environmental groups precisely because its record in this area has often been shown to protect the interests of industry and of the military. Thus one wonders whether, along with the bureaucratic and academic models for environmental protection and cleanup imported from the U.S. Lithuanians will also be developing the sort of citizen watch groups which have cropped up in the United States.

So it appears that as in the newly forming capitalist business ventures, so too in science education are Lithuanian-Americans and their American paradigms instrumental in rethinking Lithuanian science. To get a picture of Lithuania's tribulations in communications technology, Adamkus suggested I call Algirdas Avizienis, a Lithuanian-American who was the former chair of the Department of Computer Science at UCLA and had just spent the last three years in Kaunas helping his compatriots set up a new university. I phoned Avizienis in June 1993, catching him a few days before his next departure for the Baltic region. Avizienis had lived the first eight years of his life in and around Kaunas. His father had been an officer in the Lithuanian army which was later dissolved by the Soviets. After the Second World War, the family lived in an United Nations refugee camp in Germany for a few years and then, refusing repatriation to Lithuania, emigrated to Chicago. Avizienis was seventeen when the family arrived in the Windy City—the largest Lithuanian enclave in the U.S.—and went to work making mattresses on the south side of Chicago for a year until he got into the engineering program at the University of Illinois, in Urbana. By the early sixties when he was doing his Ph.D., computer science departments had not yet been formally established. "We just had physicists, electrical engineers, and mathematicians all working together." Avizienis' wife, who was also born in Lithuania grew up in São Paulo, Brazil, "the biggest Lithuanian center in South America." She teaches French at UCLA. Between them they have a handful of languages: Lithuanian, Russian, German, English, and French.

The main university at Kaunas, Lithuania's second largest city and the seat of the government during the brief period of its life as a nation-state between the two world wars, was closed down, Avizienis tells me, by the Soviets in the 1950s. In 1990 Kaunas sponsored a conference bringing together Lithuanian nationals and Lithuanian faculty members at universities in the U.S., Canada, Italy, and other emigré countries in order to discuss building a new university.

> After two days of debate we decided to open the university again. Initially I was going to work on informatics, but I was invited to be the first chancellor of the university so I stayed on for three years running it. We just graduated the first 180 students this year [1993]. When I left in 1992 we had 1500 students and it was running reasonably well. We had some resistance from conservatives, but I must say, Lithuanians were the fastest moving reformers in all the Baltic states. They just abandoned the Soviet model. They even passed a law saying that degrees given under the Soviet regime would have to be verified because there were lots of degrees, especially in economics, which were absolutely unscientific.
>
> We pretty much took the curriculum from the standard U.S. curricula. As a matter of fact, I should tell you that the university requires that students in their last three years be able to attend classes taught in English. We had about twenty non-Lithuanians on the faculty last year: ten Americans and three or four Canadians, primarily in the English department. We advertised for people who would be willing to be paid in local currency. We had a fine fellow who was the chair of our English department from Spearfish, North Dakota, who had his degree from Yale. So, the English requirement was for the whole university no matter what you were studying. The first three hours every morning were devoted to English study. We gave everyone a proficiency test when they entered and then broke it down

> to different groups with no more than fifteen in a classroom. One hour every day, five days a week until they reached the point where they would have to pass the graduate and TOEFL exam. The new main undergraduate curriculum we patterned after the Harvard core curriculum. Graduate studies were pretty much the University of California model. It's a typical North American model.

Why the American model, I wondered. Because a good number of the planners taught at American universities? "Generally speaking European universities are also taking on the American model for their graduate programs. I've sat in as the outside member of a number of doctoral examinations in France, so I know that at least in the sciences they are leaning very much toward the American model: larger committees and more advanced study before the dissertation. In Lithuania they were left with the Soviet model. When we decided to adopt the American model no one had any countermodels, they didn't know the European model, and besides, as I said, Europe is in a process of change. Sweden, I know, has just decided to modernize their graduate studies. The American model was adapted there, too."

Listening to this description of the Americanization of higher education in Eastern and Western Europe, I noted that Avizienis was referring to something happening more in the sciences than in other advanced fields. Why would this be so?

> Well, things are becoming more standardized in the sciences. And also because in Lithuania it is the scientists who have had the most exposure to what is happening in the West. Many of them had actually had the opportunity to visit the United States. We didn't meet with much resistance there. In the other academic fields there were no faculty who knew how their subjects were taught in the West. When we looked for faculty members we discovered that there were no people trained in such things as sociology. Political science had been pretty much the domain of Marxism, which was really a joke—like ROTC at an American university where you're just given orders and you march. No, the biologists and the physicists had done quite well. They had plenty of contacts in the West. The weakest point is really the social sciences. In the humanities, Lithuanian literature and language were studied very seriously, not at the university of course, but at the Academy of Sciences.

Of course, Lithuania's educational legacy lay in this dual structure about which Rimantis had also spoken. "The universities were the teaching institutes where research was strictly your own business if you were so motivated. The Academy of Sciences was where you had a few graduate students working on their theses." What then happened to this duality under the new system? "Lithuania did a drastic thing. They disassembled the academy as an organization that ran large institutes and employed thousands of people. They declared that the academy was to be an honorary organization with membership for distinguished scientists. We encouraged our best research scientists to start teaching at the universities. In fact, two-thirds of all our full professors at the University of Kaunus are all senior scientists from the academy. So the institutes then became independent, state-supported research institutes directly responsible to the newly formed National Science Council.'' I recognized that this was the institute directed by Rimantis Slizis. ''Yes, Rimantis is the head. He is a real reformer. He worked hard to put these changes into effect. He is having more of a problem now with the new government, which is very conservative. The new president is the former head of the Communist Party, but he was the one who went along with the reforms, the one who gave the cathedral back to the Catholic Church. I actually know him. He's considered to be a reasonable person, an engineer by training.''

The fellow who preceded the engineer as president of Lithuania was a professor of literature. Did Avizienis think this change of professions in the highest office would make things easier for the sciences? ''Well, [the new president] is reasonable, but for example, I don't think he's ever been in the West. He finally visited Chicago in 1991, I think.''

The criteria of ''having been in the West'' intrigued me. As Avizienis said, of all Soviet Lithuanian citizens, scientists were by far the most likely to have traveled to the West. This seemed to go into the creation of scientists as an affinity group based not only on the obvious factor of a similar way of thinking, but also, in the case of these Lithuanian scientists, on the experience of their conference travel. For Soviet scientists did such professional meetings functions like pilgrimages creating perhaps the same sense of community as journeys by the far-flung medieval Christians to Rome (or, for that matter, by contemporary American Catholics to Mejagory, Yugoslavia)? I am reminded of something Benedict Anderson wrote concerning the imaginary aspect of a community whose identity is to a large extent formed by such travel:

> the strange physical juxtaposition of Malays, Persians, Indians, Berbers and Turks in Mecca is something incomprehensible without an idea of their community in some form. The Berber encountering the Malay before the Kaaba must, as it were, ask himself: ''Why is this man doing what I am doing, uttering the same words that I am uttering, even though we cannot talk to each other?'' There is only one answer, once one has learnt it: ''Because *we* . . . are Muslims.'' (Anderson 1991:54)

Perhaps the experience of Lithuanians, French, Japanese, Americans, Pakistanis, and Russians mingling together in a Brussels conference hall creates one sort of cohesive force when all the participants are there in Belgium speaking the English of the scientific community: ''We are scientists,'' they might feel, "members of an international community." When they return home, however, is there a different sort of imaginary solidarity produced which now depends more on the context of their place of origin? In Lithuania those scientists who had "been to the West" were primary players in constructing a new educational infrastructure which affected much more than just the sciences. This begs the question of how we are to understand an institution like Lithuania's National Council of Science. In what ways is it a nationalist institution? With its emphasis on English as a required language in science education and its borrowings of a core curriculum from American universities, what notion of nationalism is operative here? With all the rumors of a rise in what the Western press often portrays as nineteenth-century-style nationalism in Eastern Europe it is interesting to note how the identities of the newly forming nation-states of the former Soviet Union are in fact being fashioned.

I was also curious about the new conservatism Avizienis spoke of. What does it mean to be a conservative in the present circumstances? "By *conservative* I mean going back to the status quo, the way it was during the Brezhnev years. You know, many of the people who recently went into the government had already been members of the government ten years ago during Brezhnev's term. For example the minister of education and culture. These are not bad people, you know, but they are very limited and they are bound by their experience. [Beets and radishes.] So we may be entering a period of stagnation again, which is very disappointing. I'm glad that we managed to do what we did [from 1990 to 1992] because it wouldn't be possible now. It would take so long to get permission to do anything. Back to bureaucracy, back to the old ways of doing things which means collecting signatures and getting stamps."

Rimantis had talked about the problems created in science education by the sort of ethnic division of labor imposed from Moscow. I asked Avizienis to comment on the specialization—electrochemistry—of Lithuanian science. "Oh, yes, the central planning was pervasive. What they called universities were in fact sort of faculties for producing specialists. It was a five year curriculum, eighty percent of which was determined from the day you walked through the university doors. There were a few electives in the fourth and fifth years which you could choose out of a limited set of courses. But this method of taking a group of kids and putting them through the same courses together was in fact very limiting."

At this point I told Avizienis about a conversation I had had a few weeks previously with some of the Russian scientists who had recently been brought over to help create the University of Minnesota's new Institute for Theoretical Physics and had now been teaching in the U.S. for a couple of years. (Part of the exodus of Soviet scientists one heard of in the press within months of the collapse of the Soviet Union [Burdick 1992].) The scientists I spoke with were Russian Jews whose travel to the West for scientific conferences and such had been severely curtailed during their careers in Soviet research institutes due to their Jewish background. None of them had been members of the Communist Party. I had asked them whether they experienced any difference in doing and teaching science—in their case, physics—in the U.S. All of them agreed that in their field there was something one could perhaps call a "Russian" way of doing physics. It stemmed from the strong influence of Lebedev, an internationally known Russian physicist who cultivated a particular way of pursuing answers to mathematical problems. It was a style, they explained, that involved a more "intense," "sustained," and "thorough" interrogation of the problem and in this respect they felt a marked difference from their American colleagues. As far as teaching was concerned, one of the physicists told me that he was quite frustrated by some of his American students who had transferred from other universities. Despite their transcripts showing all the prerequisite work, one could tell from their work that they had not been exposed to some essential things. To this professor the students were not focused enough on their studies, they were taking too many extra classes and the classes they took were often not standardized enough to be able to know what in fact they were being taught.

Avizienis found these complaints humorous. The Soviets

> needed specialists, they didn't need people who could think. That was the explicit privilege of members of the Party. The others just had to learn their trade, like in a feudal system. Like court musicians, composers like Haydn and such. You know, that is how I think the scientists were trained in the Soviet Union. They were treated well as long as they knew their place. Of course, there were some scientists who entered the ranks of the party but they were generally speaking people who couldn't make it in the sciences and found other ways to build their careers. I was a visiting professor in Lithuania back in 1974 and I had a chance to see how things worked. Back in 1974 I couldn't approach anything in my field because computers were primarily destined for military use and therefore the designers—I'm actually a designer myself—were not allowed to talk to me. So I decided I was not going to teach them to design computers if they could not talk to me. So I just enjoyed myself that year doing some theoretical work. But this last time I went back all the facilities were totally accessible because all the secrecy had totally broken down.

All the secrecy? "Yes, in Lithuania there was nothing left after the declaration of independence. The government just abandoned the factories and there was no attempt made to take anything out of Lithuania in the way of computers and equipment. They had some manufacturing facilities for computer chips in Lithuania and these people could now talk to me as much as they wanted to. These facilities are in trouble now because they make chips which are very competitive in an open market and the Soviets are about ten years behind. Actually, we discovered that they were even further behind than we thought.''

I couldn't help thinking here of how pervasive the sense of belatedness is when comparing the West with the rest. And not only in the sciences where it seems "natural'' that such a temporal, linear model of progress should dominate our thinking. This temporalizing appears in purely aesthetic domains as well. I remember, for example, how in 1979 when I first walked down the streets of Budapest as an American exchange student socialized in terms of fashion seasons my first thoughts were, "these people are clearly ten years behind.'' Thus the sense of how cultures in fact differ from one another can be obscured by an almost unconscious attitude that there is one global culture whose temporality supplies us with a time line on which we can not just compare but actually measure ourselves in relation to all others. Our global economy only makes this sense more acute.

This sense of there being a correct and precise way of positioning ourselves in a global culture, especially as it concerns the sciences, has its ironic side as well. I could see this when Avizienis explained to me what was happening with the Soviets attempts to acquire computer software. "We realized that they were copying everything. They were stealing software, pirating it. They had a tremendous shortage of manuals and they were just copying it. It was government policy to just copy the integrated circuits rather than designing their own chips.''

But why were they so far behind if they were copying contemporary models from the West? "This is what is called reverse engineering,'' Avizienis explained.

> It's a tough job, you know, looking at microscopic patterns and trying to deduce what is going on. In 1981–82 the complexity of the chips made in the U.S. became so high that they just couldn't do it. They produced copies but those copies just didn't work, they just couldn't make it work. They didn't have the tools for the design. They could make individual copies perhaps but they could never have a production line going. There was a big factory called Sigma in Vilnius which was all geared up to produce a copy of the digital equipment preparation VAC3000 machine which was introduced in about 1982–83. So by 1985–86 they were going to do it. The Lithuanians were quite ready but the copying of the most important central processor chip was supposed to be done by a group of specialists in Russia. But they could never come up with it. So the Lithuanians

> were sitting there for two years with the production line set, all the circuit cards made up, everything was there except the heart or the brain of the computer. Couldn't be copied anymore so the Lithuanians were stuck with a nonworking facility and then that really collapsed. It had three thousand employees when I first visited them. They were trying all kinds of things. They even had talked to the Digital Equipment preparation people from Massachusetts, trying to convince them to set up a local manufacturing facility. They said they were quite impressed but then again Hungary and Poland and everybody else was after them, too. So that didn't work out and I think last I heard they had to just shut the facility down because there was nothing that could be produced that could be sold.

This was not only due to the time warp which the practice of reverse engineering was bound to encounter in the global economy but also due to the ethnic division of labor in the Soviet management style. I told Avizienis that another of my father's cousin's, Gintautas Zintelis, the chairman of the computer science department at the Kaunas Technical University, had told me a similar story. During my visit in February 1992 Gintautas had taken me on a tour of Kaunas, a beautiful city whose main street is a one-mile-long paved walkway lined with paste-colored baroque buildings. Just across the river he pointed to a TV factory where the two thousand employees were now sitting idly waiting on parts made in other republics. "Oh, yes, I know Gintautas. He is now the Minister of Communications. So you see, the way the economy was set up, there were no backup sources. I found out all these things after independence because people could finally speak about it, and they were bitter. They somehow had the delusion that they could compete, produce something that could be sold in Europe, but no way."

Of course, the story of the growth of the computer science industry in the United States is precisely the story of the downfall of the hegemonic management style associated with big science industries like aerospace and nuclear power, domains whose cold war structures of secrecy, unaccountability, and overspending are under serious public scrutiny even as they implode from financial ruin. The computer industry in California's Silicon Valley, for example, may well have been spent into being by the U.S. Department of Defense, but the rush to personalize the microchip and get it to the marketplace was the result of the highly antimonolithic efforts of Steve Wozniak and his garage-band style of production for the Apple computer in the late 1970s. This style of production and management has set the pace in Silicon Valley where once, back in the early sixties, the government-sponsored missiles and space business was vying with the auto industry as the country's top employer. Despite Ronald Reagan's massive national reinvestment in such large-scale, centrally organized engineering projects like Star Wars, since 1989

California has lost close to 200,000 aerospace jobs and is predicting the disappearance of 420,000 more by 1995 (Beers 1993). Soviet-style industry, like its American counterpart in what is known as the military-industrial-academic complex, has perhaps met a strong adversary when it comes to computer science with its potential for small-scale, personalized entry into the global market. This general shift in the organization of industrial production, from the massive in-house factory to the far-flung manufacturing of individual components which will only be brought together in one place at the time of assembly, is often referred to as post-Fordism. Whether such a description is also useful in understanding the transformation from a socialist to a capitalist economy is open for debate. One does hear stories of potential economic success about Zelenograd, one of Moscow's satellite towns also known as the "Silicon Valley of the Soviet Union." This is where much of the electronics work for the Soviet military was done, but it has apparently also been a center of renegade activity (Knorre 1991).

Given the porous relationship of computer science education and the computer industry in the U.S. I wondered how Avizienis had set up the informatics department at his new university in relation to a global market?

> In Lithuania, computers were only studied in one place, at the technical university in Kaunas. Their department produced the computer engineers from all of Lithuania. Vilnius only has applied mathematics. So we decided not to compete with the technical university, which trains people in engineering, but rather to train people in more general computer science, programming, artificial intelligence, and so on. Some Lithuanian programmers are quite good. They have hopes of getting a contract from the West to produce software. One thing they've succeeded in doing is using very modern, very advanced typesetting software and the Lithuanians are producing mathematical books for companies in the U.S. and England at half the price of their competitors in Western Europe. For the time being they have this little niche on the market because mathematical typesetting is extremely demanding the equations have to be proofread and they have enough good mathematicians willing to work there. So they're looking for opportunities like that and we're training people more in the areas of advanced programming languages, foundations of artificial intelligence trying to get them to do work where they could do primarily the intellectual work using small machines not needing factories or large computers—Lithuania can't do that. We want to depend on the brain power, and to use modern software networks."

As my father's cousin Gintautus told me, Lithuania's attempts to enter the global market through computer science mean that it must also create new communication channels since the networks established by the Soviet government have worked to make Moscow the single gateway for all computerized data entering the Soviet Union. From telephone lines to television channels, all routes lead to the former capital (Starr 1990). There is some talk of creating a "Via Baltica," a superhighway running from Berlin to Helsinki through Gdansk, Vilnius, Riga, and St. Petersburg, forming a developmental axis which would begin the work of creating new physical facts in Eastern Europe (Lepkowski 1992:13).

What does it mean to de-Sovietize Lithuanian science? Since the fall of the Soviet Union I've heard many people in Eastern Europe talk about de-Sovietizing their apartments. This is an aesthetic statement and usually means getting rid of the stark, angular furniture first mass produced in the Eastern bloc countries after the Second World War. You know the style. It is quite similar to what was advertised for the "modern" home in postwar America: a combination of Bauhaus simplicity and dull uniformity. A Lithuanian-Canadian woman I met in Vilnius in 1992 explained to me how she was de-Sovietizing the apartment she and her husband had just obtained in the much-coveted medieval area of Vilnius (through less than democratic connections, she admitted, since as an employee of the new government she could choose among a number of flats from which former Communists had recently been evicted). They had just finished laying new wooden floors on which—thanks to the high exchange rate of their Canadian dollars—they would throw oriental rugs purchased from merchants who with the collapse of border restrictions were now arriving in the Baltic states from the former southern republics. This would all go quite nicely with the "antique" peasant tables and wardrobes they had just acquired at the flea market. In short, the decor she described seemed to resemble the living room of a turn-of-the-century European of sufficient means, able to collect exotic objects on his or her travels.

De-Sovietization, then, meant a more ethnically and historically diverse array of objects. Such taste certainly raises issues of class. It reminds me of what happened when my own family moved from a working-class to an upper-middle-class neighborhood in the late sixties: the fifties style cherry-wood china cabinet with its high-modernist uniformity was exchanged for the suggestively aristocratic French provincial.

Like their compatriots redoing their apartments based on a newly developed and highly eclectic aesthetic judgment, Lithuanian scientists in the process of creating a new infrastructure for de-Sovietized scientific work appear to be drawing from an interesting but circumscribed variety of sources. To speak of the creation of a Lithuanian science seems contradictory in at least two respects. First, how can nationalism operate as a powerful organizing force in a field like science which generally prides itself on universal values and strict objectivity? And second, what is specifically Lithuanian about the international array of advisors and educational prototypes which inform the new scientific institutions in Lithuania? This raises stimulating questions about what nationalism is all about in the shadow of the former Soviet Union. Curiously, it was *homo sovieticus* who was supposed to provide a model for universal man over and beyond his various national identities. Does the de-Sovietization of a society, then, mean a de facto turn to national consciousness?

One way of responding to these questions is suggested in Benedict Anderson's book *Imagined Communities: Reflections on the Origins and Spread of Nationalism.* Nationalism, he claims, is better conceived of "not as an ideology" but "as if it belonged with 'kinship' or 'religion,' rather than with 'liberalism' or 'fascism'" (1991:15). Neither reactionary nor progressive in and of itself, nationalism, from this anthropological perspective, is more like a cultural artifact which will vary according to the cultural system in which it is found. And that cultural system, he suggests, is a system of differences making nationality a relational term which—like gender—derives its meaning not from intrinsic properties but from what it opposes.

This idea that national identity depicts ineluctably on what it defines itself against means that what is now emerging as Lithuanian is inevitably being forged in what Lithuanians understand as their nemesis: the Soviet state. Thus, at this point in their history, Lithuanian science will struggle to mean non-Soviet science. But, as the preceding pages have suggested, this clearly means more than the wholesale adoption of Western European and American models for science education, research, and technology. It also refers to the possibility of a demilitarized science, the sort of scientific practices we might see emerging wherever institutions whose identities were established during the cold war are now seeking transformation.

My first night in Vilnius my grandmother's sister welcomed me with the family photo album in her hands. The first picture was of her father, my great-grandfather, laid out in his coffin with his many children gathered in mourning around him. As the portraits turned to snapshots and color entered the faces, I was stunned to see a picture of myself at age five standing with my siblings in our Easter dresses in front of the yellow house, red tulips blooming behind us. My grandmother must have sent these pictures. There was even one of my father dressed in his army uniform from the Second World War.

Later that evening, Voldis's son Audris showed me video footage of the bloody events of 13 January 1991. Audris had participated in the citizens' attempts to protect the television station and the parliament from being taken over. The footage he had was pieced together from many home video cameras of friends and foreign journalists who attempted to document what was happening that night. They aimed their lenses at the hospital corridors and morgues where the wounded were taken (eleven dead and six hundred wounded) and captured image of bodies, swollen and naked, with frantic nurses and doctors asking for family background as they hooked up tubes and monitors. Some of these pictures made their way via satellite to various foreign capitals. The more decorous ones put Vilnius on the map the next day in newspapers around the world and many say that such visibility helped stop the Soviet takeover (although these events were eclipsed the following day by the onset of the Persian Gulf War, recalling for many how the details of the 1956 Hungarian uprising were similarly covered over by the beginning of the war in the Suez). This is the footage that Audris says he will show his children. A new form of memory for the family album.

It is often said that the centralized Soviet government exercised its control vertically, employing involuntary use of vertical lines of communication. What eventually eroded the power of the party bureaucracy was not so much political but rather technological; information technology was finally a source of power which eluded attempts at vertical control. Clearly, the center no longer holds. In the cold war world the two superpowers tried to set up two "centers of action," but this did not hold. Given the remapping that is going on at all levels—politically, economically, scientifically, culturally—the map of "Europe" (and of many other places) is up for grabs. In this swiftly globalizing world, is it still useful to think in terms of centers? Perhaps so, since there are indeed places where more "action is happening" than elsewhere—in other words, greater concentrations of science-making, of finance and cultural capital, and so on.

During my visit to Lithuania, I was surprised to hear talk of another center. Speaking of his hopes for the future of the region, Rimantis Slizis remarked, "Actually, the center of Europe lies only a few kilometers from Vilnius." While this is not the perspective one gets from the Maastricht Treaty, perhaps this view of Europe from just east of the Urals will someday inform a new map.

Acknowledgements

The author wishes to thank her relatives—Voldemaras (Voldis), Vanda and Audronis (Audris) Laukaitis, Gintautas Zintelis, Genovaité (Gen) Milunaitis Podgalsky and Albert Milun (Algirdas Milunaitis)—without whose help and hospitality this piece could not have been written.

Notes

1. The sorts of research programs that developed in such a climate demanded "a knowledge of microwave electronics and radar systems rather than alternating current

theory and electric power networks; of ballistic missiles and inertial guidance rather than commercial aircraft and instrument landing systems; of nuclear reactors, microwave acoustic-delay lines, and high-powered traveling-wave tubes rather than Van de Graaff generators, dielectrics, and X-ray tubes'' (Leslie 1992:9).

References

Anderson, Benedict. 1991. *Imagined Communities: Reflections on the Origin and Spread of Nationalism*. London: Verso

Armstrong, Scott. 1993. "Editorial Policies for the Publication of Controversial Findings.'' *International Journal of Forecasting* 8.

Beers, David. 1993. "The Crash of Blue Sky California: The Aerospace Industry Is Dying and With It a Way of Life.'' *Harper's* (July).

Burdick, Alan. 1992. "Coming In from the Cold.'' *The Sciences* (March–April).

Gerber, Michele Stenehjem. 1992. *On the Home Front: The Cold War Legacy of the Hanford Nuclear Site*. Lincoln: University of Nebraska Press.

Knorre, Helene. 1991. "Russia's Silicon Valley.'' *Nature* 353 (3 October).

Lepkowski, Wil. 1992. "Poland Struggles to Forge New Policy for Science.'' *Chemical and Engineering News* (8 June).

Starr, S. Frederick. 1990. "New Communications Technology and Civil Society.'' In Loren Graham, ed. *Science and the Soviet Social Order*. Cambridge, Mass.: Harvard University Press.

Leslie, Stuart W. 1992. *The Cold War and American Science: The Military-Industrial-Academic Complex at MIT and Stanford*. New York: Columbia University Press.

Andrzej Staruszkiewicz, Physicist

Koczanowicz: Our interview will appear in an annual, focussed on the current fin de siècle. They are interested in the contemporary social and cultural situatedness of science. We could proceed in various ways, but perhaps we can best start chronologically: childhood, adolescence, family history. Let's start with your family. I have heard from people who are much into it that European physicists come from a totally different background than American physicists. In Europe physicists are most often from the middle class whereas in America they are rather from lower classes. You exemplify a European physicist because your family can be described as middle-class.

Staruszkiewicz: Right, except that it means something different than in America. My father was a teacher, while this famous grandfather of ours was a very wealthy man, so my father was in a way a *step down* compared to his father.

Koczanowicz: Your grandfather was an official?

Staruszkiewicz: He was an official in Lvov, sort of head of a department in a revenue office. This was, nota bene, a very high position and very highly paid. So by Polish standards he could be described as middle-class.

Koczanowicz: Do you know anything about your family's earlier history? Does it conform to the polish model in which the majority of the intelligentsia comes from the gentry?

Staruszkiewicz: My aunt Tawcia used to hint at a possible nobility of our background, but I don't believe it. I know very little about my family's earlier history. What I know for certain is that my great-grandfather, that is, my grandfather's father, is described in the Church documents as *faber ferri,* which, I suppose, means a smith. I am not sure though if it was so in his case. If so, if he really was a smith, which is not absolutely certain, then it is an excellent background, since many very outstanding people had a smith among their ancestors. This is a fact.

KOCZANOWICZ: I've never heard about it.

STARUSZKIEWICZ: It is a documented fact. Chopin's great-grandfather was a smith. Heisenberg's grandfather was a smith. And there is a good number of such cases. The sociology is that if some lower-class people advance in to the middle class, it happens very often through a smith.

KOCZANOWICZ: Because being a smith has a mystical sense. It is someone like a wizard.

STARUSZKIEWICZ: Yes, this Felsztyn, the city from which ultimately the Staruszkiewicz family comes, because it was as late as the grandfather's generation that they moved to Lvov, that was a poor region and even for such a poor country as the Eastern Galicia at the time. He, this great-grandfather, must have been a prosperous man if he could afford to educate his son. By the way, if you are interested in anthropology in the physical sense, you will have perhaps noticed the way abilities are inherited. I have my grandfather's school certificates, and from the first to the last form he is the best student. In the Austrian schools they mentioned not only grades but comparative scores. And grandfather had both excellent grades and the first-place scores in all his forms. So, if any abilities are hereditary, they must come from my grandfather.

KOCZANOWICZ: So, your grandfather was a university graduate. In what?

STARUSZKIEWICZ: Probably in law.

KOCZANOWICZ: Your father was a historian and I believe he started on an academic career?

STARUSZKIEWICZ: My father was a historian. Before World War II you had to work without pay at the university before you were employed. My father did not do it. He always reproached his parents for it. They were wealthy people, they could afford to send him abroad, but for some reason, they did not want to do it.

KOCZANOWICZ: Did your father ever talk about his studies?

STARUSZKIEWICZ: I have some of his transcripts and university certificates, all of them with excellent grades.

KOCZANOWICZ: Your father had a doctor's degree?

STARUSZKIEWICZ: It is a little more complicated. My father started his studies in Lvov and got into a conflict with a historian by the name of Stanislaw Zakrzewski. I don't know what it was about. As a result, father moved to Warsaw and completed his studies there, with a doctoral degree, which was not a rule at the time. But his what we nowadays call master's thesis was so good that Professor Marceli Handelsman, a well-known historian from Warsaw, recognized it as a doctoral dissertation. And so father received a doctor's degree, Handelsman being his academic advisor.

KOCZANOWICZ: He is a famous figure—one of the founders of Polish historiography. By the way, let me say he was killed during the war. The case is very unclear, it is supposed that he was a victim of a conflict between the left and the right wing in the home army. Also his Jewish background might have been important as well.

STARUSZKIEWICZ: There is an anecdote about it. He was a disciple of another famous Polish historian Askenazy, also of Jewish background, and when he applied for a job at the Jagiellonian University in Krakow, Szymon Askenazy told him that two Jews in one chair of Polish history was a bit too much and so he moved to Warsaw. The fact that father made his doctorate under Handelsman's supervision is certainly a good recommendation. Father had a good chance for an academic career, but he forfeited it due to his parent's narrow mentality.

KOCZANOWICZ: Your father was a high school teacher?

STARUSZKIEWICZ: He was a teacher all the time, until he fell ill with multiple sclerosis during the war. It was an incurable illness. And that is why after the war, when he had problems walking, he worked at city archives in Bydgoszcz.

KOCZANOWICZ: And your mother?

STARUSZKIEWICZ: She had a degree in classics and was qualified to teach Latin and Greek, but after the war the communist authorities had a negative attitude toward Latin and cut down on the number of Latin classes as much as possible, so mother used to teach two hours of Latin a week and fill the rest of her teaching with as much of the humanist knowledge as encounters with books. My parents had an immense library and I regret it so much that it dispersed after my mother's death. I had a very small apartment then. I did not know what to do with the books. I think it is a terrible thing that books collected by the Polish intelligentsia go down the drain because there is no room for them in our homes. It is a tragedy. So I think that it was exceptionally valuable for me to grow up among books.

KOCZANOWICZ: It is not typical of Polish physicists. How would you estimate their mental culture?

STARUSZKIEWICZ: This may vary. Generally, I would say it is rather poor, by the way, just like here (in the U.S.). Physics is a fundamental science. It has been a corner stone of mental culture ever since the seventeenth century, but only if practiced at a high level. Average hackers of physics are at a rather low level; their mental culture, in general, does not impress me.

KOCZANOWICZ: However, many outstanding physicists had a certain philosophical bent.

STARUSZKIEWICZ: Yes, but only the outstanding ones; a step below you won't find this sophistication.

KOCZANOWICZ: Coming back to your family history, 1945 in Poland is sort of an ultimate foundation of it all, the advent of communism. How did your parents react to it?

STARUSZKIEWICZ: I don't remember that any more.

KOCZANOWICZ: Many a trauma of Polish intelligentsia takes its beginnings from that moment.

STARUSZKIEWICZ: I don't know if you realize that I had rather weak contact with my parents. Because I left for a medical treatment in Zakopane in 1948. I stayed there for six years; so until I was fourteen I had no contact with my parents. I returned in 1954 when my father was already fatally ill and he died soon after. So you see I can only speak about the time between 1945 and 1948, but I don't remember it at all. In a sense I had no contact with my parents in this deeper sense, formative or mental, just because of this stay in Zakopane.

KOCZANOWICZ: Did you take it well? Sometimes children find it very depressing.

STARUSZKIEWICZ: Ann, my wife, thinks I am very much beyond the norm [*laughs*]. This may be a result of separation from parents, I can't rule out such a possibility. Myself, I find this stay in Zakopane very positive for the formation of my personality. I don't know how to describe Zakopane, but it may have been something of a Magic Mountain. You see, the atmosphere of this sanatorium, it was exactly a Magic Mountain, of course removed to socialist Poland, for children, not for adults [*laughs*] but all in all it was something like that.

KOCZANOWICZ: Did you feel bad there?

STARUSZKIEWICZ: No. I remember it as a fine time, very nice.

KOCZANOWICZ: The teachers were good?

STARUSZKIEWICZ: Were they good? I remember one thing—when a teacher explained to me that for a body to become weightless it has to find itself at an appropriate point between the earth and the moon, which is absolute nonsense in terms of physics. It is a disgrace to say something like that. So, I'm not sure if the teachers were always of the best kind. Besides, I suppose I was a problem pupil even though I felt comfortable there. What problems? Not conflicts, I was a so-called good pupil and my teachers as a rule expected a lot from me, things I was not always willing to do. I remember a competition in which you had to make a drawing on a historical topic and describe it. So let's take a Greek temple, make a drawing from memory, or maybe not from memory, I don't quite recall, describe it, and so on. At first I couldn't wait to start it, I loved this idea, but then I lost enthusiasm and submitted my drawing unfinished. This irritated my teachers badly. Our form master made of me a hideous example of all Polish national vices, saying I had a lot of talent, but also was utterly lazy, lacked of character and persistence. I still remember it. Why? Because from the position of who I am now, it is a completely false judgement. Namely, if now, as a physicist or an academic professor I should compare myself with famous physicists, I would estimate my situation in just an opposite way. My talent in a purely physical sense, for example, a capability of mathematical argumentation—this is a good test of technical talent—is very modest. If in this sense I would ascribe to myself anything, then it would be, first, systematic hard work, that is what my teachers didn't find in me, and second, what is most important in science, a certain intuition as to what is banal and not worthwhile, even if practicable, even if you can earn your academic degrees on it.

KOCZANOWICZ: And you have a feeling for it?

STARUSZKIEWICZ: I don't know if I am right, maybe I just flatter myself, but I feel I know what is banal and negligible. Mathematics is a science which favors a physically understood aptitude, mental aptitude for solving specific technical problems, mathematical argument, calculation. This aptitude, in a scale which is required of me now, that is required of academic physicists, I would estimate as very mediocre.

KOCZANOWICZ: You don't find such confessions embarrassing, do you, as in Poland especially praised are those talented but lazy.

STARUSZKIEWICZ: This is how I was perceived in my primary and high school. Paradoxically, my academic career is a clear negation of this stereotype.

KOCZANOWICZ: It was not your problem that you did better in certain subjects?

STARUSZKIEWICZ: No, it was not. I could do anything, I was one of the best pupils. Nothing was really difficult for me.

KOCZANOWICZ: What next? After high school you decided to study physics.

STARUSZKIEWICZ: Why physics—it is hard to say. I wanted to study at the Fine Arts Academy and took some drawing lessons working towards the exam, but I thought then such lessons were not enough and I'd never get through the entrance exams. So my decision to do physics was in a way incidental. I can't say why I chose physics.

KOCZANOWICZ: And not mathematics.

STARUSZKIEWICZ: I must have realized that my purely mathematical aptitude wasn't good enough to practice mathematics professionally. You see, I participated in an inter-high-school mathematics contest. I got to the penultimate regional stage of it, and then I fell through. If I had got to the ultimate stage of it I'd have considered studying mathematics, but this contest probably made me give up math. Mathematics is quite often something like music, you need some purely physical aptitude for it, an aptitude for argumentation, association, analogizing, which I don't find in myself.

KOCZANOWICZ: If you were to relate to your student days, could you name some peculiarities of physics in Krakow, some tradition, some local areas of interest?

STARUSZKIEWICZ: As for tradition, a student has no contact with such a science that would address a tradition. From my studies, I remember two old professors: Niewodniczanski and Weyssenhoff. Niewodniczanski was the greatest personality. In 1964, when I was an assistant, he built a new institute of physics. There was also Weyssenhoff, who was a theorist and under whose guidance I wrote my master's thesis.

KOCZANOWICZ: And your doctoral dissertation?

STARUSZKIEWICZ: I did my doctoral dissertation, well, I don't quite recall, probably under Rayski. From my studies I don't remember having realized Smoluchowski's greatness as a scholar from Krakow who was a world-class scholar, a top class scholar. Instead I do remember I owe much to Weyssenhoff. But why? It is a peculiar case. Anyone who met Weyssenhoff would say he was a hopeless teacher. Like this book of his, an electrodynamics handbook he wrote in such a diffuse and unpedagogical way. You can hardly learn any physics as a craft from it. He defined for me my master's thesis, "The Evolution of Views on the Laws of Conservation of Energy in General Theory Relativity," so rather a historic-bibliographic and not a physical theme. Yet, I believe I owe a lot to Weyssenhoff. What exactly? He came from a very wealthy family. His father was a well-known writer. His mother came from a Jewish banking family. Extremely well-off people. They were wealthy before the First World War and I guess something of these riches survived the war because Weyssenhoff traveled quite extensively. He did his doctorate in Switzerland during the First World War. There, in Switzerland, he met Einstein, and in fact rubbed shoulders with the founders of modern physics. And he used to talk about these people, and in an aura of them as great people while scholars at the time of my graduation were no geniuses. If you showed Weyssenhoff an article in the *Physical Review* and asked him his opinion, he would take it into his hand like a piece of dirty rag and say, "He is really no genius" [*laughs*]. I am very critical of several things that are now being done in physics and I may have inherited it from Weyssenhoff, who really mixed with these great people and judged all others by their standards.

KOCZANOWICZ: Nowadays you often learn about ideological pressure in science. Have you ever experienced anything like that?

STARUSZKIEWICZ: It depends what you mean by ideological pressures. Well, of course, communism did exist at the university. I remember well, when I was institute vice chairman for education, our staff meeting would be supervised by a local organization's party leader. Well, it may sound unpopular if I say that these party people I knew from the Institute of Physics were in fact harmless guys. Yes, in this sense you wouldn't feel pressured.

KOCZANOWICZ: And when you were a student?

STARUSZKIEWICZ: Generally, I have a poor memory for biographic-historical events. I have always kept a fond memory of my philosophy teacher from that period of time, Augustynek. He was a party man, but he read philosophy for third-year students. Remember, my third year was in 1959, in Poland the time of full-blown communism. A lot of people in Augustynek's place would make something totally indigestible out of philosophy. I remember well what Augustynek lectured on. These were quite decent lectures, a bit of history of philosophy, a bit of philosophy of nature. I remember there was not a trace of ideology in it. He was in the Party and I don't know what he was doing there but as an academic teacher he was an absolutely positive figure.

KOCZANOWICZ: Among Soviet physicists there were heated discussions around the theory of relativity.

STARUSZKIEWICZ: This is a pathology. All these things: Party's ban on the theory of relativity, quantum mechanics, genetics, all this is a social pathology. Nota bene, this problem deserves a close analysis since it is really strange how in the twentieth century such phenomena were possible. Because ultimately it is a sheer pathology for politicians, that is, people called to solve political problems, abstract problems that are difficult even for the professionally trained to solve them. I think in Poland this pathology did exist but in a less harmful form. If you like I can tell you about Mrs. Eilstein. [Helena Eilstein was then a professor of philosophy at Warsaw University. After 1968 she left Poland and worked as a professor in England and the U.S.A.] I don't know why but when I was a young man, probably just after my graduation (I was twenty-one), I was invited to a conference organized by Mrs. Eilstein on the philosophy of nature. It was 1961 and the conference had to have Marxist accents. What were they about? Mrs. Eilstein in the end presented us with sort of conclusion. She said that contrary to Soviet Marxism, Polish Marxism's greatest achievement was respecting the Aristotelian principle of contradiction. And this is what she called the greatest achievement of the Polish Marxism—a return to Aristotle! I remember it shocked me then, because I thought that to get free of such primitive sort of rubbish like this whole dialectic is a moral but not an intellectual achievement. To call it a scientific achievement, and of an entire scientific school at that, is ridiculous and may be that is why I have remembered it so well.

KOCZANOWICZ: But some Russian physicists would join these discussions concerning the theory of relativity, for example, the famous physicist Vladimir Fock.

STARUSZKIEWICZ: He was a very good physicist. He defended the theory of relativity, about which he had a bit unorthodox views. Let me say again he was a prominent physicist, his views on the theory of relativity were those of a professional and not a psychopath. You see, the Communist Party in the USSR banned the theory of relativity, quantum mechanics, and genetics—the three most important scientific discoveries of the twentieth century.

KOCZANOWICZ: Plus cybernetics.

STARUSZKIEWICZ: This is just by impetus. Also, I wouldn't compare cybernetics to these discoveries. One would have to be really unlucky to blunder three times in a row, but I wouldn't want to analyze it. There is no sense in it, it is sheer social pathology.

KOCZANOWICZ: Let's revert to your master's thesis. How do you find it now, from the distance of time?

STARUSZKIEWICZ: I would say it is a total failure. And objectively, it is a failure, because this whole problem was already described by Wolfgang Pauli in his famous work, "The General Theory of Relativity," in the early twenties when he was twenty-one. I have told you already that Weyssenhoff was a poor teacher. He gave me this topic and did not mention Pauli's book to me. And writing my thesis I did not know of Pauli's work. When I later came across it I got terribly frustrated because what I wrote was a miserable fragment of Pauli's book which he had completed at age of 21, exactly my age of graduation. Objectively, it is easy to explain, Weyssenhoff did not tell me about this book while at the time Pauli was an assistant to Sommerfeld, who was at the hub of the world of physics and where people the world over sent their reprints so in his office you could have found anything from the entire history of theoretical physics in its original source. Thus, Pauli's book is perfectly documented, there are a few hundred bibliographic entries which he had at his disposal, whereas I did not even learn about the existence of Pauli's book from Weyssenhoff. So my master's thesis is rubbish, I would even say, a disgrace.

KOCZANOWICZ: You did not feel dealing with history of physics is a waste of time?

STARUSZKIEWICZ: No, I didn't. I don't regret this work. It was typical of Weyssenhoff to aim at major problems. Well, the problem of conservation of energy in the general theory of relativity is a fundamental scientific and physical problem, unsolved yet. So I encountered this problem quite early and I don't regret this work. However, this work would have been far better if I'd consulted Pauli's book. Because I spent all my time establishing that which Pauli had written, in this sense it was a waste of time.

KOCZANOWICZ: Is such historical knowledge beneficial for physics?

STARUSZKIEWICZ: Oh, yes, I am sure it is.

KOCZANOWICZ: But I suppose not everybody thinks so.

STARUSZKIEWICZ: Most modern theoretical physicists don't seem to realize the historical roots of their problems and do not appreciate them. My belief is the knowledge of the history of physics is essential and very helpful.

KOCZANOWICZ: Can we say that an analysis of a cause of others' failures and successes enables you to work out strategies of problem solving?

STARUSZKIEWICZ: It helps in many ways. Knowledge of history sometimes helps you avoid blundering, realize that certain problems recur, for example, the problem of conservation of energy. Besides, well, I don't know, I am absolutely all for a good knowledge of the history of physics. I think students should be taught it. I guess this has already begun in a way. I mean, earlier you could only very rarely find biographies of the good, only of the most outstanding, physicists such as Newton; as a rule there were rather superficial, not based on source studies. Now even a couple of run-of-the-mill physicists have been made the subject of biographies. So physics seems to mature in the sense there is a need to read biographies of famous physicists.

KOCZANOWICZ: Can you nowadays talk of schools of physics? Because this term seems quite in vogue now.

STARUSZKIEWICZ: This term is frequently abused. It's hard to say whether this group of physicists in Krakow can be labeled a school. The term *school* is abused because certain prominent people would like to be treated as founders of schools. When I think of some prominent physicists, for example, Sommerfeld did found a school; a number of his students were immensely successful. Nevertheless, not all outstanding physicists are founders of schools. Newton left behind a void that lasted over one hundred years. I would refrain from associating any scientific greatness with a school it might have initiated.

KOCZANOWICZ: What is a school in physics? A certain way of viewing problems, of solving them? It is understandable in philosophy, psychology, and so on, but can you talk of a school in such an advanced science as physics?

STARUSZKIEWICZ: Certainly, in the twentieth century there were schools conceived in this way but I personally wouldn't attach much importance to them. It is as you said—physics is a way of thinking which is so perfect that it in a way directs you in the right way that. Well, I am not sure though. Let's stick to the facts—there were very outstanding physicists who did not found schools and there were those who did. I don't know whether that which exists in Krakow can be called a school—I'd rather say it can't. Anyway there is no single person to whom you could ascribe it. You can certainly ascribe a creation of an Institute of Physics to Niewodniczanski, but only physically, because he built it. He was a powerful man who created conditions for the institute to exist. In terms of organization you can surely attribute the emergence of physics in Krakow to Niewodniczanski.

KOCZANOWICZ: Can you talk of any personality in Polish physics that would have a major influence on its evolution? For instance, Infeld?

STARUSZKIEWICZ: I don't know. Maybe Warsaw physics could be labeled Infeld's school. I would say Warsaw people see it this way, that several people came from Infeld's school. I would see no equivalent to Infeld in Krakow.

KOCZANOWICZ: And coming back to Smoluchowski, has his tradition survived in Krakow?

STARUSZKIEWICZ: Smoluchowski was a great personality who died very early of dysentery, at the age of forty-something, during the First World War, and he left no successors.

KOCZANOWICZ: And when you were completing your studies, did you feel a contact with the world physics? Back in the fifties and sixties Poland was a relatively isolated country. How did it feel in science?

STARUSZKIEWICZ: It is difficult to say. I'm this type of person who learns basically from books and not from people, so in this sense I can't answer your question as there are people who learn otherwise—rather from other people.

KOCZANOWICZ: But books are not accessible either, for philosophers it is a problem that nowadays, for financial reasons, they don't have access to books, periodicals.

STARUSZKIEWICZ: The question is whether it is a disadvantage.

KOCZANOWICZ: Yes, that's a different point.

STARUSZKIEWICZ: When I graduated, I came to the conclusion that I really didn't know anything about theoretical physics. Therefore, in a bit of a showy manner I threw away all my notes and sat down to a two-volume course of mathematical methods in physics. I think I was right to do it. In this way I gained certain knowledge of mathematical methods in physics, except this handbook was written in the thirties. It is still excellent even though some people wouldn't feel specially keen on it, it would be good if those who now talk so much about strings and various topologies understood mathematical methods at the level of this handbook. I am the type of person who learns from books. And if I do have a problem I try to solve it by myself. That's why I may not notice that I lack some important piece of information.

KOCZANOWICZ: And why did you come to the conclusion that you had learned so little at the university?

STARUSZKIEWICZ: Because I took a glance at a handbook on mathematical methods and did not understand what was in it.

KOCZANOWICZ: What would you say is the source of this?

STARUSZKIEWICZ: It is rather simple problem that mathematical methods in physics, that is an absolute minimum which every theoretical physicist should master, are an immense area of interest. And you can't be taught all that during your university studies. I don't remember what I was taught from mathematical methods as a student, but anyway at some point I realized I knew nothing and then I sat down to this handbook. I don't regret it.

KOCZANOWICZ: And it was then that you completed you doctor's thesis? Soon after that?

STARUSZKIEWICZ: My doctor's thesis is a peculiar story. I was never really doing it. I wrote it all in two weeks.

KOCZANOWICZ: How did that happen?

STARUSZKIEWICZ: It was like this: I wrote my thesis on the theory of gravity in three-dimensional time space, which as late as twenty years after being published was recognized properly and I must say is now quite extensively quoted and by the most prominent theorists. The story of my doctorate goes like this: Bialas, now Professor Bialas, presented this work of mine at an international school in which participated also the outstanding American physicist J. A. Wheeler, to whom this idea appealed very much and who, for some twenty years, was the only person to quote a number of times this thesis of mine. The news that I'd written a work which impressed people at the conference came to Warsaw. So I was invited to Warsaw and I gave a lecture there in the presence of Infeld himself. Trautman, now Professor Trautman, a member of the Academy of Sciences, asked me if I had a doctor's degree. I said I didn't and then he said I should present my work as a Ph.D. thesis. So I padded the original version a little to make it about forty pages long. I did it within two weeks. Academically I had no advisor. Trautman proposed to evaluate my work positively and wrote to Rayski in this tone. I did this work at the age of twenty-three and defended it at twenty-four in 1964.

KOCZANOWICZ: You have said this work is now extensively quoted. Is it so because the subject is so fashionable?

STARUSZKIEWICZ: At present there are dozens of publications titled "The Theory of Gravity in Three Dimensions" elaborating the subject which I indeed, it is no exaggeration, originated. So far there have been well over a hundred works on this topic. From the eighties alone there have been about fifty references to my work. From this I infer there must be about 200 works devoted to the subject I introduced.

KOCZANOWICZ: Do they all mention your name?

STARUSZKIEWICZ: No.

KOCZANOWICZ: Maybe some want to create an impression they invented everything themselves?

STARUSZKIEWICZ: I do not know. The point is that in physics one can create a complete theory. For example, Einstein did this for the general theory of relativity—full stop. Yet since that time there have come into being thousands of works on the general theory of relativity and it does not contradict what I have said before. There is something like a physical theory and it is sensible to say that someone completed it. For example, in 1912 Einstein didn't have the theory of relativity; in 1916 he did. You can never define it exactly and say what it means that a certain theory has been formulated. Whereas one can keep exploring certain details endlessly and so for example within the framework of the general theory of relativity there are going to be published very profound studies as long as the world exists. Which does not change the fact that this theory was ultimately formulated by Einstein in 1916. Likewise I perfected a model correctly in 1963. This model is very interesting; it has various astonishing features and therefore it is intensively studied.

KOCZANOWICZ: And is it the most frequently quoted work of yours?

STARUSZKIEWICZ: Absolutely. It is perhaps the only work of mine which has received some citation. I believe I've written much better works which have passed totally unnoticed.

KOCZANOWICZ: Why?

STARUSZKIEWICZ: I don't know.

KOCZANOWICZ: Coming back to your career. It looks as if you worked completely on your own, without a school or a master.

STARUSZKIEWICZ: Since the very beginning I have been left to my own resources and I find it advantageous. Nobody really cared about my work. I have a rare privilege to have always dealt with things interesting just for me.

KOCZANOWICZ: You think the situation is usually different.

STARUSZKIEWICZ: Radically different. Especially here, in the U.S., it's totally different. The United States has worked out a suicidal system of promoting physics. The system of grants that exists here is good in medical sciences. If you have to test a medicine's efficacy you take one thousand patients and apply it to them and then you test results. The outcome is always positive, medically it can be either positive or negative, but scientifically it is always positive. If in theoretical physics there are such subjects, you should stop dealing with them as they are not interesting. The American system of grants promotes subjects that are fashionable. Such subjects needn't be interesting for a young scholar. This system forces writing to flatter a wide public of reviewers. There exists an entire art of anticipating who is going to be a reviewer and writing to please his taste. I don't know, I may be old-fashioned, but I think it is a totally suicidal system.

KOCZANOWICZ: And in Poland?

STARUSZKIEWICZ: In Poland this system has just been initiated, in ten years the situation will be exactly like in America.

KOCZANOWICZ: And what was it like before now?

STARUSZKIEWICZ: I can't speak for others but personally I had perfect conditions for development because nobody was interested in me. I wrote my master's thesis under Weyssenhoff's supervision at the time he was already retired. And he was a man who never looked after anyone; he wasn't a man to supervise his students.

KOCZANOWICZ: It is a wonderful situation. I experienced something of this kind, too.

STARUSZKIEWICZ: I find such conditions for an academic development ideal, even though it was in Poland, deep under communism, despite lacks in access to literature. I may be wrong because someone from America, reviewing my work, might think I have not done much. It is none of my business. I believe the system elaborated in America is suicidal, just suicidal it is completely clear by now.

KOCZANOWICZ: Can you imagine something better?

STARUSZKIEWICZ: You should see a difference between those who set requirements, whether it is generally some authorities or your direct superiors. The American system is two-fold: people giving money require something from grant directors, who in turn require productivity from performers. I was lucky that nobody wanted anything from me; neither authorities nor superiors. I believe that other people at my age were not in such a situation. I mean authorities might have really not required anything, as I have personally participated in various verifications at the university as a director and a dean, this was a complete nonsense. It was clear nobody cared about anything. In this sense the authorities did not care about anything, unless they wanted to destroy somebody for political reasons. However, if there was no such need they really did not care about anything related to research.

KOCZANOWICZ: As a result, the average productivity in communist Poland was, they say, one article in ten years.

STARUSZKIEWICZ: In Krakow your immediate superiors might have been interested and promoted a certain area of problems. After all, it is understandable. If I am to supervise or review some work, I can't promote subjects in an area I am ignorant about. Automatically, the entire system of science prefers to repeat activity similar to what is already known by seniors scientists. I was lucky that by a coincidence I did my doctoral dissertation in two weeks and that nobody was really interested in it. In America the most important quality of the system of grants is that it exacts publications on a definite subject, which may be not the most essential and which may not correspond to your true interests.

KOCZANOWICZ: Because it is easier to get a grant for something fashionable?

STARUSZKIEWICZ: Of course. I think I've done something more important and profound than this three-dimensional gravity. But I don't think I would get a grant for it here because this area of interest is not familiar to people who give grants.

KOCZANOWICZ: So it is a self-maintaining mechanism.

STARUSZKIEWICZ: In the United States there is a man whose name is Clifford Truesdell. He adapted such terms taken from Orwell as *plebiscience* and *prolescience*. The first refers to studies whose results, although highly predictable, are not certain while the other's results are absolutely certain. The system of grants very surely promotes plebiscience if not prolescience. Everyone wants studies in physics to be good, original, and comprehensive for everyone. In all probability, these three conditions can't be fulfilled all at once; sometimes everyone recognizes the importance of the new theory, like quantum mechanics, but this happens rarely.

KOCZANOWICZ: At present Poland is introducing of a system of financing science based on the American system of grants. How would you estimate the first period of its functioning?

STARUSZKIEWICZ: As far as physics goes, the grants system in Poland has one advantage as compared to that in the U.S. In my opinion and to my knowledge of grants in physics, committees deciding about grants take into account only foreign publications, and this is a good criterion in Poland. A Polish physicist, as compared to an American one, is for a variety of reasons handicapped publishing abroad. I am an editor of a scientific journal and I know that when an article arrives from an exotic country and signed by an exotic name, you immediately get cautious. It is an instantaneous reaction, I can imagine it is a common thing everywhere. For this reason, a Polish article, in order to be admitted in a Western journal, has to overcome more obstacles than an American one. Often it is worse technically, linguistically. It may have such clearly noticeable defects. If there is any doubt these will be good enough reasons to reject this work. As a matter of fact these committees' motivations are quite sensible. Their criteria may select the best. Such criteria do not exist in the humanities at all, so I really don't know what decides about assignment of grants in this area, since I can see no verifiable criteria there. The criteria physicists apply are purely formal and hence anonymous, objective, and verifiable. And this is its value.

KOCZANOWICZ: Do you think management of science under communism was essentially different from the system of grants?

STARUSZKIEWICZ: I should think so, although I have to say that I was never a member of a money-giving committee, so I may be wrong. That I was a dean or institute director in a way means nothing. I imagine I was even a member of a committee that assigned grants, except, somehow I never participated in it because sessions were so boring that I felt I had no time to waste there. So in this sense I had nothing to do with financial assignments, but certainly these were not grants, it was something else, I don't know what they called it, in my opinion, as for theoretical physics it was, as it were, a deliberate extra financing of those who deserved it. The committee worked on the assumption that salary wasn't enough to survive and that those who deserve it should be given a bonus. In this respect I believe the committee functioned in the right way.

KOCZANOWICZ: You have said that at least in physics the system of grants functions correctly. Do you think this pathology that appears in the United States is going to reach us, too?

STARUSZKIEWICZ: So far, the Polish system of grants is based on the criterion of publications abroad. In a few years our people will have learned English, will travel abroad, and will all have computers. So only their names and affiliation will make them different from Americans.

KOCZANOWICZ: Your arguments can be easily refuted if one thinks that in the so-called socialist countries not much has been done, that the system of grants has yielded concrete results in science.

STARUSZKIEWICZ: In Poland not much has been done but Soviet theoretical physics was much better than American.

KOCZANOWICZ: And you think it is a result of the steering of science?

STARUSZKIEWICZ: You are asking about the secret of a very high level of theoretical physics and mathematics in Russia. One could ask how they have good mathematics and theoretical physics in the country that has no medicine, where you can't get a tooth filling. I guess the answer is very simple. Namely, in the Soviet Union all, but absolutely all, existent paths of a social or personal advancement have been destroyed, so if in the U.S. quite frequently not the most able people in a given generation practice science, in the USSR it's been only the best people in a generation to do it; and there we have the results.

KOCZANOWICZ: It is quite obvious, but let's come back to the system of grants. Your arguments against it can be refuted by these I've mentioned.

STARUSZKIEWICZ: You can't compare Poland and the United States, it is a matter of different proportions. The United States would have to commit massive suicide in order to step lower than Poland in terms of science. However absurd their system they are bound to have some results. Whereas if we are to compare the U.S. and the USSR, the results will be very ambiguous, as I would say Soviet theoretical physics and mathematics are much better. They are simply more intelligent.

KOCZANOWICZ: What did your international contacts look like?

STARUSZKIEWICZ: I have stayed abroad longer twice, in the U.S. and in West Germany. My stay in the U.S. was facilitated by Infeld. He was Einstein's assistant at the same time as Bergman, who was later professor in Syracuse. I got there just on Infeld's recommendation. However, by Polish standards, my international contacts are nonexistent. An average Polish physicist has lived longer abroad than in Poland. Polish physicists' contacts began long time ago. I remember some legends about people refused passports, summoned by the militia. It was very colorful but in the Jagiellonian University people used to travel abroad long ago. All my colleagues have stayed abroad much longer than myself.

KOCZANOWICZ: Don't these wide international contacts incur the risk that Polish science will be derivative, that for example, it will aim to solve, say, American problems.

STARUSZKIEWICZ: It is hard for me to say. There are no American problems in physics. But I understand your question. I would always do things interesting to me. I can't speak for others but I do recognize an abstract hazard that if you practice physics as you do it in Poland, you are at risk of being derivative. It is hard to say whether this risk is real.

KOCZANOWICZ: And the journal you are an editor of, does it have an international appeal?

STARUSZKIEWICZ: Quite unexpectedly it is quite wide ranging. It can be found in all major American university libraries. In this sense it is an international journal.

KOCZANOWICZ: But how is it valued?

STARUSZKIEWICZ: It is an American mania to work out ranking lists. There is also a ranking list of scientific journals. It is based on certain criteria, I don't quite recall them all, but the number of citations is decisive here. I don't know if it is an absolute number of citations or divided by the volume. Because obviously the only sensible criterion is the number of citations divided by the volume. The number of quotations and also what they call "lasting influence,'' that is, how long a given work has remained in circulation. So my work on three-dimensional gravity has scored a lot because as far as this lasting influence goes, it has been referred to many times in thirty years. Well, certainly on this list, *Acta Physica Polonica* can't compete with the *Physical Review.* However, it occupies a relatively high position on this list. Optically, it would be somewhere at the end of the first third. Which means first, quite decent, but second, surprising to me. Beside these great international journals there are minor national periodicals and *Acta Physica Polonica* has been higher on this list than, say, its counterparts from Holland or Italy. It is an astonishingly good position for a Polish journal. Of course, there are Polish mathematical journals of high prestige such as *Fundamenta Mathematicae,* but they are not really periodicals as they have long production cycles. Apart from them, *Acta Physica Polonica* is the most frequently quoted Polish journal. I'm a member of its editorial staff. At present we have grave problems because we lack money but we hope that if it has been coming out for sixty years, it will survive.

KOCZANOWICZ: This shortage of money for science is an interesting thing because opposition programs used to point to the small outlays for science as one of the worst vices of communism. But now that the former opposition has become the government, outlays for science are even more miserable.

STARUSZKIEWICZ: It is Poland's misfortune that the opposition treated its programs in an instrumental way and disregarded the necessity to take responsibility for them, as it did not take into account the possibility of assuming authority.

KOCZANOWICZ: But after all, under communism Poland was a poor country and yet somehow it managed to find this money.

STARUSZKIEWICZ: Mechanisms have changed. Now it is hard currency, theoretically convertible into dollars. Then it was paper money. So there are also changes for the better, too, for example, we receive journals regularly, unlike before, probably because then they lacked hard currency.

KOCZANOWICZ: You held some administrative positions in these hard times of martial law and after.

STARUSZKIEWICZ: But I was elected director and dean by my colleagues.

KOCZANOWICZ: How did you tolerate it?

STARUSZKIEWICZ: I don't remember anything sensational. A few of the people from our institute were arrested. I even had to go to the security police to get them out. I had to sign something but I don't remember what [*laughs*]. What I am saying may be not representative because, first, I am not a sociable person, and second, I rarely appear at the institute, so certain things may have escaped my attention. I don't remember any dramatic circumstances, apart from such obvious facts as that journals did not arrive, and some visiting controls, so beside these trivialities I can not remember anything special. When I was dean I had a very serious conflict with a local party organization. This was because our Institute Council decided to employ Professor Czyz from the Nuclear Research Institute. The council offered him a professorship and this was protested by the local Party organization, for reasons never explained. I had a series of violent negotiations with the leader of the University Party Committee and it was there that I hit a barrier. I suppose there was a veto by the security police and that this local Party man just worked at their command, he never explained any reasons. By the way, he was a university professor of law so our conversations were very polite. I used to say, "Professor X, please tell me what it is all about and I will try to explain the situation," but he would never tell me, no response, just no and that's it.

KOCZANOWICZ: So at least in physics the Party did not pressure science too much.

STARUSZKIEWICZ: Yes, I think so.

KOCZANOWICZ: Summing up these bits of stories about Polish science, what would you say about its organizational system, now and in the past?

STARUSZKIEWICZ: I believe it has always been hopeless. I'd rather not talk about Polish science in general, because the Krakow Institute of Physics is after all something atypical of Polish science. I don't have any wider experience of Polish science at large.

KOCZANOWICZ: I mean, also these recent attempts at reforming science, like the introduction of grants or periodical assessments of particular institutes, assessments which condition financing.

STARUSZKIEWICZ: Certainly the Institute of Physics can't be considered representative. According to this latest assessment we got an A and it was obvious we would get that much. There are no right criteria for assessment of science, this should be made clear. There are and there will be none. The only right criterion I know of is this maximal automatic one. Such as the one we've talked about before, a number of publications in foreign periodicals. It is in a way automatic and therefore it is relatively good and certainly separates absolute rubbish. However, I would have serious doubts whether it makes positive selections. The introduction of the system of grants in Poland is a worse nonsense in Poland than in the United States where there is bureaucracy which estimates the proposals, while in Poland it is done in an amateurish way, without such a basis. Besides, there is still another element. The United States is an immense country and you can theoretically count on finding a reviewer who, on the one hand, is an expert at the problem, and on the other, will not try to trip you up maliciously or competitively. Poland is a small country and finding competent and impartial reviewers is very difficult. I reviewed some grants projects, they were proposals from my best colleagues, so I graded them positively as they deserved it, but you can always find problems. It is generally a wider problem of introducing of American models onto a foreign ground. For instance, lately there has been a debate on the need to submit a thesis to qualify as an associate professor (the so-called *habilitacja*). When our university's professor went to the ministry to discuss this problem, the vice minister's answer was, "Do you know what you call 'docent' [assistant professor in Polish] in English?" This was to be an ultimate argument in this conflict.

KOCZANOWICZ: Is there any model of science you could accept?

STARUSZKIEWICZ: In my opinion the closest to the ideal would be the German science in the second half of the nineteenth century. If we would like to set up an optimum model of science, we should find out what caused the explosion of the German science at that time. You will notice that previously the German science couldn't compete with the English or the French, and practically within a few decades it became much better. I have analyzed the problem of selection of professors at German universities at that time. At least in physics, these choices were almost always correct, verified by their later careers. You certainly can't say that about contests for professorships in the United States.

KOCZANOWICZ: And in Poland?

STARUSZKIEWICZ: As we had no such problem before. Positions used to be created according to needs. This problem has cropped up only now that the number of positions is limited. You can imagine the correct steps that need to be taken, although for practical reasons they are difficult to enact. For example, one should absolutely prohibit employment of your own graduates: this however would be impossible if we take into account, say, the housing problem.

KOCZANOWICZ: So far we have talked about external conditions of science; however, I have detected some traces of your anxiety about the status of modern physics. What would this crisis in physics concern?

STARUSZKIEWICZ: As a matter of fact, for sixty years physics has revolved within a framework of the same notions. There are no essentially new notions.

KOCZANOWICZ: What was the recent invention?

STARUSZKIEWICZ: Quantum mechanics; 1926, 1930. In 1930 the real creativity in the field of physics, but conceived at this highest level as creating of notions, came to the end. At present physics takes advantage of this set of notions that was created in the twenties and there are no signs of going beyond these notions. These are subtle things. Of course, there are the results of the work of an immense number of people. For instance, at present there are mathematical methods of physics that did not exist before. I'm talking only about theoretical physics because experimental physics is a separate issue. The latter always advances, as it were, automatically. It is a matter of technological advancement, so it is a separate question, even though there is a crisis in it, too. Deleting this superaccelerator in Texas from the U.S. budget is a symptom of a crisis which resulted from an exclusively extensive development of experimental physics—more money, more processing powers.

KOCZANOWICZ: Are there any differences between the U.S. and Europe in ways of practicing physics?

STARUSZKIEWICZ: I can see no differences between the U.S. and Europe. It may be a difference in scale, media, these secondary factors. Nevertheless, I see a fundamental difference between people of the old school and modern people. And to the disadvantage of the modern. This generation of the twenties and thirties has somehow now successors.

KOCZANOWICZ: Why?

STARUSZKIEWICZ: There has been a giant advancement, and in this connection all the energy of the successors was focused on putting those achievements to use. The next step forward might be objectively difficult because, let me put it this way, we don't know how high it should be, we don't know the obstacles to overcome. It is like from this joke about someone in the mountain who spent a whole night sitting at a deep ditch because he thought it was an abyss and it was rational of him because he didn't know what was in front of him. As we really do not know what we are facing, we can only make suppositions. We don't know how high this threshold is. That no one has managed to do it is a result of ignorance, or a negative effect of social factors such as the system of grants, or the fact that theoretical physics is practiced by hundreds of people who do not communicate with one another. The founders of quantum mechanics were in touch with one another on a daily basis. There were no faxes, no computers, maybe mail worked better than nowadays. It was all old-fashioned but they were all in contact with one another on a daily basis. A good part of quantum mechanics was made on postcards which traveled among its authors. Whereas nowadays someone like me has no contacts with most people occupied with the same field of study. I have no contacts maybe because I'm not sociable, I travel rarely; but I suspect that people who think they are here in the center of events don't have contacts either, simply because it is impossible.

KOCZANOWICZ: Do you have any predictions for the future?

STARUSZKIEWICZ: I can tell you about possible scenarios, but I have no predictions. These scenarios are as follows: (1) Optimistic. There appears a man or a group of people who make a breakthrough. He will be the messiah everybody expects. This is not very likely. (2) An extremely pessimistic option is that this dissatisfaction and frustration, clear results of which can be noticed now, will effect an enactment of a scenario in which people will abandon science, just as they abandon industry in favor of ecology. I guess it is a very strong chance people should be encouraged to take advantage of, because there is no way to fight an ecological crisis except to tell people that if they want to live in a clean environment they have to produce healthy food in their gardens and not buy it in a supermarket. There is no other way out. (3) All in all the most probable option is that science is going to evolve but by way of diminishing steps. That is there is going to be progress toward a certain terminal point that is not a real terminal point. Imagine you want to get to Houston but you move at a speed that constantly diminishes. Then it may happen that you will move infinitely, until the end of the world but you will get only halfway. The same may happen with science, progress by diminishing steps consists in a subjective feeling of progress but this progress won't go toward say, the Platonic purposes science aims at. The ultimate physical theory is a certain Platonic being which exists and toward which we have been moving, but here we have got a few possible scenarios. Either we reach it, or we don't reach it as it is in infinity, whereas we move at a finite speed which makes reaching this infinity impossible, or lastly it is not attainable because science evolves by way of diminishing steps, and doesn't approach any real destination. All in all this last scenario is the most probable. Plebiscience. Science practiced by thousands of people with the results of research being more or less predictable. You can make an experiment. If you know somebody who assigns grants in Poland, ask him if he will be astonished by the result. I bet most of these people can predict the results. And this according to Truesdell is plebiscience.

KOCZANOWICZ: What consequence would such a crisis of science have for Western civilization and culture which, after all, was founded on a union of philosophy and science as well as on a concept of technological effectiveness of science?

STARUSZKIEWICZ: I don't worry about Western culture and let me tell you why: maybe just because, even though I'm a physicist, I view it from a historical perspective. Science is basically a tradition that began from Pythagoras, that is as early as 2600 years ago. Out of that just 300 years can be called Western culture. So you are talking about 300 years which are but a small fraction of 2600. And now, if you are asking about Western culture's perspectives, you should juxtapose this 300 years with that 2600. And consider what happened to nations which now create what we call Western culture: England, Germany, France, Italy. Well, the histories of these nations were different, and what we call Western culture is not this whole period, it is a combination of culture brought from Rome with science and technology. And this is but 300 years old. There is no use considering what will happen to it because there are too many unknowns. There is enough coal for 500 years and yet coal is so terribly destructive of normal environment that it seems that burning of this coal is impossible for environmental reasons since this would mean suicide. Such is a temporal perspective. In the next 100 years something dramatic has to happen. What mankind wants to do is physically impossible; when you ask what the entire mankind wants the answer is simple: live at the level of North America and Western Europe. This is impossible.

KOCZANOWICZ: Many people tend to accuse science of this situation, they think that the model of technologically applicable science is bound to produce the results you have described.

STARUSZKIEWICZ: I always protest when you talk about science in this context because it means looking for a scapegoat. People have to think twice. People. That is, governments, societies, public opinion, press, parliaments. People have to consider whether they really need this. I like the comfort in the U.S. very much, but I believe that without lowering it substantially, without depriving people of showers and running water, we can still have it at a fraction of present costs. It would be enough to rationalize it all on, say, ecological principles, so that food is produced locally and not imported from one continent to another by jet planes So you see, it is not science that has to reconsider itself—as I believe theoretical physics has nothing to do with it. Theoretical physics has quite a clear purpose for itself, except it does not know how to attain it. This purpose is another great cognitive step forward, such as quantum mechanics. It is obvious that this step is necessary as in the existent theory there are contradictions, these are notional contradictions, so you don't need to be a great thinker to notice that something is missing, that the world image presented by theoretical physics is not coherent. The need for the next step does not come even from a conflict with experience. The paradox is that there is a correspondence with experience. There are no signals that modern theoretical physics would find it difficult to describe experience. No. Here everything agrees. Whereas this image is internally incoherent and this is important. If all founders of quantum mechanics, Einstein, Bohr, Schrödinger, were in accord that this step was necessary, it was because the world picture created by quantum mechanics on the one hand and by the theory of relativity on the other is incoherent. It is like that. Once in my lectures I compared the world pictures of nineteenth-century and twentieth-century physics. The physics in the nineteenth century was something like an old photograph. You can hardly see details in it but somehow it does make some whole. Whereas modern physics is like a broken mirror. Particular pieces give ideal reflections but the whole is incomprehensible. When you look at a broken mirror, you know you have to collect pieces together some way. Except it is terribly difficult to put them in order.

KOCZANOWICZ: Is it possible at all to put them in order?

STARUSZKIEWICZ: It is Einstein's credo that you can do it.

KOCZANOWICZ: God doesn't play dice.

STARUSZKIEWICZ: No, God is not malicious. It is an assumption that doesn't have to be true. It is an assumption that ultimately can be false. Except it has turned out correct so far. It is not that I know it, because one can't know it. It's a credo that the universe is rational. It is not absurd. Only there is a real difficulty no one knows caused by what. All these factors I've talked about, this system of grants which is suicidal, these are all circumstances that reinforce the basic fact that the direction of research is unknown. Einstein for ten years had deliberated on a single fact, namely, the equality of heavy and inertial mass. It is a model of how one should work. Not everyone is capable of such an investigation of a simple idea.

KOCZANOWICZ: We observe the same phenomenon in philosophy, the same problem of putting together of a broken mirror, except in philosophy many people believe one should discard the very idea of a mirror as a metaphor useful for a description of the relation between cognition and reality.

STARUSZKIEWICZ: To my way of thinking, the evolution of modern philosophy is essentially combined with the evolution in physics. I mean, almost all important impulses come from physics. Look at it diachronically beginning with Descartes. Descartes was someone you would now call a mathematician. But he considered himself someone you now call a theoretical physicist. His ambition was to create a holistic world picture. The lasting achievements of Descartes, that which has remained valuable, is mathematics. Descartes's physics is valueless. Although it was his main ambition. Descartes, Leibniz, Newton, Kant, as a matter of fact, all basic ideas were created under the influence of scientific revolution.

KOCZANOWICZ: Would you say the crisis in philosophy is a reflection of a crisis in physics?

STARUSZKIEWICZ: The crisis in philosophy derives from the crisis in physics as we seem to have run out of ideas physics delivered for three hundred years. I also believe that the crisis in philosophy has been also caused by another phenomenon, a sociological one, concerning the population of philosophers. You certainly know more about it but if you look at university populations of philosophers, you'll see that people inspired by science (such as Thomas Kuhn or Karl Popper) make up a minority. The majority of academic philosophers are traditional philosophers. What do I mean by traditional philosophy? Philosophers practicing philosophy in the style of Hegel or Marx, or even the British analytical school. I think this traditional philosophy has lost any reason for its existence. Hegel might have believed that what he was doing was sensible, but Tischner, who philosophizes in Hegel's style, should realize that it is often nonsensical, that it is literature and not philosophy or science. This is my reply to the problem of Hegel's philosophy. I like some papers of Nietzsche very much. I think they are intelligent, but should we press them too much and look for their positive value, the answer would be no. Because they are based on nothing—they are Nietzsche's speculations from the time when he was no longer a professor and went mad. They are private thoughts not based on any research, or on any critical methods. They are essays. Nietzsche is good essays, Hegel is poor essays, Marx, hopeless essays. All this is literature, not serious science or philosophy.

KOCZANOWICZ: So what would be decent philosophy in your conception?

STARUSZKIEWICZ: It is a problem. As early as in the twenties Max Born said that in his opinion the role of traditional philosophy that was in Greece and that was in Europe until the end of the nineteenth century was taken over by theoretical physics. Of course, only as far as epistemology and ontology are concerned, because ethics and aesthetics are separate questions. About these disciplines physics has nothing to say. Ethics and aesthetics still legitimize vast areas of philosophy, unlike ontology or epistemology. The latter have been taken over by theoretical physics. Science does not articulate value judgements, their origin, and so on. The contrasting of reality with mind is for physics too crude a distinction because the whole quantum mechanics is a problem of a relation between measurement and interpretation. Reasoning in categories of such a distinction doesn't explain anything because we know already that it is much more complex. If quantum mechanics, an excellent physical theory, collapses under such a simple thing as a relation between observer and observed system, then everything is more complicated and can't be reduced to a simple contrasting of the world and our image of the world. Science views it in an infinitely more accurate way. Science views the world as a set of atoms and particles. Science views the brain as a set of neurons and synapses and this is all infinitely more complicated so that reasoning in categories of a simple distinction is too crude. It doesn't contribute anything.

KOCZANOWICZ: This is also the viewpoint of many contemporary philosophers, most clearly defined perhaps by Richard Rorty. European philosophy is identified with the problem of representation, a "mirror of nature." Discarding this idea equals discarding of European philosophy as it has been.

STARUSZKIEWICZ: Even now physics struggles against things which result from this distinction. You will agree that the problematic of quantum physics can't be formulated in this language. So you see how crude this language is. The basic problem of quantum mechanics is that you can't describe the act of observation in a manner understandable to all. I would say that traditional European philosophy is very valuable. Descartes's book *Rules Governing the Mind,* for instance. There was a time I used to read this book again and again, I think it is a work of genius. You can find a number of such great works, so I wouldn't dispose of European philosophy as a whole. But you have to realize what Max Born said that as regards ontology and epistemology traditional philosophy can't add anything to what theoretical physics says.

KOCZANOWICZ: And has it ever been capable of adding anything?

STARUSZKIEWICZ: In the seventeenth century, certainly, yes.

KOCZANOWICZ: Isn't philosophy the stunt man of science? Philosophy jumps onto those areas not occupied by science. Understandably, philosophy can be less rigorous in this way of reasoning.

STARUSZKIEWICZ: I have no recipes for the future but in the seventeenth century this role of philosophy was real. As I've said before, Descartes emphasized his own preoccupation as that of a physicist. Incorrectly, since his physics is worth nothing. Yet, his mathematics is excellent and his philosophy is excellent, too.

KOCZANOWICZ: So you support such a neopositivistic concept of philosophy?

STARUSZKIEWICZ: I wouldn't label it neopositivistic. I have always suspected neopositivists of a wish to impose a worldview upon science. And quite an absurd world view of that. Physics has not for a moment been neopositivistic in its history. Physics have always applied the principle of the most effective means. If a phenomenological study, such as positivists preferred, yields results, good, if not, if the method of mathematical hypothesis, what they actually opposed, yields results, good too. Physics, simply, always applied the method of mathematical hypothesis and the phenomenological method, whereas neopositivists, for reasons unknown to me, attempted to dispose of, to delegitimize this method of mathematical hypothesis, the method of speculation. It is totally absurd, because you can't prohibit speculating. Speculation may be meaningless, but this is what scientific criticism deals with, testing of speculation. Think of Boltzmann. He committed suicide, we don't know why, but supposedly because he had pronounced the atomistic hypothesis all his life and had constantly been under attack by people positivistically oriented like Mach, on the ground of creating nonexistent beings. So maybe when he got old, he took this criticism so badly that he committed suicide. So fatal was this positivistic criticism, absurd since the very beginning because the point is not to discard speculation but to speculate in a critical way. The point is not that beings should not be created, but created so as to move real knowledge forward.

KOCZANOWICZ: You have said that physics replaces ontology and epistemology. I doubt if this is wholly possible, for instance, what to do with the notion of God?

STARUSZKIEWICZ: For a long time physicists have used the concept of God. You have to distinguish God as a quasi-scientific hypothesis from God as a product of culture. It is a separate question if we talk about God as a product of culture, or about the God of some tribe. Physics doesn't relate to such topics. But if we treat God in quite a brutal way as a scientific hypothesis, then it belongs to physics' interests. In the sense that He either exists or not. If He does and if physics moves in a right direction at a finite speed, then it will eventually reach its destination. And if not—the problem with God is like with curvature of space. The proposition that God exists is verifiable, while the proposition that He doesn't exist is unverifiable. It is exactly like with the curvature of the earth. The proposition that the earth is round is verifiable because the Magellan voyage around the world proves that it is round. But if the earth were flat, the proposition that it is flat is not verifiable because you can't reach infinity you always have to retreat and then there arises a question: maybe it curves further on? The same situation is with God, the proposition that God exists, if it is true, is verifiable. Such is my thesis. But the proposition that He does not exist, if it is true, is not verifiable, because it is negative. No negative propositions are verifiable.

KOCZANOWICZ: Who would be this God of physicists? It would have to be an element of reality accessible by physical methods.

STARUSZKIEWICZ: I don't know. This something that is traditionally called God means some personal factor, existing somewhere, maybe beyond time and space but still affecting physical reality. God may be, however, something quite different. In fact, so conceived, God concerns nobody but physicists. If people are interested in God in a commonplace sense, it is a totally different God. This commonplace God can be an element of a common psychic space of a nation or a tribe, but it may be also something like man's character. Every man has a character. It is difficult to say what it is. It is not a physical body that can be grasped, but it is something definite. So in this sense the God people perceive out there can be an element of that which is psychic. It may be an objective factor that exists in a human psyche and which controls this psyche somehow.

KOCZANOWICZ: It is interesting what Catholic philosophers would say about it, even those who deal with the philosophy of science. Say Bishop Zycinski?

STARUSZKIEWICZ: Bishop Zycinski wouldn't agree with it. He wrote the book *The God of Abraham and the God of Whitehead.* You can guess all from the title itself. Zycinski's thesis is this: the God of Abraham and the God of Whitehead is one and the same thing, except it functions and reveals itself in a different way. I wouldn't say I would object to it. I simply suspend my judgment because I just don't know. It may be as Bishop Zycinski claims, except he claims so having no proof of it.

KOCZANOWICZ: I think that a consequence of your reasoning is that it can't be the same God, because Abraham's God is, as you put it, an element of that which is psychic, isn't it?

STARUSZKIEWICZ: Possibly, but I say again that I don't know. You know what Einstein said. He claimed that it is obvious that there exists a God of physicists as a source of the world's rationality whereas there is not a personal God who would be interested in our lives. Because Abraham's or Israel's God is interested in a concrete man, whereas Einstein did not believe just in that. He claimed that probably there is no such God. Bishop Zycinski claims there is. These are beliefs, and not scientific propositions. There is no contradiction between what Bishop Zycinski and Einstein say. Einstein said expressly, there is God as the source of the world's rationality and the source of the world's mathematicity. What people want is a God interested in their lives, because in fact it is a psychological motivation of religion to look for a father, for protection. So this God that really exists is useless for people. I have found here an extremely interesting book by Frank Manuel on Newton's religiosity based on Newton's manuscripts. It wasn't so that he founded great mathematics and great physics and then took to this discreditable alchemy, but just the other way round. Newton believed there was something to be discovered in alchemy. And he had the same problem as modern physics, he encountered too difficult a question. What he had at his disposal wasn't enough to make progress in chemistry. To my mind, it is absolutely right to say that Newton did this theological and alchemist research very conscientiously. It was no disgrace and if these manuscripts of his were published now, we would just the contrary see how great a figure he was and how rigorous his reasoning was. More specifically, what is this religiosity about? Newton often used to be involved in various philosophical disputes, for example, Leibniz's polemic with Clarke is a philosophical discussion on various very abstract topics that Newton had with Leibniz but which were provoked by Leibniz. Manuel is absolutely right in saying that Newton hated such disputes, that he considered them useless and inconclusive. In the *Principia* Newton wrote, "I do not pose hypotheses," which wasn't true in purely physical terms. He posed numerous hypotheses, except he conceived the posing of hypotheses in a very modern way. You are allowed to do it but you have to control them and check what they mean. Similarly, Newton hated all abstract, abstractly philosophical discussions he used to be involved in. Leibniz accused him of godlessness on the grounds he supposedly discarded God, made God unnecessary. It was Leibniz's thesis. Newton's religious views were very simple—what is there in the Holy Bible and Revelations is a fact. The Holy Bible is a certain fact, just like the moon is a fact. Newton was very sober-minded and critical of this fact. He believed the Holy Bible indeed relates certain revealed truth except transmitted in a very imperfect language adapted to the needs of the people whom it originally addressed. In this connection the purpose of this biblical exegesis is an understanding of this objective truth which can be found in the Bible. Those works of Newton that are mistakenly thought to be the most discreditable concern St John's apocalypse, what it predicts, and its chronology of events. Newton used the same scientific machinery for philosophy that he used for nature. And in an extremely critical, ingenious way he carried out this exegesis, presenting various explanations of what it predicts. Altogether. Newton was a religious man in such a fundamental-historical sense, that he regarded the Bible as factual certainty which must be rationally understood. Physics is an objective ontology. There is no doubt about it. There is a God of Abraham and God of physicists. Bishop Zycinski claims it is one and the same thing. Einstein claims there is God of physicists, and no God of Abraham. I suspend my judgment on this matter because it is of no use to argue if there are no relevant data. I believe that part of scientific thinking is that one should know what one says and one should suspend judgment in matters on which one doesn't have data. I know that the God of physicists exists. Let's ask a concrete question: Is the God of physicists and the God of Abraham one and the same thing, is there any relation between them? My answer is, I don't know. Physicist as a physicist can't pass opinions about it, it is impossible.

KOCZANOWICZ: It is a fact that the majority of scientists in Poland are believers and I would say people in the natural sciences are more often religious than people in the humanities. I had a logic teacher, a famous professor who was atheistic in the style of, I would say, the Vienna Circle. He often remonstrated against this fact, he considered it a disgrace to science that so many scholars would go on pilgrimages.

STARUSZKIEWICZ: To my mind this religiosity of scholars in Poland that you have mentioned is the religiosity of the same type as Newton's. Not so fundamentalist—you know yourself that in Poland nobody reads the Bible. I understand Polish religiosity in this sense that the Church is something already existing in Poland. It is a fact. Religiosity in Poland consists in either positive or negative attitude to the Church, to the Catholic tradition, to the fact of ritualism. Polish religiosity consists in having a positive attitude to these facts. I do not think it is a matter of some philosophical motivation, I don't think most people, even scholars, would consider this fact in this way. I've never been on a pilgrimage to Czestochowa because neither my health nor time allow me to do it, but I wouldn't see anything wrong if I did it. I can see no contradiction between my going on pilgrimage and my scientific views.

Note

This interview with Andrzej Staruszkiewicz is, I hope, a self-explanatory piece. I would like, however, to clarify the points which were raised at the editorial meeting for *Late Editions*. To begin with the general question about the philosophical implications of physics and the neopositivist controversy, I wish to mention that this was never a real issue in Poland. The philosophers of the Vienna Circle believed at least for some time in the possibility of introducing a very strict program in philosophy and science which would reformulate the theories in terms of observational sentences. We had our own tradition of analytic philosophy, the so-called "Warsaw-Lvov School" which was formed originally under the influence of the Austrian philosopher Franz Brentano. (The founder of the School, Kazimierz Twardowski, wad his student.) People who belonged to the school, such as Alfred Tarski, Jan Lesniewski, and Kazimierz Ajdukierwicz never shared the neopositivistic illusions as to the possibility of translating all terms into the observational ones. Therefore the remarks on neopositivism in the interview deal rather with the abuse of certain aspects of the physical investigations and the historical case is an illustration of the possible peril of such abuse. Andrzej emphasizes this point in his papers on the philosophy of science. In his opinion, neopositivism is a dead issue and some physicists, like Schrödinger, have always had clarity on this point. Mathematical speculation is the basic method of theoretical physics but it must be critical, that is, always checked for logical consistency and agreement with the experiment. Also, Andrzej maintains that Max Born's assertion that theoretical physics takes over the function of philosophy (of epistemology and ontology) is as true now as it was in the twenties.

I should also add some information on my interlocutor. Andrzej Staruszkiewicz is a professor at the Institute of Physics, Jagiellonian University, Cracow. A great portion of the interview was made in the United States in Columbia, South Carolina, where Andrzej was a visiting professor at the University of South Carolina. He was invited by his former student, Pawel O. Mazur, who is now a professor at this university. Andrzej spent a couple of months there and came back to Poland in September 1993.

Andrzej's most quoted work mentioned several times in the interview is a theory of gravity in a (hypothetical) three-dimensional space-time, based on the same principles as Einstein's theory in the real four-dimensional space-time. The mathematical properties of the theory, first described by Andrzej in his Ph.D. thesis, are remarkable and have attracted the attention of numerous theoretical physicists in the eighties; among others, Gerald t'Hooft, a very influential Dutch field theorist, who wrote several papers on three-dimensional gravity.

SHARON TRAWEEK

13

BACHIGAI (OUT OF PLACE) IN IBARAKI: TSUKUBA SCIENCE CITY, JAPAN

City on a Hill

Most of the Japanese intellectuals I meet in the large cities of Japan and at campuses of universities in the United States are utterly disdainful of the very idea that a university has been built in the Ibaraki countryside. They also think that building an "American-style" landscaped campus is silly in Japan. They point out that the entire city of Tsukuba is artificial, planned by government officials, and seems quite boring and American. I counter that there are city planners all over the world and that Tsukuba resembles all these new cities, like Novosibersk in Russia and Irvine in California; that they resemble each other more than any one origin.[1] We talk about the postmodern architecture at Tsukuba, La Jolla, and Irvine; we agree that the Renaissance-inspired buildings and plazas of Isozaki in Tsukuba and Michael Graves near La Jolla are monumental and Eurocentered (however ironically) and that they border on the colonialist when built far from the Capitoline piazza.[2] We prefer, for example, the more playful "regional vernacular" building by Frank Gehry at Irvine.[3]

Quite a few Japanese intellectuals have told me that they think Tsukuba was designed by right-wingers in reaction to the student demonstrations of the 1960s and 1970s; they suggest that anyone who joined the University of Tsukuba faculty in its early days might well have been right-wingers, too. They recite the rules supposedly in place at the university: no public gatherings of very large groups, no bull-horns allowed on campus, and so on.[4] These intellectuals are uninterested when I tell them that many of the people I know in Tsukuba had spent a great deal of time in other countries working in scientific laboratories, accepting the offer to return to Japan and work at laboratories in Tsukuba with many mixed emotions, but apparently few political thoughts about Japan. I add that since 1976 I have heard stories from several people about rather intense political conflicts among researchers and faculty in Tsukuba; some of these "political" activities seem quite insidious to me (stalking and harassing individuals) and others sound very principled (challenging the Ministry of Education to give a regular faculty appointment to a Korean-Japanese).[5] The campus hardly seems apolitical to me. Nonetheless, I am sure there is a striking contrast between the Tsukuba campus of today and the campuses of Tokyo University in the 1960s and 1970s, just as there are on campuses all over the world.

I mention to my interlocutors that the new American universities of the 1960s and 1970s, such as the campuses of the University of California at Irvine, San Diego, and Santa Cruz (where I was a graduate student in the late 1970s and early 1980s), also were seen by activists at the more established campuses like UC Berkeley and UW Madison as being deliberately sited in rural areas, small towns, or remote suburbs, and their architecture planned to circumvent large-scale political activity. I point out that from their inception these schools have had a much greater proportion of women and other under-represented groups among their graduate students and faculties. I add that as political action among American students gained momentum in the late 1980's, students at these sorts of schools have been quite active. I told them that I also think that such schools were much more accepting of the newer, nontraditional fields of study and that the senior faculty had much less power than in traditional universities. I argue with the Japanese intellectuals that I believe these new campuses built in the 1960s and 1970s all over the world, whether Irvine, Tsukuba, or Nanterre, resemble each other at least as much as any specific and singular origin; I add that as a resident alien I found Tsukuba University rather welcoming.

One Japanese physicist told me that the idea of Tsukuba began with the University of California at San Diego. He said that he was very impressed with how scientists in the United States had imagined the best sort of university for educating scientists and then just got it built. (I have heard this version of the establishment of both UC San Diego and UC Irvine from several American physicists and biologists. While a visiting faculty member at UCSD in 1990 and actively participating in a seminar at UCI, I realized that many faculty at UCSD and UCI find these stories ludicrous; the scientists are not fazed by the incredulity of their colleagues in the humanities and social sciences.) The Japanese physicist said that in the 1960s he had become intrigued by the American higher education system and particularly the liberal arts colleges where so many of his American scientist friends had been undergraduates. He eventually traveled by car across the United States, visiting many of those schools. He said he was especially impressed that students on those campuses had, simultaneously, active intellectual lives, active social lives, and even active athletic lives, all centered on the college as a community; it seemed so very healthy to him. He also liked the fact that students did not choose a specific area of study before enrolling in college, but took a general course of study and majored only in the last two years. He noticed too that the physics faculty at those American schools were not isolated within their departments, and that they tended to have many interdisciplinary discussions. He said it became his dream to build a school in Japan like what he saw at UCSD and on his trip. (Since the Meiji restoration in 1868 many Japanese scholars have established schools according to some model found in North America or Europe; in fact, there is an even older history in Japan of building schools based on models found in China.) I remembered that in 1968 many student activists in Paris had wanted to change French universities, based on autonomous disciplines, toward what they regarded as the more open American curricula.

I asked this senior physicist how he managed to build his dream school in Tsukuba. We had been talking in the lobby of Tokyo hotel and he had been so expressive and even cheerful up to this point in our conversation; I was surprised that he suddenly became taciturn and withdrawn. I felt that inadvertently I must have deeply offended him. Since 1976 I have heard many stories in Japan about this man. Several physicists in Tsukuba told me they believe that only a person with a very forceful personality could have built Tsukuba and its laboratories; they acknowledge that having a forceful personality in Japan and being willing to use it to accomplish self-defined goals certainly leaves that person open to scrutiny, gossip, isolation, and perhaps hostility. (There is also some gossip about his political and religious views.) Most physicists seem to think that the largeness of vision and creativity of this man easily compensated for any errors in his judgment, while pointing out that many Japanese feel much less sympathetic to such people than Americans would.

Erehwon

When I first went to Tsukuba in 1976, I was not sure how to get there.[6] I found the village of Tsukuba on the map, located on a small railroad line branching from the Joban trunk line at Tsuchiura, about halfway between Tokyo's Ueno Station and the city of Mito to the northeast in Ibaraki Prefecture. I saw on the map that the village was at the foot of a mountain with two peaks named Tsukuba. A few months later I visited the temples atop Mount Tsukuba and discovered not only that the higher peak is male and the lower is female, but also that an emperor of Japan once visited the mountain and wrote a haiku about it. The main temple on the mountain (all important mountains in Japan have important temples) has an impressive large wooden structure and old, steep stairs up the mountain. Many people go there to buy an amulet that prevents traffic accidents and then for the view they take a funicular up to the peaks. There is also a huge wooden farmhouse near the top which has been turned into a restaurant serving the peasant food which recently has become so fashionable in the big Japanese cities.

Ueno Station, the embarcation point in Tokyo for Tsukuba, is located in a working class area known as *Shitamachi* [literally, downtown]. I quickly learned that this area was considered quaint by some foreign tourists. Some guide books and the older American and European Japan hands consider it "the real Tokyo."[7] It includes the workshops of many traditional craft industries; one can find there too a street of shopwindows filled with those precise plastic replicas of prepared foods, displayed in vitrines by restaurants all over Japan as their storefront menus. *Shitamachi* was destroyed by fire after the earthquake and firebombed at the end of World War II. The main black market area in Tokyo during the American occupation, now a bargain shopping street, is near Ueno Station; nowadays, it was whispered to me, Ueno Park, near the station, with its zoo, many museums, and large open spaces (so very unlike Japanese gardens), is a gathering spot for local *yakuza*. I eventually learned that Ueno Station and the destinations of the Joban line trains from there were considered *terra nullius* by most intellectual Tokyoites I met.[8] I already knew that the Mito family, based in what is now the capital city of Ibaraki Prefecture, had been famous for violently opposing the Meiji Restoration in 1868 and, of course, losing. I came to realize that Ibaraki Prefecture was poor and had been poor, at least since 1868. It appeared that I would need to change trains at Tsuchiura, which I recognized as being near the famous airfields used by *kamikaze* [literally, divine wind] pilots during World War II.

KEK [*Ko-Enerugie butsurigaku Kenkyusho*, pronounced "kay-ee-kay" by foreigners and "ko-eh-nay-ken" by Japanese in the Tsukuba area] was in 1976 the only laboratory operating in Tsukuba. It consisted of several small research buildings around the newly constructed 12 GeV proton synchrotron accelerator and a smart white multistory building with large windows for the offices of the theorists, experimentalists, administrators, and staff. The accelerator had once been designed to be the equal of Fermilab, outside Chicago, but the Japanese government had cut the expected funding considerably. The physicists, coped by building what they hoped (rightly, as it happened) would eventually be merely the "booster" for a much more powerful accelerator. Visitors like me stayed in a trailer located near a makeshift sports field; I walked in midsummer around the games to the office building through thigh-high grasses. There was an old car that all the visiting researchers used to get to and from the train at Tsuchiura; by checking with the different experimental groups it was possible to borrow the car to do grocery shopping in one of the five nearby villages or to visit a local restaurant.

At that time there was only one restaurant fairly near the lab; it was at a rather rudimentary golf club, located on cheap land at the foot of Mount

Tsukuba many hours from its desired customers in Tokyo—and about an hour from its likely customers in Tsuchiura. It was a place with large windows and stucco walls: the same "modern" architecture my parents chose for our house in 1948. About five miles south of the lab were a few high rise residences for the first researchers and faculty in Tsukuba; the buildings were unlandscaped and surrounded by mud during the rainy seasons. The university was under construction. Some of the Japanese physicists were annoyed by all the space and the absence of urbanity; they often yearned for the bars of Tokyo. Years later a little crowded district of restaurants, barbeque spots, and bars with red lanterns did appear in Tsukuba, bordered on two sides by six-lane boulevards.

The Old Villages

I once had a very amusing experience trying to buy rice in one village near the lab, finally learning that uncooked rice is not called *gohan,* the generic word for the plain white cooked rice usually served at breakfast, lunch, and dinner. (The words for those meals all include the word *gohan.*) My interlocutors and I pantomimed the entire process of planting, transplanting, harvesting, winnowing, packaging, distributing, selling, and cooking rice before we could establish what I wanted. These circumlocutions were necessary, I later learned, because they had heard foreigners ate bread instead of rice. They were incredulous to be meeting a foreigner who ate rice. Someone eventually drove me to the correct store and then returned me to my car, thinking I could never understand his directions: we were surrounded by small rice paddies for miles in every direction, except toward the mountain. (It was because of the fertility of this Kansai plain that the Mito family had once been so rich.) The second director of the lab told me that during those early days they had visited the elders of each of the five local villages near the lab, assuring them that the lab would hire young women in equal numbers from each of the villages so that any marriages to laboratory staff would be equally distributed among them.[9]

I was surprised that that director had realized he needed to visit the village elders; he was surprised at my surprise. Of course, I should have remembered. It seems everyone in Japan has a *furosato* [hometown] they visit each summer during *obon,* a period of remembrance for those who have died within memory of the living. In most industrialized nations cities have quintupled in size in the space of one generation; in Japan that mass migration from rural areas happened after World War II. No wonder my urbane intellectual friends were shocked by Tsukuba; their symbolic map of Japan included only mountains, the sea, declining hometowns, and huge cities. In fact, by the late 1980s the tourism industry had begun to cultivate nostalgia, encouraging visits to particularly picturesque villages as generic *furosato* with the common advertising theme, in English, of "my hometown." The government, too, had begun a popular program of providing funds to villages that would use the money to enhance some distinctive feature of the community, not necessarily long-lived, in order to stimulate tourism to the village.[10]

It seems that while most of those urbane intellectuals' families had lived in the cities of Japan for generations, many of the physicists' families had not. The physicists did not say much about their hometowns, but if I asked a specific question they would talk. Once I asked if it were difficult to maintain those machines that transplant rice and I got a flurry of stories.

Repro Expo

By 1986 the Tsukuba of the laboratories and the university was designated a city [*shi*] to distinguish it from the town [*machi*] of Tsukuba and it had become an expo site. (Tsukuba is pronounced "tsku-bah" by locals, "tsu-kubah" by Tokyoites, and "scuba" by foreigners.) By then it had over forty national research labs, many high rises, a major hotel and plaza (designed by the internationally acclaimed architect Arata Isozaki), a supermarket, underground utilities, a huge department store with the only curved escalator in the world, innumerable restaurants (including Denny's, Kentucky Fried Chicken, and MacDonald's), an important university, several aerobics clubs, a few colleges, and 150,000 new residents.

Tsukuba has had bad press. I dreaded returning from Tokyo on the last Joban line train at night from Ueno Station during 1986 and 1987 because it was always jammed with drunken men, smelling badly of alcohol; they were unfailingly cheerful, as is the custom among the vast majority of inebriated men one sees everywhere in big Japanese cities after about nine o'clock at night.[11] Nonetheless, I would forget about this while enjoying myself in Tokyo, and then take that same train, yet again. I never failed to startle them; how could I be missed, the only woman, the only foreigner, five feet eight inches tall with red hair, our bodies touching in the packed mass of humanity rocking slowly with the train's movement for nearly an hour. No one ever offered me a seat. I had thought that a large woman would not be attractive to Japanese men; I did not grasp at first the implications of Japanese men's life-long, intense, complex, and intimate relationships with their mothers. One man would always begin talking to me, all the others listening, asking where I was going. If I pretended not to know Japanese, someone would always inquire in schoolboy English. I would answer that I was going to Tsukuba and my bleary-eyed, leering trainmate would then ask if I had found it depressing there. The newspapers and television news shows had been making a big deal out of the suicide rate in Tsukuba, announcing that the city was too new and too strange. I thought I would have been considered evidence of the problem, rather than an object of concern or an informant.

The modern tollroad from Tokyo was completed by 1986, so that when the traffic was light (in the middle of the night and early Sunday morning) I could drive between the lab and someplace in Tokyo within two hours. The city of Tsukuba had been platted in a narrow grid, about a mile wide and eight miles long, with KEK at its most northeastern point, so it was quite easy to venture back into the land of the villages and rice paddies. I occasionally drove around the prefecture, exploring, and sometimes would get misled, especially after dark, by following signs home to Tsukuba and then find myself in the village, rather than the city, of Tsukuba. Stopping at remote gas stations to ask directions, I always found friendly and cheerfully inquisitive village people. One of the young foreign researchers at KEK was able to rent an old traditional house in the compounds of a farming family alongside the newer prefabricated house the farmers had built for themselves out of the new money available from providing the newcomers with what they wanted.

Five years later the traffic signs throughout the prefecture pointed to the Science City [*kenkyu gakuen toshi*]. Express buses stopped at labs several times a day and arrived at Tokyo Station, the center of political, financial, and mercantile Tokyo, in about two hours; multiple round-trip tickets were available to commuters at a discount. By 1991 Tsukuba Science City had expanded, and real estate prices had escalated quite beyond the reach of the researchers and faculty; they were beginning to buy small houses three railroad stations beyond Tsuchiura, toward Mito.

Returnees

There's now a tiny Mexican restaurant in a residential area of Tsukuba, run by a Japanese man who spent some years in Mexico. Nearer the lab is a popular and lovely English tea room and close by is an interesting northern Italian restaurant; there are no Italians or English there. I bought imported fabrics and sewing patterns in a small shop owned by a Japanese woman who had lived in northern Wisconsin. But I knew that Tsukuba had become not only cosmopolitan but also sophisticated when I was taken by some young American physicists to a traditional rustic Japanese tea house serving peasant foods, run by Japanese artists, refugees from Tokyo.

While I was at Tsukuba in 1986–87 I began to think about my childhood, mostly because my "informants" began then to ask me so many questions about myself (they seemed to find it fascinating that a single woman in her forties could also be a professor, especially at MIT or Rice University, with a Fulbright Senior Researcher Grant).[13] I grew up near Los Angeles at the margins of the old empires of the Atlantic, Spanish America, and Anglo-America, and a continental empire, the United States of America. I grew up at the beach, at land's end betwixt the ocean to the west, a mass to the east of aircraft factories, first built to fight the Japanese in the Pacific theater of World War II and then used to win the race with the Soviet Union to the moon, a refinery to the south named El Segundo (*El Primero* was elsewhere, of course), fueled by tankers depositing their energy into a pipeline stretched far into the Pacific, and a sewage treatment facility to the north named Hyperion (after one of the ancient Greek titans, son of heaven and earth) which dumped waste through another pipeline into the same ocean. The air contracted rhythmically with the roar of jets leaving the Los Angeles International Airport while the smell of oil and offal rose from the miles of sand dunes where we post-Pacific war children played "Japs and GIs," hunted with bows and arrows, and dug tunnels to China. Sometimes we watched adults make romantic films there about adventures in Arabia, before the birth of new empires and new Asian wars made the pulsing roar of the airplanes too frequent to make good movie sound tracks. I thought about that at Tsukuba as my friends showed me the massive new industrial plants along the Ibaraki coast, another part of the government's economic development plans for Ibaraki Prefecture. As we drove to the coast we passed the old airfields near Tsuchiura where the *kamikaze* pilots trained during World War II. My friends laughed when I said I felt like an Ibaraki native.

Pioneers and Natives

At home I was taught a romantic adventure story I now think of as our family's version of social Darwinism: the docile and infirm and dull-witted stay put and the brave and fit and bright move on. I was told that our family had been pioneers for the past thousand years and that to fulfill the family heritage we too had to become pioneers, as our parents had, and our grandparents, and our great grandparents, and back one thousand years. Their stories were my childhood stories: to California in 1941, to the Oklahoma land rush of 1906, to another edge of an older Spanish empire in east Texas nearly a century earlier, to Tennessee from a "crowded" Virginia, to southern England from southern Denmark. Did others come to the American south from the Netherlands via western France? Were those stories really true? About ten years ago I drove with my mother through Virginia and Tennessee toward Oklahoma and Texas and beyond to California. Seeing the Virginia and Tennessee countryside for the first time, she kept remarking that it looked just like the stories she had heard as a child about when the family lived there in the 1700s and 1800s.

We had heard only a few of the stories of "the locals" married along the way. Mind you, "the locals" referred to, variously, the French, English, Spanish (and, later on, Mexicans), Native Americans (Cherokees, Comanches, Choctaws), Cajuns, Germans fleeing the revolutions of 1848 to Texas, and African-Americans. So who are the we in these stories? Or, as my first mother-in-law-to-be asked me in Manhattan (the borough of New York City, not the southern California beach I had visited nearly daily through my childhood), What are you? Unfamiliar then with the habits of American northeasterners, I found the question quite a challenge and began, patiently, to try to answer. As I spoke of several nationalities, ethnicities, religions, and classes, I slowly realized that she thought I was being insolent; my fiance intervened, suggesting that I give a one-word answer, so I shrugged and said, in that case, I guess you could call me a *gringa,* although it is not quite right, and politely asked her the same question she had asked me. She and her son now were horrified that I had drawn no conclusions and had attached no particular meaning to their family name.

I thought of that while watching debates on television in Japan about archeological research at the sites of former residences of the imperial family: many of the artifacts suggest the hitherto inconceivable conclusion that the family might have been from Korea. Since 1868, but especially since 1945, children have learned in school that all Japanese are alike, physiologically, racially, culturally, and linguistically, and different from everyone else, including Koreans. On that assumption the government retests all imported pharmaceutical drugs, for example. The physicists I met expected that, as an anthropologist, I knew all about these sorts of differences, and would ask me to explain to the other foreigners about how Japanese brains were structured differently. When I explained that, increasingly, physical and medical anthropologists were concluding that race was a political term and not a scientific one, they were quite unconvinced. When I added that many anthropologists had also concluded that Japanese were not necessarily alike, nor significantly different from other Asians, genetically, most were shocked.

While traveling on the train with physicists from Tsukuba to some physics conference, I would ask if they thought everyone on the train looked alike. They would readily point out what they regarded as the basic types: narrow aquiline faces from Sendai and the northeast, round flat faces from Kyushu and the southwest, and men with a five o'clock shadow were thought to be partly Ainu from Hokkaido in the far north.[14] When I would ask if anyone had come from Hokkaido, several people would readily answer yes, clearly surprising their colleagues. With further questions they would cheerfully recount family stories of emigration to Hokkaido many decades ago. I had learned to ask where people in such families had been born and buried, because for many it was clearly very important that they had been born on the main island of Honshu and that the family continued to keep ashes of the dead there, too. Others were very proud of the fact that the family had chosen Hokkaido as their home; they always knew the date of the transfer of the family's legal residence and funerary tablets, usually several decades after the first emigration. Even in those families many had taken the precaution whenever possible of going to Tokyo for the delivery of their children.

In the elementary school in my hometown we learned that for a brief but very important historical moment California had been an independent republic, with a bear flag: we sewed one and learned to sing "California, here I come, right back where I started from," imagining that we had come home from a diaspora. The United States of America came to California in 1851, but we did not celebrate that anniversary. The biggest event in town each year was the *fiesta:* we all dressed like *Californios* (who were not natives, either) and we carried the flag of the republic just as my cousins in Texas paraded with the flag of that short-lived republic and my cousins in Oklahoma knew all about the Sooners of the 1906 land rush. Our elementary school teachers, inspired by John Dewey, taught us California history by having us learn Native American ways, followed by the ways of all the *conquistadores,* in chronological order: the Spanish, the Mexicans, the Russians, and the Americans. We carved *kachinas* (although Native Americans in California never made them) and heard stories about how the Indians came from Asia, visited some of Serra's missions, and heard stories about the harsh rule of sword and cross, wove *serapes* and cooked *tortillas* and sang songs about *cinco de mayo,* played with animal pelts and heard stories about the pioneers from Irkutsk on Lake Baikal in Siberia, and built miniature log cabins and heard stories about the Donner party and the Gold Rush of 1849. We were very proud of our frontier history and we thought that the edge of empire was the right place to be; when I got home I find everyone there still does. But, of course, it was not a frontier to the Native Americans who once lived there.[15]

Foreigners

I thought about all this in Tsukuba as the foreign physicists became friendly with the Ibaraki people working at the lab and were learning to speak a bit of Ibaraki-ben, the local dialect. Some of the younger physicists, Japanese and American, were marrying Ibaraki women who were working as secretaries at KEK. Most of the Japanese physicists cultivated a disdain for the locals which reminded me of all the faculty from east of the Rockies I have encountered at California universities who have the same attitude toward their native Californian students. When I was an undergraduate, we students were fully aware of their disdain, and, in return, called them "carpetbaggers." I wondered what the Ibaraki people thought of the suddenly arrived scientists and engineers; gradually I began to hear that they seemed to like the foreigners much better.

The number of foreigners at the lab suddenly mushroomed in the mideighties when an international collaboration from about twenty universities around the world was formed to do experiments at a new KEK facility named, acronymically, TRISTAN (Transposable Ring Intersecting Storage Accelerator in Nippon). According to the then laboratory director-general:

> Our accelerator plan at KEK was nicknamed . . . "TRISTAN" after the passionate story of . . . Wagner's opera, with the love and dreams for our science research, particularly for hunting quarks in Nippon. The first conception of this idea was about a decade ago, and now our TRISTAN [has come] on stage. An opera is really a team work of singers, instruments players, the conductor, the stage manager and many more important people setting the drama behind the scenes. So is the construction of a large accelerator complex such as our TRISTAN. We could only make TRISTAN's initial operation successful, with the excellent cooperative work of our colleagues. . . . There is a famous story about . . . Wagner's idea in composing this opera with is based upon his original musical drama. He wrote in a letter to Franz Liszt, "Because I have never tasted the true bliss of love, I shall raise a monument to that most beautiful of all dreams wherein from [beginning] to end this love may for once drink to its fill". It took about ten years of Wagner's work before . . . "TRISTAN und ISOLDE" was first performed at Munich in 1865. . . . Now, we are very glad to publish this TRISTAN construction report on the occasion of the dedication ceremony for TRISTAN on April 7, 1987. Needless to say, our physics program is just about to begin. Taking . . . Wagner's opera, this [corresponds] to the beginning of . . . Act 1, when Tristan and Isolde are about to depart . . . from a quay [to meet King Mark]. We don't know at present how the highlights of the succeeding acts will develop. We will continue to make our best efforts so as to be able to taste the true bliss of our love, and leave the rest to Heaven.[16]

By the time TRISTAN was completed in 1987 the laboratory had expanded significantly from its beginnings over a decade earlier, acquiring not only those cooperative colleagues and the TRISTAN facility, but also large numbers of scientists, several new buildings, and many bureaucrats from the national government who managed the very large amount of money that had been bestowed upon the lab. With this expansion there were many new roads on the laboratory site. One day some Japanese physicists said they were trying to think of names for the roads and since they thought I was familiar with the history of physics, they wanted me to recommend some names, I suggested that instead of using the names of Europeans and North Americans they should use the names of important people in the history of Japanese science and the arts. This suggestion met a stony silence; they preferred a more "international" approach.

Entrepreneur

In this enriched and enlarged lab research groups no longer shared a car. One man who was born in Ibaraki developed a new business around the foreigners visiting KEK and their desires for automobiles: he rents us cars. Now in his fifties, he shifted from repairing farm equipment and farmers' cars to renting to foreigners some older, dented cars most Japanese would not use. Registering, insuring, and getting the regular official auto inspections are both complex and very expensive. Only recently have foreigners at KEK begun to do this for themselves. Some of them have purchased the cars they once rented and continue to get their cars serviced and get advice about the bureaucratic procedures from the same man. Tsukuba now has the highest automobile-to-resident ratio in Japan and the highest automobile accident rate. The man who once rented just to foreigners at KEK now sells, leases, and repairs cars with his son. Three years ago next to his old house and the garage he built a very large, new, traditional-style house for his extended, multigenerational family; it is lovely with a great deal of fine woodwork. He has also acquired a new hobby: flying small planes with his old friends, one of whom has a nearby sushi shop, decorated with a large propeller. Most of the foreigners have visited this man, his family, and his friends, some on many occasions, and he visits them; I have only met one Japanese physicist at the lab who knew him.

I thought about him as I listened to physicists at KEK talk about the *tanshin funin,* the "married bachelors," who had apartments in Tsukuba but commuted to Tokyo and farther away every weekend because they did not want their children to go to school in Tsukuba. A few of the Japanese physicists who came to KEK in the seventies decided to settle in Tsukuba and raise their children there. I have heard their stories of working with the local schools and city councils and of the difficulties the outsiders and the locals (children, women, and men) had in learning to work together. Now, twenty years later, they can laugh together about the difficulty they all have in working with the newcomers moving to Tsukuba. (The suburban expansion of Tokyo finally reached Tsukuba and beyond. Inflated real estate prices in Tokyo first fueled development to the south and east of Tokyo; in the mideighties, development and commuters finally moved north and then finally toward the northeast and Tsukuba.)

Missionaries

When the cities of industrializing countries suddenly implode with emigrants, new religions are born, meeting those emigrants' desire for some semblance of community and even continuity. Japan is no exception and many new religions emerged after World War II; the most successful ones have had charismatic leaders, often women, with syncretic cosmologies forged from Buddhism, Shinto, and Christianity. Unlike Korea, only about one per cent of the Japanese have become Christians; the proportion has remained constant for many decades. Nonetheless, a much larger minority of physicists at Tsukuba are either Christians or attended Christian schools. They are impressed by Joseph Needham's thesis that the scientific revolution did not happen in China because there was no tradition of a single law-giving god. They are also drawn to the Christian message of forgiveness for the banished; to be excluded, or even out of place, in Japan is to be invisible. Many of the Japanese physicists at KEK are returnees, urged by the leaders who built KEK to reverse the "brain drain" of the 1950s and 1960s by returning to Japan to do physics.

Two American Protestant Christian missionaries (a very conservative Baptist and a rather more progressive Methodist) and their families arrived in the Tsukuba area in the mid-1970s; another missionary and his wife arrived from Germany in the 1980s. (Each of their congregations were predominantly Japanese, as was the membership of the local Roman Catholic Church.) I met the older Baptist missionary in my local grocery store, introduced by a Japanese friend who was a member of his congregation. I had heard enough from her to conclude that the pastor had an extraordinarily conservative reading of Christianity; he seemed to have heard enough about me to not make much of an effort to missionize me. I met the young German missionary and his wife because we had been invited to the same event by Japanese who were friends and we were quite noticeably the only non-Asians present in the large audience of mostly middle-aged Japanese women. Held in the Dai Ichi Hotel which had been designed by Isozaki, the event was sponsored by the Soroptimists of Tsukuba, an organization of some prestige. It was a lecture and demonstration by a Kabuki actor who specialized in women's roles. As he lectured he undressed, applied his makeup, and dressed for a "performance," concluding his lecture as his dressers placed the final combs in his hair and the last twists in his *obi* belt. Middle-aged women provide the primary audience for the traditional kabuki and bunraku dramas; they are especially enthusiastic fans of these cross-dressing artists.[17] I thought it was hilarious that our Japanese friends had invited a foreign missionary and a foreign ethnographer to this particular event, a highly gendered performance of deconstructing and reconstructing a performer of a traditional art in a postmodern building before an audience of articulate, successful members of an international women's organization in a modernist city saturated with people who had lived everywhere, successfully.

I met the Texan Methodist missionary and his family at one of the Fourth of July celebrations at KEK. Having mentioned the various denominations of my cousins in Texas and the ones who had become ministers, the missionary made a joke about his family and mine, thereby acknowledging simultaneously that we could place each other perfectly, according to Texas social codes, and that those codes were a bit far-fetched from the fields of KEK.

The missionary and his wife had become active in forming many community groups in Tsukuba and were especially involved in planning an international school. They also worked behind the scenes to mediate many awkward situations. I remember when one foreign family arrived in Tsukuba and enthusiastically enrolled their son in one of the local schools. His hazing began when he changed clothes for sports one day and the other boys noticed that his underwear was not white; his problems persisted and then suddenly escalated into an altercation in which one of the Japanese boys was hurt. The Japanese parents expected an apology and more, none of which the foreign family was willing to provide. Fortunately the mediators were able to solve the problem constructively for the parents, the boys, and the school; the next year the same boys became good friends and the parents could laugh about the problem.

The American leader of an international team of collaborators at KEK and a friend of the Methodist missionary once got into an automobile accident in Tsukuba village while he was showing visiting relatives the area. The police chief became intrigued by the American and his Japanese language skills. Later he explained to the American that he had to visit the hospital room of one of the people in the other car and apologize, even though everyone involved knew that the American had not caused the accident. All were impressed that the American was not only good natured about this performance, he also took a gift. After the American finished filing the final papers on the accident, the police chief wished the American and his team success in their search for quarks.

Volunteers

Aside from these foreigners acting as mediators in Tsukuba there is an important volunteer organization of Japanese women who have lived abroad, called the Rainbow Club [*Niji no Kai*] which helps foreigners. Most foreigners first learn about them through a pamphlet which outlines the basic resources in Tsukuba, providing information about the hours, operating procedures, and special services of government offices, such as schools and post offices, telephone companies, banks, buses, physicians, and pharmacies. When I would get sick I would use this guide to find different doctors in the local villages, so I could combine my visit to a doctor with learning about another village in the area. The Rainbow Club also provides many Japanese language courses for visiting researchers and their spouses, some of which I have taken. Most of the members of the Rainbow Club are married to senior researchers and faculty members working in Tsukuba.

These very capable women remind me of the women I have met around Fermilao and the other major research facilities in the United States and Europe which were built at some distance from the community resources they regard as necessary, such as excellent schools with the latest educational techniques and technologies. These formidable women immediately set about building those resources by mobilizing parents and gaining access to local, regional, and national government agencies, foundations, religious groups, and women's organizations. I remember visiting Waxahachie in east Texas where the now-defunct SSC (superconducting supercollider) was being built for high-energy physics research and I toured the center of town with its several houses designated as listed on the register of national historic places. Talking with the proprietors of an antique store and a restaurant I heard about an important cluster of Waxahachie women who had first gotten the town's better houses listed as historic and had then approached the Texas Film Board about listing the community as an appropriate site for filmmaking. Eventually, *Tender Mercies* and *Places in the Heart* were made there, among other films. Having learned to deal with the special needs of filmmakers, they thought they could certainly cope with scientists and worked hard to get Waxahachie named as the site for the SSC. When the senior scientists' spouses appeared and focused on the schools, they found their counterparts in the Waxahachie women.

Women in the United States, especially those who are well educated and upper-middle-class, have long had significant and powerful positions in their local communities. They bring those same skills to the communities around the new science cities and it is impressive how quickly the senior researchers' spouses are able to establish the resources they want. Many members of the Rainbow Club had spent considerable time in the United States; while their spouses were doing research, their children were born and educated in public schools, and these women joined their American counterparts in the local community organizations. Similarly, the spouses of senior foreign researchers in Tsukuba were also active in voluntary organizations, particularly those associated with the Christian churches; some established a widely distributed English-language newsletter for foreigners called *The Alien Times*.

Hanami in Ibaraki: Plum, Cherry, and Iris Blossom Viewing Time

Mito is about an hour by bus and train from Tsukuba, to the northeast. In the early spring of 1991 I went to Mito with two women who lived in Tsukuba; the purpose of our trip was to view the famous plum blossoms in a renowned garden at the former estate of the Mito family; I was also eager to visit a museum there which houses an important art collection of objects once owned by the Mito family. (The Mito family were known as important patrons of art and scholarship.) Another time I went to Mito with a friend from KEK to see the inaugural exhibit at a new contemporary art museum and performance space which had been designed by Arata Isozaki. Beside these aesthetic attractions in Mito there are other arts to be found in Ibaraki, of course. There are two particularly well-known folk pottery centers in Ibaraki, Mashiko and Kasama, about an hour by car directly north of Tsukuba, each of which holds popular festivals and welcomes tourists all year. Outside Kasama is an artist colony established several decades ago by a family which had owned one of the first galleries in Tokyo to show Western-style paintings by Japanese. This family had also founded an excellent museum of nineteenth- and twentieth-century art in Kasama. While showing Kasama to an American friend who was quite familiar with Japan and Tsukuba, she remarked that it was surprising to find such impressive artists and art collections in a prefecture stereotyped as a haven of right-wingers and the *yakuza*. My friends in Tokyo seemed to think I had been rather clever to find these redeeming features of Ibaraki.

I was astounded when I went "back east" (to New York) for the first time and heard people say that they had more "history" than "out west"; it took me a while to realize that they did not regard the Native Americans, Spanish, and Mexicans as having a history in North America worthy of the name. We had different origin stories: when I would tell them that the oldest dwellings in the United States were built by the Native Americans in the twelfth century and the Spanish in the sixteenth, they thought I was untutored; I thought they were provincial. They grew up facing the Atlantic, in the newly powerful provincial margins of the old European empires; we grew up facing the Pacific, at the volatile edge of many old and new empires and with our backs to those "back east," which is what we called every part of America that drained into the Atlantic. They thought we were a joke from the coast; they called us lazy, shallow, and erotic—the usual stereotypic for those expected to stay in their place. We had different stories to tell from the margins of different empires.

Once I went cherry blossom viewing with a group of KEK physicists very near the lab. We watched a group of older men fishing in the pond near the trees; the men were all wearing *tenugi* bandanas on their heads, scooped neck undershirts, baggy khaki pants, *tabi* socks, and wide elasticized bands around their waists. Japanese concerns for their stomachs getting cold match the intensity of American anxieties about our heads getting cold or the French preoccupation with their livers. I thought the men fishing under the trees were picturesque, but my Japanese friends wanted to find another spot where they did not have to look at old fishermen. To the west of Tsukuba, toward the coast, is a village famous for its canals. That part of Ibaraki was once known for its marshes; small boats used the rivers winding through the marshes for transporting agricultural goods from Ibaraki to Tokyo. During the season for viewing irises a group from the lab drove to that picturesque town, had lunch, and strolled along the canals. Middle aged women called to us every few steps, urging us to cruise on the canals in one of their small boats. I noticed that the young male physicists seemed especially uncomfortable with these women's rather forceful entreaties; I was told later that they were stereotypical Ibaraki women: unattractive in every way.

Hanabi: A Traditional Fireworks Festival near Tsuchiura

A man who works in the guardhouse at the entry to KEK lives in a village famous for its traditional fireworks. He invited a couple of us foreigners to the village festival one summer's night. Small children became enchanted with the sight of us, walking straight toward us and tugging at our clothes and asking us to bend down so they could look at our faces more carefully. Their parents would retrieve the children and invite us to join the line dances so popular at Japanese summer festivals. Most of the Japanese were wearing traditional thin cotton *yukata* robes with narrow *obi* belts. There are traditional basketweave and geometric patterns for men and flowers and dragonfly designs for women, usually blue and white, but some are in purple, yellow, or green. The last few years some large Japanese companies in the big cities have encouraged their young women employees to wear *yukata* to work on certain summer days and some of the fashion houses have begun to offer *yukata* designs. Most Japanese wear *yukata* only at home and at the local *matsuri* summer festival. When the fireworks started we were surprised to see that they were not thrown into the air, but had been strung on wires which were arranged in various patterns. In fact, as the patterns fired up, one after the other, a story unfolded of a journey to a faraway place.

At an American Anthropological Association meeting not long ago I was giving a paper about my research at KEK and afterward a prominent senior anthropologist introduced himself to me as someone who had once done fieldwork in Tsuchiura, studying schooling there. Since then we have had many conversations about our work and our affection for Ibaraki, even discovering that we had gone to the same shop for, yes, bread. This past summer we continued those conversations by electronic mail while I was in Tsukuba doing fieldwork. He asked me to look up a friend of his in Tsuchiura; I could not find her but I had a lovely day walking through the older sections of Tsuchiura with a Japanese friend of mine, frequently encountering a small parade of children, young mothers, and older men who were guiding their *mikoshi* float through the neighborhood toward the festival grounds. It was hot and humid, as it always is in Japan in the summer, so we stopped at a old *kura* storehouse converted to a cafe serving traditional Japanese-style cold sweets. As we pored over our guidebooks and maps, someone at the next table gave us directions, so we began to chat. She knew people at KEK and someone in her family was now visiting friends in Texas.

Fourth of July

The Fourth of July is always an important day at KEK: years ago the American visitors began holding a barbecue near one of the playing fields and inviting everyone at the lab, their families, and anyone else who had befriended them in Tsukuba. These sorts of inclusive, informal gatherings are rather unusual in Japan and many people come each year to enjoy the spectacle. The American physicists encouraged each nationality represented in their research group to provide some traditional barbecue dish from their country for the party so the food was very good. I liked to invite a former student of mine who had gotten two degrees in nuclear engineering at MIT; after she returned home to Japan to work she felt that KEK was the only place that she had found in Japan that reminded her of MIT.

She used to come see me at KEK on her way to Hitachi, a town to the north of Mito, where her employer operated a nuclear power plant. She was a great hit at the Fourth of July parties; when the directive came from *Monbusho,* the Ministry of Education, Science, and Culture, to hire more women, two group leaders asked me for her address because they wanted to recruit her. At that time the senior physicists at the lab were eager to cooperate with any new initiative from *Monbusho* because they wanted *Monbusho* to help them bring graduate students to KEK. However, the universities were trying to prevent physics graduate students from getting degrees for research at KEK with KEK physicists. The traditional academics' notion of the correct training for young minds was to read canonical great articles and replicate canonical great experiments.

The rules were changed and certain of the new national research institutes and laboratories, such as KEK, were allowed to have graduate students, including foreigners. In the early 1980s the Japanese government had resolved to increase significantly the proportion of foreign students in Japan, especially those from Asia. Their model is the American Fulbright program that so many Japanese academics, bureaucrats, business people, and politicians have received. Foreign students in Japan have increased, but so have the problems they face. Each university was supposed to take about the same number, but that has not occurred; instead, Tsukuba University has taken a much larger share, apparently because the students face less resistance there than at other places in Japan.

The summer of 1993 I had conversations with some foreign students getting graduate degrees in physics at KEK. They were citizens of Brazil, China,

France, Germany, Greece, India, Mexico, and the United States. The Brazilian was a great-grandchild of emigrants from Japan. Some loved Japan and others hated it; some were doing well and others were not. The one thing they had in common was, intriguingly, that none had formed a strong friendship with a Japanese physicist which they thought would last well into the future. Some of these foreigners had managed to form lasting friendships with Japanese outside the lab, however. Many had also formed close ties with other foreign students and were already planning to try to visit each other in the future. Almost all of the older foreign physicists had the same experience.

Home on the Border

Two senior Japanese physicists at KEK will retire next year. When asked about their plans, one said he is already busy establishing a college for the study of health and radiation where he hopes there will be many foreign students, especially from Asia. The other physicist says that he would like to see the UN university move to Tsukuba. Tsukuba is a collage of natives, pioneers, foreigners, missionaries, returnees, immigrants, emigrants, locals, and aliens. It is someplace that some persist in believing is nowhere because it is not a traditional place. I have lived there a total of about three years, first in 1976, then in 1986–87, and again in 1991 and 1993. Recently I began to feel that Tsukuba was beginning to feel too settled. I decided to visit the new science city under construction between Kyoto, Nara, and Osaka, especially since a meeting of the Society for the History of Science in East Asia was meeting there. I was a bit disconcerted to see that the new apartment blocks and shopping centers and nice roads were already in place, although only two laboratories were operating. Our conference met in a fancy new hotel located in the same building as the administrators of the Kansai Science City. One of the physicists on the administrative board arranged for some of us to visit the two labs which were affiliated with private companies. Kansai Science City administrators and national policymakers have decided, as they say, not to rebuild Tsukuba, but to build a new kind of science city. They want to focus on building a community and emphasizing its relationship with the rest of the region and the rest of Japan. Much is made of the contrast with Tsukuba. I wish them well, but I began to feel a little annoyed that they were dissing my town.

Tsukuba has become a *furosato*. I noticed in the summer of 1993 that a television program based on travels to interesting places in Japan was doing a show on Tsukuba. They featured a man and woman who had retired to Tsukuba and had established a nice, simple theater for old movies. They carefully painted faithful enlargements of the original posters and advertised the films to their neighbors. I thought the show could have gone on and on. What about those people at the Bach Grove who built a lovely place in the woods to house a magnificent organ and then held frequent concerts. What about—but by then I had begun to think of retiring to Tsukuba, my favorite noplace, at the margins of new empires.

Acknowledgements

I would like to thank George Marcus, the editor of this volume and this series, for including me in the workshop from which this paper emerged. I would also like to thank John Singleton of the University of Pittsburgh for our many conversations about Ibaraki, some on e-mail while I was in Tsukuba; he has conducted years of fieldwork in the nearby town of Tsuchiura. Finally, I want to thank my hosts and the many people in Japan who talked with me about Tsukuba.

Notes

1. There are many studies of such places, usually by architects, urban planners, regional development specialists, economists, and policy makers. For a cultural studies approach, see Doreen Massey, Paul Quintas, and David Wield 1992.

2. Isozaki has written compellingly of being hired by the national government to design Tsukuba City Center. He defends his deliberately dislocated and ironic rerendering of the Capitoline piazza in his designs for Tsukuba in "Of City, Nation, and Style," reprinted from *The Japan Architect* (January 1984) in Masao Miyoshi and H. D. Harootunian 1989: 47–62. The Japanese version appeared in *Shinkenchiku* (November 1983). He notes too that "several points in the whole plan . . . allude to Velazquez' *Las Meninas*" (60).

3. On the politics of two forms of postmodern architecture, see Kenneth Frampton 1983:16–30.

4. Masao Miyoshi (1991 227, 280.16) discusses these issues in a chapter entitled "Conversation and Conference."

5. Nonethnic Japanese, [for example, Korean-Japanese, Chinese-Japanese, and Ainu], groups such as the Burakumin (traditionally those families which engaged in the work of human cremation, butchering, leather trades, and maintaining temples), and "foreign workers" (including Brazilian-Japanese and Peruvian-Japanese) face enormous discrimination in Japan. In many prefectures nonethnic Japanese cannot, by law, hold government positions, which includes teaching at public schools. Teams from the predominantly Korean-Japanese high schools are not allowed to compete in the extraordinarily popular annual national baseball competitions held every summer.

6. Isozaki notes that while Tsukuba City Center was being built it did not exist, administratively speaking; it was sited within the boundaries of Sakura village. He argues that "if I were to say that Tsukuba Center Building is the expression of the absence of something, I would not be taken seriously. Yet actually, the central concept of the design is that very absence" (Miyoshi 1991:49, 60).

7. Theodore Bestor 1989 and Edward Seidensticker 1983 have written about Tokyo neighborhoods.

8. A huge proportion of the population in Japan, as in the United States, consider themselves to be middle-class. To my knowledge very little has been written on this ideology in Japan; two interesting books have been written on the upper classes (Matthew Hamabata 1990 and Takie Sugiyama Lebra 1993).

9. Several "village studies" have been written by American anthropologists. I am particularly fond of the two studies of *Suye Mura* written by Robert J. smith and Ella Lury Wiswell (1982) and John Embree (1939).

10. Jennifer Robertson (1991) has written about this phenomenon in another district near Tokyo.

11. There is no public debate in Japan about alcohol consumption, although there are quite strict laws about driving under the influence of alcohol. The physicists I know seem to always respect those laws, even after drinking only one or two beers. Many of the larger restaurants in Tsukuba have large vans or small buses which can be reserved to pick up the members of a dining party and then deliver them home after dinner.

There is also no public debate about the pervasive pornography depicting violence against women. For example, thick booklets (two to five hundred pages) on this subject in comic book format are for sale in almost all convenience stores and book stores in every kind of neighborhood. On subways and trains everywhere I went in Japan, including the Joban line between Tokyo and Tsuchiura, these books were being read by perhaps a third of the boys and men between ten and forty years old who were traveling alone.

There is a little public debate about tobacco consumption in public places; between half and two-thirds of Japanese men smoke. The only smoke-free public spaces I have encountered are subway cars and certain designated train cars.

12. On the "reverse culture shock" experienced by Japanese returning to Japan after long sojourns abroad and the discrimination they face, see Merry White 1988.

13. Portions of the next several paragraphs, now heavily revised and expanded, were first published in Sharon Traweek 1992.

14. On the symbolic construction of bodies in Japan, see Emiko Onuki-Tierny 1987.

15. As Ursula Kroeber LeGuin points out, "After all, California was not empty when the Anglos came. Despite the efforts of the [Spanish] missionaries, it was still the most heavily populated region in North America. What the Whites perceived as a wilderness to be "tamed" was in fact better known to human beings than it has ever been since: known and named. . . . An order was perceived, of which the invaders were entirely ignorant" (1989:82).

16. Tetsuji Nishikawa (1987:iii). Notice that the name Tristan is not always capitalized; in this title the name is not an acronym. The unusual grammatical constructions are in the original. For a discussion about postcolonials' relationship to English and postcolonials' transformations of standard English, see Gayatri Chakravorty Spivak 1993:67.

17. Yukio Mishima (1966, 1989) wrote compellingly of his own obsession with

these *onnagata*. Jennifer Robertson (1992) describes the Takarazuka troupe of young women actors, their stylized performance of gender, and their extremely enthusiastic young female fans. Milly Ivy has written about middle-aged male entertainers who tour the provincial cities of Japan, each performing as both a man and as a woman in both Japanese and "western" dress to cheering fans of middle-aged women who stuff large sums of money in the actors' clothing during the finale.

References

Bestor, Theodore C. 1989. *Neighborhood Tokyo*. Stanford, Calif.: Stanford University Press.

Embree, John. 1939. *Suye Mura: A Japanese Village*. Chicago: University of Chicago Press.

Frampton, Kenneth. 1983. "Towards a Critical Regionalism: Six Points for an Architecture of Resistance." In Hal Foster, ed. *The Anti-Aesthetic: Essays on Postmodern Culture*. Port Townsend, Wash.: Bay Press.

Hamabata, Matthew. 1990. *Crested Kimono: Power and Love in the Japanese Business World*. Ithaca, N.Y.: Cornell University Press.

Marilyn Ivy. "Discourses of the Vanishing: Modernity, Phantasm, Japan." Typescript.

Lebra, Takie Sugiyama. 1993. *Above the Clouds: Status Culture of the Modern Japanese Nobility*. Berkeley: University of California Press.

LeGuin, Ursula Kroeber. 1989. *Dancing at the Edge of the World: Thoughts on Words, Women, Places*.

Massey, Doreen, Paul Quintas, and David Wield. 1992. *High Tech Fantasies: Science Parks in Society, Science, and Space*.

Mishima, Yukio. 1966. *Death in Midsummer and Other Stories*. New Directions.

———. 1989. *Acts of Worship: Seven Stories*. Trans. John Bester. Kodansha.

Miyoshi, Masao, and H. D. Harootunian, eds. 1989. *Postmodernism and Japan* Durham, N.C.: Duke University Press.

Miyoshi, Masao. 1991. *Off Center: Power and Culture Relations between Japan and the United States*. Cambridge, Mass.: Harvard University Press.

Onuki-Tierny, Emiko. 1987. *The Monkey as Mirror: Symbolic Transformations in Japanese History and Ritual*. Princeton, N.J.: Princeton University Press.

Robertson, Jennifer. 1991. *Native and Newcomer: Making and Remaking a Japanese Town*. Berkeley: University of California Press.

———. 1992. "The Politics of Androgyny in Japan: Sexuality and Subversion in the Theater and Beyond." *American Ethnologist* 19:419–42.

Seidensticker, Edward. 1983. *Low City, High City: Tokyo from Edo to the Earthquake*. New York: Alfred A. Knopf.

Smith, Robert J., and Ella Lury Wiswell. 1982. *The Women of Suye Mura*. Chicago: University of Chicago Press.

Spivak, Gayatri Chakravorty. 1993. *Outside in the Teaching Machine*. (Routledge, 1993)

Traweek, Sharon. 1992. "Border Crossings: Narrative Strategies in Science Studies and Among High Energy Physicists at Tsukuba Science City, Japan.'' In Andrew Pickering, ed. *Science as Practice and Culture*. Chicago: University of Chicago Press.

TRISTAN Project Group. 1987. *TRISTAN Electron-Positron Colliding Beam Project, KEK*. Tsukuba, Japan: National Laboratory for High Energy Physics. Report 86–14.

White, Merry. 1988. *The Japanese Overseas: Can They Go Home Again?* New York: The Free Press.

Science Beheld

These contributions represent two cases of lay response and attitudes to twentieth-century science; one is artifactual and the other is embedded in issue politics distinctive of contemporary United States life. This is the general arena in which much cultural studies of science has been thus far developed. Mario Biagioli's tour of the Jurassic Museum, itself a defamiliarizing, ambiguous assemblage on science and technology, includes attention to an exhibit of letters exposing the cosmological visions of otherwise anonymous persons addressing astronomers and physicists. Kathleen Stewart's sensitive ear for local voices of reaction around Las Vegas to the construction of a disposal site for nuclear waste records a precise sense of widespread contemporary faith in and paranoia about science and its consequences for everyday life.

Kathleen Stewart

14

Bitter Faiths

1

In 1989 I spent a year in Las Vegas trying to track the mirage of discourses of "science," "government" and "the environment" then swarming around the contest over whether to site a national nuclear waste repository ninety miles north in the desert. The U.S. Department of Energy (DOE) was proposing a "final solution" of deep geological repose for eighty thousand tons of "civilian" or "peaceful" high-level radioactive waste now sitting in pools and casks at a hundred sites around the country. The waste would be transported cross-country and buried deep in the bowels of Yucca Mountain on the edge of the Nevada Test Site. Without such a solution it would be extremely difficult to start up any new nuclear reactors. But the state of Nevada, like other states before it, and eighty percent of its citizens, like other citizens before them, opposed it. DOE had a "credibility problem" and "science" and "technology" had been called in to settle it once and for all. The stakes were high since a solution here would also give hope of solving the problem of military radioactive waste and could revive the potential for state and industrial uses of nuclear powers of all kinds.

Today, two decades after the initial inception of a "final solution" for nuclear waste, and billions of dollars later, the issue is still unsettled. But the battle in Las Vegas, as in other contested sites before it, has not been unproductive; it has produced not a solution but myriad twisting, proliferating discursive effects as the signs of "science," "government," and "environment" have been volleyed back and forth between politicians, hunters, ranchers, off-road vehicle users, wilderness proponents, populists, survivalists, Indians, whistle blowers, nuclear test site workers, scientists, antinuclear activists, and UFO enthusiasts (among others).

From my vantage point as an ethnographer of contested discourses, the many different positions voiced on "the repository" or "the dump" come to define themselves in relation to radically other positions. The bonds between hostile discourses and competing ideological stakes can grow tight and complex as their contact is prolonged and their fates intermeshed. Take for instance the way that some anti-nuclear activists adopt the DOE's "science-speak" or "technobabble" in the effort to catch their enemy at its own game. It is almost eerie to see a young man who lives, in vigil, in a tent at the gates of the test site stand up at a DOE hearing dressed in Native American garb but armed with a copy of a huge, hypertechnical DOE report to debate minute technical points until even the DOE technicians themselves protest the unnecessarily unwieldy language.

Or take the complex mixture of bitterness and faith voiced by the "ordinary citizen" who opposes the "dump" and yet feels powerless and overwhelmed by the sheer technical and political scale of the issue. The resultant "bitter faith" in science, technology, and big government is but one example of the many discursive sediments that have settled out of the fray. Here I focus on this bitter faith to demonstrate some of the intensity and complexity of a popular scientific imaginary and to begin to imagine the kind of cultural politics that gives rise to it.

Bitter faith is a view to power in a grand, apocalyptic scheme of things. Within its purview, power rises from the ashes like a sphinx, at once destructive and magically creative. I think of it as a kind of panoptic scan that traverses a range of positions such as the bitter pragmatism in a call for law and order, conspiracy theories aimed at the government, beliefs in UFOs, and other magical superenergies, or that brand of American pragmatism that still fetishizes the technological fix by still, despite everything, holding dear the image of the lone inventor tinkering in the garage (and on home computers). In some cases it is a faith that begins and ends in a sense of personal helplessness or victimization and demands redress—a "solution"—from the very forces that are the sources of its alienation; a faith by default. It may be aimed as a threat to the powers that be meant either to unseat them or to purify them and enlist an ultimately redemptive, salvific power. It may arise in the border zone between an attitude of free speculation and inventiveness and a sense of dangerous free-fall without a net. But wherever and however it arises, it enmeshes "science," "government," and "the environment" in the far-flung reaches of a popular imaginary and its powerful sediments. Whatever else it is, it is a fiber in the mesh of a scientific imaginary.

The day I arrived in Las Vegas I drove around town with a realtor looking for a place to live. She was retired military and had moved to the desert with her husband. He liked being away from it all and spent his days landscaping with cactii. She liked the town and gave me some tips on the best shows and the best buffets on different nights of the week—"you'll get the hang of it."

I told her what I was doing there, looking at the different positions people were taking on the proposed repository. She said "Yeah, well. You have to be realistic. Okay, there's contamination all over the place. If it has to be cleaned up, regulate it more, but do it! *Just do it.*" I recognized her irritable, impatient realism about "the dump" from other comments I had heard on the street. In the midst of widespread, strident public opposition to the dump and federal "dumping" on Nevada, public spaces were also filled with bitter one-liners: "just get it over with," "we're tired of all the politics," "if the government wants to do it, you *know* they're going to do it," "we should at least get some money out of it for roads and schools." Antidump positions centering on antinuclear sentiment or NIMBY (Not in My Backyard) politics, found themselves combatting these bitter faiths as well as DOE discourses and other pronounced prorepository positions.

So I probed the realtor's comments. She said she had friends in government who had told her that it was all just politics on both sides of the battle. She just wanted to get the whole thing over with. "Why don't they just recycle it? Build a plant over there and recycle it. *Whatever.* I'm *sick* of all this waste." Others, too, were spitting out wild speculations on inventive technological alternatives with the irritation of swatting at a buzzing fly: "Why don't they just stick it on the moon" (or in outer space or at the bottom of the ocean); "Why don't they just recycle it? I heard some guy talking about it on TV and it sounded good. And I never heard another word about it—it's been covered up like everything else."

I told the realtor I'd read that the recycling process she'd heard about actually produces by-products far more deadly than the original waste. She grimaced.

> Look, people are parasites, we live off each other. Look at the service industries in Las Vegas, there's a 7-11 on every corner. Look around you! That's the way it is—people living off each other. *Be realistic!*
>
> So you see I'm pronuke. No, I'm not. I'm really not. But you've got to be realistic. Life is not fair. My brother—twenty-eight years old—went to work one day and never came home. He was the manager at a fast-food place. They think a previous employee who he had let go waited in the bathroom and he was there alone counting receipts and he sliced him all to pieces. It was a *real mess* when they found him the next morning. He just got up and went to work and look what happened to him. And I never saw him again. That was it! I've never gotten over it. It's not fair, but you've got to be realistic.
>
> It's like these accidents out at the Test Site—they're always the workers' fault. Because they say like I'm just cooking potato soup, I'm just mixing up some stew 'cause they do the same thing for so long, that's how they think of it. You gotta control 'em and clean up their mess after 'em.

Having probed, I was speechless at the response—as I would be many times again in the months to come. Here was a bitter faith in law and order and technology that arose as an angry demand. Centered in horrifying personal experience, it scanned the ground of a degenerate social order and a numbingly routine human nature. It constituted a "faith" more unshakable, perhaps, than any unambivalent faith in science (or government) as a good in itself.

I saw the bitter faith again in the couple whose teenage son's body was found in the bottom of a chemical drum in the factory where he worked as a night janitor. Haunted by the image of their son's brain turned to mush overnight, they were consumed by conspiracy theories and their unrelenting efforts to expose the company and its chemicals through painstaking research into chemical compounds and OSHA regulations. They had uncovered case after case of disease, disability, and death caused by the same chemicals, and the same company, across the country and in Mexico and Canada. They bitterly hated "the system" and the lawyers and local politicians that they dogged to "do something" even now, years after their son's death. I can remember sitting on their living room floor poring over hundreds of pages of data and newspaper clippings that documented the company's crimes against truth and humanity. I remember, too, how Carol calmly stood from the couch, walked over to me, put her foot down about an inch from my knee, and twisted it back and forth only once. "It was a black widow" was all she said before continuing with the epic saga of her research.

I saw the tensions of bitterness caught in faith again in a group of marginal small businessmen whose businesses had been destroyed by an explosion at another chemical plant in town. The group was started after a federal hearing for emergency relief in which Bob, the leader-to-be, said, "These people are pissed off, their minds are on the brink, ready to crack up. Because they lead people on with false hope, and put 'em deeper into debt. I've got people calling me up, looking for a ladder. We're gonna make a bigger explosion than Pepcon."

Their career as a group was a violent emotional oscillation between sudden wild hopes and bitter fury. They tried to enlist scientists and engineers to help them get to the bottom of things. They kept records of threats and political corruption. They threatened exposés. They went door to door to residences collecting evidence of damage and victimization. My favorite story was the one of a young man who had been raising boas, tarantulas, gila monsters, and other dangerous exotic creatures in his apartment to raise money for college. With their deaths in the explosion he claimed losses of $15,000 and his only chance at an education.

The group—"just a bunch of pissed-off Americans"—held "secret meetings" with chemists and engineers for "fact-finding" about the chemical that had exploded and the valves that had failed. At one of these meetings they decided to build an ammonium perchlorate plant of their own to "show" the government and the banks and to compete with the company that had destroyed their businesses. They said this would "get them where it hurts." They said this was "doing something positive." Then they were bitter when the banks would not fund the idea. I remember a secret meeting with Bob standing in a parking lot on the edge of town with the expanse of the desert and the mountains surrounding. He told me his story and his philosophy for hours as we stood through a spectacular sunset and then in the dark leaning against his pickup. He said sometimes he had visions of a future America so corrupt that he would take his boat to the lake and float in it all day thinking about putting a gun to his head.

> But I'm not crazy. I'd never do that. But I don't like what I see. Every time I've been knocked down I've come back stronger. Used to be you could look at that mountain and say I'm going over there but now every time I try to go somewhere there are roadblocks come up in front of me, and I can't understand for the life of me why the government can't step in and set up better programs. They've got the know-how, they got the power, why can't they do better things for the people? I mean, I come back to the state I was raised in—you know, the hunting, the fishing, an area I was proud of—and you got nuclear tests, you're lookin at disposal sites. Nevada has become a waste site. In my opinion, and it's only an opinion, I just hope that the grace of God looks down on the people of Las Vegas. Because that's about all the people we've got, that's the only person we can turn to now because our government damn sure ain't the place to go. And we don't know what avenues to pursue now. But I'm telling you right now, we're gonna put our heads down and we're gonna bear into this and we're not gonna give up until we get our answers. And that's a promise.

The next day he received a promise for special financing from a state politician and was once again filled with gratitude and hope until the funding disappeared and he realized they were trying to corrupt him, destroy him, run him out of town.

When people in Las Vegas say they have no choice but to "let the experts decide" such a big and technical issue as nuclear waste disposal, they stir the sediments of helplessness and irritation, faith and fury, hyperpragmatism and wild leaps of the imagination. The fibers of the mesh of a scientific imaginary tense and strain and grow visible as they unravel in a panoply of voices.

One voice offers a claim of helplessness. Listen, for instance, to the voice of a somewhat frail-looking older woman wrapped in a sweater: "I see the clouds and I don't know how much of it is dust versus dangerous pollution. I don't know what that is. I know the television says what the dangerous point is and so on, but I don't know what that means. I mean, if I'm an asthmatic I guess it's worse than if I'm not an asthmatic or it doesn't—.'' This is not just helplessness but a reasoned accusation leveled at her government. Her voice, calmly centered in a sense of moral right, common decency, and common sense speaks as if explaining to them what they must not know.

Another voice expresses excited faith in a great scientific project with untold benefits—''The repository will be a 'great experiment' creating jobs and other useful by-products.'' Yet it ends in the deadening pragmatism of trite one-liners—''like how the space program came up with Tang.'' Or a blind faith in scientific progress expresses a dream of redemption that some good will come of things: ''What's done is done, so let's make the best of it.'' Listen to the words of a middle-aged man standing up for the government after repeated attacks: ''We already have the Test Site; this can't be any worse than that. We're used to it; we're not afraid. The government and its scientists know what they're doing and have public safety at heart. Back in the fifties they didn't know about radiation but now they do. It's important to the country.''

Or there may be a hope to redeem a pure science and technology from the spectre of big government secrecy and ''lies.'' In the fifties ''downwinders'' walked their property with Geiger counters: ''We had Geiger counters and we would talk to the AEC people and say, 'Well, it's kinda hot today,' and they'd say no it's not and we'd get our Geiger counter and compare it to theirs. Well, they'd get mad and go off and that burns me because they thought we were stupid and I guess we were for a while until we got educated.'' Test Site worker: ''I don't go for all this secrecy. People should have a right to go out there to the Test Site and take their little Geiger counters with them and walk around.''

There are the wild speculations over the future opened up by the dictum to secure the waste for ten thousand years. In the free zone of the unpredictable, terrifying free-fall meets limitless possibility. Listen, first, to the sense of unpredictable free-fall or chaos voiced by an older man calling for more responsible, rational government research and decision making at the same time that he feels that the DOE is incapable of such seemingly simple and common sensical measures: ''You can't predict the future. In 10,000 years Nevada might be a lush agricultural area. By that time maybe people won't be able to read the signs on the dump anymore and they might plant right there. So then you'd have nuclear food.''

Then listen to the response from another man who is excited at the thought that anything is possible: ''Or maybe somewhere down the line this stuff we're storing will have some kind of fantastic value and people will be coming from all over. A negative thing could turn out to be positive.''

The insistent pragmatic faith that ''seeing is believing'' stacks up against the paranoia of invisible threats. In the words of one Test Site worker, he will believe that radiation has ill effects when he sees. What he *can* see right now is the "paranoia" of the antinuclear activists: "My father came to the area in the fifties to work at the Test Site. We watched. We'd go up to the Tonopah Highway and watch the mushrooms and we still have our eyes, we don't have cancer, and I'm fertile. Been here for forty years. But on these antinuclear videos they're talking about jelly fish babies."

On the one hand, there are the Test Site workers who joke about the guy who wore a gas mask to work, "as if that would protect him! We call him Dr. Strangelove." On the other hand, there is the worker who quit when he asked for information about radiation releases and "all they would say is, Don't go in that tunnel."

There are the whistle blowers reporting that instruments at the Test Site are strategically placed to demonstrate least contamination or that the DOE put the monitoring equipment at the Fernald, Ohio, plant at the bottom of a steep embankment so that it did not detect the releases passing overhead. There was the Fernald technician who was unable to identify the instrument he had supposedly been reading and recording for nineteen years. There was the hearing on a nuclear test in which the DOE reported that there were "no detectable releases" of radiation but an investigative reporter who had read the phrase in the written report and thought it strange, discovered that the equipment itself had been accidentally destroyed in the explosion. "And *that,*" she said, is what they mean by no *detectable* releases." There was the antinuclear activist who stole a little boy's amputated cancerous leg from a lab near Fernald and smuggled it into Canada where it tested at very high levels of plutonium.

There was the story of the man who remembers his participation in the atmospheric tests in the fifties as a proximity to power. His wife attributed his skin cancer of thirty years to his exposure to radiation. But his story portrayed only the drama of men included in an awesome project. They would wait in the trenches and then walk to ground zero to check the equipment. The sand melted into glass. "That's how they contain the stuff in the ground. It forms a natural kind of glass. You'd be walking on it crunching underneath." He described how the massive towers would disintegrate in a flash, the huge balls of fire. They'd build platforms underground and then watch the whole area collapse. They didn't know much about radiation then and it was really interesting working there. "They had trenches and they'd tell you don't pick up your head. If you went into the bad areas you wore a badge and if it turned purple you got out fast."

There was the chilling story of the woman whose husband unwittingly brought contamination into her house. A woman whose husband became, for her, no longer her husband but a carrier of deadly disease—an intruder.

There were the endless meetings like the meeting of the Rotary Club in a small town in the desert where an antinuclear video was presented showing a range of possible transportation accidents. At the end of an hour of violent debate there were three testimonies from parents making a plea to protect their children and future generations. Then a final testimony from a man who had visited the Test Site as a consultant many times over the years invoked faith in progress, science, and problem solving at the heart of the American way:

> Well I'm not sure what year it was, somewhere around 1900, a man called George Westinghouse came out with something new he called the transformer and people criticized him and said if we use that our hair will fall out and our babies will be born deformed. And now we have transformers up everywhere. There's probably one right next to you in that VCR. But there is a thing called faith. And if we take a look at history we have certainly done a lot of things to humanity with blowing each other up with wars and everything. But there is also a thing called faith. We have created a problem. We have always had problems and we have also always created solutions. We have experts. I'm sure if there are problems they'll be solved. But we *are* problems solvers. I have kids, too, mine are a little bit older than Jim's, maybe, but I'm sure my kids will become problem solvers too.

The DOE works hard to manage "public perception" as it twists and turns on the slippery slopes of pragmatism and faith, problem-solving inventiveness and irritated dismissal, bitter anger and dreams of redemption in the end. The DOE uses slick public relations campaigns, "seeing-is-believing" demonstrations, and expert testimony. They run tours of the Test Site and Yucca Mountain. They give out souvenir packets of "nuclear waste"—small lumps of carbon mounted in plastic. They recruit local science teachers as volunteers at the EPA atmospheric monitoring stations in the desert around the Test Site. At their public "update" meetings they show videos of nuclear waste transportation casks being rammed into brick walls at eighty miles an hour without any significant release of radiation. They display models of the repository and massive hypertechnical reports measured in pounds—one quarterly report was twenty-seven pounds, the next was thirty-three pounds. They keep the overhead projector busy with charts of DOE organization, oversight committees, and planned quality control systems. They goad DOE scientists to translate their scientific preoccupations into idioms appropriate to the cause of public perception management. "The public" watches, restless in its seats, and shifts its own idioms from side to side to meet the political challenge.

The DOE says that radiation is as natural as the sun. A projects officer projects a slide of him resting his hand on a cask filled with twenty-four tons of "spent" high-level nuclear waste. "If I stay there for about fifteen minutes I'll get about a millirem, about as much exposure as if I was on a cross-country airplane flight. If I move away from it, get away ten feet, probably stand there eight or ten hours before I get one millirem. We rely on engineered features, robust containers for transportation safety.'' But an antinuclear physicist in the audience points out, as she did in a book that the DOE has read fastidiously, that the two kinds of "so-called radiation'' have nothing to do with each other; one penetrates the body and kills its cells; the other doesn't.

The DOE says there are foods we eat every day that are naturally radioactive like bananas and beer—beer, they say, is extremely radioactive. But someone in the audience points out that the beer referred to is the Milwaukee beer that was contaminated by uranium pilings that were dumped in the Great Lakes.

The DOE says nuclear energy is best for the environment. They say Mother Nature is amazing; when you put something in the ground, it stays. They say archaeologists found glass that had been perfectly preserved underground for ten thousand years, still green like a pop bottle. We could put nuclear waste in such glass, too. Everyday life and pop bottles will last forever. They have a story about what they call a naturally occuring nuclear waste dump found in Africa where the waste never moved for two million years. But when they call on a DOE scientist to support the story, the scientist talks like a scientist. He says there are hypotheses, uncertainties, empirical problems. The geologists are the worst; they say their science is not a predictive one at all, let alone for ten thousand years. They think a mountain is fluid and the life of rocks is structured by contingency, accident, and the unpredictable confluence of forces.

The state of Nevada's Nuclear Waste Projects Office scans the DOE reports finding "Freudian slips'' in which the DOE repeatedly refers to "burying'' the problem instead of "studying'' it. They leak dissenting documents from DOE scientists. The "Szymanski report'' proposed a dynamic geologic model that pictured the possibility of sudden dramatic surges of the water table up into Yucca Mountain where the waste would be stored. A U.S. Geological Survey memo, signed by seventeen DOE scientists, called the Yucca Mountain project a "mess'' and accused the DOE of having political, not scientific, priorities: "In subjugating the technical program to satisfy Department of Energy political objectives, we may succeed in making the program comply with regulations while being scientifically indefensible.'' The General Accounting Office found that "of DOE's 127 facilities that handle nuclear materials, 124 of them have documented radioactive and hazardous contamination of the groundwater and the soils around these sites.'' A Nevada official reported that Secretary of Energy Watson criticized DOE for "implying . . . that we just need to make a judgment call. That science is really never going to tell us what we need to know at Yucca Mountain . . . saying . . . the issue of whether it will be suitable or safe is largely a state of mind. So you can see . . . there is a great deal of concern about whether . . . science is what DOE has in mind or not.''

On the one hand, opponents of the "dump'' may dream of turning science and technology to righteous purpose. On the other hand, citizens invoking ultimate issues of life and death may refuse to be made "guinea pigs'' for scientific experiments whose importance pales by comparison to their concerns. Listen, for instance, to the self-assurance in this woman's voice as she invokes a human authority over the whims of scientific hearsay:

> Well, my name is . . . and I represent myself, my children, and my grandchildren, part of an endangered species—living, breathing, healthy people in the state of Nevada. . . . And all this research is just scientific hearsay at the moment because nobody really knows what that repository is going to do to us or anyone else, and if it comes, we will be a very big scientific study—if we stay alive.

There are people who back-talk the discourse of graphs, probabilities, and obscure technicalities with a "bottom-line'' discourse of lives and bodies. Again a woman and mother, speaking with the authority of future generations:

> I think if nothing else could convince me that this project is a mistake, it is your little graph showing the mountain with the water table underneath, telling me that you guys are going to study and study to see how long it will take for that radiation to reach that water table. I don't want to know how long it is going to take to reach the water table. I want you to guarantee me that it is never going to reach that water table. [*Applause.*] Until you can do that how can you say this is a suitable site? What's a suitable site?
>
> DOE: The regulations do define that and that is if a mountain will isolate the waste for ten thousand years, that is defined as a suitable site.
>
> CITIZEN: Well I don't want my great, great, great, great, great, great, great, great-grandchildren unable to drink their water because you guys decided it was a suitable site because it would take ten thousand years to get there.
>
> DOE: And there is a basis for that. The radiation decays away to a level where it is no more harmful than natural uranium which is in the ground right now. Natural uranium exists around the West. So that is our basis for the ten thousand years.
>
> CITIZEN: And I think that you telling me that is like them telling us in the fifties sure, sit on your rooftops and watch these explosions, because they are not harmful. [*Applause.*]

Once reference is made to the earlier Atomic Energy Commission's now-proven history of secrecy and lies, people may move in for a more direct attack on the methods of the DOE, the same agency under a new name. In this case, another middle-aged woman, also speaking as if from a higher moral ground, took up the questioning: "Could you please tell us what is an 'acceptable death rate' for your experiments? Why does anybody have to die?

> DOE: These codes, computer codes, do not predict deaths. What they do is look at potential consequences or impacts that could occur from one activity so that you can compare them against other measures like natural background radiation or other activities. There is a misunderstanding about the meaning of the term *death* used in this manner. Those do not predict deaths of individuals. What those do is look at rates of exposure to very large populations. They are predictions of statistically very small things that cannot be compared with saying you're going to die.

Then there were snickers from the audience and finally an older man asked a question almost in the form of a joke that pushed the DOE's claims for a probability science into the realm of the improbable and ludicrous.

> The international astronomical union claimed a thirty foot diameter asteroid barely missed Earth. They noted, had it hit earth it would have destroyed a city. What if it hit one of the hundreds of nuclear plants scattered about the earth? A Test Site geologist *told* me that a high-level nuclear dump would not be safe from a hit from an asteroid. [*Audience laughing.*] The question is, if an asteroid hit Yucca Mountain dump, will it be safe? [*Really laughing now.*]

And as I was leaving the meeting that day I overheard a bitter joke that referred directly to the agency's change of name from the AEC to the DOE. "Sounds like DOE is waking up to the whole 'credibility' problem," said a woman. "Yeah," said a man, sneering, "time to change their name again."

Over and over you hear the desire to return to some kind of "common sense" in the face of imminent disaster, to somehow clear the "smoke screen" of politics and obscurantist technicalisms and get back to the basics of the really real. In the words of a newly retired engineer who had been actively following the news of nuclear contamination around the country,

> Now the scientists say one thing. The scientists say—What in the hell do they say? They say a lot of things. But what are the net results? You know this country, everywhere they've been, including the Test Site, is polluted forever, with plutonium. I asked you at the last meeting how many catastrophic, serious, real stuff, accidents happened handling nuclear materials. You said you didn't know. Well, I told you it was over 3500 at the first meeting, in Savannah River, [you] said there was a little plutonium, not enough to cause any harm. At that very moment, there was a crack in the floor, and pure plutonium was leaking down the Savannah River. Now it goes everywhere. You can tell me, and your engineers can tell me what you are going to do to make it safe but you can't

> make it safe. It is utterly impossible. Man has opened the world to uranium, converted it into plutonium, one of the most deadly things known. Now the only thing you've got left to do is use common sense and do away with it. What have you got? What's the net result? If it's so damn safe why do you spend billions and billions to make it safe? You are spending more to make it safe than you get out of the nuclear power produced. So what the hell good is it?

There is a fear that no one is in control and that power could fall into the hands of conspiracists and terrorists either because our government is ineffectual or corrupt or, in the opposing view, because people are trying to interfere with government control. The issues of control twist and turn complexly in the debate. First, in the words of a middle-aged male Test Site worker, power has to be concentrated in the hands of experts precisely because there are dangerous people out there:

> As an elderly gentleman got up a few minutes ago and was reading from a magazine concerned about what crazy dictators were going to do and stuff like this, I'd like to answer his question if I could. You have nuclear plants all over this country. You have extremely dangerous radioactive material in relatively unsecured sites with a few rent-a-cops watching them. That's all you got. He's worried about what's going to happen. Well, there's a lot of terrorist organizations around this country that would like nothing better and if they ever got into that instead of trying to keep it in a place where it's secure, I'll tell you, near those big cities it would make Chernobyl look like a backyard barbecue. [*Applause.*]

Then, in response, the words of a woman who doubts that the DOE is capable:

> Well, for the women and mothers in this state, we are fighting for lives and for the health and well-being of our children and for the future. To even transport that stuff across the United States is a ludicrous idea. There is no easier target for a terrorist in the world. They are going to know which trucks are filled with that stuff and we are not. Because I'm sure it is not going to be marked so that all of us driving along the highway are going to know what it is. But they'll know.
>
> DOE: Those transportation shipments will be marked, and they are marked today. We are always shipping high-level waste in the country, even today.
>
> WOMAN: DOE has a terrible track record of lying to us and I think it's ludicrous that you say you will train people to handle any accidents. It's ludicrous. You are shipping this stuff across the middle of nowhere, across this country, and you are going to get these sheriffs of little towns who don't know anything and they are going to get hurt.

In the end, the DOE has reverted to an effort to "circumvent the human factor" altogether. One day they displayed an empty nuclear waste transportation cask so that people could "see it and touch it for themselves." We listened to various presentations showing how the DOE is using systems and technologies to circumvent the human factor. They projected a cartoon which showed a pickup truck piled high with barrels of nuclear waste. The truck was stopped at the entrance to a New York City tunnel and a group of "rednecks," with beer bellies hanging over their jeans, were standing around in traffic looking at the bridge clearance sign—11 feet. The truck stood eleven feet six inches high with barrels piled higher yet. One barrel had fallen off and was rolling down the ramp into the tunnel. It was the moment of decision. One of the men was saying, "Go for it!" The DOE argued that such popular perceptions of nuclear waste management are pure fictions and showed slides of the DOE organization and its high-tech transportation casks. Then they sent us out to the parking lot to see one for ourselves. The cask was housed in a semi-trailer after its first experimental cross-country run. We touched it and listened to the hum of its "containment systems."

Then I went up and talked to the truck driver sitting in the cab. He was wearing a cut-off Harley Davidson tee shirt and earrings in one ear. He wanted to talk about his high-level security clearance, which he was very proud of—a G-clearance, better even than a C-clearance. Then I asked him about the computer keyboard sitting on the passenger seat. And he told me a story. As he was driving away from the origin point in New Jersey they told him to log in every 15 minutes so (hypothetically) they would know where this hypothetical waste was at all times. He had never used a computer before so he responded as if it was a joke: "Yeah, sure, I'll call you when I get there." But it turned out to be easy to use—nothing to it. Except that the first time he logged in to the satellite that was tracking him they sent an urgent message back asking if he had been rerouted. He punched in that this was the route he'd been given and received an alert that he was on the wrong route and a second urgent request that he specify if he had been rerouted. So he shut it off. Because if he'd had real nuclear material on board he certainly wasn't going to listen to these people; anybody could have gotten control of the satellite—terrorists or nuts. There are terrorists out there who'd love to get their hands on this stuff. So he drove cross-country, with no system tracking his locations, and followed the route he'd been given by his dispatcher.

When he finished his story I went back inside to the hearings and listened to more talk about how the DOE had bypassed the human factor. But the reactions of a scientific imaginary proliferate unchecked as they twist and turn on the wings of government and technology, pragmatism and hope, life and death. It is an imaginary that emanates like a cloud from any productive event.

Take, for instance, the signs of such an imaginary buzzing around a tour of Yucca Mountain and the Test Site.

2

As we drove the sixty-five miles through the desert to the Test Site our tour guide entertained us with near mythological stories about the places we passed: the cast of characters living in a nearby ghost town with a few trailers and a store; the history of a grim looking, near deserted air force base; a prison that was, he said, just like a luxury hotel; the high-tech air force bombers flying overhead on testing missions; the place they call the "widow maker": "A month didn't go by that there wasn't at least one serious accident. There used to be no speed limit out here. It's a challenge just to see how fast the vehicles can go. So when there was an accident it was a real grinding affair." The traces of a mining town: "There were fifteen thousand people there but it lasted only one year and then the town disappeared. In 1958 when they came in here you could still see the roads of the town. There was an old Model T and lard cans. A guy interested in the history went and found the assessor."

When we passed through the gate of the Test Site the tour guide made jokes about the protestors on the side of the road. He said he had recently discovered that they were actually paid to do this—"It's their job." As we piled out of the bus to go through security clearance an antinuclear activist on the tour went up to the guide to contest his accusation. "These people sacrifice to come out here. We're just as dedicated to what we believe as you people." As their discussion continued, a man standing next to me said, "She won't let up," and another man, passing us, said, "That's the great thing about this country. You can have both sides on the same bus." He seemed very happy about this.

Outside the security trailer a sign attached to a barrel half filled with sand read, "Discharge all firearms here. No weapons or cameras." We were given radiation sensing badges to wear. Then we passed through entrance gates that looked like a border crossing and men in uniforms boarded the bus to check IDs and take another head count.

We cruised slowly through two housing settlements where some of the ten thousand Test Site workers sometimes live. The one called Mercury has a movie theatre, cafeteria, steak house, laundromats, and long lines of newspaper dispensers and water fountains ("They drink the water out here?"). It has the largest vehicle fleet in the DOE—2,400 unmarked white pickup trucks. The camp at area 12 is much more remote—a collection of dusty silver trailers in a bleak moonscape desert about fifty miles up into the Site.

We had a free lunch in an army barracks cafeteria filled with Test Site workers.

In between, the tour guide filled us in on points of interest scattered over the desert of the huge reserve. Rock Valley was a special study area for both animals and plants. It had three fenced-in areas which were catalogued for their plants, squirrels, lizards, and snakes. A three hundred and fifty-foot tower contained the cesium source to irradiate the area. One species of lizard had become extinct but in all other species there was no detectable change in numbers or reproductive capacity. "The experiment itself is highly successful." There were the bleachers of "news knob" where the press watched the atmospheric tests in the fifties. At one time they were testing nuclear powered planes in one area but they spewed junk all over and you couldn't land them. "The selection of the Test Site was something of a miracle. It has closed basins so the water never leaves its area. It migrates out with a test but then it slowly migrates back in."

The Armad facility is where all the sensitive high-level radioactivity work is conducted. "They have machines that can do something real powerful one minute and then turn around and pick up an egg and place it in a car without cracking it." At the core sample testing facilities for the Yucca Mountain project technicians dressed in white coats showed us boxes full of ordinary looking rock samples and complained about the elaborate "quality assurance" systems they were developing. "It has nothing to do with good science but you need it to go to court."

We passed ghost towns of bombed-out simulation worlds. One is a Japanese town used to simulate the damages sustained in Nagasaki and Hiroshima. The bomb tower, now dormant, houses a still operative elevator. "I was in it going up and the wind was blowing and the safety devices took over about halfway up. And I really appreciated Mother Nature taking over because then we came down. But even on a calm day the top of that thing sways eight to ten feet."

A bombed-out 1957 "survival town" displays the effects of the bomb on various building constructions—small, large, brick, and frame houses, row houses and motels, a service station. Each is fully equipped with everything but electricity and plumbing and fully furnished right down to the cereal box on the kitchen table. "Each one cost $35,000 and to me in 1957 $35,000 was just about all the money there was in the world and knowing they were being built just to get destroyed kinda hurt." The bombed town makes an eerie landscape of traces of buildings and collapsed steel bridges that were never meant to connect anything in the first place.

Paiute burial sites are cordoned off and if a Test Site worker is found removing anything from them he is immediately terminated. A metal horse head marks the place where a man was killed by being kicked in the head by a horse. There used to be a wooden horse head with an epitaph marking the site but some unthinking person removed them. There used to be experimental EPA dairy cows roaming the range in one area. One of them—"Big Sam" had a window cut in his side so you could reach in and get the food out. He was on display at fairs.

At the Climax mines—"where all the lizards glow green"—we were taken in small groups down a long shaft for a tour of the nuclear waste cylinders set in cement in the floor. The miners activated a switch to lift one out of the ground for us and laughed as we backed away.

Back on the bus, as we drove out to Yucca Mountain, people talked about horror movies set at the Test Site. An iguana, hit by radiation, turns into a monster. Aliens drop down in the desert and live in an abandoned mine shaft where the radiation kills off a parasitic bacteria that has been threatening their civilization. They need the radiation to stay alive but they have to temper it so that it will not kill the human bodies they take over and inhabit. Two brothers sitting behind me reminisced about coming out to this area "caving" as kids. They had heard there was buried treasure out here and that anyone who dug it up would be cursed. One man who had dug up some treasure in one of these caves had ended up spending eight years in jail. They themselves had buried a silver box with a gold chain in it. But when they came back years later they were unable to find it. They remembered it was three steps north and four or five steps east of a Yucca tree but when they retraced their steps they ended up on the edge of cliff—"but of course our steps were smaller back then, too."

At Yucca Mountain everyone got out of the bus and walked around looking at the ground. I asked the bus driver what he thought about all this nuclear waste business. "I firmly believe there are UFOs. There are ships that have come down, they're keeping some of them here." A neighbor who worked for the government had a signed affidavit from the governor that they had destroyed the ashes. Khrushchev banged his shoe on the podium and the Trilateral Commission got rid of him. Trotsky's revolution was financed by our biggest bank. The less educated people can't read or write hardly at all (they would self-destruct). Organized religion keeps the truth from the people. East and West religions—no one has the truth in itself. This had really opened up his eyes on the nature of man and the planet. Man is ignorant of what goes on because of the powers that be. UFOs could return the waste to the sun, to the source. Powers that be are keeping the UFOs from us because they would have a little box or something that would give you all the electricity you needed for your house. It's the utility companies in cahoots with the government. We have to get together in brotherhood. Something will happen at the turn of the century.

A woman tried to get on the bus with a souvenir rock and the tour guide told her she couldn't have it.

"Why?"

"Well, this is a government project. They'll probably have to map out every single little rock and then when they're done they'll have to put them all back exactly the same way."

On the way down the road back into the Test Site signs marked the sites of accidents—"Jesse's curb," "Billy's hill." We stopped at a famous crater where a piece of earth the size of a three-bedroom house had flown out and landed on the lip on the other side of the crater. That was the hot spot. The bus driver called me aside to talk. He wanted to tell me more and to ask me what I thought about all this. Speechless, I told him I had heard this sort of thing before and that there must be *some* truth in what he was saying.

As we were leaving, the tour guide pointed out a place on a mountain they call the monastery that was built as a kind of transcendent site from which to photograph the atmospheric testing. It sits high above the desert, a monument perched on a precipice, while a prolific scientific imaginary scans it in passing.

Mario Biagioli

15

Confabulating Jurassic Science

The Museum of Jurassic Technology is not exactly a museum if by that term you mean a large, corporate looking, copiously staffed institution populated by million-dollar exhibits.[1] Nor is it a community museum dedicated to local history and culture. Rather, it resembles a mix between a boroque cabinet of curiosity assembled by a gentleman and visited by his friends (and by the appropriately introduced visitors who happened to come through town), and a large installation piece—the art form previously practiced by its founder, David Wilson. It is officially presented as "an educational institution dedicated to the advancement of knowledge and the public appreciation of the Lower Jurassic and is in cordial cooperation with all similar institutions throughout the world" (Worth 1964: cover 2).

It occupies a gallery-size space in an unconspicuous building on a generic block in a nondescript Los Angeles neighborhood. If it were not for a medium-size banner marking its entrance, you would probably not notice it or think that somebody had rented a shop and turned it into something you cannot quite figure out. Maybe a private club trying to keep a very low profile. Maybe one of those Los Angeles upscale restaurants that, in order to (pretend they) cater only to "those who know," mimeticize themselves while making sure you do understand it is a restaurant incognito.

In these days, museums, public exhibitions, amusement parks, libraries, and archives are attracting much attention from academics. Much literature has focused on the museums' many roles as trend-setters of artistic taste and cultural values, as organizers and enforcers of taxonomies of knowledge, as providers of national historical identity and, more in general, of comforting representations of "us" as different from "them." Science museums too have seen their discourses analyzed and often deconstructed—especially the naturalizing representations of racial differences and gender roles embedded in many natural history exhibits, or some science museums' attempts to shape national identities through the celebration of a nation's scientific and technological "monuments" and heroes.

While studies of contemporary or relatively modern museums have been often critical of their social and cultural role, early modern collecting institutions have been usually depicted as less pernicious: as instantiations of older notions of evidence, natural order, taxonomic anxieties, and gentlemanly curiosity about nature (and everything else). Renaissance or baroque cabinets of curiosities seem more likely to remind one of Borges's fantastic "Chinese" taxonomy of animals than of the racist worldview embedded in the older exhibits at the New York Natural History Museum. Although we know of renaissance kidnappings of American natives and of the "reconstruction" of their villages in European cities (practices that were quite congruent with the logic of collecting behind apparently benign cabinets of curiosities), historians have tended to see older museum practices as somewhat distant from that which produced tragedies like that of Minik, the so-called New York Eskimo.

Yes, we know that the "curiosity" or "playfulness" embedded in those early museums, galleries, or *wunderkammern* reflected processes of social distinction and control. We also know that, behind their composed facades, museums functioned as potlaches through which noble and princely collectors were challenging each other's power to spend and to drag the "other" in from the most remote parts of the world (or the past). However, the cabinets of curiosities' effectiveness as canonizers of both exoticizing and naturalizing discourses about differences among humans or of specific views of scientific evidence and knowledge has appeared dismissable when compared to our modern museums and their role within so-called mass culture.

It almost seems that, when confronted with the exotic and unfamiliar culture of early museums, the historians' contextualizing skills have sometimes jammed (though not in uninteresting ways). It is common credo among self-conscious historians that if the past looks too familiar, then it may be that we are normalizing it. Possibly, our presentist glosses are erasing its historical specificity. The antidote to this often unconscious colonial attitude about the past is to do our best to defamiliarize it, to make it strange to us, to get to realize how "other" it is. Although this defamiliarization of the past is, of course, good intentioned, it sometimes becomes difficult to draw the line between recovering the otherness of the past and exoticizing it.[2] This is particularly so in the study of early modern cabinets of curiosities, museums of natural history, and their apparently benign imaginaries.[3] In short, while studies of more modern science museums have often been critical of the scientific imaginary they embedded—one that is deeply involved in the simultaneous spectacularization and naturalization of problematic master narratives (white, patriarcal, colonial, and so on), some studies of early modern museums which started with other-friendly, noncolonial commitments have sometimes resulted in a reification of the exotic imaginary of early modern collectors.

Equipped with this reading of recent literature on scientific museums and collecting, I walked into the Museum of Jurassic Technology to realize very quickly that an interpretation of this site in either the "critical" or "exoticizing" registers would not have taken me very far. Its exhibits neither present science as a success story nor confirm its status as a hegemonic narrative. This is not a museum dedicated to great scientists, their discoveries, and their contribution to harnessing nature in the name of progress. While the category of knowledge (of some sort) is central to the museum's discourse, success and progress are not.

At the same time, this is not a museum of quack science or of irreproducible results. That would betray a "reactive" posture—one that would accept "real science" as a privileged term of reference. But here we are in the realm of Jurassic technology, not of "real" science. However, if this place is not about cold fusion, it is not about science fiction either. The point is not to imagine the future as our science may shape it, but rather to reimagine the present as "Jurassic."

Unlike other museums, this is not an institution bent on the reification of collective memory. Nor does it provide a self-reflexive critique of how museums tend to do that. This place neither canonizes nor deconstructs memory. As we will see, its founder does not quite believe that memory exists to begin with. What you visit is a workshop on the elusiveness of memory and on the pervasive and constitutive role of the imaginary in constructing it.

However, the imaginary of this museum does not get caught in the exotic—as it usually was the case with both late renaissance and more recent museums. True, the visitor is exposed to a good dose of prima facie exotic materials and narratives: tales of scientists exploring the Amazon (heroically, of course) chasing mysterious bats endowed with X-ray vision; "an extraordinarily curious horn which had grown on the back of a woman's head" which is said to have been first observed in the seventeenth-century Musaeum Tradescantianum; or an eleven-millimeter-wide fruit pit on whose front we are told it is carved, among *many* other things, "a Flemish landscape in which is seated a bearded man wearing a biretta, a long tunic of classical character, and thick-soled shoes" while its back depicts "an unusually grim Crucifixion, with a soldier on horseback, Longinus piercing Christ's side with a lance, the cross is surmounted by a titulus inscribed INRI."[4]

However, these narratives are neither presented as true statements nor are they poked fun at as tall tales. They are simply attached as extended captions under these objects (a corrugated fruit pit or a hairy and weird-looking horn) which, if separated from their captions, have nothing inherently fake or authentic about them. Therefore, it is not the authenticity of the object that certifies the truth value of the narrative (or vice versa). Rather, the status of both the object, the caption, and of their mutual connection is left ambiguous.

It is up to you to decide whether the exhibit is a joke on the credulity of the person who wrote the text reproduced in the caption (if he/she ever existed), on the curator who reproduced this tall tale, on you who for a moment felt that maybe there was something plausible about the carving on the fruit pit, or on the silliness of early modern cabinets of curiosities which appreciated so much this type of specimen and narratives. Or maybe you are not going to feel that it is a joke at all. Or you may get "analytical" and think that this may be an interesting example of how the play between object and its caption does not only construct an aura of authenticity, but it can also destroy it as easily and effectively.

And, of course, the way the exhibits are set up does not overdetermine your response.[5] Everything (from the display case to the way the caption is printed and presented) is very serious looking (though you may think that, were these things real, the security level here would be awfully low). Whatever option you pick, the exhibit is going to provide you with resources to support it. For instance, the story about the carvings on the fruit pit sounds pretty outrageous but, actually, this pit looks unusually corrugated. Maybe there are microscopic carvings on it—carvings you can't see from where you stand but might become visible with a lens or a microscope. Or maybe they were not produced by some ancient artisan but by some new high-tech procedure. These days people manage to etch anything on a silicon chip. And remember, there was a picture in the *New York Times Magazine* of a microsculpture of Pope John Paul II with the usual red garb, holding the cross—and the entire thing was fitted in the eye of a needle (Burdick 1993).

There is no parking lot in front of the Museum of Jurassic Technology, though there is a bus stop. Often the door is open. Sometimes you need to ring the bell and wait. As you step through the door, what you see is neither a science museum nor a cavernous space that tells you, "sacred ahead," nor a well-lit gallery in which a few "select" pieces are hung at respectful (and value-enhancing) distance from each other. Actually, the way the space is internally subdivided and dimly lit prevents you from seeing much except the "receptionist" behind a desk with the usual computer sitting on it and a few interesting-looking scientific instruments (or parts of them) whose function is far from evident (and not only to you). A minimalist museum shop is behind the desk.

Unlike museums shops where you can count on finding the usual coffee-table books, the postcards, and maybe some designer item, the objects on sale here may not fulfill your expectations. While they make (some sort of) sense if you look at them at the end of your visit, they may seem either weird or a bit too subtle when you quickly scan them before going through the exhibits. In many ways, the taste of the objects in the gift shop is emblematic of the museum itself—a mix of subtlety, lack of explicit cultural canons and related marketing strategies.

The place conveys the feeling of being serious, professionally designed, and meticulously kept. While the style (and the scale of the locale and especially of the exhibits) is not that of a corporate museum, the place does not have anything of the alternative art space, either. The feeling is not that of informality but rather of subscaled formality. Also, the small (early modern?) size of the exhibits, their visitor-activated nature, and the pervasive adoption of recorded descriptions delivered through phone receivers enhances the privateness of the experience. You "plug into" the exhibits as much as "look at" them.

A few feet past an official-looking but smallish panel engraved with all the names of the museum's trustees, officers, donors, friends, one gets to a very dimly lit and moderately labyrinthine area where the exhibits begin. A professional-looking multiprojector slide show, complete with serious voice-over, is projected on a very small niche (about eight square feet). An elegant plaque identifies the small screen you are looking at and the small bench you are sitting on as the Judy and Stuart Spence Multimedia Theater.

The slide show recounts the history of museums in general and of the Museum of Jurassic Technology in particular. It begins with Noah's Ark, represented (following the seventeenth-century Jesuit polymath Athanasius Kircher) as a neatly taxonomixed floating museum in which each animal (and its opposite-sex companion) occupies the appropriate taxonomical box. The narrative that follows is pretty much historically correct until about the industrial revolution. At this point the proposed genealogy of museums begins to deviate from accepted historical narratives as the Jurassic is introduced as a period which took place somewhere between the eighteenth and twentieth centuries (to the surprise of anybody who had any geology in high school). As I was following the voice-over, I expected to hear "modernity" anytime, but was told that, instead, this was the "Jurassic."

As images of buildings ruined by bombings and large underground repositories flash on the screen, we are told that the Museum of Jurassic Technology has been established, destroyed, and heroically reconstructed to preserve the knowledge of the Jurassic and to foster the appreciation of its technology.[6] Although the music swells as the narration celebrates the museum's quest, nothing makes you feel that what you are going to see is a celebration of scientific and technological progress and of the ways in which it has shaped our lives. Although the Jurassic is presented as taking place at the time history textbooks tells us that the industrial revolution was emerging, it does not seem to have much to do with the culture of modern science and technology.

The Jurassic seems to refer to an imaginary space parallel to but sharply distinct from modernity. It is informed by technology, but not one identified with progress. As all museums, this one too is an institution which celebrates (in some peculiar sense), but it celebrates a period and a knowledge that never quite happened (or was systematically forgotten). However, the tone is not of nostalgia for a far-gone knowledge nor of hope for a different technology. The Museum of Jurassic Technology does not try to make grand statements or to promote scientific and technological utopias. Yes, it claims to promote "research into life in the Jurassic" but it does so with a sense of humor.

Actually, it is not very respectful of its subject matter, which it literally pushes around. For instance, while at the beginning of the introductory narrative the Jurassic is presented as a historical period which overlapped almost perfectly with the beginning of modernity, a few minutes later it is suddenly transformed from an era into a country. Moreover, what is flashed on the screen of the Judy and Stuart Spence Multimedia Theater as a geographical map of the Jurassic is actually one of Egypt in which its north and south regions have been relabeled Upper and Lower Jurassic (Fig. 1).

These overlaps (between modernity and the Jurassic) and inversions (from the Jurassic as a remote geological era to a geographical area, or of "upper" and "lower" from chronological to geographical markers) are a recurrent theme of the exhibits of the museum. Interestingly, these displacing moves are not particularly sophisticated. As I will discuss later, they are more like childlike puns than subtle slippages of discourse. They seem to be reminders that well-crafted strategies of persuasion aimed at convincing you of being in the presence of "deep meanings" are not the point here. At the same time, these fairly crude displacements don't make you pull back and dismiss the entire thing as a simple joke. There is something apparently so serious about the entire place that it makes you feel that there must be more to it—and, I think, there is.

What follows the introductory slide show are a number of small exhibitions which resemble a conceptual artist's interpretations of the genre of experiment and of scientific didactical display, reports of historical events, and narratives about marvelous objects. As in any respectable science museum, the visitor is guided through the exhibition by a series of very scientific-looking audiovisuals with very serious (and appropriately long and often pedantic) voice-overs. The style of the displays is very minimalistic though elegant. Great care is put in their formal aspects, and even the lighting devices are carefully crafted to resemble those clamps used in labs to hold instruments, viols, sensors, pipes. The layout of the museum does not force you into a preestablished path, nor does it make you feel you are lost in a labyrinth. Then, the small, quasi-microscopic scale of the exhibits makes you feel that, maybe, there is something precious about what you are observing and trying to figure out. While there is often a conspicuous gap between the scientific

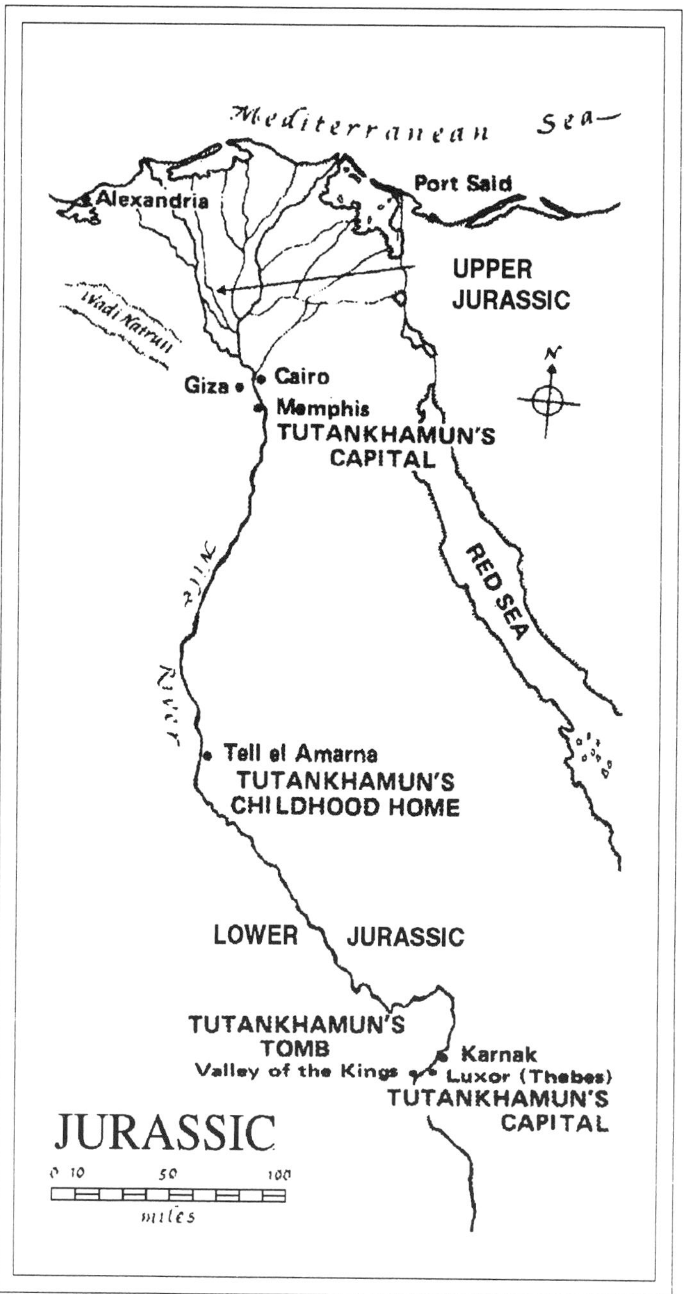

Fig. 1 Map of the Jurassic. (All figures in this chapter were provided by and are reproduced by permission of the Museum of Jurassic Technology.)

seriousness of the description and the banality or unintelligibility of what one sees, the viewer is given no strong clues about the possible presence of irony. Professionality seems to reign.

A little hammer keeps banging on a brass sphere every few seconds as the tape tells you about the process of crashing necessary to produce synthetic precious stones.[7] By now, the hammer, the brass sphere, and the cam which drives the hammer are all worn out, and brass dust is piling up at the bottom of the display case. The rest of the "instrument" connected to the brass sphere being banged on looks real but, again, it is not at all clear whether this is a replica of the original device through which synthetic gems are produced or some sort of sculpture which happens to resemble an instrument. In any case, the ongoing banging conveys a sense of obsessive repetition that, nevertheless, does not add much to the understanding of the "scientific" narrative being heard (which, in fact, is quite self-explanatory). Sense and the labor behind the wear do not display any apparent connection. Normal inference seems to fail.

Next, you see some glassworks as you hear a very plain narrative about "purification by sublimation."[8] While the setting of the exhibit suggests that the glasswork being shown has some *hinc et nunc* quality about it, what you see is a simple still, a viol, a beaker, and few other plain glass items which don't make you feel you are witnessing something particularly high-tech or historically groundbreaking. Later on, you find a display labeled as Under Repair, but one does not understand whether this display was actually about something scientific (something that "made sense") but broke down, or whether there was ever any "scientific sense" in this display even when it "worked." The "instrument" looks like a microscope which collapsed into and crashed the object whose subtle structure it was supposed to make visible. But it could also look like a machine to measure the promptness of one's reflexes which got broken because one's reflexes were off scale—either too good or too bad. In short, is this a "real" broken instrument? A pun on destructive experimental practices? A joke on laboratory instruments that never work and routinely embarass the technician? Or a reflection on the opaque distinction between a sense-making machine that doesn't make sense because it's broken down and a machine that never made sense but, by having the label "Under Repair" on it, encourages the viewer to assume that there was sense in it (somewhere, sometime) and that one can reconstruct it provided one knows how to reconstruct it.

There is very little spectacular or entertaining about this museum. No gigantic displays, impressive laser shows, scary dinosaurs, high-tech sound effects, or fancy reproductions of historical experiments. However, one does not get the sense that this is a transparent joke on science and its celebration through science museums. The meticulous care inscribed in the displays does not feel like a trick—a pun on "scientific professionality." What one feels is neither sarcastic laughter nor celebratory gestures toward great science. Rather, it feels like the work of a buff of peculiar forms of science.

After these small "introductory" displays, the exhibits become larger, more intricate, and more Borgesian in tone. A "tropical rain forest" theme begins to emerge as one moves through a series of exhibitions about failed great technological feats, opera singers, a depressed Chicago neurophysiologist turned philosopher of "obliscence," mysterious bats, and ants with chemically induced suicidal tendencies who all populated, tried to build bridges in, went through, sang in, or could not sleep in the rain forest. In some cases, the live paths of some of these people and animals happened to cross, if only momentarily, in the rain forest.

In a *Fitzcarraldo*-like narrative, a German-born engineer, Wilhelm Sonnabend, attempts the very ambitious construction project of a

> suspension bridge over the Parana River in Northern Argentina designed to span the seven hundred meters distance between the Argentinian and Brazilian banks of the river just at the point of the Garganta del Diablo, the largest and most spectacular of the Iguaçú Falls. The bridge which was to have been the longest foot suspension bridge ever built was designed to pass within ten meters of the crest of the massive falls, which would have afforded the visitor a spectacular if possibly terrifying view of one of nature's most extraordinary achievements. (Museum).

However, the bridge collapsed as it was being constructed (knocked down by an "unseasonable storm"), and Sonnabend is said to have moved to Chicago (with the help of Charles F. Gunther, a Chicago-based eccentric millionaire collector who had met the engineer as he "was visiting South America for the purpose of enhancing his already renown collections"). Apparently, "Gunther had heard of the demise of the Iguaçú bridge project and his taste for the extravagant led him to want to meet the person responsible for such a daring project." The visitor is now given a chance to experience the sublime of such a terrifying view by gazing at a two-foot-wide model of the falls (with real but not so turbulent water going through it) which can be conveniently watched either from the Argentinian or Brazilian border.[9]

The exhibit follows Sonnabend to Chicago where, we are told, his engineering practice thrives and his son Geoffrey becomes a professor at Northwestern University. But not all is well, as Geoffrey "suffered a mental and physical collapse" as a result of an "exhaustive and largely inconclusive research into memory pathways of carp." His loving mother Viviana (an Argentinian native who had lived in Iguaçú as a girl) felt the Amazon would help Geoffrey's speedy recovery and decided to take him back to Iguaçú, where her own father had previously recovered from a similar ailment. Viviana's move back to Iguaçú was a personal memory trip—one in which she was accompanied by her son who had collapsed in his struggle to understand memory. Iguaçú was a memory spa.

One evening, in Iguaçú, Geoffrey is said to have attended the performance of Madalena Deloni (whose life and career are the subject of a separate wing of the museum). Deloni, we are told, was a famous New York-based opera singer, whose voice, according to a *New York Times* critic, was "steeped in a sense of loss"—a feature which the museum's curator sees as possibly connected to her problems with short-term memory. After listening to Madalena Deloni that evening Sonnabend could not sleep. But during that wake he had a ground-breaking vision that led him to solve the problem of the nature of memory which had eluded him during the research on carp and had eventually caused his psychological breakdown.

The tangible result of this revelation was *Obliscence: Theories of Forgetting and the Problem of Matter,* a three-volume work he is said to have published ten years later. In this work (which I have not been able to locate), Sonnabend presents memory as the effect of the intersection between the cone of consciousness and the plane of experience (fig. 2).[10] Significantly, Sonnabend's revelation was that memory is, actually, nothing but loss or, more

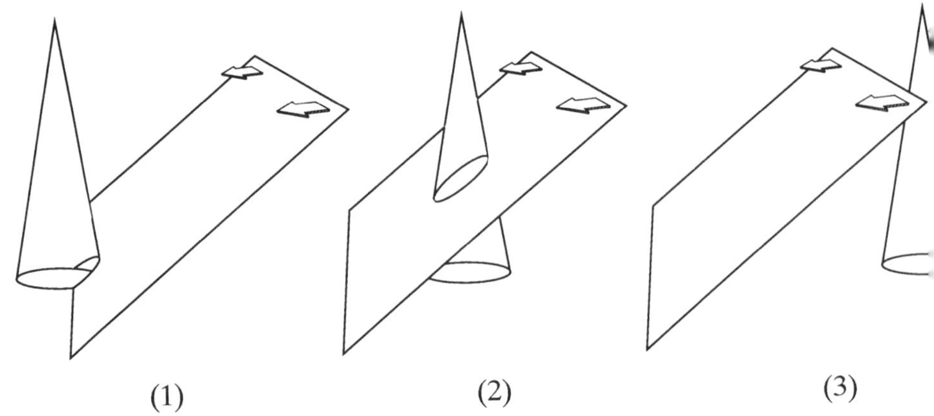

Fig. 2 Sonnabend's memory diagram as reconstructed by Wilson: (1) being involved in an experience; (2) remembering an experience; (3) having forgotten an experience.

precisely, it is the result of our desperate attempt to deny that memory does not exist:

> We, amnesiac all, condemned to live in an eternally fleeting present, have created the most elaborate of human constructions, memory, to buffer ourselves against the intolerable knowledge of the irreversible passage of time and the irretrievability of its moments and events. (Worth 1991:1)

The Deloni-Sonnabend exhibition suggests an interesting homology between Sonnabend's theory of memory as confabulation and the discourse articulated through many of the exhibits—especially those curated by Wilson.[11] In fact, Wilson's paraphrase of Sonnabend sounds like an epitome of his museum:

> What we experience as memories are in fact confabulations, artificial constructions of our own design built around sterile particles of retained experience which we attempt to make live again by infusions of imaginations—much as the blacks and whites of old photographs are enhanced by the addition of colors or tints in attempt to add life to a frozen moment. (Worth 1991:1–2)

A few feet later, the theme of forgetfulness is reinforced by a minimalist Proustian exhibit. Beside a little dish with three madeleines (one is partially eaten) there is an empty cup of tea. A short passage from *A la recherche du temps perdu* tells the well-known story of the vivid episode of involuntary memory experienced by the author after biting the tea-soaked cookie. But more important is another quote from Proust that Wilson has chosen to include in the exhibit:

> But when from a long-distant past nothing subsists, after the people are dead, after the things are broken and scattered, taste and smell alone, more fragile but more enduring, more unsubstantial, more persistent, more faithful, remained poised a long time, like souls, remembering, waiting, hoping, amid the ruins of all the rest; and bear unflinchingly, in the tiny and almost impalpable drop of recollection.

Sonnabend, Proust, and Wilson overlap. All we've got are vivid "memory effects." And these are mental artifacts not just because what they refer to is long gone, but because what triggered them to begin with were fleeting, perishable, and almost immaterial things.

However, Wilson does not seem to indulge the tone of deep nostalgia of Proust's text. Three transparent plastic pipes (one for each madeleine?) are connected from the inside to the plexiglass screen separating you from the cookies and the tea. Right below the endings of the pipes there are three buttons. When you push them, a bout of air comes out of the pipes toward you. You try to smell. Maybe you are going to smell the madeleine. Maybe you'll have a vision. Or maybe this is a cheap trick—like those scratch-and-sniff cards you used to get at some movies. Instead, no smell comes out of the pipes, no matter how hard you push those buttons. No memories can be reproduced, except the ones you produce by confabulating. Forget longing.

Wilson's apparent fascination with Sonnabend's theory (which might be Wilson's own) and with the allegedly memoryless opera singer, Madalena Deloni (whose performance at Iguaçú is said to have triggered Sonnabend's theory of memory as an epiphenomenon of loss) provides an interesting register for reading the museum itself. Wilson's narratives around the various objects on display may be read not as contextualizations or celebrations of "real" scientific facts or discoveries, but rather as "confabulations" produced to "buffer ourselves against the intolerable knowledge of the irreversible passage of time and the irretrievability of moments and events" (North 1991:1).

Maybe the exhibits are nothing but pretexts for producing memory. Maybe it does not matter whether (or how closely) the exhibit refers to "real" events or authors in the history of science and technology.[12] In and of themselves, these objects and narratives about them are not "memory." What's important is that they help Wilson (and the visitors) confabulate. It is in this sense that the museum is not an institution that canonizes a given historical narrative about science, but rather a place in which different memories are produced as part of a collective performance which involves visitors, objects, and curators. And, these confabulations feel very private because of the setting in which you experience the exhibits. You watch them standing or sitting in a dimly lit spot in which you are barely aware of other visitors (if they are there at all) and hold a receiver that carries the voice-over.

The fragmentary material on display points, quite literally, to the impossibility of any viewer's or curator's achieving "total" experience and reconstruction of the scientific objects he or she observes. Also, the museum shows that memory is private and multiple at the same time, and that the imaginary is constitutive of any knowledge one can produce. Sonnabend (or Wilson) does not deny the existence of reality and of our experience of it, but argues that our discourse about it (and about ourselves as its experiencers) is bound to the imaginary of our confabulations. It also suggests that science and memory as confabulation are bound together and that, therefore, the museum is the "perfect" laboratory for the production of that kind of temporary and multiple knowledge.

However, the cognitive predicament described by Sonnabend/Wilson is not that of the "postmodern condition." The instability and fragmentariness of our knowledge about nature or anything else is not presented as reflecting a historically specific reflection about the workings of language and the meaning (and politics) of interpretation. Our problem (or maybe just our predicament) is that, simply, memory does not quite exist. This museum is neither modern nor postmodern—it is simply Jurassic. In a sense, this is a self-conscious "memory palace" which knows itself to be just a theater (maybe an installation piece), and not a theatrical device instrumental to arranging and maintaining "real" memory according to some scientific "order of things." David Wilson is a late-twentieth-century skeptical Raymond Lull—somebody who does not dream to reconstruct the Garden of Eden, the Great Chain of Being, or some sort of "natural order."

Finally, the theory of obliscence may indicate why a number of these exhibits, while being about forgotten or would-be scientists, are not meant as redeeming gestures but just as *presentations*. August Verneuil, the forgotten inventor of the technology to produce synthetic gems, or the various people who wrote to the astronomers at the Mount Wilson Observatory to present their unorthodox cosmologies (which I will discuss in a moment) are not portrayed as unsung heroes or victims. Their work is presented without a comment, without a redemptive gloss. They are not celebrated, they are simply given a chance to be "confabulated about" because that's all anybody can do and expect to be done to his or her own work.

The last sizeable exhibit of the "Amazonian cycle" is about a very peculiar bat, the Deprong Mori, or "piercing Devil," of the Tripiscum Plateau. It consists of a number of disparate objects which include a model of an Amazonian natives' dwelling, diagrams of bats' sonic emissions, and the model of a radial mazelike structure. In and of itself, the exhibit is neither spectacular nor self-evident. It begins to make (some sort of) sense only after you lift a receiver and listen to the usual deadly serious voice-over:

> On his return in 1872 from anthropological fieldwork with the Dozo of the Tripiscum Plateau of the Circum-Caribbean region of Northern South America, Bernard Maston reported having heard several accounts of the Deprong Mori or Piercing Devil which Maston described as "a small demon which the local savages believe able to penetrate solid objects." . . . A typical report would describe a Piercing Devil entering a local savage's dwelling by flying through one of the thatched walls proceeding across the living quarters and exiting through the opposing thatched wall. (Worth 1964 : 1)

At one these reports involved "six Mori flying, in rapid succession as if in formation, through a domicile measuring eleven by thirteen feet, during the

Fig. 3 Museum model of a Dozo dwelling. The arrow indicates the trajectory of the six Deprong Mori.

inhabitants' evening meals'' (fig. 3). (At this point a spotlight illuminates the model of a Dozo dwelling in the exhibit.) In another passage of his field notes, Maston allegedly wrote that "a Mori is said to have penetrated the outstretched left arm of a five years old child. The arm apparently evidenced no lesions or wounds of any kind'' and the child was said to have later developed "the ability to heal warts, blood blisters and other superficial skin disorders'' (2).

Nobody seemed to notice Maston's reports (if they ever existed) for eighty years until Donald R. Griffith, a specialist in echolocation and author of *Listening in the Dark: Echolocation in Bats and Men,* retraced Maston's steps in the Tripiscum Plateau and, after eight months of strenuous research, formulated the hypothesis that the Deprong Mori was actually a bat of the species *Myotis lucifugus.*[13]

Griffith's ground-breaking hypothesis was that

> the *Myotis lucifugus* has evolved a highly specialized variation of the standard echolocation capability. This tiny bat had developed elaborate nose leaves which act as horns to focus the orientation emissions into an extremely sharp and precise beam. Furthermore, the orientation beam of the *Myotis lucifugus* is not the typical sonic beam, but is emitted as part of the electromagnetic spectrum falling generally in the range of the extreme ultraviolet.

More precisely,

> Griffith reasoned that the *Myotis lucifugus* employed variable frequency echolocations as do other bats, however the *Myotis lucifugus* applied variable frequency to its electromagnetic (rather than sonic) emissions. The key idea in the Griffith hypothesis was that as the *Myotis lucifugus* emission increased in frequency the emission actually crossed the thresholds from the extreme ultraviolet into the X-ray thereby allowing the bat to fly unharmed through solid objects.[14]

In a switch resembling that between the Jurassic as geological era to historical period and then to geographical area encountered in the initial slide show, here Wilson engages in a number of "category hoppings." From sound to ultraviolet (extreme ultraviolet, actually) then to X rays. He also switches from X ray as a technique through which people (and, allegedly, animals) can *see through* things, to an ability to *go through* things. He obviously likes it: "The research team was understandably enthusiastic about the Griffith hypothesis" (Worth 1964:7).

And I liked it too. The literalness of these switches and analogies (like the one between Viviana Sonnabend's memory trip back to the Iguaçú with her memory-crushed son who is eventually able to understand memory (though only in the sense that it does not exist) thanks to an opera singer whose voice "steeped in a sense of loss" (said to reflect her short-term memory problems) has something "excessive" about it. Like the childlike puns mentioned before, these are not "bad" jokes. Rather, they are jokes about "good," "subtle" jokes—jokes by people concerned with being either serious, witty, or simply professional. It seems that Wilson is ironically appropriating the taste (or maybe the obsession) for literal analogies typical of the culture of early modern cabinets of curiosity—though in a carnivalesque register. However, Wilson's "carnival" is as low-key as the rest of his museum. It does not want to turn science upside down (that would be a sort of "utopian" gesture, one suggesting an "alternative"), but just to tease science and those who have absorbed the discourse of seriousness and professionality it has slowly accrued through the centuries.

The same controlled irony emerges also in his mention of the quasi-magical power allegedly acquired by the arm of the child "visited" by the Deprong Mori. The child cannot perform full-fledged "miracles." He can only "heal warts, blood blisters and other superficial skin disorders."

Also, the narratives of these exhibits refer to each other, helping to construe the museum as a self-referential microcosm of cross-confabulations. For instance, the *Myotis'* emissions belonging to that "part of the electromagnetic spectrum falling generally in the range of the extreme ultraviolet" is a direct reference to another exhibit precisely about that: the extreme ultraviolet.

> Since the discovery of the extreme ultraviolet in 1891 by the German physicist Johann Wilhelm Ritter, many of its effects have been well researched. Less completely understood however are the regenerative effects of the extreme ultraviolet. Between four thousand and two thousand angstroms, the ultraviolet is known to promote the healing of wounds. Beyond two thousand angstroms a phenomenon known as ultramundane or plasmotic postvisualization becomes apparent. Plasmotic postvisualization is the ability of vision in the extreme ultraviolet to reconstruct life forms from inanimate remains.
>
> If we, like many species of moths, certain insects, and infants of our species were able to see into the extreme ultraviolet we would observe not only existing life forms (as we do in visible light) but preexisting life-forms as well. Infants see further into the ultraviolet than older individuals, some as far as 2120 angstroms, where partial plasmotic postvisualization can be observed.
>
> In this exhibit, the lens of the video camera has been fitted with a narrow-band extreme ultraviolet filter (1110 angstroms) and the display case is lit by fluorescent tubes emitting a large amount of energy in the ultraviolet. The resulting video image demonstrates plasmotic postvisualization in the extreme ultraviolet. (Museum)

What you see in this exhibit (as you hear a howling) is a canine skull placed in a transparent case, resting on a little patch of dry, sandy dirt resembling desert ground. But if you look at the skull through the eyepiece at the center of the case, you realize that the howling you hear does not come from a "real" coyote but from a man imitating one. In fact, through a system of prisms and mirrors, a black-and-white videotape of a howling man is made to overlap with the image of the skull in the case. It is not that you see doubts. The effect is that of seeing both the skull and the howling man in the same place at the same time.

Again, we find a number of "category hoppings" here: From death to life; from the healing features of the ultraviolet to its alleged ability to the rescue from death; from the invisibility of the extreme ultraviolet to the visibility of the invisible through it. And, again, the irony. Extreme ultraviolet vision cannot resuscitate, it cannot "heal" the dead: it can only make the dead visible—*represent* it as alive. Maybe a high-tech version of the story of Orpheus and Eurydice. But, in any case, the dead we are allegedly able to see is not the "real" dead. The dead coyote visible through the "extreme ultraviolet" is not a coyote but a man. And this man is neither alive nor dead—he "exists" only through a video camera. Therefore, the alleged reconstruction of the dead through the extreme ultraviolet is just a technologically aided confabulation. Everything is just memory, yet a memory that is imaginary (even when helped by "science").

Similarly, the carved fruit-pit exhibit resonates quite directly with the

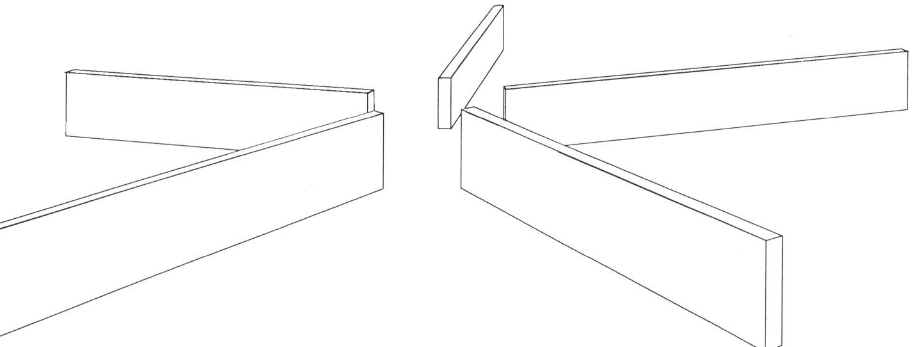

Fig. 4 Griffith's radial pentagonal snaring device.

exhibit of the microsculptures of Hagop Sandaldjian temporarily housed at the museum.[15] Beside Pope John Paul II in full regalia contained in the eye of a needle, the exhibit featured, among other micropieces, "Mount Ararat etched on a grain of rice, and a Lilliputian Christ of glue and gold dust, crucified on a single, bisected hair from Sandaldjian's head."[16] An exhibition dedicated to nanotechnology is currently being assembled. Allegedly etched fruit-stones, visible microsculptures, and high-tech nanotechnology share the same space and refer to each other blurring the boundaries between technology, art, and confabulation.

Back on the Tripiscum Plateau, Griffith and his assistants were facing serious problems in their study of the X-ray *Myotis lucifugus* as it kept flying effortlessly through the scientists' nets. Fortunately, Griffith came up with another brilliant idea. He decided to construct "a snaring device consisting of a radial pentagonal construction with walls of solid lead eight inches thick, two hundred feet in length and twenty feet high, like five spokes of an enormous lead wheel" (Worth 1964:7; fig. 4). After this cyclopic detector was put in place (how this *Fitzcarraldo*-like feat was achieved the narrative does not say), the team began to wait. And waited, and waited for more than two months until

> in the early morning of August 18th at 4:13 a.m. Griffith's instruments recorded the event that the team had been awaiting. The number three wall had received an impact of the magnitude of 10^3 ergs twelve feet above the forest floor, one hundred ninety-three feet from the hub of the wheel. The team was hopeful that one of the "piercing devils" had strayed within the radius of the wheel.
>
> Unable to wait for dawn's first light, Griffith and his assistants packed the cumbersome equipment of a specially designed portable

> X-ray viewer across the rain forest in the black of the night. Precisely at the spot indicated by the seismometer, Griffith's X-ray viewer found, at a depth of 7-1/8 inches, the first *Myotis lucifugus* ever contained by man, eternally frozen in a mass of solid lead (fig. 5). (8)

[A red LED goes off in the spot in the model where the impact is supposed to have taken place].

In some way the "containing" of the *Myotis* (which, because of its "X-ray capabilities," ends up "eternally" preserved in lead instead of being simply

Fig. 5 Rendering of the Deprong Mori as eternally frozen in a block of solid lead.

smashed on the wall) is an appropriate metaphor of memory as confabulation. As with Proust's smell and taste, the stability of the bat's "memory inscription" is made possible by the very elusiveness that characterized its existence. It is almost eternal because it almost got away. At the same time, the "permanence" of the *Myotis'* inscription is not permanent at all because, in the end, Griffith's epic is the produce of Wilson's confabulation. But, of course, Griffith's feat is not presented as a confabulation, but as "real" science. Actually, as an exemplar of "good" science:

> The juxtaposition of the work of these two men [Griffith and Maston] demonstrates in bold relief how, over the last century, progress made in the application of the scientific method as well as developments in scientific technologies have led to the triumphs of modern scientific inquiry. (Intro.)

The most recent exhibit at the Museum of Jurassic Technology is a selection of letters to Mount Wilson Observatory from 1915 to 1935.[17] Its title, "No One May Ever Have the Same Knowledge Again," is taken from the first letter in the exhibit by a woman from New Zealand, Alice May Williams:

> I want to tell you I am not after money & I am not a fraud. I believe I have some knowledge which you gentlemen should have. If I die my knowledge may die with me, & no one may have the same knowledge again. I have gone through frightful things still I go through it & I am beginning to get knowledge. I would write down & tell you what I no. But I would sooner wait till I hear from you. Because you are both strangers to me & my letter may go astray. When one writes one needs peace and quietness. (Simons 1993:3)

Even today, many departments of astronomy, physics, and astrophysics receive letters from people outside or at the margins of the scientific community—letters which often propose uncanonical cosmological views or scathing critiques of beliefs deemed canonical by the profession.[18] These letters are rarely answered. Much more frequently, they are either thrown away or posted on bulletin boards to be laughed at by graduate students and faculty—a laughter that contributes to confirming the scientists' professional identity as opposed to that of "quacks."

The Mount Wilson Observatory was a unique institution which, immediately after its founding, gained a conspicuous place in the popular imaginary thanks to the size and power of its instruments and the many discoveries achieved by George Ellery Hale and other astronomers. These letters indicate that people seemed to attribute not only scientific knowledge but also remarkable power and wisdom to the astronomers at Mount Wilson, who are often represented not as cut-throat scientific operators but as priestly figures—the priests of the biggest scientific temple of the time.[19] Most of the authors of the letters to Mount Wilson do not seem to have had much of a scientific background. Several displayed only an approximate control over the English grammar and often lived in rural areas without access to professional scientific networks.[20] A good portion of these letters are from midwestern farmers.

Before the exhibit was completely installed, I was given the transcriptions of some of these letters by Sarah Simons (the cocurator of this exhibit). Probably she thought that, being a historian of science, I might know what to do with texts like these. At first I tried to read these letters as traces of twentieth-century popular scientific imaginaries. But it soon became apparent that, in general, such an approach would not take me far, largely because I had no extended body of letters by (or contextual information about) any of the writers.[21] "Thick description" seemed beyond reach.

Alice May Williams seemed to be the partial exception. In her three letters to Mount Wilson she provides fragments of her everyday life and of her views on the cosmos and about another world she believed to exist above ours, beyond the sky. She mentions a treatment she is undergoing as well as her vexation at being "interfered with" by people who are puzzled by what she says. She derives her knowledge on cosmological matters from beings from Mars with whom she communicates and visits when she is in a "half-sleep trance." Sometimes they speak through her when she is in that state: "They sung right through my mouth to show how they sing on gramphones," or, "They can talk right down through my mouth like the gramophone no one else can hear them." Sometimes, being their mouthpiece creates problems:

> I do hate it saying who no, through me at times so people can hear, & it wont stop. It keeps working me up to say who no all the time, in the house, but not in the street my husband does carry on & he wont stop he thinks I am saying it on purpose.[22]

But if she does not seem to mind the disruptions brought to her daily life by the trances necessary to connect with the Martians, sometimes these communications turn hurtful:

> If those people from mars comunicated with people of this earth people wouldnt be able to do their work properly. They have got to put wireless on your hearing I think that is what I got. It is terriable at times as somethink else interferes with their wireless & harms people. (14)

Maybe the "interference" which sometimes comes through the "wireless" communication with these beings is connected to the treatment she mentions she is undergoing—possibly in a psychiatric hospital.

Whether she might have incurred in psychiatric treatment because of her unconventional cosmological views or whether these views are the result of a psychological condition is, of course, impossible to say. However, her uncommon cosmology seems to have been developing since she was a child—one who had unusual (though apparently nonpathological) experiences:

> Always from my childhood I have been mad on Planets. When I used to live on the lighthouse I used to live out side a great deal. & I watched everythink going on in the sky taught myself & also had a great deal of enlightment . . . that is what I want to teach you gentlemen that everythink is different to what anyone of this world of today thinks. That sky is all different. (12–13)

The "wireless" and the "gramphone" allow her not only to receive messages from her Martian friends, but to call them up, too, especially children:

> I have only got to call through my gramophone & I get shown all sorts of things. I beleive the babies [from Mars] can hear me writing this. I think one wants to talk & show me hisself now. The children are nice. When I start reading someone reads it right out to me. (7–8)

And these are not the only technological means through which communication takes place. Movies work, too:

> When I go to the pictures Thoughts people acting in the pictures talk to me & yarn. Anyhow somethink talking inside of me enjoys it self at the Theatre & chats to me. So I am not lonely. (8)

The cosmos populated by Alice Williams and her friends from Mars is a complex, labor-intensive, high-maintenance organism which—like Alice Williams herself—should not be "interfered with."[23]

> The sun comes out by day and doses his work by day. He draws the clouds and the sunspots you can see on the sun are that very black cloud, which is called a verandah, at certain seasons it is thrown out & spread over the sky for certain work going on underneath so it won't be seen, & to keep people from harm, on a very fine day it is drawn inside one of those moonhouses to give a very fine day, so that is what a sun spot is. Black clouds.
>
> Some of those planets you can see are like stations to do god's work. Venus is thickly blanketed with clouds, that is a place that stores the clouds. Jupiter be I suppose a storeplace for rain snow & hail. Others to keep air, wind, sunfire, wild fire, star fire, moisture, electricity, gases. It is worked by a human spirit world of human spirits that run over the top of the world & wind. The Planet Mars is inhabited by human spirits like us can talk eat & drink and wear

> clothes, but have great powers. . . . They are kept to do work overhead. They also work our wireless gramphones, machinery, Moving Pictures Talking Pictures and all that sort of thing. All that sky is worked spirital. (5)

The cosmos is shaped not so much by physical laws as by a lot of "spirital" labor and willful decisions by God and his Martian assistants. The cosmos works very much like any community, factory, or household. Terms such as *work* and *worked* (by which she might mean "wound up") are dispersed throughout her letters.

The sun is wound up to do its work during the day ("The rays of the sun is worked").[24] Its task "is to give nice warm days for pleasure & enjoyment to grow our plants & to keep us in good health" (8). The moon too is "worked" to do its work at night: "When the sun goes down that moon lights up to do its work to draw in the clouds or moister." As with streetlights, the moon's "shift" ends at sunrise: "[if] you go out again in the early morning you will see the moon put out" (12). Comets too are part of this cosmic economy: "Those comets you can see are somethink that doses their work in the sky I should think so" (8).

As some "construction work" is taking place in the cosmos, a "curtain" (usually made up of clouds) is drawn shut so that humans don't see what is going on and don't get hurt by possible "debris." When the job is done, the cloud curtain is lifted and stored in some planet (usually Venus), and the humans receive a nice sunny day.[25] Pleasure on earth is the result of a job well done in the cosmos.

It is this view of the cosmos as an ongoing work site that makes her very skeptical about the uniqueness and centrality of the sun in our planetary system. Alice Williams seems to see the sun and the moon as "workers" within a very local economy. Basically, the sun is that source of heat and light which shines (sometimes) right above her head. That the sun does its job well does not mean that it should be believed to be at the center of our system. Consequently, she is truly surprised by what Dr. Petit of Mount Wilson writes her about the solar system, and asks,

> Do you mean to tell me you got the nice warm sun out in California overhead & the same sun out in the world of N.Z. & then Australia, they are all different worlds of this earth world. Do you mean to tell me that, whole live worlds revolves around one sun in the air. It would be terriable. (8)

That she does not "think any live Planet revolves around the sun" is also connected to her belief that planets are not planets in our sense. They are not bodies orbiting the sun but rather God's spiritual agents which help him "work" the world in a number of different ways (12). One of the functions she attributes to them is to act as "warehouses" for clouds, rain snow, hail, air, wind, sun fire, wild fire, star fire, moisture, electricity, and gases. That some planets appear at specific times of the day (like Venus which becomes visible around sunset) indicates that they also help in the transitions between day and night, that is, when the sun and the moon need to be either turned on or put out.[26] Similarly, what we see as the moon is not always the planet Moon. When we perceive it during the day, what we actually see is a "fuel tank" that feeds the sun so that it can do its job:

> You see sometimes a round place like the moon that works the sun fire around. That place is one of the places [that is, "stations"] to keep the sun fire in to make the sun. It is worked spirital. I could show you myself if I was near you. havent you noticed in the day times at certain times you will see the sun on one side & you look in another you will see either a half moon or full moon in the day time. That isn't a planet. (12)

Moreover, some of the celestial objects we perceive as planets are actually holes between the terrestrial and heavenly worlds. It is through these holes that God reaches down and, so to speak, "works" the world machinery.[27] And it is through similar holes that the people from Mars—God's "repair crew"—come down toward our world.

These holes are in sky—one of the several "curtains" which populate Alice Williams's cosmology. As we have seen, some of them (made of clouds and sometimes even of sunspots) are temporary shields. Others (like the sky itself) are more permanent and are there to prevent us from seeing another world, that is, heaven:

> God made two live worlds, one he called heaven. The people are made of heaven, Therefore heavenly people. The other world he called earth. The people are made of the earth. Therefore earthly people. One was made of heaven, & the other one of earth. Both similar to one another, I believe heaven as all different countries like down here divided with the sea, & a sky overhead. Because the two worlds got put the sky to separate the two worlds. The sky is a covering to the other world. The sky is called firmament. That is the reason the people of heaven cant talk to the people of earth I believe that sky opens and closes on certain periods, When you see all cloud covering the sky right up & over. Those clouds are called. Blinds, shutters, & verandahs. Sometimes that sky opens underneath & you cant see it on account of those clouds covering the sky. The sky opens slowly down like a lid of a box then closes up quick. (4)

The distinction she sees between the two worlds is not as categorical as, for instance, that which Christians assume to exist between heaven and earth.

Although the two worlds are made up of different material, look different, and are populated by different beings, she believes that some level of communication and even travel is possible between these two realms. As she says on a number of occasions, sometimes holes are open, curtains are drawn up, and one can see heaven—though briefly.

More important, there is a liminal class of beings which mediate between the two worlds: the Martians. "The Planet mars is inhabited by human spirits like us can talk eat & drink & wear clothes, but have great power" (5). And they are also directly in touch with god, for whom they "work" the world. They are "material angels."[28]

Their mixed heavenly-human status is reflected in the external appearance and internal content of their spaceships. These machines look like planets and, like them, help God "work" the world. In a sense, planets and Martian spaceships fall into the same taxonomical box: they are God's "brokers" and their similar appearances reflect their similar function in the economy of the cosmos. In fact, because they have lights inside (and their windows are round and "look like the face of the moon"), they are often mistaken as planets or stars. Finally, Martian spaceships mediate between the two worlds by literally moving up and down like elevators through planetlike holes: "I beleive the people on the Planet Mars have machenes that run down in the air elevator style . . . & they look exactly like the moon" (6).

However, the morphological similarity between spaceships and planets has its dangers. "I think this is the reason they [the people from Mars] dont want people down on this earth to shoot rockets up into the moon. Because it might be one of those machines with people in and harm the people inside" (6).

But if their spaceships look like stars from the outside, they look like homes inside, confirming the hybrid heavenly-human status of their passengers:

> They have glass shutters that look like the face of the moon. Inside they are rooms with beds tables and chairs. . . . One time I seen three brothers in one of those machines one looked like a clown they had like pipes set on a table in the centre of the room & they were alight like gas, a bed on each side. another time I seen a man sitting in a easy chair reading a paper & smoking a pipe he was dressed in a blue serge suit & had fair curly hair. Another time I see two people a man & woman very tall fair people they were singing a moon song. (6)

However, the people from Mars are not all-powerful. In many ways they are like us humans. When their sky planes come down toward us, they cannot physically reach us because they fear that the holes through which they have gone may close behind them, shutting them out in the earthly world. When this happens, "they may have to wait a long time before they can get back to their own world" (8).

Interestingly, although Alice Williams simply states as a fact that the spaceships from Mars *do not* reach the earth (but can get very close to it), it rather seems that they *cannot* get down here. A gap has to remain between these two worlds, a gap that can be bridged only by her (with the help of their grammophones, "wireless," and moving pictures). These Martian technologies (which look very much like ours but are clearly more potent) allow for something in Alice Williams's mind to connect with their world and, indirectly, with the heavenly world above the sky. Also, it is because they cannot reach us (and therefore communicate with everybody) that Alice Williams can maintain her special status as their "oracle"—one equipped with the right "wireless" device.

She perceives films, radios, and grammophones not as technologies of entertainment but rather as something inherently connected to mental processes. They are "media" only in the sense that they mediate between two worlds. And it is important that they produce "representations," not embodied performance as, say, in theater. What comes through radios, grammophones, and films are voices and images, not physical bodies. It is the "disembodied" nature of the output of these technologies that fit Alice Williams's need for mediation between the two worlds. As we have seen, her cosmology does not contemplate the possibility of "close encounters" with the people of Mars (or with God) while she is alive in the terrestrial world. It is only her mind and soul that, with the help of these technologies, can travel and see those nonterrestrial realms—especially the mixed human-heavenly world of Mars.

Through these technologies Alice Williams makes sense also of the "somethink talking inside of me," of the voices she associates with experiences like "when I start reading someone reads it out right out to me," or of invisible people talking through her. The existence of radio, film, and the grammophone (and their role in cosmic communication) allows her to believe that these internal voices have an external origin and are distinct from hers. Given her cosmology and her views about how God "works" the universe and the beings who populate it, such an interpretation is not ad hoc. To her, these technologies are not communication aids, nor are they external to the mind: God himself has joined bodies and souls through a "wireless."

> In my sleep trance I come alive off my sleeping body on to my soul. To show that when we die, we have another live body. Our body down here is linked together, with a wireless to our soul body in the other world. The spirit of the body is our mind our thoughts. When we die our mind our thoughts fly to our sould body & we come alive for our judgment. The body dies in this world & lives in the other world a world god as set aside for us. . . . But the world of heaven is a live world. (7–8, 11–12)

Therefore, she sees the "wireless" connection she has with the Martians as just an instance of an ontological link connecting bodies to souls in general or, more precisely, terrestrial bodies to their heavenly doubles which are waiting for them beyond the sky.

Alice Williams's views on heaven reflect a mix of common and uncommon religious beliefs. While we find references to the last judgment, it seems that the souls and minds of the people who have ended their terrestrial life immediately join their celestial bodies, and that these bodies are not a resurrected version of their old bodies. Rather, everybody has a "soul body" waiting in heaven ready to be "coupled" with one's mind and soul and begin a new life right away. The picture she offers is not one of collective last judgment and resurrection, but rather of travel to another parallel world populated by parallel bodies. And, again, the Martians are in between. While she cannot connect to her "soul body" beyond the sky through them, she can at least see the heaven with her technology-aided mind.

But if Alice Williams spends only a few lines discussing the "real" heaven above the sky, she describes in great detail the hybrid world of the Martians. Although the Martians participate in God's designs by keeping the cosmic machinery going, they don't quite stare immobile at the eternal light emanating from God. They eat slices of tinned corn beef, new potatoes, and "all sorts of foods & soups," they sing moon songs, wear woolen tights, smoke pipes; they take pictures of each other, their babies cry and scream, their rooms are sometimes furnished with "a glass case full of lovely things." And they use money too (6–7). This, rather than paradise, seems the world she wishes to join.[29]

To conclude, her letters seem to express a desire for connections that have been made illegitimate by accepted scientific beliefs—though she does so by appropriating some of the technologies which are seen as emblematic of that enterprise. In this sense, they resonate very well with the museum's reconfabulation of science in order to question (or maybe just suspend) the emblematic status it has been given within modern culture.

But, as I said, Alice Williams's letters are an exception. Most of the other material I was given was of a much more fragmented nature. Sometimes what I had was just a short note from which I was not able to reconstruct the writer's "cosmology" or get any nonfragmentary sense of his or her culture. The most disturbing aspect of my interpretive failure vis-à-vis these other letters was the inability to avoid stereotyping their authors. As lack of information prevented me from reconstructing their thinking patterns in a somewhat holistic way, they often remained "weirdos" or even "crazies." Unable to develop satisfying narratives about their culture, I found myself stuck with my presentist, science-informed perspective. I also felt unable not to see these letters as unsurprising replays of beliefs, moves, and anxieties that, as a historian, I had already encountered in older cosmological discussions.

For instance, a few letters propose unintelligible arguments against Einstein's theory of relativity. Others, instead, focus on recent (and not so recent) astronomical discoveries which they perceive as reducing the centrality of man's role in the universe and his ability to trust common sense knowledge—arguments that reminded me of the anthropocentric anxieties so common during the Copernican revolution.[30] Analogies between early modern cosmological anxieties and those expressed here were also exemplified by a few letters which attempted to refute not only Einstein but Copernicus as well (101–3). In short, much of the content of these letters seemed a mix of cultural "leftovers" and reactive moves against established science rather than "positive" expressions either of popular culture or of original (in the sense of relatively independent) cosmologies.

Even the transparent drift of cold war tropes into cosmology did not seem to result in anything particularly surprising. For instance, a later letter (1958) appears to have been triggered by the launching of Sputnik and is concerned with developing a new view of the cosmos so that interplanetary travel (something at which, at that time, the Soviets seemed to have a better shot at) would turn out impossible. The cosmos has to be rearranged to make sure the forces of evil would not climb from hell into the heavens.[31]

Another feature among these letters that I saw as a replay of early modern themes was the concern with grand cosmologies stressing unity and harmony (instead of expansion, chaos, collisions, and so forth). The usual common denominator here are vaguely defined "electric" forces which are everywhere and never sleep.[32] But while these authors used scientific-sounding categories like "electricity," these moves seemed little more than a relabeling of older notions of action at a distance (ether, pneuma, and "magnetism"). Similarly, other letters display pervasive sexual overtones in the cosmologies they propose particularly when processes of creation, attraction, union of opposites, and cohesion are involved (41–43, 65–66, 95–96). But, again, these reminded me of pre-Socratic and early modern cosmologies as well as of alchemical beliefs.[33]

In short, my "professional" approach to these texts did not allow me to find them very "interesting." I had heard some version of that story before, and (probably because of the fragmentarity of the material), I could not manage to say much more than that.

My reading of these letters changed radically when I viewed them no longer in a transcribed and photocopied form but could see them as framed in the exhibit at the Museum of Jurassic Technology. Here they are presented without a gloss, without any introduction or comment. They simply hang there in their simple but serious frames, in a small and somewhat crowded room. The setting looks more like a nineteenth-century art show, with paintings hanging close to each other sometimes in several rows one above each other. But if the setting does not "sacralize" these letters by not placing them at a respectful distance from each other, they are neither posted on a bulletin board as they would have been in an astronomy department to be laughed at. Nor are they proposed as "historical documents" valuable only for their textual content.

They are not strings of words to be interpreted by a historian of science or popular culture. They are not stripped of their material and aesthetic qualities: their drawings and watercolors and the business or private stationery on which they were typed is there in front of you, as are the telegrams on which they were sometimes inscribed. They have their "aura," but one which is not necessarily mythical because you can't quite imagine what mythical narrative they may belong to. If you want to project an aura on them, that's fine, but the exhibit seems to remind you that that would be your own private business.

The setting indicates that they are not displayed as "other" to established science. They cease to be "funny," "cute," or "nutty." And they are not presented in the nostalgic register—as traces of a "popular science" eventually squelched by the scientific establishment. There is no suggestion that this was "great science," one the reader should recover from oblivion and reintroduce it in mainstream scientific culture. Also, nothing suggests that we should engage in the reconstruction of the "author" behind them. At the same time, it is not that they are displayed "as they really are." They do not stand "on their own terms." The museum frames them—though neither in the contexts of dominant or alternative science. Instead, the context which envelops them is, again, that of the museum itself: a setting of and for confabulation where hermeneutics is suspended. They stand to be constructed like anything else in here. They are not memory, but you may want to use them to construct yours. Or maybe not.

Notes

1. The Museum of Jurassic Technology is located at 9341 Venice Blvd., Culver City, CA 90232. I would like to thanks Sande Cohen for the useful comments on a previous draft of this paper.

2. In my view, this ambiguity (which I take to be constitutive of the historians' predicament) is made more visible by the so-called new historicism which, because of its methodological orientations, puts it under the spotlight. While endorsing the anticolonial, antinormalizing motivations behind neohistoricism, its concern with "fragments" and "wonder-effect" (which "surprise" us and jams our normalizing glosses)

may lead, unwittingly, to reexoticizations—though of a more "reflexive" type than those usually produced by colonial discourse. In particular, Greenblatt's notion of "wonder" seems a double-edged sword which can both decolonize the other and re-exoticize it at the same time (Greenblatt 1993).

3. There may be dynamics of academic self-fashioning at play here, too. Like baroque gentlemen who constructed themselves as distinct individuals (as different from pedants) by assembling what they saw as curiosities (that is, singular or exceptional specimens), some historians with vaguely neohistoricist sensibilities may have ended up fashioning their "distinctivness" (as intellectuals as opposed to pedantic scholars) by collecting "curious" museums in their narratives.

4. The entire description of the "Horn of Mary Davis of Saughall" is as follows: "We were then shown an extraordinary horn which had grown on the back of a woman's head. It was exactly like a horn, except that it was thinner and browner in colour. It is somewhat of a curiosity and it appears that men-folk bear their horns in front of such women theirs behind. It was noted on a label that it originated from a Mary Davis or Saughall in Cheshire *an. aet. 71 an. Dn. 1688.* No doubt it will have been mentioned in the *Transactiones Angl.*, or in the *Hist. nat.* of Chelshire. The horn was blackish in colour, not very thick or hard, but well proportioned. (Early visitor to the *Musaeum Tradescantianum,* The Ark)."

The full caption of the "Fruit Stone Carving" is: "Almond stone(?); the front is carved with a Flemish landscape in which is seated a bearded man wearing a biretta, a long tunic of classical character, and thick-soled shoes; he is seated with a viol held between his knees while he tunes one of the strings. In the distance are representations of animals, including a lion, a bear, an elephant ridden by a monkey, a boar, a dog, a donkey, a stag, a camel, a horse, a bull, a bird, a goat, a lynx, and a group of rabbits: the latter under a branch on which sit an owl, another bird and a squirrel. On the back is shown an unusually grim Crucifixion, with a soldier on horseback, Longinus piercing Christ's side with a lance, the cross is surmounted by a titulus inscribed INRI. Imbricated ground."

Both texts are provided by the museum as transcripts of the exhibitions.

5. For instance, the "Horn of Mary Davis of Saughall" is surrounded by plainer and definitely "real" horns of deer, goats, and other horned mammals. A replica of this "magic" horn is also sold as an amulet in the museum shop; it costs $25.

6. An abridged version of this narrative in reproduced in the brochure "The Museum of Jurassic Technology and You."

7. This exhibit is about the work of August Verneuil.

8. This is the transcript of the audio section of the exhibit: "Sublimation is the process by which a substance passes from its solid form directly to a vapor without passing through a liquid state. Sublimation is generally a property of relatively non-polar symmetrical compounds. Many aromatic compounds including anthracene, naphtalene, parasubstituted halobenzenes, and many aromatic amines and phenols sublime quite readily.

"Purification by sublimation involves subliming, or vaporizing a substance in one part of an apparatus and condensing it in another. To be successful as a method of purification, it requires that either the sublimate or the impurities sublime, but not both unless their sublimation temperatures are reasonably far apart. Many substances do not sublime at normal atmospheric pressure (1 atm, 276 mm Hg) but do sublime under specific conditions of temperature and pressure'' (Museum).

9. But if you watch the falls through an eyepiece attached to the third side of the display—an imaginary vantage point suspended between Argentina and Brazil—you can actually see the bridge which, instead, remains invisible from the two "real'' points of view.

10. Sonnabend, *Obliscence: Theories of Forgetting and the Problem of Matter* (Evanston: Northwestern University Press, 1946). I have not been able to locate this volume in the *National Union Catalog Pre-1956 Imprints* or in the *New York Library Dictionary Catalog*.

11. This may explain why Sonnabend was unsuccessful at finding the memory pathways of carp: carps don't confabulate.

12. Seen in this light, three otherwise quite obscure exhibits gain some contextual meaning.

The very first exhibit you encounter as you enter the museum is an alleged map of the medieval battle and siege of Pavia. The caption under this drawing says that many representations of the battle had been produced through the centuries in paintings and tapestries but that, eventually, it became clear that nobody who produced those representations had any direct (or even secondhand) knowledge of what that battle looked like. In short, all the "memories'' we have of that event are nothing but confabulations referring to other confabulations.

Similarly, the exhibit about Verneuil's process for producing synthetic gems may point to the problematic distinction between "real'' and "fake'' objects. As part of the exhibit there are two crystals of a synthetic gem which slowly rotate in front of the viewer's eyes (to make sure you see them from all possible points of view?). One is labeled "natural,'' the other, "artificial.'' However, while they differ in shape, the two crystals do not display any detectable difference.

Finally, the exhibit about "purification by sublimation'' may be a direct reference to objects melting into air (like the memory about them). But maybe I am trying to confabulate a bit too hard myself.

13. There is a Donald R. Griffin (not Griffith) who published a book with a title very similar to that attributed to Griffith's work: *Listening in the Dark: The Acoustic Orientation of Bats and Men*. While Griffin's book discusses in much detail the behavior of the *Myotis lucifugus*, it does not mention Maston, the Tripiscum Plateau, or the *Deprong Mori*.

14. Worth 1964:6, 7. The book by Griffin (not Griffith) is silent about electromagnetic, ultraviolet, or X-ray emissions by the *Myotis lucifugus*.

15. There are other examples of the cross-referentiality among the exhibits (or elements of them). For instance, Gunther, the rich Chicago collector who was attracted to Sonnabend's daring and unusual project of building a bridge over the Iguaçú Falls, may be an "embodiment'' of the sensibility of the museum itself. Similarly, the "rain forest theme'' provides another web that links several exhibits, as does the concern for the microscopic and the ongoing fascination about the boundary between invisibility and existence developed through a number of exhibits. There is also a "Noah's Ark''

theme which runs from the introductory slide show, to the exhibit of the "magic horn" of Mary Davis of Saughall, to another exhibit toward the exit which displays a small model of the ark (probably derived from Athanasius Kircher) which slowly rocks over an invisible diluvial ocean.

16. I have not seen the exhibit, and I am relying here on the review in Burdick 1993.

17. Unlike most other exhibits which have been arranged by Wilson in collaboration with his wife, Diana Drake, this has been cocurated by Sarah Simons and Wilson.

18. I would like to thank Owen Gingerich of the Smithsonian-Harvard Observatory, and Mark Dickinson, of Berkeley's Department of Astronomy, for all the material they have made available to me.

19. While many of these letters are critical of and often bitter about what they perceive as the elitism of the scientific community and the censorship involved in the process of peer review, they seem hopeful that the "sages of Mount Wilson" can right these wrongs.

20. A sampling of more recent letters show that this is no longer true today. People who resort to this form of letter writing today often think of themselves as scientists (though unappreciated ones), and try to gain through this informal communication the credibility that has been denied them by established scientific journals. It is not rare for modern letter-writers to have reputable scientific credentials or, at least, to have some technical background.

21. Moreover, the body of texts I was given represented a small selection of extant letters, a selection which reflected the curators' choice both in terms of content and chronological distribution.

22. Simons 1993:3–4, 6, 14, 8. Also: "At times somethink works my mouth to talk out loud & I have got to be carefull of her [her housemate] hearing as she thinks I am mad, & makes all sorts of fun of me to people" (3–4).

23. "The moon is a sphere and it works the clouds by night; it is not a Planet, & should not be interfered with" (5).

24. Simons 1993:8. Also: "The sun is worked from day to day to do its work" (12).

25. "At certain seasons [the curtain] is thrown out & spread over the sky for certain work going on underneath so it won't be seen, & to keep people from harm, on a very fine day it is drawn inside of one of those moonhouses to give a very fine day" (5).

26. This, at least, is my "confabulation" about the passages at 12.

27. Again, this is my interpretation of the fairly obscure text at 12.

28. As she puts it, "They are kept [by God] to do work overhead" (5).

29. At times, she seems to suggest that she would like to join the Martians' world not only in spirit but in with the human body as well. For instance, she claims that "they can draw you to them" (9); and that "I was told when I heard people talking & seen them in a trance to go up in a aeroplane as far as I could straight up in the air as one of those machines might be about & those people might try to get me" (6).

30. A certain T. C. Bates, a "little old twenty acre clodhopper," writes about Einstein and asks, "But what about our man Einstein? What does he mean? Is he just kidding us or is he crazy?" (9). He then invites Hale to bring Einstein to Mount

Wilson so that, after being exposed to the evidence provided by those great instruments, he would see the truth and desist from his silly claims. But that would not be an easy task: "I hear that Einstein intends to visit your observatory in the near future. Hope he will, but he, with our help should be careful that he does not go off at a tangent, for, if a straight line is crooked, a curved line is straight. If a straight line returns upon itself a crooked line must go forever. This is my philosophy. Now the line from Europe to the observatory is crooked and if he is not careful he will not arrive at the observatory" (90).

31. This letter was eventually excluded from the exhibition.

32. (34, 50–55, 60–63, 102–3). One author who thought of himself as a Christ-like figure wrote to the sages of Mt. Wilson telling them, "This is to certify, That I have found the Key To all Existence" and that he trusts his doctrine "to you of the Astronomical and the scientific world" as if they alone could understand him (55–59). According to this prophet, "electricity is the spirit of all existence" and electricity, nitrogen, and etholeum are the trinity of life (60). For instance: "Etholeum—The base of all existence—it is One with Electricity and there is no place where It does not exist. It is the conduit of The Light between all of the planets and thru the telephone and the radio and without it There would be no Earth Because there would be No sound to be transferred between" (59). He then concludes, "Because you have used the talents of your minds . . . and Have not hid them in the and consequently you are [worthy] to receive my words. In witness where I set my hand and affix the [seal] of [. . .] By signing my name - Edward" (63).

33. Some letters seem to have been written by people who traced the source of their psychological conditions to the malignant actions of "waves" emitted by powerful centers (maybe like Mount Wilson itself) to disturb their minds through various forms of interference: "All branches of absent treatment through the mental airways in the abstract with applied phenomena must be checked up on irrespective of what name or alibi the process operate under for the real danger lies in the abstract with applied phenomena because the process itself is an invisible contact with mentality in the effects only are seen which leaves individual liberty of mentality and fought at the mercy of applied phenomena I am not surmising or theorizing" (93).

34. It is interesting that when I inquired with David Wilson, Sarah Simons, and Diana Drake about who was the curator of the exhibit, I received only vague answers. While everybody, of course, acknowledged the labor they had put in the project, nobody was willing to be represented himself or herself as the "curator," that is, the "author" of the exhibit. The line I kept getting was that the exhibit was part of the "museum."

References

Burdick, Alan. 1993. "Pope by a Hair." *The New York Times Magazine*. 19 Sept.

Greenblatt, Stephen. 1993. "Kindly Visions." *The New Yorker*. 11 Oct.

Griffin, Donald R. 1958. *Listening in the Dark: The Acoustic Orientation of Bats and Men*. New Haven, Conn.: Yale University Press.

Museum of Jurassic Technology. 1993. Exhibition transcript.

Simons, Sarah, ed. 1993. *No One May Ever Have the Same Knowledge Again: Letters to the Mount Wilson Observatory, 1915–1935*. Los Angeles: Museum of Jurassic Technology.

Worth, Valentine, ed. 1964. *Bernard Maston, Donald R. Griffith, and the Deprong Mori of the Tripiscum Plateau*. Los Angeles: Museum of Jurassic Technology.

———. 1991. *Geoffrey Sonnabend—Obliscence: Theories of Forgetting and the Problem of Matter—An Encapsulation*. Guide leaflet 7. Los Angeles: Museum of Jurassic Technology.

DISCUSSIONS: EXCERPTS FROM THE COLLECTIVE EDITORIAL MEETING

The following excerpts have been selected and edited from the collective editorial meetings for volume 2 of *Late Editions,* held at Rice University, 8 May 1993. At this annual day-long meeting, each participant in the volume served in rotation as the assigned editorial critic of another participant's piece. The commentaries tended to be concerned with adequate framings for the materials presented, and a cumulative sense of "what these materials were about" emerged from the repeated attempts of commentators to see connections between the pieces they discussed and others. This contingent discovery of "governing" concepts from the comparative resonance of constantly reading across the different materials that has given such a spark to our editorial meetings is indeed the way in which we hope that the volumes of this series might be assimilated by their readers themselves. Likewise, in the spirit of the series, excerpting such discussions to give a reader who has just gone through the pieces a flavor of the themes and concepts that connect them is more appropriate then the more conventional discursive, distanced, and unifying conceptual summary common to academic works, presented as introduction or afterword. —EDITOR

"I think of weapons scientists as sort of like an onion"

HUGH GUSTERSON in discussion of Diana Hill's piece:

I will talk about five things that I got out of the interview. First, this constant theme that goes through the interview about a utopian idealism for peace, which I presume might strike some as counterintuitive but which is highly authentic. The second is this notion of U.S. and Soviet weapons scientists as sort of buddies in a very ambiguous way. They are rivals but they are also buddies who get along with each other and enjoy showing each other their equipment. There is a lovely piece in the middle of the interview which is pure Robert Merton ideology about the difference between science and politics. Hill keeps trying to return to this notion of science as the opposite of politics, and he is constantly undermining it without realizing it. Third, I really liked the way this excitement and anxiety about secrets kept saturating the interview. There is this notion of transparency, you could substitute the word intimacy for transparency at some points; there is this notion that the two sides can really see what the other side has. And there is both excitement and anxiety about that. There is also the anxiety that if they disclose too much they will lose their jobs. This is both an existential and practical or professional anxiety. The fourth thing that I really liked was this hidden thread that keeps coming to the surface of the tension between the U.S. military and the scientists. From a distance, it seems as if the military industrial complex is always one thing. But once you get very close up, it becomes clear that there are all kinds of tensions that keep on surfacing between the military personnel and the scientific personnel and between their ideologies. So the interview becomes a way that you can pull open the contradictions inherent in the militarization of science itself and display them for public view. You might want to trace that back to the very beginning, to the relationships between Oppenheimer and General Groves which was incredibly charged with both hostility and moments of intimacy and collaboration. The detail you present of the joint clarification experiment is beautiful. It is very vivid. You can sort of picture yourself there in this restaurant with these guys talking to each other. So those are the things I really liked in it.

Let me make some remarks about how you might sort of deepen the contextualization. Having spent three or four years working on this issue myself, I had a lot of context I could bring to it, but I think most people wouldn't really be able to read it in the way that I read it because they don't know exactly what is going on here. So you might want to explain what the Threshold Test-Ban Treaty is. At the very end of the interview you talk about this 150-kiloton limit. You want to move that much earlier on, I think, and explain what the significance was of limiting these explosions to 150 kilotons, which is about ten times the size of the bomb dropped on Hiroshima, but by nuclear weapons standards it is a very small bomb. When the Threshold Test-Ban Treaty was first negotiated there was huge debate about whether it actually signified anything or not, which is very obliquely pointed out in the interview. A lot of people thought that it was just window dressing. You know, who cares whether they are allowed to test bombs of 150 kilotons? That would be the political context in the seventies, but in the eighties the debate around the Threshold Test-Ban Treaty becomes much, much more charged. You don't talk at all about the other kinds of treaties that might have been signed. In the 1980s, there was this huge grass-roots social movement that is not mentioned in this interview anywhere, the

American antinuclear movement, and the Soviet antinuclear movement which is also not mentioned, and a huge antinuclear movement in Europe as well, pushing to end all nuclear weapons everywhere on earth. And the way the debate got set up in the 1980s, the weapon labs were saying you couldn't verify an absolute ban on nuclear testing but you can verify this particular treaty that we have already negotiated. There are a whole bunch of other experiments going on that the interview doesn't mention, joint clarification experiments, that were not being done by governments at all, but by geologists and people like that who are working citizens groups, the National Resource Defense Council in particular—they weren't allowed to go to the Nevada test site. So your father was positioning himself on the left, but actually if you put it in this context he is on the right because he is part of a set of exchanges that are undermining the possibility of the Comprehensive Test Ban Treaty.

RABINOW: Isn't one of the things we ought to do all the way through is at least pause when we say right and left, who is situated where. Obviously what you just said is an absolute that could be reframed, too.

GUSTERSON: I think of weapons scientists as sort of like an onion. The very core of the onion are the warhead designers, and your father is not a warhead designer. And then you have these other people that contribute on the outer layers of the onion in various ways. And the people who are working on the verification of arms-control treaties are one of the outer layers of the onion. If you work even a little at Los Alamos you get included into a particular kind of discourse which is what your father uses—of the Soviets as being sort of both rivals and buddies, of this anxiety and excitement about secrets—those are all the things that you find typical of weapons scientists. And he has security clearance. And to me, the main rite of passage in being transformed into a weapons scientist is subjecting yourself to investigation by the FBI.

"There are lots of images of battered bodies but they aren't getting us anywhere"

KIM LAUGHLIN in discussion of Hugh Gusterson's piece:

My comments will come from a bit of a different direction because I have no expertise at all in the substantive area that you work in. I have framed my response in terms of questions that I think we share in doing this: What are we after with an anthropology of science? To what end are we serving our own careers or scholarship?

Regarding Sylvia's biography, self-presentation, commitments, motivations, and so on, I think it would be nice to push on the question of to what extent she is or is not a subject of the enlightenment project. To what extent does she acknowledge the cross-cutting identities that you document her sharing, as a Japanese-American, as a woman in physics, to what extent her identity as a physicist is dominant, to what extent she thinks that she must choose amongst her identities so as to present a sort of collected, sovereign, composed self to her boss. And she may handle this differently in the different milieus in which she circulates. Another question regarding personal orientation is the question of how she regards innovation. I am particularly interested in that because one thing that is happening with the international trade treaties regarding patents is the assumption that we aren't going to have any technology if we don't provide financial remuneration by the patent process. So we might use our work on scientists to question the assumption, because it is not considered a questionable concept even by the Clinton administration.

In terms of the social organization of science, I think that one thing that we are asking is to what extent community is dependent on consensus. You tell of a virtual integration within this group of physicists through the testing procedures. I think it would also be nice to perhaps talk about in which domains within the lab difference is left intact, tolerated, in what areas of inquiry or perhaps subject matters. Clearly you talk about how people place themselves differently on the political spectrum within, and I think that while that does herald a negotiation of difference, one thing that one has to be careful of is suggesting that "Oh-la-la, scientists really are different!" But it is still a consensual methodological agenda where again politics is separated from the science. They work together as scientists and they leave their politics at the door.

Another question about social organization is the ways Sylvia perhaps participates in translation activities? For example, policy debates. You speak of her involvement with young high-school women that she is helping to be budding scientists. How does she present science to these young women? What are her mechanisms of presentation, both conceptually and rhetorically? Also in those situations it would be a nice place to ask to what extent science is considered a monolithic describable entity. There is a particular tension in a translation situation because the goal of sense-making sometimes compels a reductivist description of what we do.

Then there are questions of scientific practices and self-perception of them. Specifically, Sylvia clearly likes rationality and I am very curious to know if she thinks that the rationality that she practices is the same thing as the rationality she thinks occurs with these politicians that keep their fingers on or off the bomb. Also, I think it would be interesting to have some specific material on how Sylvia constructs her descriptions of the world tropologically. Specifically what I mean by that is the question to what extent she relies on empiricism as a basis of understanding, and not only within her conceptual process but within her representational process. That gets you into questions of, in critiquing oppositional people, the people outside her gates with puppets. Clearly she enjoyed their putting on a puppet show for her but did she think they had a chance in hell of intervening in the process if they didn't have their facts laid out on a graph that could be appended to a court document? What is her awareness of tropes of discourse? Perhaps through different efficiencies in different forms? Or does she assume that there is one discourse that crosses them all? And this would go back to the issue of translation. If she does participate in translation activities how does she conceive of difference or strategize working with it?

Third, I think you make the very good point that the battered bodies of Hiroshima are remembered differently. Mike [Fischer] has been very helpful on my Bhopal material in saying that you have got to start with the fact that there are lots of images of battered bodies but they aren't getting us anywhere. And so if you can start with the idea that there are multiple images in the first place, then you can work with the problems and places of maneuverability among them. But if you say, okay, she carries a specific imagery of the battered body of Hiroshima, what is it? Then possibly you could get to that by pushing on her concept of the authentic. You have that section where she doesn't like the film narrating Hiroshima because it is not native Japanese speaking, and if that is her criterion of the authentic in victimization, how does she conceive the expression of victimization in Chernobyl? Bhopal? What constitutes a legitimate representation of suffering, of crossing the boundaries between sufferer and he or she who watches the sufferer?

Regarding the question of the agenda of science, what project does she see herself undertaking, and to what extent does she think that progress will come by technological and economic advance? I think that continues to be a very practical question and I know that in negotiating environmental issues in the third world, that is the basis of congressional decision making. Her approach to weapons scientists seems to be more a survivalist orientation than innovative. The story she tells was about making sure all the bolts haven't rusted, that they are still in good shape, they aren't going to blow up at the wrong time. She doesn't talk about innovation, production of new knowledge. Maybe she does. I have heard in casual conversation with people things like there hasn't been anything interesting in this kind of science [weapons science] since the fifties. I don't know if that is true or not, but how does she conceive of the newness or the status quo maintenance? How would she define obsolescence? Is it when a machine or a piece of technology doesn't work any more? Is there an efficiency argument? Is it use value, exchange value, efficiency value? Because clearly one of things that is very much up for grabs in politics which is largely defined by science is what is value? To query her on that I think would be helpful.

You also say that she calls herself a Humphrey Democrat. I am curious to know what that means in the 1990s. Surely this does involve questions of resource allocation of which funding for big science is a part. And then to push it more conceptually, does she work from this assumption of an economy of scarcity or of excess? How does she conceive of wealth? Does she conceive of it as a static thing residing in machines themselves or does she have a sense of its circulation? Particularly the sense of the kind of work she does within the geography of imperialism. Because you know a lot of the arguments you hear about high technology is that it makes circulation cheap which usually implies a distribution argument, a fairer sense of distribution. But of course just because you can telecommunicate very cheaply between Taiwan and here does not mean that he or she who pulls the puppet strings is any more egalitarian in that decision making.

Now for some issues regarding the form rather than content of your paper—how we need to write differently in responding to science in the present context. I admit that my bias is toward a negotiation between politics and scholarship and some may find that extreme. You discuss nuclear items as weapons but you don't mention their disposal or the organizational issues they connote. For example, the very protest slogan that nuclear power necessitates authoritarian governance. And in limiting your discussion to the weapons issue I find that it is very similar to what we fought very persistently against with the Bhopal disaster of encapsulating an event as isolated in space and time, where disaster has a very conventional definition as occurring in this circle for two hours. I think that to be helpful information for political work specifically you have got to extend it. Within anthropology I think it is interesting you make comments about how studying scientists we can avoid the exoticizing gaze of studying the peasants. Personally, I think we have to be very careful that we don't exoticize scientists. I think that we do need to still study villages, but we study them differently and without the walls. And so this sense that continually breaching encapsulation of identities of events, and so on, I consider a shared project. It's nice that it's shared. It is certainly a shared agenda with the political activists I worked with, and so it's nice to say we are for once doing something the same.

Another encapsulating maneuver is the implication that science is somehow exempt from processes of social change. There is no mention of a changing science for Sylvia throughout her career. You start with saying, What do those protesters think of the Russians? Well, the Russians have changed a lot just in the last three years. Does she have a different conception of who the Russians are today as opposed to a couple of years ago?

Last thing, your information on Sylvia's purposeful pursuit of the Peace Museum in Hiroshima, going off to Russia, that is a very nice way to humanize her. You include an excerpt from a letter where she mentions that you seem surprised that she cries or she's upset when she sees a road accident. I think that probably the general public doesn't assume that scientists like road accidents, but the sense that how they deal with their placement in the world—questions of the venues by which scientists seek an affable position in relation to their societies I think are important. And its interesting that she, like in going to a nuclear protest, she troped it as an experiment. That was where it went off, was where it ceased to be that.

Finally, as a question of form, you say you take as axiomatic Latour's sense that a postmodern strategem of objectivity is a reflexive one that concentrates in making rather than presentation of interpretation. I think that particularly with our concern for where our scholarship is circulating we have to be highly attentive to our presentation as well as our interpretation, because although identified by many with a tendency to exoticize, we [anthropologists] do want to make sense. On the other hand we don't want to be complicit with the logical structures that we critique. The issue of different audiences, we have lost the dream of the mass audience. It used to be that you write for academia or you write for the masses. To what extent can we strategize diverse audiences without that nostalgia for a mass space or an unpluralized space. But still, this isn't where I would place the postmodern take on scholarship.

GUSTERSON: Where would you place it?

LAUGHLIN: This is again from a political position, dissemination was the critical issue for us [activists in Bhopal]. There is clearly a problem of information availability, the facts. I mean that is one of the things that I think constitutes victimization. On the other hand, the circulation of the information, I guess, is certainly an economy of excess rather than rarity.

Where you just see another starving child in Somalia and it's like, Oh, Christ, where do we move from here? There is a numbing that goes along with the commodification of these issues, particularly through visual representation. How can we play with narrative devices not only to strategize representation with sensitivity to a pluralized audience, but also to track knowledge claims, especially if we don't want to track them by recourse to a canon? Tracking them tropologically is a way to still remain accountable or remain aware of what we are doing. I know that certainly within the politics I participated in the best way of critiquing existing political programs was not an ideological analysis of whether it was right or not but attacking the efficiency of its representation. I just think that is a practical imperative.

GARY DOWNEY: It seems to me the central message of this piece is in its title—becoming a weapons scientist. And that is in becoming a weapons scientist one learns to draw a sharp distinction between science and politics. In becoming a weapons scientist you learn to shift your attention to the technocratic agenda, "what works best," as you describe it. It seems to me that shifting your attention to that agenda presupposes accepting a world in which there is a very sharp distinction between scientific and political activities. Drawing that distinction is one of the reasons that there is confusion among some weapons scientists about those people outside the gates saying, "I share politics in traditional terms with some of those people outside the gates and yet they should be outside the gates and they shouldn't be in here and I should be here and not out there." I think a crucial ideological difference, in your terms, between the weapon scientists inside and those outside has to do with accepting that distinction as a presupposition of everyday activity.

GUSTERSON: These people [weapons scientists] had become so isolated by the time I arrived. I mean there was a time in the early eighties when they regularly attended parties in Berkeley and San Francisco, but then came huge demonstrations—1500 people were arrested in one day—and they started to think twice about accepting invitations to go outside Livermore. So I was sort of forcing them to still engage in translation with the outside world. But if her process of professionalization was one of sort of progressively closing these windows and pulling the curtains and being just in this world, my process of professionalization was one that involved systematically putting my own beliefs in jeopardy in a way that she started off by doing but progressively ceased to do more and more. So when you asked about a reflexive sort of climax at the end, for me it ended up with a situation—I remember last year when I was living in New Mexico, I remember being told by an antinuclear activist in New Mexico, that the guys at Livermore designed this weapon, they put it down the hall, it was all ready to go, and Bush canceled the test. Simultaneously I felt two reactions. One was "Hurray, we got the bastards," and the other was "I wonder if that was someone I know, who had been working for two years on this." And that was sort of a different place in which we ended. I don't see myself as reifying this dichotomy between science and politics at all. I see myself as putting it in jeopardy. And I see myself as undermining all kinds of perceived notions of politics in the journey that I went through and I am sort of writing to former comrades in the antinuclear movement, putting their politics in jeopardy in a way that is not done very much.

On "Bubble Boys"

MARY-JO DELVECCHIO GOOD in discussion of Livia Polányi's piece:

You pose the bubble boy's big science, bomb science against what I would call biopower, which is what I think your father and uncle were involved in even if they were inventors. Because if you are talking about angioplasty and things of that sort that they are engaged in, I mean you almost belittle that, and yet, do you really want to? Big science has become in some ways passé science and the bioscience and the biopower that your uncle and your father are engaged in is in many ways big science. It is one of the larger domains in which we invest our GNP. And so that kind of tension, I would like to see if you are going to do anything with this at all.

DOWNEY: One thing that really struck me was a connection in both your [Livia's] piece and Paul's that presented industry as a refuge. And again it calls to mind for me the distinction between science and politics that permeates the relationship between science and the academy. The academy presupposes the same distinction between science and politics such that we do science and not politics but what that really means is an intense political component. We have to have neutral politics, a politics of neutrality. This *is* a political position. If one then becomes perceived as moving to one side or another and becoming, hence, explicitly political, then that is inappropriate and grounds for removal or not hiring. In this era and this case, you are gone. And people flee to industry where that sharp distinction has already been blurred and hence people feel a greater sense of freedom in an industrial environment, which your piece describes beautifully.

POLÁNYI: I think there is a line in my piece which is almost identical to your point—they'll [industry] let us do what we want. It was so interesting to get that.

RABINOW: I smiled all the way through that too. We want to talk more about that, absolutely.

DOWNEY: That is a different story about capitalism.

RABINOW: It is a different story, and more complicated and interesting. The bubble boy stuff struck me as well. I would just be a little careful about not essentializing this into some universal gender thing, which I don't believe for a second. It is not only that we are moving into an era, particularly in the social sciences, where not only more women and people of color are occupying higher positions but there are a massive number of highly visible bubble girls and people of color positions. I mean, Cornel West's frequent flier miles are probably as high as someone at the DOE. So there is no gender or race thing inherent in that.

GOOD: This is not how I read the way this bubble boy image is created. It is a sense of commitment and passion in one's work that entirely consumes you so that you don't go home from the lab. It is not only that you make it big because obviously you have other ways of making it big or becoming even more powerful. When people talk about whether or not, at least in the work that I do, how much they are going to invest in the lab, they say, "I haven't been able to stay until midnight to keep an eye on my experiment. I don't want to live that way. I'm leaving the lab." It is sort of an overwhelming consumption of the self. Now people who are really successful can manage the time and go home from the lab. And those are the people who often rise to the top and become powerful. Being a bubble boy doesn't necessarily mean becoming a powerful person. I think it's more a sense of whether or not your life is internal to the lab, like Sylvia's, maybe.

RABINOW: Well, I think the biggest surprise for me was the place of industry. Because industry is this sort of black box, as it were, in the academy. Just sort of inherent corruption. And increasingly the corruption narrative, to the extent I am going to tell one, is mainly from the academy out. So industry then becomes a kind of more transparent domain in which the motivations for doing science are more transparent than in the academy, among other things. And then also where the hierarchy relations and gender relations are clearer than in the academy.

DOWNEY: It seems to me the distinctive contribution that comes from an anthropology of science is not only to document the multiplicity of visions across what are called the sciences, but also to examine the power relations between them. Each positions itself by means of these relations. It seems to me what is critical in lots of these papers is looking at relations between institutionalized technoscience, and a variety of other activities. So you look at the relations between academia and industry, or between basic science and so-called applied science, or between, in this case, Western science and Soviet science, or the bubble boys and others. The piece on medical science [Del Vecchio Good's] I think is crucial in that respect because the person involved is not a bubble boy. Most work in science studies and technology studies ignores such relations and misses their importance.

SHARON TRAWEEK: I was just struck by three of the papers, Diana's, Livia's, and Michael's, dealing with their parents as scientists, dealing with this military industrial complex sort of thing, and life in the family coming through, reverberating between the lines quite powerfully, and the various moral choices that people made about what kind of positioning they took with respect to those horrible dilemmas. I think this very much reverberates with other papers here around what Gary calls the industryuniversitygovernment world. But I think this personal dimension, with children of these people becoming intellectuals who work very much on uncovering secrets, is a very powerful motif here. And what Michael mentioned, the reverberation of all the people saying, "Oh, yes, well, I have a lover," or "I have a family member . . . ". I have certainly encountered lots of people, children of these people, lovers of these people, who subsequently become academics who uncover secrets.

A Decolonized Nation and Its Recolonized Science

DOWNEY in discussion of Kathryn Milun's piece:

This is one of very few pieces that go beyond Western technoscience to look at relations between it and others. In this case it is a characterization of two different visions that coexist simultaneously in Lithuania, a Soviet vision of science and then an attempt to replace the Soviet vision of science by explicitly adopting an institutionalized Western vision of science. And so we end up with a hybrid space, a hybrid nation. And you give a fair amount of documentation about how this is a hybrid both in the orthodox political sense, because the new president is a former communist, but also in the scientific sense. And that is what I want to focus on because I think it is a central contribution.

I think that the story about peer review is a crucial part of this story. I think that Lithuanians say, "We have this problem with peer review in the Soviet system since we are all here in one institute, we are all localized. How can we have genuine scientific, that is to say, scientific but not political, peer review? And a way to do it is to bring in lots of Western social scientists into our institutes." The irony there is really important. They are getting rid of the Soviets, right, but in the process of getting rid of the Soviets, they are bringing in a lot of German, American, and French scientists to populate their institutes in a proper way to introduce a neutral politics. And the presence of the Swedes is also important. "What do we do about this nuclear power plant? Call the Swedes. Get the Swedes in here to help us." So the implicit colonialization that is taking place through the transference of an imaginary, and the irony that it produces I think are important to bring up.

MICHAEL M. J. FISCHER: Well, there might be an important role for some comparative footnotes. For instance, the peer review stuff comes up in Leszek's account, too, in the Polish case. It is close but is slightly different because it operates through the international reckoning, through what journals you publish in. So there is already a global hierarchy that is working. In the case of the nuclear power plants, it is not just the Swedes, all the Scandinavian countries and Western Europe have relations with particular aspects of these nuclear power plants. So it is not just the Lithuanians saying we have got to get the Swedes to do this. It is a larger set of people.

DOWNEY: Yes, okay, a more global hierarchy is also working here on the Lithuanians, but still they have to figure out how to position themselves. Lithuania has become the smallest country in the world with a nuclear power plant. "We're stuck because the Soviets imposed it. Now what do we do?" Elaborating the relations between Soviet and Western visions is an important way of describing the ambiguities they now find themselves in, stuck between a vision they want to exorcise and another one that introduces new problems. They have no clear alternative. Lithuania has to deal with other countries within its borders. The way of achieving active national liberation from the Soviets is to become thoroughly embroiled with other nations.

FISCHER: Yeah, for instance there is the Armenian comparison, too. They could make the choice to shut this thing down cold. Armenia did, and they are suffering terribly.

I thought the really interesting sentence in the previous discussion [of peer review] was that statistic, about twenty percent of British peer review, or whatever it was, was based on non-Brits. I don't know where that comes from, but if you could track that down it would make much more interesting the relationship between national liberation, the attempt to set up something autonomous as a romantic, utopian ideology that can't exist. The other kinds of strategies to set up something that would be a national scientific establishment, but that would, of course, have relations with the outside world, and how do you protect the inside from the outside. How do you negotiate that sort of thing?

RABINOW: Actually, I think that it doesn't have to do with the past entirely. A student at Berkeley is writing a really interesting thesis on the Ukraine, built around Chernobyl and the establishment of exactly these norms about power and the legitimation of the state via health and science, and the restructuring of the entire territory of the nation and the discourse of the nation around, not only environmental cleanup per se, but the sense that what the state is there to do is to establish norms of health. Therefore you have to start studying people, you have to start bringing in the equipment to set up a normative distribution, ranges of things like that which is utterly connected to communication technologies and new spatial organizations. So it may be that something happened in the past but what it is going to mean is going to be very future oriented. The state certainly didn't have any legitimacy in the past, and not much representational imagery. So the project of constructing that now passes very much through science and medicine.

"The word is leverage"

DOWNEY in response to a discussion of his piece:

I see these people working very hard to blur the boundaries that separate them, motivated by objectives that transcend their self-interests and their organizational interests. But yet that motivating force is national chauvinism. The whole goal here is to beat the Japanese.

I like the idea of blurring these boundaries among government, industry, and universities in the pursuit of global interest. What I try to do in individual conversations, some of my interviews, and one invited talk is to provoke them beyond national chauvinism into a more global sensibility. But while sitting there, I keep wondering, I mean constantly wondering whether or not I am seeing something that has real promise.

My own goal is somehow to force multinational corporations, multinational capitalism, to accept into the corporate person a sense of citizenship. I picture the world today as consisting of a population of nations, a layer of nations, overlain by a population of multinational corporations. The corporations basically live in the world of nineteenth century America, a world of laissez-faire capitalism. There is no collective regulation of those multinational enterprises. They get to seek the maximization of self-interest with the confidence that if everyone maximizes self-interest it will work to the benefit of the whole. Well, the experiences of the late nineteenth century suggest otherwise. What I picture is the development of global robber barons through the progressive concentration of wealth in fewer and fewer hands. How can we get past that?

I think it is naive politically to concentrate on imagining a world without capitalism. One pathway might be to set up the global analogues of the Sherman Antitrust Act and federal regulation, but such a strategy presupposes a different system of world government than the intensely nationalistic United Nations. It is worth the effort, but good luck. Another alternative might be to find ways of transforming the concept of the corporate person by adding a sense of citizenship. In such a world, every corporate actor would take it for granted that it bore direct responsibilities for global welfare, leaving it unable to appeal passively to hidden-hand mechanisms.

In the ACSYNT Institute, I see a variety of people motivated in part by a sense of citizenship. But for most of them, this citizenship is intensely nationalistic in content. So I wonder, am I participating in something that actually is leading toward the goal I would like to see? Or am I watching the military-industrial-university complex at work and finally realized as a monolithic whole? The level of collaboration that began to take place among government, industry, and universities during the 1980s is both striking and frightening, far exceeding what took place in the sixties and seventies, yet with little public debate. I mean, these people are working together more tightly than ever before. If we blur these boundaries, we might end up with one huge dominating organization. So what is going on? Which one is it?

LAUGHLIN: Do you think there is less public debate now? Or just more that should be debated?

DOWNEY: There is less public debate on that issue.

GOOD: The public has the same goals, doesn't it? Of defeating the Japanese or competing with them. The public has the same goals as the company.

DOWNEY: Oh, yes. The people I have been working with just take it for granted that this is legitimate activity. They go home feeling very good that they are each stepping beyond a limited perspective, adopting a more global perspective that stops at our national boundaries.

LAUGHLIN: But there are fractures in that. Because the labor movement, some environmental politics, is really questioning this agenda of a free-trade imperative and things. And to what extent do these people acknowledge that? For example, would they differentiate something like corporations that behave worse or better? Or is the corporate structure seen as a monolithic structure that serves our end and it operates the same in every case?

DOWNEY: The vision in the ACSYNT Institute is not the same as the free-trade imperative. The free-trade imperative seeks the nation's support for corporate objectives by means of governmental action, extending the maximization argument that what's good for IBM is good for the country. The vision in the ACSYNT Institute is reframing the nation itself by blurring the boundaries among constituent institutions: government, industry, and universities. That is, together they pursue the national interest directly rather than conceptualizing it as the collective outcome of everyone pursuing self-interests. Although this reframing has indeed involved a shift from political to economic terms, the strategy holds the potential of also reformulating the nation as a citizen of the world. The free-trade imperative offers no such hope.

Participants in the ACSYNT Institute do indeed distinguish among corporations. I was very surprised to learn that some of the older participants had made career commitments to alternative energy during the 1970s, until the Reagan election, while one remains strongly committed to sustainable agriculture today. So I believe by participating in the ACSYNT Institute they are indeed trying to remake their corporations in a way that improves them. Without giving up the goal of profitability, they are thinking in terms of citizenship.

HILL: Its surprising to what extent the anti-Japanese thing will go. Because from what I was hearing my father talking about Los Alamos and them trying to envision their future, they talk very much about cooperating with the Soviet scientists because the Soviet scientists are very cheap, a fraction of the cost of American scientists. And it's one way for the government to stay competitive, because right now they can't be. Government scientists are too expensive for industry to use. But if they teamed up with the Soviets and sort of took advantage of that, then that would sort of put them back in the—

DOWNEY: That is exactly the sort of argument that is used here. The word is leverage. You leverage your funds. In the ACSYNT Institute, each firm invests thirty thousand dollars a year, which is a pittance. And so the engineers interested in the institute can go off and get approval for thirty thousand dollars to enter into a relationship that upper management might reject as too radical if it were on a larger scale. But because it's only thirty thousand dollars a year, they let it happen. The parent organization of the ACSYNT Institute, American Technology Initiative, lures new members by telling them they can leverage their funds and get benefits that far exceed the investment. So conceptually, at least, it's a situation where everybody wins.

RABINOW: Have you read Robert Reich's book *The Work of Nations?* Reich's basic argument is that there was only a short period of time in which capitalism was beneficial to nations. We are no longer in that period. And that therefore the question of social justice has got to be raised in a different form because multinationals have no national social justice questions. Therefore, a short ending to the book, what do we do? He doesn't have much to say.

DOWNEY: And that justifies an activist government agenda which is what the Clinton administration is all about.

RABINOW: Right, potentially. But your arena is one in which its not pure capitalism, its not quite the same.

DOWNEY: Right. There was a transitional period in the 1980s where it became appropriate for governmental agencies to participate in collaborations as equals, but not to steer them. Not to be an activist selecting projects.

RABINOW: Well, how transitional it is, I think, is a question. All the university patent policies or whatever are leading in that direction, too.

DOWNEY: Sorry, okay, yes. Transitional in the attitudes of government.

RABINOW: But this may be a strong component of governmental attitudes for a while. So maybe a threshold into what we are going to be seeing for a while.

"Biosociality"

RABINOW in discussion of Mary Jo Good's piece:

There are three elements that I thought were really quite fascinating. First this biopower, or what I am calling biosociality, element. As some of you know, my rap on this is that we are moving from a point where a lot of talk about molecular biology and genetics is ideology or culture. Sociobiology has nothing to do with truth in the biosciences. They are never going to find the gene for altruism, there is no gene for altruism, and so on. So we are moving out of that to what I am calling biosociality where a whole form of identity, both individual and group identity, and a vast array of cultural, political, social, therapeutic institutions, practices of all sorts, are emerging very rapidly around truths. So I am interested in that Foucauldian sense of seeing how truths have emerged, been produced, and then circulated. And your paper seems to me to be one of the clearest examples that I have ever seen of this hypothesis of mine that such things are going to go on. AIDS is one of the major initiators of this, but clearly breast cancer and Alzheimer's are going to be, and then many other smaller diseases are going to be the next set of arenas. So breast cancer, your data there are extraordinary, because in fact what is being produced are truths. Your doctor and thousands of other people like her and millions of women who are engaged in this trauma and in these treatments are very concerned with the truth and there are truths to be produced, and to be verified, and to be experimented with. So you have a whole range of identity productions around truth on the one hand, and then you have a very moving and complex experimental situation, in which you have at least initially in the AIDS epidemic, and this will be significant and true in breast cancer and in Alzheimer's as well, an arena in a relatively privileged group of patients who are going to be extremely knowledgeable about what has happened to them, but who are also then willing to subject themselves to experiments. This is basically what is going on here, in which we've seen a really voluntary self-experimentation in which, historically speaking, there is a massive transformation of a justification of experimental protocols and the right to experiment. Just briefly, in the eighteenth century you could experiment on animals and on prisoners. You experimented on prisoners because they were outside and had forfeited their right to be part of humanity and you return them to humanity via that route. Two hundred years later we are taking upper-middle-class people, in this case women and in the AIDS epidemic largely men, who are subjecting themselves freely, in the name of autonomy, which is both an older value, but really, basically, a 1980s presidential commission on ethics is where it was really articulated, and in which they are creating a whole range of new subject positions, to use that language, but in which the field of truth and power is not really one of domination at all but of subjectification of various sorts, and then of this biosociality where the creation of identities is forming around these truths.

And that leads to a smaller point and then a third point on this. One, the multidiscipline areas of the science involved are extremely interesting. Watching this multiplicity function I think is going to be very interesting because it is politically and socially high on the agenda but against the grain of medical education and of molecular biological approaches. In comparison, the sort of science and society debates around the genome project in particular, to my mind are very abstract and without interest. The really fascinating place in which new scientific collaborations are being articulated are in situations like the one you are working at where the new knowledges are being situated, new subjects are being created, and power relations then are being articulated there that will then spread throughout other arenas of the class structure, of the racial situation, of what have you. So I think you picked a really interesting laboratory situation in that multiple sense of laboratory. I have always been more interested in self-colonization than in colonization of others.

Then, finally, there is this question of autonomy and the role of experts. Because in this autonomy culture, both in bioethics but in the structure of hospitals now, a patient is expected to be, particularly upper-middle-class patients are expected to be as knowledgeable as the physician, and in many cases in the AIDS epidemic, more knowledgeable than the physician. You have a kind of complicated, interesting end point of the autonomy movement, because if the physicians are themselves enlightened, Cambridge area, New York-correct types, they will abdicate as much responsibility as possible in the name of patient autonomy, and that then leaves them in—and your doctor brings this up a number of times—in a situation in which, what is the role then of the physician? Is it some role of basically subtechnical specialists in a clinic which has many technical specialists and in which the decision-making power and the authority to proceed in this biosocial experiment then is left to individual patients? It is a very strange situation. So I think we are going to see a lot of negotiation of the role of authority of physicians which we are likely to continue to see as the Clinton and other managed plans go through. I think it is going to raise for us the question of how to critique the excesses of medical power and then wonder if we are not seeing here an instance in which we might begin to call for more medical authority in the way that this subjectification goes on, that is to say, in the way that these decisions are being negotiated. They strike me as maybe having tipped awfully far in one direction. So there is sort of a false autonomy involved as well.

But then this raises the question of where these physicians learn these emplotment strategies? In other words, and I am not necessarily criticizing them, although I myself am fairly allergic to the sort of soft core therapeutic style—

GOOD: It's not soft! When they are working with someone, they are quite directive.

RABINOW: Well then it's different because a friend of mine is just going through this and she was offered so many different options without anybody coming down anywhere.

GOOD: That is when you have no narrative. That is when you have fragmentation, the fragmentation of the narrator, at least in the way I am looking at it. You know, where a physician cannot do proper therapeutic emplotment. It becomes a treatment narrative that is fragmented, so there is no directionality to it. And that is true of so many people who have had breast cancer experiences from our patient interviews, there is just fragmentation of care all over the place. You don't see coordination. These doctors are talking about trying to pull together a unity that will lead to coordinated care, but the question is whether the patients feeding in and out of these units will in fact experience this integrative, coordinated treatment.

"Relatedness, intimacy, transparency, kinship, love, and sex are part of the story"

TRAWEEK in discussion of pieces by Rabinow and Tom White, and by Leszek Koczanowicz:

I am really intrigued, and not just for these two papers but for lots of papers and of course for my own work, too, I am really concerned with two scenes which on first gloss may not seem related. One is this emergence of this new milieu, new forms of social organization around innovation. Everything from Gary's paper and Joe's industries, and universities and government rearranging this and reconfiguring that relationship, whether because, in the case of Leszek and Andrzej that the country's infrastructure is falling apart, or in the case of I think with Tom's work, new social forms, just like Gary was talking about. So I want to come back to that. I think that is actually an issue in several of the papers—Michael's, Kathryn's, mine as well.

On the other side, this theme of relatedness. Whether it is fathers and daughters, mothers and sons, or maybe some fictive kinships. I am really struck with Hugh's paper that the name that he gave to his informant was the name of his advisor.

GUSTERSON: Ah, you spotted it, who is Japanese-American.

TRAWEEK: Of course. So this sort of dopplegänger of Sylvia the informant. I read it twice just to see it. There are all of these kinships and fictive kinships and intimacies and transparencies we were talking about, and the friendship and trust, and also especially I think in the building of community among—and I will come back to this, the friendship, the trust, the reciprocity issue, plus the building of community, in order to shape a new milieu of integration. I'm really intrigued by all this. I want to go back to Leszek's conversation with Andrzej for a second. I think that it is really striking that in this turbulent milieu in Poland so many of the old cliches of what it means to be an intellectual, what it means to be a scientist, and what kinds of institutions we work in are either being enacted or there are serious efforts toward reconstructing them.

The voice of authority Andrzej uses, the construction of his own genius, the control of his own biography and autobiography, the auto-nonautonomy, the autobiography, are really quite striking. And I would just contrast that story of genius that he tells of himself, which is what Michael talks about in the construction of genius in traditional biographies, with the story that Tom and Paul present, a story of discovery that is utterly unlike the story of isolation and autonomy and genius and self-invention, and properly so. There is that real tone of "I am proper" in Andrzej's story, self-story, "this is the proper way to do it. These other people may gain attention they may even do good work, who knows, it might be possible." But that sense of the real way to do it is that way that he thinks he is doing it. Also his invoking of powerful networks. Of course, the powerful networks are elsewhere, but nonetheless, by simply getting referred to in somebody's talk at the center of the action, he gets a Ph.D., by letter of recommendation. These powerful networks create his career and affect him. He eventually becomes quite obviously—we don't know how he ends up sitting in the United States during the interview but you gradually begin to learn that this interview is situated not in Poland, that his adroit manipulation of these powerful networks has been astute, that there is no question about it. I will never forget, one of the guys he talked to was saying, "of course the bright ones will come to us. It doesn't matter that we are in a rich country and they are in a poor country, the smart ones will come here." And also that some people you can't transplant because some people are the wallflowers, but some people are the real geniuses, really strong thinkers. And he is obviously somebody who circulates. Also invoking his intellectual antecedents through his family. It is really interesting for me how he constructs his genealogy, because he is asked a question about class and he comes up with a story about smiths. And the idea of a smith being at the origin of a crafty intelligence is of course an old one in all kinds of anthropologies and folklores. That he wants to dismiss his other class associations and also try to play with this funny classness that this smith must have been rich because he educated his sons and so forth. He is very playful in the sense of remaking his genealogy away from a class story and into a story that provides the raw materials for his own genius. He wants to cite the transcripts of his relatives to point out that they had good grades. I think it is really quite striking how familiar this story is in a site where we don't expect it to be. And particularly what is absent are all these alternative institutions that are these kind of social forms that Tom has been building or that Gary is talking about and various other people here are looking at. They are not interested in those kinds of social forms to build up recommendation. They are trying to reconstruct a very authoritative center that a lot of the rest of you, and for me studying the guys in Japan, are trying to say, "What is next? Where do we go from here rather than reproducing these really very traditional social institutions?"

I just want to go right away into talking about Tom and Paul's piece, which I think is a really interesting discussion of permeation and collaboration in various contexts. My first version of what I was going to do in my own work that I wrote as a proposal to graduate school was that I was going to study what I called collective creativity. And a lot of the guys in the lab started saying, "Sharon, you seem to know what creativity is, could you tell us?" The only thing that I could tell them was that they couldn't work alone in order to do the job. That was clear. We all agreed on that, that they had to work together. And there was this paradox that nobody gets educated to work together. I think that that is really brought out very nicely in this paper. How you get from getting trained alone to working together and what are the variety of social forms that are available to scientists for working together I think is intriguing. Especially the geopolitics, the geopolitical economy of the generation in this story, the kinds of itineraries through the geopolitical economy is really quite interesting. And there is just one line in there situated someplace that I thought was really intriguing, and I think it is yours, Tom: "Well, there is physics and then there is biology." And it suggested to me in the context in which it appears that there is a story about what kinds of forms you can work in and what kinds of careers you can have in biology. I have heard informally people talk this way. About seventy-five percent of the postdocs in the field I study [high-energy physics] leave the field at the end of the postdoc. A lot them migrate toward biology, a lot of them go into industry. And they invariably say, "I want to get out of this highly politicized environment and get to someplace I can really do some science. I am going to go into industry." I think that this is a very different kind of discourse than the elder generation uses. I haven't heard anybody over, there is some funny line here right around fifty years old, speak in this way. You don't hear this from the older people, about going to work for Du Pont. But I think that the lineage you have, Tom, through your father is really an interesting counternarrative to the ones that senior academics have.

I want to posit the notion of these weird careers versus the kind of career that Andrzej has in Poland in adroitly manipulating and receiving the benediction of these powerful networks. I want to offer a third possibility. Pnina Abir-Am in a very interesting piece has written about the beginning of molecular biology in terms of a group of Jewish immigrants in London who were thrown together partly through immigration and partly through English anti-Semitism, that nobody else in their labs was really friendly to them or talked to them and so they formed seminars in which they had really intimate intellectual discussions. In Nina's historical work that is really a very interesting moment in the history of molecular biology. I was hearing a lot of reverberation with Nina's story with the story in your paper of these exciting clusters of people being brought together. That of course goes back to the notion of industry as a refuge, and refugees, and who is a refugee from what. I just said that the elder generation doesn't participate in this story of going to industry as an apolitical place, but there is this elder generation, two generations higher, refugee story. The circulation that I referred to in Leszek's paper, I think there is a really interesting story of circulation not just in this paper but in several of the papers. It is interesting the similar sites in which California, Washington, New Mexico, Wisconsin, Massachusetts, and what these sites mean for what kinds of work are possible, what kind of social organization is possible. I think New Mexico is often in the scientific community a code for "if you are going to live in that place you do military science." It is very hard to do anything else in New Mexico. The very poignant story Diana told about her father being unemployed until he was willing to get a security clearance I think is just a sign of what it means to be a scientist in that geography. And people saying, "I don't want to stay in Wisconsin." Anyway, these circulating intellectuals of where you can go, where you can come back once you come into this country, back to this country from being abroad, where you can make these new social organizations.

So I just say that I think there is an interesting line here from political consciousness to community to curiosity to ethical questions that all comes finally together at the end of this paper. Because I think the ethical dilemmas inside these different social organizations are part of what is being referred to all along. I think in a lot of the other papers, too. I mean the ethical dilemmas within the work in the laboratory. Making up data, forgetting data, destroying each other's equipment, destroying careers, cutting tables; that is back to the kinds of competitions, the kind of careers that people are making, what are the possible careers in these different settings and what is the behavior that is out of bounds in these different settings. What is out of bounds and what's just barely over the line in a university context is different from what is just over the line or way out of bounds in these collaborative communities like Tom and his friends are building.

Everything that I have said has been posited in terms of the science and scientific organization. In a way that I think comes up at the end of Tom and Paul's paper more strikingly is the reciprocity issues. But I think the mutual construction of the careers, because now Paul's career is very much intertwined with attending to Tom's career. Also, what kinds of social organization do we have for producing knowledge in anthropology? The kind of stuff that we are engaging in here, I don't know if Tom's aware of this, how rare this kind of thing is where we get to look at early drafts of people's work and roll up our sleeves and say, "Oh, this is really interesting. And what if we could configure it this way? What direction would you take your paper and what direction would I take my paper?" Not replicating but going—I see in here that kind of conversation in various settings among the scientists I study but very rarely among anthropologists. Also, in fieldwork it is relatively rare, rarely acknowledged at least, that this kind of collaboration is going on. So I think there are three kinds of collaboration here, and three kinds of social inventions of new milieu for innovation: in the laboratory, in anthropology, and in fieldwork interaction. And again that is back to friendship and relatedness, and it's also back to industry, the government, university, global political economy. I think its really intriguing that is relatedness, intimacy, transparency, friendship, love, and sex, are part of the story of global political economy, industry.

TOM WHITE: Let me just say that, of course, in starting out with this I wasn't at all interested in appearing to be the hero of this thing. I was interested in the process by which the discovery was made because it contrasted with the stereotypical image of the solitary inventor working alone. Which I think in fact had little to do with what actually happened. But actually the best contrasting story, in some sense, is the story the inventor himself is now making, which is to recreate the solitary inventor history. And then the person who did some earlier work is now recreating in his own mind the solitary inventor thing, so solitary that he never told anybody. In a sense, the tragedy of the story is if people go on believing that this supports innovative work when in fact it is the very process that crushes it, you are not anywhere.

JOE DUMIT: Peter Galison's story about how the great discoverer in physics has shifted from the one who does the experiment in some form or another to the manager, the one who runs the experiment. He still gets called the discoverer, but in this odd management notion of "vision." They saw how to direct the team that would do it. Maybe one thing would be to try, as Sharon is hinting at, to play at this different kind of story. And also play with this sort of fear of reproducing Tom as the great manager with this amazing insight vision. And also toy maybe with this sort of ecological notion that the various forces that go into this aren't going to be disentangled by any one sort of obligatory passage point somewhere along the line.

"The conditions of innovation at a particular historical conjuncture"

RABINOW in discussion of Sandy Stone's piece:

This is yet again another story of innovation and a sort of incandescent moment of creativity. I mean, you say in this short period of time at Ashibe basically everything that Silicon Valley was going to do took place there. We could obviously have some thicker contextualization about how that came about and about why it was that this perfectly ordinary organization produced this. It may well be mainly about the state of play of the software and machines at that point but I think we would like to know more about that because otherwise it is a little bit like the journalism, "Another miracle in the Silicon Valley. All these nerds got together and created virtual reality and every one. . . . " You know, "genius, genius, genius" sort of stuff, which I know that's not your intent. But it could be read that way. So there is that. The conditions of innovation at a particular historical conjuncture. This is another theme that comes up in a variety of different places, certainly in Mary Jo's work and in mine and in yours, as opposed to some of the physics moments which are past—that quick coalescence of creation after which then people work on the details for a long time.

Then you raise a concern at the end which dovetails with a number of other people's interests, and maybe Gary's in particular, of here's a failed instance where a potentially, if not liberatory, at least a productive imagined set of potentials doesn't seem to have happened. Both in terms of kids then going out and playing sexist war games rather than another range of possibilities, and the people themselves building these things being a bunch of sexist warmongers when they could well have been something else. You lay out that decline sort of graphically, but explaining it more and explaining what alternatives other players elsewhere might be trying, again setting it in a work environment and a product that was better than this one, might be a good counternarrative that would cheer us up a little bit.

A LOOK BACKWARD: *PERILOUS STATES* REVISITED

JAMES HOLSTON

16

INSURGENT URBANISM: INTERACTIVE ARCHITECTURE AND A DIALOGUE WITH CRAIG HODGETTS

Shortly after the Los Angeles uprising of April 1992, I began a series of discussions with architect Craig Hodgetts in which I proposed to consider what kind of urbanism could emerge from such devastation. As an anthropologist, I wanted to explore possibilities for a new kind of architectural production of the city. I suggested that the future of architecture is not architectural at all, in the autonomous sense of the word. Rather, I proposed that architecture and planning had to become interactive with other disciplines (like anthropology) and other technologies (like desktop publishing) which could innovate new modes of communication, narrative, and modularity. I had been thinking about this proposal for a long time, but the L.A. rebellion put it into a different perspective: I had never been in a city burning before. My own neighborhood around La Brea Avenue, from Pico to Sunset Boulevards, was firebombed, and my experience of the uprising as catastrophe explains the grim mood with which I initiated the dialogue.

I saw the violence as a unilateral declaration, cynical on all sides, about the mythologies of American society and about the powers of the modern to change it for the better. As I watched the televised torching of Samy's Camera Store on the corner of my street, I was not mobilized by that access to the simultaneity of images, voices, and scenes from the many worlds of Los Angeles to imagine the construction of a better society, in the sense of an alternative future, emerging from the montage of film clips. Rather, I saw such figuration as emblematic of the exhaustion of that very hope of modernity and its technologies to reinvent the social. I saw them, in other words, as indicative of the collapse of one of the central paradigms of American modernity: the idea that it is possible to plan society, manage the social, and build an alternative future on the basis of planning initiatives which embody that future as an imagined construct. By the latter I mean embody as an absent cause, present nowhere in the world but only in plans and their technologies which are supposed to colonize the old through their internal force and create the new in relation to which they then appear as natural offspring. In short, I refer to the modernist idea of inverted development (see Holston 1989:31–144) in which an exemplar, enclave, beachhead, fragment, or blueprint of radiating change creates the new on the basis of the values that motivate its design.

For several centuries, in both the Americas and Europe, this imaginary of planning focused especially on the city and the problems of mass urban society. It stimulated the great public projects of nation- and state-building which defined modern experience: projects, for example, of urban development, national citizenship, standard health codes, mass education, and home ownership for the working classes. These projects were aimed especially at the urban masses and attempted to forge new kinds of cultural forms, legitimating historical narratives, collective associations, and even personal habits. Through such state-building initiatives, modernizing political authorities imagined and legitimated new kinds of public spheres, with new subjects and subjectivities for them. With unprecedented publicity, the L.A. uprising exposed the exhaustion of this ideal of modernity in American society, not by televising the fires as such but by showing two sorts of social and spatial preconditions: on the one hand, the growing balkanization of city life into fortified, privatized, high security/high risk enclaves in which discrimination produces difference and difference tends to result in discrimination; and, on the other, the hard fact that many Americans exhibit little sympathy for the urban deterioration which results and little interest in reversing the process.

I sent Craig some comments, written in this dire tone, to initiate our discussion. They asked: "What kind of urbanism can we imagine emerging out of this exhaustion? If we reject the urban status quo, what kind of insurgence might it be? Surely, conflagration is one type. Another is *Blade Runner*'s dystopian cybercity of mile-high corporate towers and acid rains. But can we imagine an insurgent urbanism that reorganizes the social and reinvents the modern without relying on either Armageddon or futurology? Can we describe an architecture interactive with other disciplines and technologies which could innovate new modes of communication, narrative, and modularity, and thus become interactive with its users?

"As an anthropologist, I would first want to pursue these questions ethnographically. I would try to identify sites of change in the city in which such reinvention is embedded in social practice, probably untheorized. I would try to use ethnography as reinvention's register, as a means of showing it as a possible alternative to the status quo, taking shape as possibility in the complex social life of the city on the one hand and in the practice of architectural and planning professionals on the other. I would hunt for situations which engage, in practice, the problematic nature of belonging in American society and which embody, as narratives of the city, such problems as the decline of citizenship, the exploitation of democratic institutions, the privatization of public resources, the hardening of class positions, the hostilities of multiculturalism, the deterioration of public space, and the dominance of nonpolitical activities in the public sphere (for example, consumption, leisure-time associations, and sports).

"How would you, as an architect and urban designer, go about it? What would an insurgent architecture and urbanism look like to you? How would you conceive of an interactive architecture and planning which express the contemporary narratives of social life? How would you use them to reinvigorate the public realm of cities? What kinds of design strategies and technologies would you try? Would they include user participation? Do you think that they can be applied to social change, or that such propositions only reiterate a failed modernism? Can you describe an urban planning that is incremental and not totalizing in the modernist sense? How would you rethink architecture's intervention into public spaces governed by suspicion and amnesia, by the priorities of shopping, by the segregating forces of police, zoning, and aggressively narrow property interests?

"To prepare for our dialogue, I propose that we think about three conditions which I suggest structure the production of architectural concepts about the urban landscape, in junction with the economics of regional reorganization: (1) architecture's rejection of the social engagements of the modern movement; that is, the disillusionment with the redemptive power of architecture deriving not only from the perceived failures of utopian modernism but also from the more general dissolution of the idea of the social itself; (2) the inability of the professions of architecture and planning to move beyond that rejection to develop a new social imagination; and (3) the preoccupation in postmodern architecture with aesthetic formalism, technologies of communication, and concepts of virtual reality which dematerialize the social and emphasize commodity images.

"The dissolution of the social, exhaustion of the modern, and dematerialization of the postmodern—I suggest that these conditions dominate contemporary architecture and planning as reflections of their broader eminence in American culture. I propose that we focus our dialogue on their possible counterformulations, that is, that we use them to consider what kind of urbanism might emerge from and stimulate a reorganization of the social, reinvention of the modern, and innovation of technologies of communication, narrative, and modularity."

My proposal also had the objective of bringing into dialogue an anthropologist and an architect who are, from very different perspectives, professionally concerned with the conceptual production of the city but whose professions rarely collaborate. Indeed, as a city dweller, I have always been impressed (and increasingly fed up) with the ability of the architectural professions to exclude the rest of us from the process of designing and planning the urban environment which affects our lives so profoundly. Craig is a well-known innovator of high-tech avant-garde architecture whose practice is based in Los Angeles and who is a professor of architecture at the University of California, San Diego. I invited him to the discussion because his designs seem to me concerned with the conceptual production of the urban landscape, especially inspired by theater and film technologies. By conceptual I do not mean that he conceives more than he builds (an architect's fate generally), but that his designs often incorporate a sense of cityscape and therefore suggest at least some aspects of an urban society which render them meaningful (see figs. 1 and 10).

For me, the dialogue is a standard form of fieldwork. It is essential to ethnography as a means not only of gathering information but also of developing the object of research. That object is always, in significant aspects, a collaborative construction which ethnographers talk into being with their informants. They push and pull it into shape through the ignorance, insistence, and insight of their questions just as their interlocutors do when they respond, perhaps in new ways, to unusual or dogged questioning. This collaboration is creative on both sides, and the dialogic form captures that process. To be sure, not all fieldwork is dialogue or collaborative construction. In that sense, the following version of ours is just one facet of a research project on American urbanism.

I call it a version for several reasons. First, I have edited the transcription of the recorded discussions to make them easier to read. I have added punctuation for the sake of clarity and pace, and I have also fixed up some grammatical infelicities and struck out some redundancies. Occasional clarifications are given in brackets. Second, both of us had the opportunity to prepare. I did so extensively, bringing notes and slides to our two meetings—one in Los Angeles on 30 June and the other in San Diego on 1 July 1992. At times I spoke from those notes, and in editing the transcriptions I occasionally referred to them. I also included as illustrations a selection of slides from the "slide show" which punctuated our discussion. During the dialogue, we used the notes and slides to think through a series of issues in a collaborative process of investigation. It is my hope that the dialogue captures the excitement of this process and suggests that such collaboration is essential not only to an ethnography of urban America but also to the architectural production of an innovative urbanism.

In retrospect, the dialogue seems to me to have a number of important limitations which are worth signaling. In my initial comments to Craig, I proposed that architecture and planning had to become less autonomous conceptually (and in fact less architectural) and more interactive with other disciplines and technologies as a means to innovate new modes of communication, narrative, and modularity. The dialogue explores this proposal, especially during the discussion of slides. It ends by suggesting that new interactive technologies may completely transform the profession of residential architecture, eliminating the role of the architect as presently conceived. But the dialogue does not succeed in applying that suggestion to urban planning. Instead, it often slips from a discussion of urbanism to one of "signage" or "plumage," as Craig calls it.

This slippage reveals some of the limitations inherent in thinking about the city from a postmodern architectural perspective which, having rejected modernist concerns for the social, is largely preoccupied with the building as object and its design on the one hand and with the marketing of images on the other. It seems hard indeed to move beyond that limited conception of architecture. I say limited if only because very few architects ever build buildings in ways which correspond to their obsession with the design of objects, an obsession fostered by the overwhelming dominance of design studios in architectural education and by the mystic of the profession's "star system." If a city is more than a collection of objects, then an architecture which purports to engage it must surely be concerned with more than buildings. Such interaction would require an expansion of the idea of architecture beyond a preoccupation with permanent buildings and their technologies to include public policy and social activism concerned with rights to the city as well as impermanent interventions such as those of performances, exhibitions, and teach-ins. An interactive architecture can only develop from a much broader conception of caring for, teaching about, and occasionally building the built environment.

The slippage from urbanism to signage also reveals problems of context. Almost all of the signs and images we discuss are about commodity consumption, mostly of the upscale sort. There are two types of problems here. First, although there is certainly more to urban life and its signs than consumption, our discussion tends to restrict its semiotics to commercial exchange. In one sense, as early postmodern populists like Robert Venturi and Charles Jencks emphasized, the focus on commodity image and marketing represents an unromanticized attempt to accept the commercial strip as an American vernacular, as the language of local culture. But in another sense, this embrace of Main Street also has the inherent problem of reducing any critique of it to just another recycling of images. Second, the discussion of urban consumption and the market forces behind it is further limited by an unacknowledged omission: in fact, there are virtually no shopping malls in poor neighborhoods in Los Angeles. Black retail business in South Central L.A. has been driven to near extinction not only by discriminatory banking practices but also by competition from just the kind of upscale regional shopping that the dialogue invokes as cityscape—like the Fox Hills Plaza on the ghetto's periphery—that are the focus of many of Craig's architectural examples.

Thus, the idea of "the local" is really little more than the generalization of just one context, that of the prosperous. In this elitism, our dialogue is not different from most discussions in architecture in which urban contextualism and typology are extolled: what contextualism might mean in poor neighborhoods or in center cities as a whole is rarely considered. Changing the context rather than commodifying, preserving, or reproducing it is probably more to the point in these circumstances. But change raises the so-called problem of utopia, which has all but been banished from postmodern architectural discourse. Furthermore, an interactive and contextualizing architecture would have to express what is really paradigmatic about inner cities like L.A.'s downtown: that they condense within a contested terrain the growing and complex conflicts between international circuits of capital and luxury consumption, international low-wage labor migrations, and local disintegrations of older industrial economies. The collision of these forces constitutes the context of city life today. A discussion which misses that collision leaves the future of architecture and planning in this urban terrain largely unexamined, even if it comments on their usual roles as instruments of inner-city gentrification, high-rise development, and land speculation.

Although the dialogue ends with a technological transformation of architecture, it leaves unexplored the uses and meanings of this technology in the contested space of American inner cities. The image of a new kind of architecture evoked in the final pages is powerful indeed. What it might mean for the politics of culture and what consequences it might have on conditions for a more democratic society are left for another discussion.

Dialogue

HOLSTON: Let's begin with Aldo Van Eyk's statement about the difficulty of developing a social imaginary of any convincing sort in modern architecture. He says, "We know nothing of vast multiplicity, we cannot come to terms with it, not as architects or planners or anybody else, and if society has no form, how can architects built its counterform?" (cited in Frampton 1980:276–77). I would reply initially that of course society has form and that most architects, like Hollywood producers, perceive its lowest common denominators and conceive their designs in those terms. But I take the sense of Van Eyk's statement to be that generally neither their understanding nor their design methodology advances beyond that rudimentary and uncritical level. My question then is how indeed can architects build society's counterform?

HODGETTS: That is interesting. You know I have just come back from Rome and in a certain sense I was really surprised by the reality of Rome as opposed to the image of the city that I had in my mind. I am reminded of all the Karl Popperisms of history as received and fabricated by a power elite. The history of the nonelite is perhaps excavated long after by researchers like yourself and others, but the image of urbanity, the image of successful architecture in an urban context, all those preconceptions that we have when we look at our cities, our societies, and so on, are all grounded upon what we received about the nature of the past, its successes, its failures. So there is a tendency to look at it through the fabricated gloss of Piranesi, Turner watercolors, and whatever it is you want to think of, which are recounting a kind of splendid relationship between the architectural and urban environment and the people that inhabit it. Now, you surmise that that was a very benign relationship and fairly successful, when in fact, given the kind of scrutiny we are bringing to the L.A. riot right now, or the Watts riot, it may be that this is an endemic situation, that architecture and urban design have perhaps never really been very successful at confronting all the multiple agendas of the underclass, let us say, those that lack transportation, lack enough coal to cook their dinner, lack enough water to drink, and so on.

HOLSTON: Isn't it the case that today's architectural urbanism also doesn't address the needs of those so-called power elites, or of the middle classes? They've fled the cities, fled the public spaces, fled the very idea of the urban. If the urban means in some way this presentation of an image of the best, or of the improved, or of the social in progressive development, then the feeling I get is that the middle and upper classes have fled the very idea that made that image of Rome so persuasive to so many for centuries, and that is really what I am addressing as well.

HODGETTS: I think that is a similar question and a separate one. Because if you think of the cities that really work that way, they are mostly European. They are mostly cities which came into being in the sort of little window between the industrial revolution and the medieval city. They flourished in a very special technological climate, I suspect, where devices such as the telephone, the automobile, didn't exist and yet one had succeeded in managing the infrastructure. Paris and Frankfurt and London—all of the cities that one thinks of as really fantastic cities to live in were ones where the infrastructure was managed: you had water, you had sewage, you had cancerous air, but the infrastructure was managed. But then you didn't have this further overlay to break down the human equation of the motor car and the telephone. So that the city's paradigms, the things that really controlled it, were well within an anthropometric kind of relationship to man.

HOLSTON: I also think that the important circumstances were political. On the one hand, the notion of burgher democracy, if you will, was linked directly to the constitution of city space—you know, "city air makes free"—the idea that the city and its citizens were precisely those who had overthrown seigniorial and ecclesiastical authorities, so for them their very space was politicized and represented their triumph. And on the other, a little bit later, as the absolute monarchies became liberal constitutional monarchies of one sort or another you had the raise of the capitalist bourgeoisie, of a group again claiming the city as their political space. So there was every reason to be out there in the space which served not only as the metonym but also as the metaphor for that political life and for that access to information, knowledge, excitement, and power.

HODGETTS: Yes, it was the opera house, the conservatory. All those things were jewels in the city, were places for cultural exchange. And all the exchange places in fact were in the city, that is where they were grounded. The monetary exchange, the dissemination of all information, scholarly, cultural, economic, had to do with passing through the city. I think that when we start thinking about a modern city like Los Angeles, we have to begin at the same time to wonder whether as a mechanism it is not a husk, you know, whether we are not asking it to do something in terms of our visceral relationship to it that it needs all these other ingredients in order to perform effectively; in other words, that being denuded of these critical functional mechanisms, those being extracted from the operation of the city, can we expect it then to offer us all these other goodies? I think that this is a real tough question: because the stock exchange is on the end of your telephone.

HOLSTON: But there is tremendous social conflict and social drama out there in the city and those are not disappeared—to use an Argentine kind of expression. What you say of the stock market is true, information has been virtualized in that sense, but society hasn't. That is an impossibility unless our bodies are vaporized. So there is all this social stuff out there. What's happened though is that these dramas are no longer seen as somehow essential to information, to the kind of information that you are talking about, so that many people think that they can just abandon them, forget about them, and leave it to their suburban homeowner associations to build protective walls around them. I don't think we can ask of architecture that it change the world, and agree entirely with you that L.A. cannot be something like Paris of the eighteenth or nineteenth century. But I am not asking that question only. The question is also: Isn't the exhaustion of the social a way of thinking about cities? You seem to be saying yes, it's inevitable, it's there, we can't ask L.A. to be a kind of city that it is not.

HODGETTS: Yes, I guess I am seeing it not as exhaustion but as a kind of replacement. I think that there was a very good set of operating principles for a city like—which had to do to some extent with probably a more or less monolithic cultural apparatus, which had to do with a literal need to exchange information and emotional life with those around you in order to get on with your business or your purpose or your goal. It had to do with a need to really measure very carefully the effects of various changes, both geographically and in terms of density and in terms of population because it was a delicately balanced kind of mechanism and that is why you got really quite intelligent diagrams of a city like Ebenezer Howard's things where it didn't seem ludicrous to say, Ah-hah, in order to make a perfect city all we need to do is have a park within so many minutes of each house, we need to have a centralized space, we need to have rings of alternating agriculture, and so on. Those did not at the moment of their fabrication appear whimsical. They didn't appear ineffectual, they didn't appear silly. If you look at them today and you propose anything like that, an ideal city, which was based on geometric measure or cosmetic overlays or anything like that, you would be laughed out of the room. They are silly.

HOLSTON: Inherently silly?

HODGETTS: Inherently in our culture, in our society, in terms of the ways that transactions are made in our society, and the way that productive work is accomplished, they are silly. They just have nothing to do with the complexity and the number of layers, and in fact the effectiveness of certain replacements of those things. What I think has thrown this question open is that it is clear that there is a new structure desired and no one has any idea of what it is.

HOLSTON: Exactly, that is an important point. I am not advocating by any means a nostalgia—

HODGETTS: I know you are not.

HOLSTON: All the things you just described for Ebenezer Howard, you can say were part of a modernist ideal.

HODGETTS: Easily. But Corbusier's Radiant City is a silly diagram. If you put that diagram out today, it's silly.

HOLSTON: The point is that modernist architecture clearly thought it was embodying a vision of the social, and it was a disaster. You say it was silly; I would say it was a calamity. It showed us that those utopian conceptions, Brasília, Chandigar, modernist L.A., end in perversity, the perversion of many of these very social goals.

HODGETTS: Many of the same values and goals that it was trying to establish.

HOLSTON: And ended up exacerbating, perverting them. So we have experienced enough of that to demand something different. Yet, we also know that without of a utopian factor of some sort we are just prisoners of the status quo, and there does seem to be a great, almost an overwhelming need, perhaps symbolized by those riots for something else, for some new structure as you put it. But what is it? We don't know, we have a very hard time articulating it in form. That is one problem. Another problem is that our society has, as you put it, changed dramatically its way of conducting its business. Yet I would also propose, and I wonder whether you would agree with this, that the way our society does its work culturally, economically, politically in the urban spaces and structures that we have is quite alienating and disjointed and premised on suspicion, fear, fortresslike withdrawals, or withdrawals into fortresslike structures, urban flight, a rejection of the inner city, a kind of simplification, a symbolic simplification on a cognitive map of the inner city: rich and poor, white and black. What I'm driving at is that the experience of the built environment today for most people, if one can talk on those terms, is not illegible as some people claim—that architecture does not mean anything—but rather that it is not interactive. I wonder whether or how you could imagine, if you think that is possible, an alternative architecture?

HODGETTS: What kind of architecture are you talking about?

HOLSTON: I'm talking about, well, what comes to mind?

HODGETTS: If you're talking about commercial architecture, it's very legible. If you're talking about architecture as architects, academic architects, are practicing it, it's highly illegible.

HOLSTON: What do you mean by that?

HODGETTS: The work of anybody on the stellar map of architecture is highly illegible to people, I believe. What is not illegible to them are strip malls, drive-throughs, and so on. As we talk about the inner city, I'm reminded of this weekend when I happened to be down by Marina Del Rey, which is among the most desirable locations in Los Angeles. It is highly desirable, it's near the water, sailing ships, harbors. Immediately inland there is a dense automotive strip, lots of repair shops, drive-ins and so forth, and a huge mall which I wound up at because I wanted to see the movie *Batman*. It's right on Lincoln Boulevard, a huge mall. I was really surprised. I haven't been in this mall for many years. It was probably thirty to forty percent black, almost no Asians, all very clearly middle class, all sublimely intermingling, absolutely no overtone whatsoever of strife, just a completely successful place, where people were just doing their thing, they were shopping, which is of course a great leveler for everybody. I had the same experience out in similar places in the valley, in Montebelo which is primarily Asian, and one thing which strikes me very strongly is that success as measure of harmony is very important, that lack of success—

HOLSTON: Success? You mean personal success? Economic success? Or you mean success of the mall?

HODGETTS: More generic than that, success. In other words, the atmosphere of purpose and achievement, and fulfillment, and the attributes of success, whatever those may be: successful communication on the part of the mall, a place where people want to go and not a place where people must go. It's not a commodity place. It's a place where people are going about fulfilling themselves—and they are not into meat and potatoes. But there is a strata there which has nothing to do with race, but has much to do with class, but nothing to do with race, and I believe it's true even of the sequestered communities you are talking about. I don't think it is appropriate to call them places of white flight. That [African-American] fellow that Peter Ueberroth [co-chair of Rebuild Los Angeles] has just put on his planning commission [to rebuild L.A. after the uprising] lives in Brentwood [a very rich westside enclave].

HOLSTON: I don't know, I don't agree with that. I think that white flight pioneered these enclaves and that the number of blacks in them is very small.

HODGETTS: Well, mainly because they can't afford it, but not because there is any other reason.

HOLSTON: I think that historically this is not true. For example if you look at the by-laws, or deed restrictions—

HODGETTS: I agree, that is true, back in the fifties and sixties that is absolutely true. I'm proposing that it is not true today. I'm proposing that Eddie Murphy is a big star, Arsenio Hall is a big star, there is an endless list of successful people of all races in the society today, and that the class system is far more important, I think, in terms of all of the stratifications we're beginning to see in cities than traditional racial segregation. It's really about—like it or not—the flair that energizes a neighborhood, whether it's a small entrepreneur or a—it has to do with style, I think, at some level. And I think what is different about Los Angeles now as a prototype is that it is probably more precariously dependent upon what might be considered superficialities of style and of—I'm thinking theatricality but that is not the word—of *gesture*.

HOLSTON: Presentation?

HODGETTS: Of presentation, than probably any other city in the world, and probably any other city until this point in time. It is highly susceptible to the manipulations in that strata.

HOLSTON: I've often thought that this is no doubt true, but I think that Angelenos like to overplay this a bit. If you think of Paris, and if you think of the world expositions, the fairs, the playing of images. Think of Manet's paintings. T. J. Clark (1984:60–66) has a wonderful comparison between two paintings by Manet, one done in 1862 [*La Musique aux Tuileries*] and the other in 1867 [*L'Exposition Universelle de 1867*]. The earlier one is of a group of what appears to be high society people, but one is not quite sure, presented on the canvas in an apparent jumble of figures seated in the gardens of the Tuileries. There is no depth perspective in the painting, and the point is that if you don't already know the social code, the canvas is quite illegible. Then he does a painting five years later of the universal exposition of 1867, from the viewpoint of one of the new promenades. It is all about perspectives, long perspectives, and city life as a spectacle revealed by these perspectives. There is the promenade itself, you see out to the horizon through the city with all its monuments; there is a man in a balloon drawing the city from above; there is someone with binoculars looking up at the man in the balloon; and there are people parading on this new stage of perspectives in the latest fashions presumably bought at the new department stores. And all of that is completely legible. That scene, it seems to me, describes Los Angeles as well, and indeed most major cities.

HODGETTS: Very much so.

HOLSTON: And what I meant by the legibility is what you mean, I think. To me, the architecture of a city like L.A. is legible. It's legible by most people who read out of it a whole variety of narratives. And clearly the stellar architects are practicing a kind of illegible architecture, and that is a kind of separate question. But I wonder if you see that legibility as socially static, as kind of brutalized, as not interactive, as not critical, you might say. In other words, the architecture that we have now in this city of presentation, this architecture hasn't quite caught up to the demands of the stage so that the buildings don't give us back very much and people don't know to interact with them.

HODGETTS: There is a real question there about whether it is the building or the content of the building; in other words, whether the configuration of the building itself is meaningful at any level, laden with content or not, or whether it's the displays and the shop fronts. There is a famous restaurant in L.A. called L.A. Maison which parks all of the pure-bred cars of the diners in its parking lot in front of it in a very orderly fashion, and the less well-bred cars are shunted off somewhere else. So the facade of the building is literally the token presence of the various people dining in the building because after all that particular restaurant is all about personality and to some degree the vehicles that they drove reflected the personalities. So the architecture is really a background for this other theater of personages. And I think you see actually on many, many, many of these "hipper" shopping districts of Los Angeles a prevalence of very highly neutralized architecture and very highly idiosyncratic displays, and idiosyncratic signage and idiosyncratic apparatuses to identify the novelty of the particular shop owner. The architecture is the foil, the background foil for this, but this other activity is going on and I suspect that this kind of transparency—that the architecture is not interposed between the user or the consumer and the merchant or between the service and the user—is a trend and a very powerful one and that the orderly context which the architecture provides in the traditional city is being dispensed with. That is not to say that it perhaps shouldn't be a presence. I think there is an interesting debate about whether the stability of the city is manifested in its architectonic form against which these rather minor variations in texture, color, individuality, play.

HOLSTON: I think that describes L.A. architecture very well. Take a street like La Brea, the buildings are clearly just sheds for all this other stuff, for plumage.

HODGETTS: Plumage, exactly.

HOLSTON: For the cars, signs, clothes, displays to occur, and they are very good as that ground for the figures, for the plumes. And it would seem to me that what that signifies is really an expansion of the concept of architecture itself, because the notion of architecture as a structure which imposes a story is really not working there but rather architecture has expanded out to include—

HODGETTS: Well, if it does.

HOLSTON: I think that some of your own fronts have grabbed the sidewalks in that way.

HODGETTS: But there is a tremendous question of ersatz and of this plumage aspect as transitory—and deliciously so, as it should be—

HOLSTON: Trendy?

HODGETTS: Next year there is something else, and it is very, very, very, trendy. In a situation like the City Walk Project, the transitory nature of this kind of spontaneous architecture, or of architectural expression I think is probably fairer to say, is being memorialized. It's being inflated in scale and is being memorialized in far more permanent construction materials than would be possible on Melrose Avenue, let's say. It is amusing to me in a way because it is precisely the transition which the Doric temple went through on its journey from being a wooden column porch to finally a marble temple with nails still evident but obviously fossilized in place, and I think that there is something very like that happening in the sense that there seems to be a constant need to stabilize on an architectural urban level the sort of, you know, the ebb and flow of reality. Just kind of seize it and memorialize it: "Oh, it's good now, stop changing." And of course that is the moment one could say that establishes a city's long-term durability. I guess what I'm saying is that there must have been a point in the history of London or the history of any city when there is an "Ah-hah, half-timbering and little leaded glass windows."

HOLSTON: Was it 23 April 1753?

HODGETTS: Right. But there was a point in time when the citizenry realized that they were on the apogee of this development curve and that they had somehow to congeal it.

HOLSTON: What do you think makes that moment?

HODGETTS: If you look back in time I think that you see that there are—we were just talking about England—there are ways in which one thinks of Christmas and England and Brooks Brothers and a whole set of pieces, cultural pieces: like plaid, burnished brass, dressing, and so on, and the picture is very coherent; it's an incredibly coherent picture, and it's memorialized and celebrated. There is one in Greece.

HOLSTON: What I hear you saying is that what gels is an architecture of content which somehow captures a story, an identity, a narrative which has a coherence—like Christmas time, there is a story in which people see their individual stories played out and reconfirmed—

HODGETTS: And fulfilled, really fulfilled.

HOLSTON: That leads me to the question: what about the new mode of urban narrative? Has it crystallized around consumption? Is there anything else? Is that it? The moments that you have been describing all have to do with consumer activity, and the comparison with the Doric temple is in one way appropriate, but in another like apples and oranges because it refers not to consumerism but rather to religion and foundation mythology and the sense of belonging to some— So what I'm asking whether our only coherent narrative is that of the consumer?

HODGETTS: Probably if you're looking at the U.S. in the 1920s or 1930s, and you're looking at what is a coherent narrative, it would be built around productivity, I think.

HOLSTON: Work?

HODGETTS: It would be built around work. The powerful images were Dorothea Lange on the farms. They were Charles Sheeler's and Diane Arbus's photographs in *Fortune* magazine. They were powerful, powerful images of productivity, whether about the energy of the Hoover Dam, whether it was a factory town with their smoke plumes, all of those were very romantic images.

HOLSTON: How were those translated into architecture?

HODGETTS: In every single place, every single place possible! That's the factory aesthetic, the factory town. You could go from one end of the country to the other collecting postcards and you would have an album that was very, very coherent.

HOLSTON: Even freeways, commuting to work?

HODGETTS: All of these things, and it was highly coherent, and it really gelled, and I think it was a driving force right through the Second World War, right into the fifties. Whether there was a segregation of work and play, they were obviously serious activities, and they were very clearly sorted on the map and those activities were actually kind of figure-ground relationships of those things. You had things like golf courses and recreational facilities and things like that being overt expressions of the culture and of the city. Suddenly golf courses weren't country clubs for the rich, they were pieces of urban furniture. You had tennis courts and football fields and high-school football fields and just a blossoming of recreational things in the twenties and thirties.

HOLSTON: In other words, it's a work/off-work, figure-ground relation.

HODGETTS: Exactly, and I think that there is a really powerful articulation of that; it's there.

HOLSTON: But European cities have that, too, it was part of the industrial revolution.

HODGETTS: Where do they have it?

HOLSTON: The soccer stadiums, the fairs and exhibitions, the parks.

HODGETTS: The parks were aristocratic parks, they call the parks in London Regent's Parks. I'm talking about a different thing.

HOLSTON: What the European cities have are the boulevards for wandering, for playing.

HODGETTS: Sure. What I was just saying is that the American ones were asserted in a different way.

HOLSTON: Big blocks of space.

HODGETTS: And not integrated into an urban network, but actually quite separated because of the motorcar and that tradition of driving Miss Daisy, of getting into the motorcar and taking a Sunday afternoon drive. The drive was recreational, the drive was to a recreational spot. It was an excursion, even here [in Los Angeles] the Red Line went from downtown up to the beach, Pasadena. Santa Monica was a beach community.

HOLSTON: And this was so powerful that it triumphed over the idea that one might live where one works.

HODGETTS: Exactly, and I think that the seed was laid right then and there, that there was a real joy in this separation of work and pleasure and it was structural and it was really built in.

HOLSTON: So that was a coherent narrative up through the Second World War, this separation of residence and work which historians of capitalism often describe as a fundamental change in daily life, one that I agree came to fruition here with the help of the automobile. What about new relevant narratives since then? How might the ordinary urban of today—not the grand projects like going to the moon, the grand urban histories that might form the stuff of a city hall mural—but how might the ordinary urban be an inspiration for a new architecture of content, an architecture which gels around a narrative which is coherent for us today? And I'm calling that coherence interactive architecture.

HODGETTS: I think that is an incredibly important question. It's just that it is very difficult to grapple with in traditional terms. We have to think first of the whole idea of services and the sense in which services replace productivity or product as a new economic force and as a new focus for people's energy. And look at the fact of being able to identify, proselytize, capitalize on, create an ethos around service which is nongeographical, not place-specific.

HOLSTON: Many services are of course very place-specific, like getting a xerox or Chinese food to go, but some are.

HODGETTS: But they can also be atomized in a way that an automobile assembly plant cannot. I guess that is what I am getting at: services can be atomized in a way that production cannot.

HOLSTON: Although flexible accumulation, post-Fordist modes of production, aim at just that.

HODGETTS: Which would be wonderful. There is a utopian vector there.

HOLSTON: Thinking about services as a dominant force in cities, what kind of urban narrative can you as an architect make out of them? But probably I should first ask if you want your designs to have narrative readings? Can they make a difference in the experience of public space in American cities?

HODGETTS: That is one of the biggest things that modernism struck out of the canonical form of architecture, isn't it? Content was systematically eliminated from architectural vocabulary and pure abstract space became the ideal of a kind of refined architecture, to the degree that even the barest minimum of signage and notation other than the abstractions of structure and enclosure were eliminated from the palette. If you look at the modernist buildings, on Park Avenue in New York for example, like the Seagram Building and the Lever House, even the tradition which was quite valid in European modernism, let's say, in the twenties and thirties—if you think of the PSFS Building in Philadelphia, if you think of the World's Fair Pavilions of the Russian Constructivist—there was a grappling with typography and with signage and a grappling with fusing some kind of content into the building fabric. I mean, it was visible, it was sending messages to you.

HOLSTON: Including a social and political content, which was stripped out when modernism came to the United States.

HODGETTS: I think it is difficult to say whether social content is the responsibility of the architect or not, and it is difficult, I think, to say whether that is something "the building" can convey. Maybe the building says, "I have many small windows and therefore I am domestic," or "I have large, undisturbed areas and therefore I am industrial."

HOLSTON: What I meant is that in the modernist case, architects thought that their buildings were embodying a political and social content.

HODGETTS: I don't think that is true, I don't think they did. I think the objective was to develop a building vocabulary which in the abstract could respond for all of that. I think that the issue is whether there was any content in that response, whether there was a level of communication, whether the buildings communicated anything about either their purpose or their destiny or their history, whether that was of abuse or of elaboration. For instance, look at a modernist building in Italian housing or third-world countries and it is festooned with mattresses hung out the windows, with clothing hung out to dry, with gardens on the balconies. It is very clear that it's a place where people are living. The sensibility in the United States was very, very parched next to that. The modernist tradition in the United States came to be one—and I think that it is probably why Sorkin calls Johnson a fascist—in which the image of the architect was sine qua non: it was the ultimate expression and not to be violated, so that any kind of emendation to the basic architectural design was actually considered a—

HOLSTON: A violation of law. In Brasília, for example, Lúcio Costa's 1957 Master Plan was law, you couldn't change the facade.

HODGETTS: In fact, in the extreme case, Mies van der Rohe at the Lake Shore Apartments in Chicago lined the exterior of all the curtains with a uniform material, so that no matter what personal decor the apartment might have, the exterior of the building would retain his detailing.

HOLSTON: Don't you think that today modernist architecture and planning are being recycled, reused to create new rules of exclusion, demarcation, and segregation, through the use of implants like gated communities which use, for example, modernist decontextualization to create isolated spaces? In other words, today's planners and architects very often use a modernist vocabulary to create isolation and seclusion, both in image and in fact. These of course respond to what people apparently want; they want to be protected, want to be isolated. What I am suggesting is that one of the narratives we currently have is that of modernism itself reinvested with a content of fear, class flight, racial tension. The idea that modern architecture, both residential and commercial, seems to put forward so clearly today is that people don't want a rich multiclass or interclass public space in which encounters are unpredictable and adventurous—even multicultural environments must be commercially orchestrated to be appear architecturally.

HODGETTS: I think you have to look at that through sort of two lenses: one is a management lens and a financial lens, which has a lot to do with a kind of conservative financial atmosphere which doesn't want to take any risks. It's sort of like trying to make a hit movie: people play to the lowest common denominator when they are actually creating their management strategy for any particular place so that they will say no, we don't want to attempt any street traffic in here because that will make our tenants unhappy. Now, at that point, they are speculating about their tenants, they are speculating about what their tenants really want and imposing their particular values as developers, leasing agents, and so on, on them. That may affect the architecture to an unwarranted degree. There are actual cases when any inflection toward the street is prohibited *a priori* by the managers. On the other hand, I think you will find that architects very often try very hard to subvert it. They try very, very hard if the fact isn't there of access, to at least suggest that access is possible visually.

HOLSTON: Some architects will do it. But certainly if you look at condo communities in Los Angeles or ones here in San Diego, like Del Mar West, those are not architected buildings in the way you mean. They are done by the kind of developers you just referred to, and we must admit that they are enormously successful. So, there are three agencies: the architects of the caliber you mentioned, the development people (including realtors), and the buying public. And it is clear that the buying public wants homogeneity. They want to be with others who are like them, who have the same life-style. If you looked across the land and tried to read the social organization from the buildings, you would come to that conclusion. I think that it is a very powerful factor. It seems to me that there is a narrative in that organization that we can read; it is clear. And I think it is one basically motivated by a desire for a kind of conformity that is fundamentally reassuring and unadventurous, but also motivated by fear and suspicion. All of which amounts to a real impoverishment of the urban environment, because at its inspiring best the urban has always been about the unexpected encounter, the small discovery that one makes. What do you think, though, might be some of the other urban narratives out there? Because my question to you yesterday was, what is public space in American society? If we don't want to commit again the modernist mistake, we have to begin with what we have, with the cities as they exist today.

HODGETTS: I think that the direction to go is toward a sort of elaborating on the palette. If you look at English factory housing, lets say, row upon row of little houses, or town houses in Manhattan or in Boston or whatever, Mikonos the Greek island, I think you find a long, long tradition of place making in which the homogeneity, the visual homogeneity of the place, creates a sense of well-being, creates a sense of social structure. In the past, I think that all kinds of different social adventures took place within that single context because there was a more, let's say, harmonious society, right? The context was very cohesive, extremely so, more cohesive than we are able to achieve in today's building vocabulary with all the different materials, all the different modes of transportation, all the different this and that and so forth.

HOLSTON: You think that there are two cohesive contexts: a kind of materials and technological one and a social one?

HODGETTS: And a social one, right. But today a lot of that has really fragmented in a very dramatic way. So even though it is offensive at some basic level to many people, I think that there is nothing unusual in people wanting to have a planned community in which all the roofs are red tile, all the walls are white, all the lawns are green, where they have a sense when they come home that they are coming home to a kind of holistic environment. It seems to me that for people who are less adventurous in temperament it is understandable. I don't think that it is necessarily driven by a hostile sense but rather by one of simply moving towards comfort and gratification rather than—

HOLSTON: And you see that as fairly neutral. You really don't see Simi Valley as definitively wanting to keep L.A. out?

HODGETTS: Well, I think it is wanting to keep whole, which is a subtle difference, but I think a difference nonetheless. I think it is wanting to be in a place which is knowable.

HOLSTON: Don't you think there are other ways to achieve that?

HODGETTS: I think there are, but I think it requires a certain maturity of perception on the part of the user. It takes a person who is relatively sophisticated not to feel threatened by the erratic nature of many cities.

HOLSTON: It strikes me that even in these gated communities, or in the more generic suburban condo developments, people rarely interact. They are in their cars so much that in terms of social experience the homogeneity becomes just an image. Once they are in the development, there is nobody around, no one to interact with—except, of course, the immigrant maintenance men who are servants of sorts. I'm amazed at how you never see residents walking, for example. It strikes me as incredibly inefficient; but leaving out the inefficiency, this brings out the problem of legibility that we talked about, about reading, reading buildings and reading public space, about image making, about this thing that I'm trying to get at called interactive architecture, which is a kind I think you are very committed to, in terms of signage, and I wonder whether you might elaborate on that concept. In the interactive moment in architecture, what is communicated? What narratives and so forth are possible beyond function (where to get in, how to cross the space)? What is there that the architect can put into the facade—if that is where it is—that allows the viewer or user to become a participant in more than a functional way?

HODGETTS: I think you want to develop a kind of transparency towards the different social groups and users. In other words, if you think about Japan, where you have a homogeneous society, it is amazing that their need to communicate about different attitudes or different activities is really quite muted. It is surprising to look at Japanese television; there is not that feeling of urgency, of immediacy, and of identity that in the United States is taken for granted. Every single conceivable product, activity, in the United States is given a hook, is given a name, is marketed, is taken out and displayed to the fullest in front of the rest of the country. Well, there seems to be a good reason for that, because we don't have a homogeneous society, we don't have a homogeneous ethic and any given novel activity needs to be on display in order to attract whatever constituency it needs. It is like a spore; it has got to put a lure out. And when I think about architecture's role in that I think it is to aid and abet that activity in our society. I'm surprised, for instance that in Frank Gehry's work, which I admire very much, there are allusions to communication, there are structures which look like they might have held a billboard at one point or another, there are structures which look like they might have had a sign ripped untimely off the facade. But at the same time the buildings retain that kind of abstraction of modernism which ultimately doesn't communicate anything except formal information.

HOLSTON: Except to those who happen to be inducted into that secret language.

HODGETTS: Right, right, but it is a secret language. I suppose that what I would promote in terms of the city fabric is an architecture which has the sort of vitality of configuration of many of Frank's buildings, but which is using that vitality to communicate and to allow opportunities for people to communicate about what is going on in the buildings, about what is going on in their lives, what is for sale, what is happening in their day-care centers, about somebody who just won a big prize, whatever.

HOLSTON: How would they communicate what is happening in their lives?

HODGETTS: Well, they could communicate what is happening in their lives by making decorative use of the artifacts they use in their lives—back to clothing hanging on a facade. It's interesting, you see, the impulse in the United States is to take everything and tuck it out of sight if it's got some utility value. Everything with some utility value is tucked out of sight, so our washers and dryers are somewhere, we don't hang clothing out. God forbid somebody should see that we wear scanty briefs or something like that! There is a concerted effort to take the real stuff off the streets and substitute a cosmetic facade. That's true for automobile service centers, it's true for old age homes.

HOLSTON: What about strip architecture? A lot of that strip architecture communicates in just that way. Think of Michelin tires and there is the Big Tire Man.

HODGETTS: Which is great. Strip architecture is fine.

HOLSTON: Are you talking about an interactive architecture in terms of a facade which somehow communicates.

HODGETTS: Or lack of facade.

HOLSTON: Can you describe that facade or lack of facade?

HODGETTS: I think you just did, the Michelin tires are perfect. At a certain level that is very, very strong. It's right there, it's in your face, and you know. You don't even have to go to the Yellow Pages.

HOLSTON: Would you say that the idea of an interactive architecture—I don't know whether that is a term that you would endorse—

HODGETTS: Yes, it is.

HOLSTON: —is one that communicates the building's or the space's situation in the city or some greater context or multiple contexts and then something of the viewer's or user's relation to those contexts through the building, readily and easily, without requiring elite or professional education; a communication that enables users to understand their place in the world by revealing something about the simultaneous orders and disorders of their lives? That is the way I am thinking about it. And the question becomes how does this dialogue between spaces and buildings and people take place?

HODGETTS: Well, it is like reinventing the wheel. I don't think it's so hard if we get over our prejudices. I think we've got terrible, terrible Puritan prejudices in this country which seek to impose order where none is necessary. In other words, our prejudices—which I think are highly tied to speculative land values, a whole capitalist kind of attitude about sustainability in terms of value—look at communication as a corrupting aspect of the city, look at all the things which densify our understanding of the city as things which ultimately make it profane and thus value less. I would suggest that the opposite is true for a healthy city.

HOLSTON: Wouldn't you agree that you are talking about several kinds of communication? On the one hand, there is the communication of a redeployed modernist architecture I was talking about that has to do with the facade of orderliness, of rightness, of correctness, that is to say, of status, propriety, property, real estate value, and of the corporate finish, the Midas touch, and the fear that goes with it. That is communication, too, so I wouldn't want to say that the architecture we have most of doesn't communicate, but we seem to be talking about another kind of communication.

HODGETTS: I would say that it symbolizes. I think that what you're talking about symbolizes all of those things but that it doesn't have content. I think that there is a difference here.

HOLSTON: I would argue that it does have content. If we did an anthropological study of how people really read buildings and public space, what they base their choices on, what they think about as they drive by on the freeways, where they go, and where they put their money, I bet we would see that there are very interesting narratives that people read off of and also insinuate into those buildings.

HODGETTS: I'm sure they do.

HOLSTON: I think that is what any architecture does, and I want to call that communication. But I sense that you have an idea about another kind of communication that uses the ordinary urban, daily life, all that stuff which you say American culture tends to hide, and what I want to ask you is how do you actually put that into a building, literally, can you describe how you would put that into the facade of a building. How you would make it interactive?

HODGETTS: It's by moving the building out of the center of the action, by putting it behind the action. It's not by "facadism." It's by letting the plumage be in front. It is by taking a far more recessive attitude as a building designer and suggesting that all of the vitality, the vital activities, really step front and center. That is not to say that there is not a very strong and really important architecture which has to do with the creation of the space and the volume within which that occurs, but then you're into the architecture of the street, the sidewalk, the daytime/nighttime lighting, the volumetric area between the tops of the buildings and the pedestrians and the vehicles, and all of those things which again, given Los Angeles, are almost deliberately designed to inhibit the kind of thing we've just been talking about.

HOLSTON: Can you give me an example?

HODGETTS: One of the reasons La Brea is one of the streets that works in the sense which we are talking about is because buildings tow the street line, there are no setbacks.

HOLSTON: You mean, they are contiguous on the sidewalk.

HODGETTS: Right, and the sidewalk is of relatively good proportion in relation to that. There is not a lot of masking with street trees and so forth and so on which tends to be soporific, tends to communicate something else in fact: let's go to sleep, let's have a picnic, let's do this and that, but it doesn't say something about urbanity. So those things all add up on a La Brea or on a Melrose as a street architecture to support a certain urban vitality. You move into other areas, like around Wilshire Boulevard and many, many other areas where setbacks are required, where street trees are required, and things like that, and you move into a far less dominant kind of urban context.

HOLSTON: What you describe in putting the buildings behind the action seems quite like the classical figure-ground/solid-void relations of a premodernist urbanism—which I would stress very much as well. I remember when you showed me the L.A. City Walk project on which you consulted (fig. 1). I was really struck with the inventiveness of the composite facade you developed. How would you describe that project in the terms we have been using? I am really interested in this because in that project I think you accepted in a very creative way the impossibility of recreating the classic solid-void/figure-ground relationship, and yet you're intervening to create something that allows the urban social context to become figural in relation to the building which is ground, which is a foil for this figure.

HODGETTS: We are trying on that project to kind of jump-start a genuine

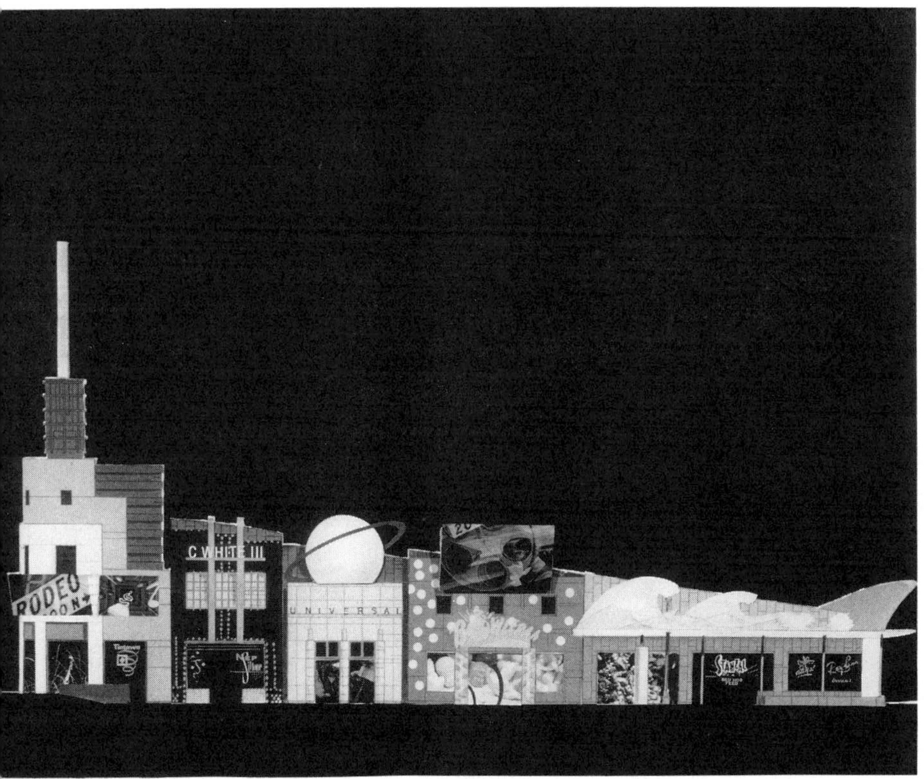

Fig. 1 Hodgetts and Fung Design Associates, Universal City Walk. Los Angeles, 1993. Photo: Hodgetts and Fung.

vitality and visual variety by providing a foil to whomever the final and ultimate tenants might be—because we don't know them in advance. And one of the things that is clear from any kind of street experience is that the preemptive notation of the street generally becomes something which acquires a patina over time—it acquires the abrasions of use—and that there is a far better relationship between the variety of tenants who might occupy that thing if in fact that patina of use is already established. It's like establishing a horticultural event. It's like planting trees and vines and so forth and so on, and you will say: well, this garden is not very well established, it has only been in place for twenty-five years or fifty years, in another one hundred years it will be well established. That kind of attitude, I think, is relevant to our architecture. Rather than seeing the successive uses of an urban structure as being somehow abusive of the basic architectural framework, one begins to see the fragments of those prior uses as "establishing the reality" of this place. As it begins to be established in memory, it becomes established in the sense that it is no longer diagrammatic. It is not the wisp of a tree planted here, here, and here, but a tree with some mold on the sunny side, mushrooms on the other side.

HOLSTON: How do you do that in the project?

HODGETTS: On that project we do that through many artifices because it was a matter of jump-starting. So we created fictive histories for each building and then we went about creating fictive users for each building and acted out in a dramatic way what each user might have done: Oh, he put a sign up here, and it was torn off when he went broke.

HOLSTON: So these are commercial stories. Would these stories be on different shop facades or on one facade?

HODGETTS: There were many shop facades, perhaps fifty or sixty, and each one had a bit of a back-story. One was a movie theater which had declined into being a porno house, and then the porno house had been converted into a swap meet, and then the swap meet was revitalized into another incarnation. And there were little flashes of all these different incarnations in the facade.

HOLSTON: In other words, the remnants of these histories were still there to be seen.

HODGETTS: Exactly, and these layers were thematically acted out by making a sketch of what it was at each stage in its lifetime and seeing what fragments were left. Finally, you have discussions about, well, all the symbols, like the brass nameplate all polished up, the sense of propriety that it has, as opposed to the sense of propriety of a neon sign, as opposed to the rather ephemeral sense of propriety of a blow-molded and vacuum-formed sign, and the hierarchy of values that all of these things portray and then how they jostle together on a street which maybe has had many occupations. There were technical ways to achieve that: you put fly ash in the paint, in one layer of paint, so that if you hose down the paint, certain parts of it flake and leave the paint underneath visible. It's very much a kind of scenographic excursion, and it's ersatz in a profound way. But on the other hand I think you could argue that the sterility of a jump-started mall which has not been given that kind of horticultural accretion never matures beyond that point because there is a great inhibition on the part of leasing agents and everyone else to break down what appears to them a very sanitary, neutral kind of order. So, we try to give a kind of dirty order to begin with, a kind of fuzzy order, in full knowledge that it wouldn't be preserved. It has a layer to be succeeded by another layer, to be succeeded by another layer. And the reason that one has to resort to that today is because the day of the small incremental development—mom and pop coming in, digging into their nest egg, putting up their shop as a groundup building next to another groundup building which another mom and pop built, midtown American landscape with the Newberry's next to the petshop, next to the multishop, each of those being a building with an authorship by an individual and with a story—that's gone, that's not going to happen anymore, we know that's not going to happen.

HOLSTON: Why?

HODGETTS: The banks are not going to lend the money for it. The parking requirements will not allow somebody to build a twenty-foot-wide building without providing parking for it, and it is economically impossible to build parking in a twenty-foot-wide lot, and so on. So there are baseline changes in the structure which make that not possible. I have suggested elsewhere that if the city wants that kind of incremental development to happen that we change all our parking requirements or we build a big parking lot, resubdivide the land into smaller bits and streamline the process so that mom and pop who want to build their shop don't have to jump through all the professional hoops, the handicap access, and so on. You know, if you have a second floor on a twenty-foot-wide shop, you have to have an elevator on it—who can afford that? Nobody. So that means no second floors unless all of a sudden you suspend that requirement. So what I'm suggesting is that there is such a network of legal and socially accepted norms, such a dense set of requirements that must be fulfilled, that without restructuring all of these things the small shop is an impossibility.

HOLSTON: And rewriting all those city codes is a very difficult project.

HODGETTS: You're looking at fifty years, probably.

HOLSTON: It seems to me that you are now talking about an architecture of intervention which is a much stronger architecture than you were suggesting a minute ago when you advocated an architecture that recedes. Now, you seem to be talking about an architecture which intervenes, but does so in a complex way trying to establish a dialogue about the process of intervention on the one hand and all the other types of interventions on the other. In other words, it's an architecture of intervention which allows for other kinds of interventions, by nonarchitects, by time, by users, by all other users, on an equal footing.

HODGETTS: Exactly, on an equal footing.

HOLSTON: And that is what modernism never did.

HODGETTS: That's exactly right.

HOLSTON: Modernism moved in and established its intervention, period, and never gave any other intervention an expression. What you are talking about is an architecture which represents human interaction with the city as one of constant intervention: splicings, cuttings, fragments, memories, leftovers, rejects. And the role of the architect or of architecture is to orchestrate all those interventions so that they are there in the design, in the building or space, in traces, so that they are available to the palette of urban experience. Does that make sense?

HODGETTS: Yes, and I think it was wonderfully stated, actually. I'm going to remember it because I think that the role of trying to nurture the environment as opposed to impose on it is really what I'm talking about. Recognizing that all these other forces in fact have to be healthy, have to have either entrepreneurial or personal or culturally satisfying realities behind them, that they cannot see the architecture within which they are living or working as adversarial, as trying to impose on them, that they have to be given central stage and giving them central stage is a complex task, hard to do.

HOLSTON: It would seem to me that the kind of interaction we are talking about requires an expansion of the idea of architecture beyond a concern for permanent buildings and permanent planned urban structures, to which the profession is normally dedicated, to include both impermanent interventions—tense, changing facades, or even the idea of change within the facade—and a commitment to the idea of the city as a social text, a commitment modernist notions of city did not allow. So that part of the architect's professional responsibility becomes caring for, learning about, and occasionally building in the city, and when the occasion comes up, then there is already a pool of resources that the architect has been involved with, all of which become important elements of the content of the design.

HODGETTS: There is a whole element of almost casting in that idea, of creating, for instance—

HOLSTON: Casting? As in film?

HODGETTS: As in film. One begins to intuit or create or nominate certain regions, certain places, certain rhythms of an urban framework with certain specific kinds of opportunities, so that you would say: on this corner it would really be great to have a flower shop; that flower shop would be terrific if there was a coffee shop next to it, because those two things would begin to generate a formation and you might begin actively, as John Jerde [design architect of L.A. City Walk] does, to search for a snappy young couple that wants to make a flower shop.

HOLSTON: He seeks them out?

HODGETTS: He seeks them out. He has the place because he is developing a shopping center.

HOLSTON: I see. We are back to the world of the shopping center.

HODGETTS: I don't think you're going to get away from the world of the shopping center. I really don't.

HOLSTON: Let me challenge the problem of the shopping center by asking this: one of the issues that the notion of an interactive architecture raises is the problem of who speaks, whose language, whose public. It may be that none of us would object to a flower shop, but it does suggest that the developer, or the architect, is deciding what goes there. But who is to say that it shouldn't be something else? How do you address the problem of whose language? Whose language does the casting director speak? What public is he addressing?

HODGETTS: I understand. Here is where the thing goes. I think there is both a moral quandary and a historical quandary because the terms of the occupation of space, of a place, have been so profoundly affected by the structures of contemporary society. These terms are really not very accessible to the man in the street. And the mechanisms aren't there; more and more the mechanisms aren't there. And I can even go further and say that our society has begun to have very narrowly defined carrier paths. Lots of specializations—I guess that is one of the words I would use—have created a society that doesn't have anything like a uniform attitude toward the environment, except in big *a priori* things, you know, like we cannot kill the planet; but there is not even really agreement there. But it's tough to see consensus about other forms. Even people who give it attention in their lives like what they like, and if they don't like it, they go do something else, because there is so much variety in the world or in the city. I think that one of the results of that is that imagination—like it or not—has begun to supplant utility in cities. In other words, pure utility, the marketplace where everybody comes and sets out their stuff and sells it, in just a neutral way, pure utility doesn't stand a chance in our society next to a festival marketplace because the festival marketplace has enough drama, enough assurance that if you drive twenty minutes to get there, you're going to be fulfilled. That twenty minutes is valued for people, and they will take that option, if it is offered, as opposed to another one. They will take the designed option.

HOLSTON: It's a hook. Things have to have the proper sales pitch and that is what you're calling imagination?

HODGETTS: Exactly. And that is true for communities, too. People move to a community because it has a clear identity.

HOLSTON: But the production and distribution of identities and spaces are produced by a market that is driven by consolidated power relations. Power produces space. The man in the street does not have power, so developers are not going to "speak his language." But it is clear that we have tremendously consolidated power structures here: advertising, real estate, banking, contractors, and so forth, which determine to a tremendous extent the images that make up our experience of the city. What I wonder is whether we can imagine other ways in which an architect who wants to create the kind of interactiveness that we have been developing might work with the problem of how to address the ordinary urban as well as the extraordinary urban—that is to say, both the foundation mythologies of our society (the festival market idea comes to mind), the big strategic forces, and also the multiple interactions that make up ordinary life and over time make up social history. How could an architect, planner, or artist work such content into the built environment? The question is really about the reinvigoration of public space in American society. I was thinking that one of the ways that we might discuss it in greater architectural detail is to look at the problem of direct address—you called it communication: How does a building or a space come to address people in their daily lives? And I thought that we might look at this problem of address through a series of slides I have put together which go from ancient Rome to the modern commercial strip.

HODGETTS: Let's go.

HOLSTON: It seems to me that one of the principal ways in which architecture has worked with the idea of address is through the monumental commemorative column, and here is a slide of Trajan's Column in the Forum at Rome (fig. 2), a literal depiction of events, a history there for all to read.

HODGETTS: It's a movie marquee, isn't it? I think it's exactly what you said, it's a literal story. You unwind it from the column. It has always been

Fig. 2 Trajan's Column, Roman Forums, Rome, A.D. 98–113.

unclear to me whether people literally read it off, bit by bit. Whether or not it supplanted real discourses, it certainly came to symbolize discourse. It says, I'm about discourse. I'm not sure whether people really read that.

HOLSTON: Do you know the history it depicts?

HODGETTS: Not really.

HOLSTON: It's about Emperor Trajan exhorting his troops to victory. But I think you're right: the actual details of the story are not as important as the idea of the story.

HODGETTS: And the idea of the story is memorialized.

HOLSTON: Yes, history memorialized. It strikes me that this technique is one of the oldest, the most direct, and in many respects most appealing ways that architecture can address its public and create a public for it.

HODGETTS: Except what is difficult about that in today's world is that it is extremely difficult to identify an event which is significant enough to deserve that. It's really tough.

HOLSTON: We are back to the point made yesterday: this is the stuff for City Hall; these are founding mythologies. But there are other possibilities of configuring histories. Can't you imagine using computers—a kind of large-scale desktop publishing—to let people make instant histories, to put up their histories on public displays? I'll come back to that; at the end of these slides I have an example of that kind of thing.

HODGETTS: You see, in a nonarchitectural framework we have something like that, you know: the animated sign outside of Earl Scheib's car place [franchised auto body shops], the things over meat counters that have little news broadcasts on them, there is a lot of it out there.

HOLSTON: They are usually one-liners, aren't they? Unlike ancient Rome, which was full of official history, we have an ordinary world filled with one-liners that are unofficial, almost illegitimate, but a world nearly completely filled, almost the inverse, isn't it? Ancient Rome had little of that but it had all the monuments. This kind of architectural discourse—I'm going to call it that although I know that there are many architects who are revolted by this kind of decoration or appliqué, as they would call it—is produced here not just by the column; look at the statuary in the Forum, it's the whole place; here we have a broader slide of the Forum, look at all the stories out there.

HODGETTS: Everything is written on.

HOLSTON: We have a literal use of building and space as writing surfaces. The building recedes, it becomes—

HODGETTS: —a big billboard. What is a triumphal arch? It's a big billboard. We may think that it has a more noble architectural form, but it's a big billboard anyway you look at it.

HOLSTON: Here we have a slide of the Hall of Justice at Bruges (fig. 3), literally cemented to the main cathedral on one side, and you see this

Fig. 3 Hall of Justice and Cathedral, Bruges, from 1376.

wonderful writing in stone, chapter and verse of the Bible played out in statuary, in the stone. You have two stories here, a religious story which absolutely abuts and supports a judicial one, and vice versa; that is to say, both worlds, the civil and the spiritual, hold each other up, and their interdependency is so legible. Another kind of interdependency that this sort of writing in stone raises is how it can put people into a global context. I think that it can be very successful at making people aware of the transnational context of the local lives they lead. A great example is the next slide, of Michelangelo's Piazza del Campidoglio (fig. 4) with its equestrian statue. Here the context is that of Roman imperialism, and I think that architects have always been very sensitive to the imperial mission. How do you see this in terms of such direct address?

HODGETTS: Well, I was just there. What they had done was to make a huge sand painting on top of the street paving, using colored sand, very carefully applied, which continued the pattern of the pavers, but imposed words and figures onto it, so that from the very same vantage point as the slide it was really again a big billboard. It was about International May Day. It was political, very strident propaganda.

Fig. 4 Michelangelo, Piazza del Campidoglio, Rome, 1537–64.

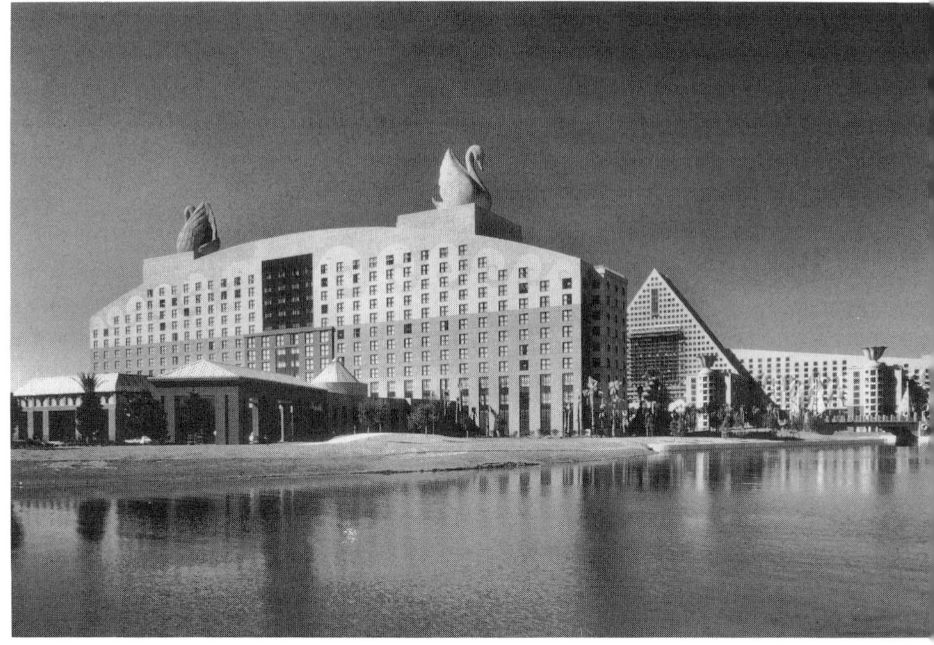

Fig. 5 Michael Graves, Walt Disney World Dolphin and Swan Hotel/Convention Complex, Walt Disney World, Florida, 1988. Photo: Courtesy of Tishman Realty and Construction Co., Inc.

HOLSTON: Michelangelo's is also a piece of propaganda. It's about the Rome of Marcus Aurelius dominating the world, occupying its center: all roads lead to Rome as the pattern of paving suggests. Here is a superimposition of a political message using the energy of this global juncture, all of which it seems to me is powerfully architectural. To me, contemporary architecture needs to expand to speak publicly about our own transnationality.

HODGETTS: However, again, what is so difficult about architecture in today's framework is finding that solitary thing—one wouldn't memorialize one's client like that.

HOLSTON: Well, I want to juxtapose the next slide in relation to just that problem: Michael Grave's Walt Disney World Dolphin and Swan Hotel and Convention Complex (fig. 5).

HODGETTS: Is that architecture? I want to gag every time I look at something like that. I think I do because it is so much about what I was just saying was impossible today. It has so much to do with the kind of singular totemic presence identifying a singular objective at the expenses of the users and of the inhabitants; and, you know, this is the ultimate in creating a gesture which obliterates all content except that of the architecture. I wouldn't look at that and say this is Walt Disney. I don't know what the symbol means, it's a pyramidal symbol.

HOLSTON: It's not Walt Disney to you?

HODGETTS: No, it's probably Masonic—Michael was Masonic and it is ultimately a Masonic temple.

HOLSTON: But are you suggesting that the urge to Marcus Aurelianize the client here, if I can say that, is similar to Michelangelo's Piazza del Campidoglio? Is that what you're suggesting?

HODGETTS: I don't think it does at all. I think that Michelangelo's piece memorializes a hero, and maybe appropriately if a hero can be identified. The Graves's building memorializes nothing. It uses all the apparatuses of important memories to memorialize nothing and to trivialize all of that, and to make it moreover meaningless because there is no trace of other content.

HOLSTON: In other words, the difference is that with the Michelangelo we have a hero and a story, a history of battles, conquests, laws.

HODGETTS: There is meaning in that. In the Graves, there is no meaning. I don't know what it means, I really truly don't. I wouldn't know if it's a hotel unless somebody told me. For instance, in my terms, I don't know whether those two apartments at the very peak are exceptional rooms. If one were interested in meaning, they would be exceptional; they would be rooms which had a character which said they were above all the other rooms. Maybe their windows would be different, maybe they would recognize the fact that the human being who was looking out of those windows, as opposed to any other window, had a different kind of charge. But they do not. I think that the meanings are obliterated in this building and this building is about symbols. It's not about meaning, it's not about content. There is no content whatsoever.

HOLSTON: And one of the other points that I think we could make is that this building is not about any relationship of the individual citizen to some kind of political or ethical community, although Graves is using symbolism of a classical sort. By contrast, Michelangelo uses the paving pattern as a scaling device to give individuals the opportunity to measure their size and standing in relation to the hero in the center. In this evaluation, he gives individuals the means to consider, in a very physical and immediate way, their relation to the political community through the memorialized hero.

HODGETTS: That's probably right, that's the humanistic thing.

HOLSTON: In the Graves, there is none of that negotiation of size between the anonymous and the great, the citizen and the city, the individual and society, the person and the culture. But here is another type of intervention, another kind of billboard, here is a slide of one of Jenny Holzer's electronic displays (fig. 6) which are on the whole more critical about the relationships

Fig. 6 Jenny Holzer, from *Truisms,* Times Square, New York, 1982. Photo: Courtesy of Barbara Gladstone Gallery, New York.

between the individual and society, between the ordinary urban and the world order that grounds it.

HODGETTS: This is a great photograph because it also says Tow Away Zone, No Turns, and the imperative signage is what Jenny Holzer has echoed there. She has done no more and no less than that. The only interesting thing is that she has brought our attention to the fact that all these other signs and all these other pieces of urban paraphernalia which we don't often look at are very powerful in the urban context.

HOLSTON: Even though her message is subversive.

HODGETTS: Well, is it?

HOLSTON: Well, that's a question.

HODGETTS: I'm quite happy that she does her work, but it does not seem to me to—what it seems to me it does best is to alert all of us to the fact that this is going on.

HOLSTON: To the power of signs, to the imperative of signs that we are flooded with all the time.

HODGETTS: Yes. It is even interesting here, this photograph favors her sign, but if we look at the number of degrees, gradients of vision, occupied by that sign in the far distance and some others closer to us, even on the issue of scale, retinal impact, and so on, it is at parity with all the others.

HOLSTON: Parody or parity?

HODGETTS: Parity.

HOLSTON: It is also a parody at some level.

HODGETTS: That is what makes it amusing. It is much more amusing than Tow Away Zone.

HOLSTON: And it is also making fun of our obsession with the signs. I don't think it is very subversive either. I think that it is trying to establish some kind of dialogue, but through shock, and that it quickly succumbs to the numbness of shock.

HODGETTS: Do you remember Op Art in New York? There was a period before that when corporate logos and identities were very powerful presences in New York, when Chemical Bank and everybody started having new and very severe black and white logos. Eight or ten years later Op Art came out which emulated those forms, and I think that is the cycle of art in New York City, and that rather than breaking new ground, basically what art in New York City succeeds at doing very often is echoing the reality and bringing the reality in focus.

HOLSTON: I would say that it's parody in a sense because parody reestablishes the norm at the end of its ridicule, and that is what this does. And I think that like many modernist gestures, its dialogue ultimately ends in a shock that grows numb. There is a certain numbness after all, you don't see it anymore than you do that Tow Away sign.

HODGETTS: That is what I mean, and it doesn't have any more content.

HOLSTON: Holzer's giant display in Las Vegas, "Protect me from what I want," is really on a par with Caesar's Palace. It does not have any more content than—

HODGETTS: But it does have more psychic impact. I think it has more impact than a fortune cookie.

HOLSTON: Another thing that is interesting is that you can change it. Here is a slide of the same Times Square display saying Private Property Created Crime. Here the phrase is more subversive.

HODGETTS: That is true.

HOLSTON: Here is an added opportunity, or an opportunity at least for the kind of subversion that the previous Money Creates Taste doesn't have. If you stop to read it, I think it achieves the kind of interaction that an effective billboard has. But is also shows some of the limits.

HODGETTS: Again, though, Holzer's billboards are rather inaccessible to the population.

HOLSTON: But in Times Square or on the Las Vegas strip, anybody can read it.

HODGETTS: Anybody can read it, but not anybody can put it up there. An architecture that would communicate would give that opportunity to everybody in Times Square interactively. There is a real populist thing. You would only allow that kind of scale of gesture to somebody with power and money.

HOLSTON: You would allow anybody to come by and have a terminal and type in their own message and put that up.

HODGETTS: Sure.

HOLSTON: You are talking about the use of ATM-like machines to make big billboards.

HODGETTS: Or shop fronts or whatever. In other words, the legislation of all this kind of stuff is only "proper" if it comes from an artist. It is something I'm completely against. I have no sympathy for it.

HOLSTON: How do you make this kind of billboarding, this kind of public statement, available?

HODGETTS: Well, it happens with graffiti, as we see here. But I think one of the reasons it happens with the graffiti is out of the massive amount of frustration that the man in the street has that, for the most part, the city is not talking back to him in a language he understands. It's talking back to him in a language which is esoteric and threatening and really not very useful to him.

HOLSTON: One of the things that the Holzer's signs bring up is the limitations of parody; that is, as in the relationship between Op Art and corporate logos, her signs mock the culture of signage, of one-liner values, by imitating them, by making more apparent and poignant their oddities and powers. This imitation capitalizes on the characteristics of the original, in this case, the display, one-lineness, voyeurism, and so forth. But in so doing, it must exhibit a secret sympathy with the original, like homeopathic magic, from which it gains its powers of ridicule. And in this sympathy it reaffirms the normativeness and coherence of the original because it depends on showing that there is something normal in contrast to which it appears comical. This ground of the normal constitutes an inescapable limitation to a public art that would be more critical or interactive. In this respect, we could contrast Holzer's signs with Krzysztof Wodiczko's projections (figs. 7 and 8) because it seems to me that they challenge the normal in a way that Holzer's do not. If Holzer's works are ultimately in complicity with the culture of signs, Wodiczko's are not. Here we have a slide of the Brandenburg Gate in Berlin onto which he projects two nuclear missiles, an American one and a Soviet one. Here is another slide showing a statue of the kaiser onto which he projects a tank and missiles filled with coins. These complex images, I think, challenge the present not in the kind of abstract modernist utopian sense, that is to say, in terms of a plan or an ideal which is not present anywhere other than in a master plan, but rather in terms of the immediacies of

Fig. 7 Krzysztof Wodiczko, Projection, Grand Army Plaza, Brooklyn, N.Y., New Year's Eve 1984, in *October*, vol. 38; reprinted with permission of MIT Press. Photo courtesy of the artist and Galerie Lelong, New York.

the global networks we live in, like colonialism, multinationalism, capitalism, imperialism, communism, in terms of the idea that the space we occupy is saturated with simultaneous meanings, in a way that Holzer never gets to. Simultaneous and decentering meanings, this is a very decentering image, isn't it?

HODGETTS: It's interesting to me that he makes such a powerful dialogue

Fig. 8 Krzysztof Wodiczko, Projection, Equestrian Statue of Condottiere Colleoni, Campo San Giovanni e Paolo, Venice, 1986, in *October*, vol. 38; reprinted with permission of MIT Press. Photo courtesy of the artist and Galerie Lelong, New York.

between the architectural surfaces that he is using and the images. In other words, there is a fusing of the sensibility of the triumphal arch or the buildings and the messages which he imposes on them. There is a fusing, as opposed to the billboard which is simply imposed. That's very interesting. On the other hand, it's so ephemeral that it is hard to distinguish it from a kind of— In other words, it's not durable. If it were durable in the sense of constantly changing and always happening, one could imagine it being a very, very powerful force.

HOLSTON: I would imagine that anybody who saw this—

HODGETTS: It would be memorable, obviously. But it is more like, "Wowie, isn't that great!" than offering the viewer the possibility of being able to dissect the content, that is what I am getting at. I think that when it is a slide and we are sitting here discussing it, it is memorialized and it is a permanent thing in the slide; in the city, it is so transitory. Unless it is there for a month, then it probably works at a city scale, people coming and going.

HOLSTON: It's a performance piece.

HODGETTS: I think it's great stuff.

HOLSTON: I have something Wodiczko (1987:42) wrote that I would like to get your reaction to: "The aim of critical public art is neither a happy self-exhibition nor a passive collaboration with the grand gallery of the city, its ideological theater and architectural-social system. Rather, it is an engagement in strategic challenges to the city structures and mediums that mediate our everyday perception of the world: an engagement through aesthetic-critical interruptions, infiltrations and appropriations that question the symbolic, psychopolitical and economic operations of the city." So he has this idea of an interruption, an infiltration that engages our taken-for-granted, our assumptions, that undercuts our official history as embodied in things like triumphal arches, statuary, and columns by superimposing an image which is almost like the skeleton in the closet around which it is clothed. One might describe this kind of work using Walter Benjamin's notion of dialectical montage. There is a lot of avant-garde film strategy here, of using simultaneity as an architectural device. Your idea of fusing the surfaces— that's the device he uses to fuse the simultaneity of world and historical experiences so that we can see them in one view.

HODGETTS: Rauschenburgian, past and present and all of that.

HOLSTON: An architect who has experimented with simultaneity as an architectural device is Jorge Silvetti in his project for the tower of Leonforte in Sicily (fig. 9). I read it as an experiment in simultaneity because as you climb the tower stairs you find telescopes and sighting devices piercing the walls which focus on places of importance in the town's history. And here is a composite drawing of various elevations, and here you see the model of

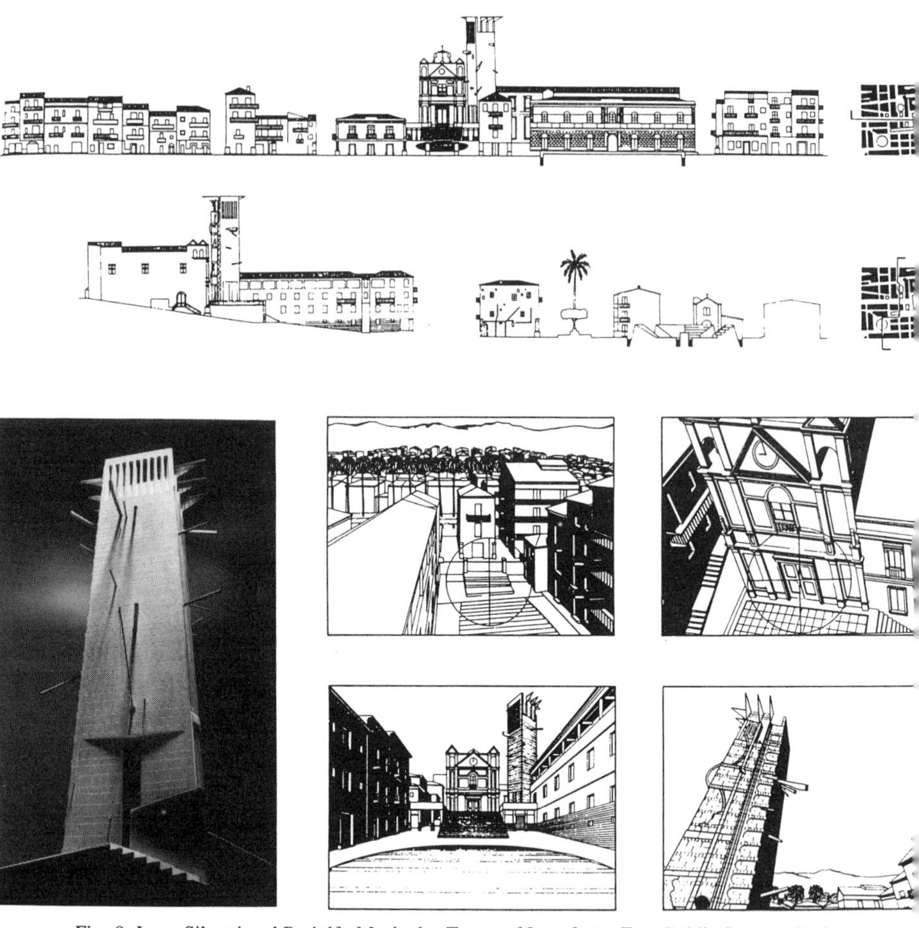

Fig. 9 Jorge Silvetti and Rodolfo Machado, Tower of Leonforte, Four Public Squares Project, Leonforte, Sicily, 1983. Photo: Courtesy of Machado and Silvetti Associates, Inc.

the tower with those tube telescopes. Here it strikes me that simultaneity is an organizing architectural device.

HODGETTS: I don't understand the simultaneity as a notion—I like this project a lot, but I don't know.

HOLSTON: What I mean is as you walk up the stairs you are invited to see various important places so you get a variety of perspectives coming into your field of experience. The view from any tower gives us a simultaneity of perspectives that we can't achieve on the ground—I guess that's one of the attractions of heights—but Silvetti's is explicitly didactic in this sense.

HODGETTS: I see it as a kind of exhibition strategy. In other words, I see it as a way of bringing the city into a single place. As we are sitting right here looking at the slide projector, in a sense it does that with the city. In other words, it displays everything within a common field, on a common ground and compresses it into a single conceptual frame—you're probably talking about the same thing—so when you are within that conceptual frame all these various aspects of the city come into play and you are thus not forced to kind of change environments or change physical relationships to it. Your physical relationship is absolutely neutralized to each thing that you see because it's constant.

HOLSTON: Turning to the last set of slides, I would like to discuss something that I think our discussion has been building up to, something that we have both mentioned at different moments as a new means to make the built environment more truly interactive—less official, static, and didactic than commemorative columns, more public, participatory, and plural than Holzer or Wodiczko. I am referring to the use of computerized technologies to compose architecture and urban space and to change fundamentally their representational processes and structures. Although I don't have slides of them, I wanted to bring up the Statue of Liberty–Ellis Island renovation projects begun in the early 1980s. Have you seen them?

HODGETTS: No.

HOLSTON: What happens in both is the creation of a foundation mythology in which migration to an imagined future, which was the United States of America, subsumes all migration histories. There are real problems with the projects: no mention of the African slaves' transatlantic passage or of the Latinos' northward trek, for example. In restoration, the Statue of Liberty and Ellis Island have become what they never were: rites of passage for all who came regardless of port, time, or condition of entry. In spite of such universalizations and misrepresentations, which appear everywhere—in the ethnic souvenirs at the gift shop and the international foods restaurant—there is in this myth making a very powerful computer technology which is not yet really interactive but which suggests the possibilities. Computer banks and video walls enable people to trace migration history. If you have prepaid and sent in your family's story—there was a national campaign to do this as a means of raising restoration funds in which I think two hundred thousand people participated at one hundred dollars each—then you could plug your name into the computer and see your migration history in the context of many others. If you are not preregistered then you merely access other people's stories. But it is still exciting to do and suggests a kind of interactive memorializing in terms of recording and projecting individual histories that one can imagine might revitalize public space, public encounters, even notions of citizenship. Another project which also uses this technology in an interesting way, though more didactically official, is Venturi,

Rauch, and Scott Brown's Benjamin Franklin Court in Philadelphia—do you know this one?

HODGETTS: Yes, I do.

HOLSTON: In the underground museum, there is an installation of telephones which connects you to the voices of the Founding Fathers. It's corny and superficial and canned, but I have seen people having a lot of fun at the installation getting a sense of what that historical moment was and also playing with the voices. As someone who has been involved with this kind of technology—I'm thinking of your Panasonic project (fig. 10) which seems spectacular—could you discuss what you see as some of the interactive possibilities of computers, video walls, holograms, and the like, in the built environment.

HODGETTS: Well, in an entertainment context, as in the Panasonic exhibit, you can think about it as it is really direct. In our case, an exhibition context, one of the notions that we've got is this sort of electronic graffiti of people's wishes and hopes and things like that which are presented in a kind of focused sound; that is, you can engrave your voice on the space. And that

Fig. 10 Hodgetts and Fung Design Associates, Panasonic Pavilion, Universal City Walk. Illustrated by Charles White III. Los Angeles, 1993. Photo: Hodgetts and Fung.

voice is sonically separated from others so as you walk you hear different people's voices who have visited the exhibition prior to that time. You hear exactly what they said and it's like a graffiti.

HOLSTON: A kind of sound museum of ordinary conversation?

HODGETTS: No, people make particular wishes, hopes. It's like a wishing well in a sense, but it gathers all these things together and kind of puts them out there. But that is an entertainment context. I think that in a context of real politics, of real events, the way to harness it is probably more on the front end, to try somehow to inventory and access individuality, individual ideas and dreams and all sort of things prior to the inception of the project.

HOLSTON: Prior? That sounds like a traditional relationship between architect and client, which is all about dreams.

HODGETTS: Well, I think that at this point you can begin to erase the architect out of the equation, because I think it's more and more possible to communicate very vividly with a person, with absolutely no stake in the world of architecture and design, and communicate with him so vividly and with tools which are so accessible that they can begin to compose, they could begin to impact a big piece of dough which is about to become a building.

HOLSTON: You mean the program—like using a kind of electronic town meeting where people would establish what they want for a particular project and that could be recorded and help the architects to figure it out?

HODGETTS: I think you can go beyond that, where the architect could be eliminated from the equation. I think there is a real possibility. Start thinking about desktop publishing and the kind of inventory of tools and the anticipation of what you might need—let's say a catalog of clip art in electronic publishing in which you need that finger pointing at something to show a price reduction. Where is the graphic designer in that? The graphic designer was well on his way to being a paste-up artist at that level and interceded in a needless way between the production of a flyer and the user, right? No real need for a "designer." I think that is the case in an awful lot of architectural frameworks, especially in architectural frameworks which ought to be commodified. If you take Christopher Alexander and cross it with clip art and with genuine access in store fronts and so forth and so on, I think an awful lot of nontheatrical design doesn't need an architect. The utility of a neighborhood, of a street that everybody uses as a neighborhood street, you don't need an architect for it. It's probably better not to use one. But what is not good is to impose on that framework a do-nothing, care-nothing, stick-in-the-wheel architect whose only interest is making money for his developer, who is without imagination, who is pressing the possibilities down to their very lowest order, below the common denominator, which is not a task of the imagination but which is a task of what Bucky Fuller used to call the chief steward. But there is an awful lot of architecture which is chief steward's work. That work, it seems to me, is better handed over to the user, eliminating the architect.

HOLSTON: How would that be? Through computers that would have banks of ready-made solutions which could be cut and pasted.

HODGETTS: Absolutely! And you can go the next step. A fellow who I visited this morning builds displays. He has a very large cutting machine which produces any shape you can draw in the computer. You punch a button and ten minutes later you have a sixteen-by-sixteen foot cutout in plywood, or anything you want, of that shape. Simple. It's not even state of the art any more. So, there is a weird imposition of the "designers' discipline" upon an apparatus that doesn't give a damn. As things are moving along, desktop architecture is absolute reality—absolute, absolute. It's where the political and economical are already coming together. It's reality. I think there are these two tracks: there is the track of the genuinely popular environment which happens with very little in the way of intersection and which is not expected to be marketable; and then there are things which are created for the market, and things which are created for the market are of a different order, a different imaginative order, a different financial order, and a different political order.

HOLSTON: For example, a community project to document its history could use desktop architecture—the computer, its imaging, its cutout mechanisms—to capture its stories in voice, in text, in drawings, in cutouts, and that is the kind of nonmarket opportunity you're talking about, right?

HODGETTS: Exactly. And that could make places incredibly rich, incredibly familiar, with that kind of texture, that kind of patina we were talking about but in a genuine sense, a very genuine sense.

HOLSTON: This technology comes out of the theater world and the film world of science fiction?

HODGETTS: Right now, because that is where the money is. The money is obviously not in the neighborhood association, so development money goes towards making systems for rock'n'roll musicians and *Terminator II*. But there is a great spin-off from that, like there is a spin-off from the space program of systems that have vast utility and will obviously be translated into other kinds of use. But I'm not suggesting this [desktop architecture] for the market track. It's a little bit like the difference between "America's Funniest Home Video" as a program, as a television program, and a dramatic film which is contrived for the purpose of being a dramatic film. They are both absolutely valid as enterprises, and in fact it's absolutely marvelous that the medium of television is transparent enough that it will accommodate both. It will accommodate both the highly focused work of not necessarily art, but certainly of consummate craft, very focused, and it will also accommodate the relatively casual activity of the man in the street and present it exactly in the same medium. I think that is pretty sensational. You don't even have to turn the dial. And I think that many of those same technologies, or many building technologies, are evolving toward that reality in the city. I think that there is where it's going to be. Just exactly in the same way that Jenny Holzer coexists with the Tow Away sign, you will have that possibility in the city. It is almost like the way in which the Gothic cathedral had all of its statuary and density of participation by the community around it in making all of the parts of it speak in the medieval city. I think you will have something like that.

HOLSTON: It's a wonderful image.

HODGETTS: I think it's happening, it's almost there.

References

Clark, T. J. 1984. *The Painting of Modern Life: Paris in the Art of Manet and His Followers*. Princeton, N.J.: Princeton University Press.

Frampton, Kenneth. 1980. *Modern Architecture: A Critical History*. New York: Oxford University Press.

Holston, James. 1989. *The Modernist City: An Anthropological Critique of Brasília*. Chicago: University of Chicago Press.

Wodiczko, Krzysztof. 1987. "Strategies of Public Address: Which Media, Which Public?" In Hal Foster, ed. *Discussion in Contemporary Culture*. Seattle: Bay Press.

Gudrun Klein

17

Kith and Kin in Borderlands

"I am here, because you were there."—A while ago I came across a variation of this line. It was a graffiti written by immigrants to Great Britain to remind their indigenous neighbors that they share a history, which might explain the foreigners' presence. Let me first locate the "I" and the "you." "I" am a German woman born soon after World War II, while "you" I presume to be Americans. I grew up in an area on the French border in which the Canadian and the American military were conspicuously present besides the French. A still-prominent feature of the landscape around my hometown are the army camps and sites marked "off limits." I spent a good part of my school years living opposite the American housing area of my home town, which is still carefully demarcated. In the 1950s the fence made it possible to look in but made access difficult, while nowadays the fence is like a garden fence and there are many openings for those who wish to enter and leave.

I remember that the fence fascinated me, but I can only speculate on what that fascination was about. I assume that it was the demarcation of difference which intrigued me. The houses of the Americans were painted pink and green, garish in comparison to German houses. The American cars were not only all different models, they were especially large compared to the still-rare German cars; the American children played unfamiliar games, using huge leather gloves and hard balls and wooden bats, tools I had never seen and was tempted to touch and would have loved to learn how to use.

The different languages, however, proved to be the "real" barrier, and like the fence it beckoned to be "overcome." We outsiders learnt to overcome the fence, as did the insiders. The baseball players needed their balls back, we wanted to explore the forbidden terrain. So, the fence kept getting holes which were dutifully mended by those who probably thought that "good fences make good neighbors." The language barriers were overcome for us outsiders only through a gradual learning process which the educational agenda of the 1950s had opened up for us. Before looking more closely at that aspect of "fences" and "borderlands," let me explore why these alluring foreigners were "there" at the time.

The area of my hometown is one example of a borderland in which struggles over political, economic, and cultural influence and domination have been played out in various ways over centuries. So, at that moment in history, the foreigners from further away, the Americans, were regarded more favorably by the people of the area than their own neighbors, the French. They, like the Americans, had entered the country as victors.

In my adolescence I assumed that the Americans were the occupying forces who had come to Germany as the visible sign of military victory over the German military forces. While it is true that the American military were the first to enter my hometown in 1945, it was the French who took over military and administrative rule in that particular area in the immediate postwar years. The U.S. forces returned after negotiations to integrate West Germany into a Western military alliance. One reason why the area was chosen for stationing, besides the military reason, was economic. The presence of U.S. forces promised jobs and a flourishing commerce after the rather restrictive economic rule of the French, which appealed to the politicians and the business community of the area.

My personal motivation to learn English was clearly greater than that to learn French, which reflects some of the intricate issues of neighbors and strangers I can only point toward here. But the desire to overcome the language barrier was most clearly motivated by a wish to participate in a culture which offered relatively spectacular things, while our neighbors appeared both more familiar and more mundane. The population on this side and the other side of the French-German border, furthermore, spoke and still speaks a dialect which everyone understands well enough, so that there is no immediate need to know French if one wishes to communicate across that particular border.

The educational agenda in the immediate postwar era fitted my personal curiosity to learn English. We were required to learn English as the first foreign language—High German being the required language which for us children was, in fact, a second one besides the local dialect, recognizable, yet palpably different, both semantically and socially. To acquire English as our first foreign language was one aspect of our education to become better democrats than our parents had shown themselves to be.

I am aware only now of what then was simply part of life, namely, that I and most of my peers grew up thoroughly immersed in American popular culture without realizing the connections between that and the military and economic aspects. Radio stations were installed through which Canadian and American GIs could listen to their own networks, and one of the perhaps intended, perhaps tolerated, perhaps even ignored effects was our exposure to such influences. While in the case of curricular choices, or concerning radio stations like the Voice of America, it is clear that the goal was direct political influence, in the case of CFN and AFN it is not.

My peers and I were exposed not only to the democratic agenda of postwar education, we also listened to American and British music and became consumers of these "subcultures" without knowing what they were. We appropriated heroes and heroines into our own lives. The "King of Rock'n' Roll" became the typical American GI, while James Dean, as a "rebel without a cause," could represent a vague sense of generational identity —a little lost, yet full of rebellious energy across borders, national and international, none of which we were conscious of at the time. What I now consider amusing and fascinating in its opaqueness then seemed "natural": these "heroes" and "heroines" like Scarlett O'Hara and others were part of my mental universe, and no doubt they all cohabited with many other now thoroughly forgotten ones of the fifties' and early sixties' music and film "menu" we were offered. The context from which they came remained blurry as was, indeed, the cultural and political context of my own environment.

While the military and the economic aspects of "our" differences were clearly in sight, yet transparent for my understanding, signs I could see yet not interpret other than on a most immediate level, the popular culture contributed to the immediacy of the lure across the fence. Through these cultural signs we were shaped and created our own connections amongst "our"selves and "you." The unreflected appropriation of that "other" culture, facilitated not least by the language learning process, but had as one of its side effects the almost complete ignorance about the "other" Germany, including its increasingly lethal border. The relative ease and affluence of "our" life-styles was taken as part of a happy coincidence of history that let "us" enjoy it without accounting for the history which shaped and in some sense created us "here" as well as "there," made "you" and "us" connect through at least partly disconnecting "them" over "there."

To talk of such shaping of identities as part of the economic and political agenda of integration into the West does not entail a claim to one neat causal link between cultural and political institutions and our personal lives, rather *I* wish to indicate aspects of ongoing processes of making meaning of our selves in which different agencies interact constantly. Nor do I wish to claim that such shaping and creating ever determines our sense of our "selves" in an entirely predictable or, indeed, sufficiently recoverable way.[1]

For the purpose of this story of kith and kin in borderlands, let me now look at some features of my mother's story, hers, and, insofar as she represents many women of her generation, that of German women of her class and generation more generally, before glancing at some of our kith in the "other" Germany. The story as I tell it here she would never tell anyone, not even herself, although, by now, she might listen—with pain—to my telling it.

My mother also belongs to a postwar generation—that of the First World

War. Her generation developed into a prewar one. My mother comes from a working-class background, and family life in her experience was rather oppressive and unhappy. So, according to her later self-understanding, she wanted to have peer company and flee from the everyday dreariness of her adolescent life. She slowly began to embrace and enact the Nazi ideology because it offered, first of all, an emotional space for identification with other young women in the League of German Girls they could come together over activities they enjoyed and which seemed "natural" for them to engage in.

That the Nazi ideology embodied an explicit social and political alternative to other political ideologies, was not clear to her at the time, but she could later formulate her dissatisfaction with some of the then-prevalent political slogans. To join the movement meant for her individual empowerment and the gain of social space in which she felt she could move more freely than either her family life or her work situation permitted. She saw only good things evolving from such activities, which, she also insisted, fitted very well what she and her peers had thought anyway. She claimed later that she and her peers were quite unaware of the concurrent actual political agenda.

She partly accepted as necessary, partly consciously overlooked the destructive, the antihuman elements of the ideology she made her own. She saw the "necessity" to exclude everyone from the community who did not fit, be it by race or by any other trait deemed unfit for the national community. She was willing to accept even the war, if it meant realizing what the male leaders of the national community promised, namely, living space, growing prosperity, and power.

Her sense of female powerlessness as the youngest daughter of a family of five, who had to work from the age of fourteen to help support her family, bred a sense of personal injustice being done to her. She would have preferred to be permitted to continue her school education, which she would have been perfectly capable of doing. It was a grudge she still holds against her parents, and against the politicians who contributed to the economic stress of the late twenties and early thirties.

What she heard from the Nazis fit the prevalent ideas of femininity which were part of a patriarchal pattern of relative empowerment of women as "the Other" in Simone de Beauvoir's sense. She wholeheartedly embraced a notion of womanhood as secondary to but useful for and revered by males. She was willing to be subordinate, if she felt she was respected for what she did. She came to value her material autonomy through her work and this pride combined with her grudge at having been kept from getting a better education was a strong motive force in her sending me to high school after the war.

Her anti-Semitism was mainly nourished by her sense of material and social inferiority, the other side of which was her readiness to look down upon and ostracize everyone who did not live up to the petit bourgeois standards she aspired to herself. Having grown up in a border area where the struggle over dominance was part of folklore and occasionally everyday reality—in the twenties there were separatist movements in this area whose members were considered terrorists and treated as traitors by the "right-thinking" Germans because they favored the French—an ideology which promised a new German identity in safety, both materially and geographically, was appealing. Personal feelings of maltreatment, of undeserved lack of recognition, of powerlessness, combined with mostly unrecognized feelings of shame and a desire to improve one's own lot fit the ideology of national shame through the defeat of World War I and was usable for purposes the consequences of which were disregarded or taken as a lesser evil since they did not pertain to one self in any immediate sense.

The appeal of Nazi ideology for someone like my mother can at least partly be understood as related to contested spaces, both socially and geographically as part of an overall political issue. But in terms of individual lives and day-to-day self-understanding, what counted was the perceived immediate increase in personal satisfaction from actions one did not look at critically and in a wider perspective, the relative material and social stability, and the hope for lasting betterment through a general increase in power and social and political prestige for all "good" Germans.

While my mother certainly found the war frightening, she was willing to serve her country; she accepted all measures forced upon her, and then enacted by her, such as finding work elsewhere when her hometown was evacuated, or the stationing in a foreign country, which she dreaded, as necessary and in some sense as unavoidable if one wanted to be what the Nazis promised: that all "good Germans" would be socially and politically powerful, not least through forming a strong, homogeneous, and safe community. To sacrifice personal well-being for a while, seemed only "right," not least because the "others" had all too often enjoyed what it was each German's "right" to (re)gain.

Such reasoning, I wish to stress, was probably never truly conscious, it may have been expressed in ritual moments of celebration—in the League activities, say—and/or in situations where self-justification seemed called for—especially during evacuation and the war. In everyday life, however, it seemed to have been nearly altogether transparent—too "natural" to be visible to someone who was not educated to be self-critical and to question authorities.

My mother, like many Germans, resented the Nürnberg Trials and the efforts at reeducation after the war—and, of course, different people harbored resentments for different reasons. She could find no fault with the beliefs she had shared with the majority of Germans. After the war, she could only see their defeat as an unjust or at least an unhappy turn of fate. A power struggle, her reasoning would have been, had she reflected upon the issues at all, had been risked and lost. In her view, the victors had no right to judge and to tell her what to think and do in the future simply because they had defeated them in the war.

Defeat, I assume in hindsight, was yet another shaming experience she wanted to forget as quickly as possible. Once defeat was inevitable, she was ready to accept the consequences and was relieved to find that at least the Americans were not especially hard, even relatively kind, to her when she was held as a prisoner of war. The French who "de-Nazified" her—together with what in her mind are still traitors, the socialists or communists who helped identify active Nazis—she spoke less well of, but on the whole she did not complain about the victors' attitude. The immediate postwar period was, in her memory, simply (another) one of extreme material hardship, a struggle for survival, in which it mattered, because it made an immediate difference regarding that goal, whether officials did or did not treat one humanely. My mother concentrated on making ends meet and tried to ensure that I could enjoy an education which would empower me more than she had ever been. She was (kept) busy acquiring a living standard which she regarded as desirable and closed her eyes, this time very consciously, to all social and political issues. As to her experience with the Third Reich, I submit, she, in her own way, enacted what became simultaneously the political agenda of postwar—soon to become West—German politics. The Allied Forces, not least the Americans, cooperated with those German political—predominantly male—leaders who were keen to "leave behind" the recent past and rebuild a "new Germany." The economic and political measures taken immediately after the war show that the desire to restore, rather than to radically renew the body politic, prevailed. A *slow* change of political, social, moral attitudes and the careful debate of alternatives regarding the social and political sphere—most notably the possible revision of patriarchal structures and the viability of economic models other than the capitalist one were clearly not in the interest of the Allied Forces and German male political leaders at the time. Those issues were to become the ferment of the student movement of which I became a part; a movement which in its turn was inspired not least by movements in the U.S. which questioned the legitimacy of aspects of U.S. social and political life and policies at home and abroad. The readiness to question publicly what had come to be known as the "economic miracle" grew out of the experience of my generation growing up in the 1950s and the conflicts inherited through the immediate past.

I happened to have school teachers who were willing to discuss the issues of the Third Reich rather than skipping those sections of the curriculum, and so my school education cast my mother and myself as antagonists in what I later began to comprehend as a much wider than simply a personal, intrafamilial generational conflict.[2] What I learnt at school did not fit with what little

I had heard about the past at home. It did, however, lay bare some reasons why silence over the recent past had reigned generally.

As an adolescent I questioned and challenged my mother with great vehemence. As the parent, she first tried to refuse to be made accountable at all, then she denied anything disastrous had happened. She refused to acknowledge the sufferings of the victims of German policies. Neither the unimaginable destruction of the Jewish people, nor the sufferings of all others who were ostracized in the Third Reich, nor, indeed, of the war victims of other nations, made her show emotional signs of remorse. In as much as she registered the voices of the victims at all, she first discounted their claims as lies. When pressed, it was the victims' personal fault or their otherwise general deficiency that had brought them to be ostracized or eventually destroyed. When much later she expressed gratitude for gestures of humaneness shown toward herself when a member of the defeated army, it seemed to me a first sign of an appreciation for understanding and warmth shown by others. This, however, did not prevent her from expressing the same coldness as before toward the victims of Nazi policies. When I ask my mother now to acknowledge her shame and guilt, she can do so at least verbally at certain moments. And only recently could she admit that she had to forget in order to go on living, but not without asking *me* what good it does me to know.

One may interpret such moments of acknowledgment as moments of recognition. Not that she will recognize the victims, but a glimpse of such recognition it is nevertheless. It is the recognition of a part of her own self she had suppressed. What began rather violently antagonistic during my adolescence has slowly evolved into an engaged conversation with the goal of trying to understand and be understood. In this process it gradually became clear to me that my mother could perhaps only cope with what had happened by denying that it had happened, or more simply by forgetting everything that was too shaming to face. And it may be that process of understanding that makes it possible for her to at least verbally agree that what was an essential part of her life was disastrous for many and lethal for people who did not deserve to die for the sake of realizing goals of homogeneity and grandeur she had espoused if not all deliberately. It would take more of such experiences of mutual recognition for her to be able to acknowledge an Other in her and his own right. I have little hope that such a change can still happen; too much seems to have been invested in keeping pain and shame under control and getting on with life. To dwell on the causes of these events, to account for her own implication in destruction, to admit the also self-destructive elements of her own attitude then as well as later, to reflect on the consequences of it all or to mourn was and still is too painful, too demanding, and she seems to have avoided that work for almost her entire life, as have many of her generation. Luckily, it seemed, no one in West Germany insisted on such work for long. In the immediate postwar period there was a brief span in which alternatives were audible and visible. They were submerged in the "reconstruction," which, on a personal level, meant for people like my mother, indeed a restoring of attitudes and behavioral patterns which had contributed to the destruction it now served to leave behind, to make invisible, to "make good."

Over the issues of the Third Reich I rather painfully learned that truth is highly complex; that, where verification is not easily possible, one has to accept that different authorities may render very different accounts of events. My mother's account, together with many other accounts I have read and conversations I have participated in over the years, has made me slowly aware that I as a member of my generation participate in other events, follow other priorities which are or may turn out to be destructive. And I, too, certainly exclude from consciousness much that I may be held accountable for any time from anyone who is victimized by my very way of life.

For my mother the feeling remains dominant that not only was she not aware of the atrocities committed in her name. She concludes that she was therefore also not personally guilty, and hence is not personally answerable, neither on a moral nor on a legal plane. I cannot and do not wish to come to that conclusion regarding the issues of today or of her past. I believe that especially through the process of unification, issues have come to the fore again which resemble issues left unattended after World War II and which need careful attention in personal as well as in economic, cultural, and political terms.

Born into the space of the Western Allies after World War II, the legacy of the Third Reich is for me to understand what it is to be a "normal human being" in my given and partly consciously appropriated context; to comprehend how "normal" human beings can become implicated in processes which are incredibly destructive to unimaginably many human beings; to recognize how far and in what way I and others are implicated in processes which may be or turn out to be similarly catastrophic as the Holocaust, even if or especially when they are not propagated as blatantly racist and exclusionist as Nazis propagated their goals or as some radical voices do now.

My mother is not alone in never really having changed her convictions, nor is her generation the only segment of German society sharing views which one may call anti-Semitic, anticommunist, homophobic, racist, and latently chauvinist regarding national identity. People with such convictions take patriarchal power structures and values for granted and look for what can be termed masculinist solutions to problems. Among those attitudes anticommunism became an acceptable shared value and has openly been part of the political and social agenda of our society since the 1920s. The patriarchal fabric is still too transparent—too visibly normal to interpret as a fence that needs to come down. Masculinist solutions are favored not only in Germany, but in many parts of the world, as are the other aspects mentioned: racism, anti-Semitism, as well as a more or less explicit chauvinist notion of nationhood. How it happens that we tend to overlook them, assuming they have either never really existed, have disappeared, or, indeed, that they are part of what is taken to be "normal" is part of the issues I am exploring here. The unification of Germany brought to the fore some of these issues again in unexpected ways. Let me therefore now tell another story, risking a look across that "other" border, once called the "inner-German" border.

In a middle-size West German town in the mideighties I became part of a rare spectacle: the theatre company of a small East German town which was renowned for its radical productions had come to play Heinar Kipphardt's *Brother Eichmann*—the play focusing on Adolf Eichmann, the main organizer of the transportations to the death camps during World War II, whose trial gave rise to Hannah Arendt's notion of the banality of evil. The play had its world premiere in Munich in 1983. Its author became well known as a dramatist through his "scenic report" on J. Robert Oppenheimer (1964), based on Oppenheimer's trial during the 1950s—a trial motivated not least by the cold war and McCarthyism in the U.S. of that period—and it seems not altogether beside the point to mention here that Kipphardt came to West Germany from what had begun to be East Germany soon after World War II.

The local theatre company of the West German town had put on its own production of *Brother Eichmann* not long before. The main actor who had played Eichmann is in the audience that night. He, too, is one who crossed the borders east to west while it was still legal to do so, before the building of the wall. Everyone is curious to compare the productions. The lead actor—lets call him Fritz—leaves the deepest impression that night. Fritz will become the focus of attention not only because of his role, but also because he is one of the liveliest talkers of his troupe. After the show some of us are curious enough to meet the relatively "exotic"—because East German—actors. We greet them as they come out of the stage door and thank them for a moving theatrical event. While they are pleased to be met with such appreciation, some are clearly reluctant to get involved in a conversation. The invitation to join us for a beer in a nearby bar is taken up by a handful of the actors, Fritz among them. It turns out to be a very enjoyable evening which marks the beginning of some friendships across the then still rather impermeable border between East and West. Our East German visitors casually indicate that at least one Stasi informant is among us, but we West Germans cannot tell which of the group he or she might be. No doubt, such danger adds to the "exotic" flavor of the occasion; in other words, we West Germans are not quite sure how serious to take such a danger—it certainly increases our curiosity and probably enhances the East Germans' awareness of their differences amongst themselves and their otherness toward us.

Over a few beers we get into a discussion not only of the politics in the play, but also how these politics are seen in an East versus West German context. Cautious responses indicate interest in further contacts. Our local "brother Eichmann" invites us all to his house the next day. We spend an unusually warm late summer day on a lovely porch overlooking a garden; we talk, eat, drink and enjoy that unexpected yet welcome East-West encounter. The late summer warmth blends curiously with the vexed issues we discuss with intensity throughout the day. Some of us debate what intricate financial arrangements one would have to make if the East German state ever truly considered opening the borders for such professional and other, more tourist-oriented, traveling. The consensus East and West seems to be, for simple practical, especially monetary reasons—lack of so-called hard currency—it is inconceivable that the East German authorities would ever risk doing that. We are assured that the Stasi informant is not present, and we note the cheerful, even whimsical-seeming acknowledgement that one way or another they all will be reported on for the sheer fact of spending time with us. Each, they say, will tell his or her own version of what happens.

Only after unification did we begin to look at the facts of how such narratives were asked for, who delivered what to whom. Whether we will ever understand what they were used for, what their effects might have been then, what they now are remains open. Through the political changes since that crucial period of connecting as a new/old unit, the sense of exoticism has been replaced by differing degrees of recognition, of getting to know each other. Not everyone in East and West Germany is happy either with that change or with the challenge of having to get to know each other, of having to face mutual recognitions. The causes for skepticism, even resentment, are different, however, and it will take time and patient attention to detail to understand this process of "familiarization."

That day in the fall I cautiously ask Fritz whether he sees any similarities between the Kipphardt play and any aspects of East German reality. His eloquence and vehemence subside, he falls silent, and when he speaks again he remains politely reserved about the issue. While trust has been established, it clearly needs to grow for an answer to be risked. That risk will be taken later, when circumstances are entirely more favorable to such probing.

The "other" Germans only gradually became more concrete, more "real," if you will, as a part of an ongoing process of understanding my self, my culture, and its various contexts. A visit to East Germany and Poland and having—if only briefly—been exposed to everyday life behind the iron curtain made the question who "they" are more urgent for me. It is worth noting that Kipphardt's title refers to Thomas Mann's essay "Brother Hitler," written in 1939, when Mann was living in the U.S. as an emigrant. In this essay Mann argues that it is more important to recognize ourselves in Hitler than to hate him. It seems now still and seemed to me then that at least the West

Germans have not come to terms with the challenge in Mann's proposal. I assumed(d) that Kipphardt wrote his play not least because he realized that the "banality of evil" which Hannah Arendt had diagnosed during the Eichmann trial was dangerously "normal" for all too many people still. I did not feel superior to Fritz in any sense, I wanted to understand how he, as an "other" saw himself in his context, and what "brother Eichmann" meant for him. The silence that had fallen between us bespoke the space between us, it indicated an invisible barrier which our common language had hardly begun to overcome.

"I am here because they were there." In June 1992 I visit Fritz, his new wife—let us call her Bertha—and their recently born son in a three-room apartment in East Berlin, close to the former German-German Berlin border—"the wall." Where formerly West Germans had to go through a series of checkpoints before they could reach East Berlin, now I simply drive through West to East Berlin and can look with amazement at the now defunct installations of the surveillance systems which remain as a sign of the times hardly passed.

I have come to expose myself to this new sense of easy connection between East and West, to experience the changes unification engenders in the space of the former border, and to renew our friendship. I am especially curious to hear what Fritz and his wife think and feel about those dramatic and unforeseen changes. On my way my eyes are caught by a Trabi—the East German equivalent of the once all-German, then West German Volkswagen. It is a wreck sitting on the sidewalk at an oblique angle to the main thoroughfare. In its askewed state it quietly bespeaks the sign above it: a colorful U.S. Western landscape of a Marlboro ad with the words "Go West" crossed out but not erased, visible, legible, open to interpretation. Now that "the West" is "here" the Trabi has nowhere to go.

During my visit we slowly and deliberately walk across the space where the wall had been the forbidding no-man's-land between East and West Berlin. One can still see the traces of the wall, some hints of the watchtowers are left, the road connecting both parts of the city is under construction. We delight in the leisurely pace measuring the distance from East to West, look forward to a meal at a small Greek cafe, and when Fritz suddenly exclaims with angry intensity: "these fools opened the wall much too quickly and only to save their own skins." We agree and have to laugh at the paradoxes we intuit as also being ours.

Fritz and Bertha admit that to be able to get rid of their old Skoda, to buy a Mitsubishi station wagon, to enjoy a greater range of wares on display, to travel more extensively, are pleasures eagerly taken advantage of as far as their very limited budget allows. Those pleasures, however, conflict with the sadness and the deep skepticism at what has happened in Germany since

World War II. We share the sense that another opportunity has been lost. Yet again, if differently, an appropriation and too sudden an identification has taken hold of the majority of our fellow Germans, of "us." The sense of belonging to what is more powerful, greater, better, at least in all matters visible has overpowered other senses. It is a moment of recognition across social and cultural differences, across now open borders, communicated in a language which, as we are still finding out, is both shared and unfamiliar.

I pose Fritz the same questions I asked him when we first met: what made you choose that play, *Brother Eichmann?* How did you feel about being the lead actor? Did you see any similarities between your situation in East Germany at the time, and if so, what were they? No simple questions, no simple answers. Painful processes of remembering, now a readiness to share. Fritz's words come out in a torrent, he is very energetic and at the same time reflective, self-critical, and ironic, often humorous in his tone and idioms, roaming associatively through space and time.

How real the horror of Eichmann's banality was as part of history did not grip Fritz or his fellow actors. They took for granted when preparing the production that everyone by now knows the terrible facts of the genocide and they were not interested in showing the beast, the devil. Instead they wanted to explore what it means to be Eichmann's brother everywhere and in the present. "During rehearsal we laughed ourselves to tears because we felt so close to him"—the similarities between the system which seemed to have been at an end in 1945 and their own were shocking and hilarious. It was a laughter of painful shame. "The pain and the shame were a powerful motive force to do the production. Eichmann didn't understand anything intellectually, conceptually, but he intuited that what he was doing was terrible nevertheless; he was ready to suffer and inflict pain on others. We wanted to change society, we thought hard about how to effect our ideals, yet there remained a gap between our actions and their results. It is eerie to realize how your thinking is nothing in relation to the horrifying immensity of your actions." On reflection I want to say, isn't our—my?—problem with Eichmann's problem, though, to have accepted the immensity of the consequences of his actions by focusing in his thinking on the limited nature of his own contribution to this immensity? During our talk all I can do is be attentive to Fritz's train of thought.

"During rehearsal something terrible happened—my voice failed me. I felt I could not possibly continue playing this role—a horrible pity for the man gripped me—such an engulfing pity overpowered me that I thought I cannot possibly continue. It had never happened before. A frightening sense of terror, desperation, a sense of the inescapable connection between guilt and innocence overcame me, held me in its sway. I felt it was simply chance that had saved me from becoming an Eichmann myself, pity turned into self-pity, and the play seemed to me to be a very German tragedy, suddenly, not just something that happened to Germans under fascism, as fascism, but something more encompassing." I can empathize with Fritz's feelings, especially with his own sense of failure regarding the claims of socialism. I spontaneously question his diagnosis that it is a German tragedy.

We cannot sort out what our disagreement is. And as before, I have to detach myself in space and time to reflect why his association leaves me uneasy and unpersuaded. If tragedy is an experience in which a choice has to be made between an incompatible, yet equally obliging (seeming) alternative, in my understanding, the Germans in the Third Reich whom I have encountered and have been talking about in this essay, did not ask such questions. Furthermore, if tragedy also requires the recognition of an inescapable dilemma and the concomitant feeling of guilt for having to choose and having to live with this guilt not proudly but humbly, that, too, does not apply. The matter-of-factness in Eichmann's attitude toward the immensity of the consequences of his contributions to the Holocaust don't fit that definition either. Unless you assume that what unified the Germans who actively and passively contributed to the destruction, who perpetrated or tolerated it, was the sense that to win greatness one had to pay the price of suffering, inflicted upon others, and, if necessary, to bear it oneself. It would seem, that suffering itself could be desired as the only means toward greatness, and, of course, that suffering was, if at all controllable, not one's own. It only became that in the course of "unfortunate" events. For men of the SS, as Goebbels's speeches, for instance, show, their own suffering—when having to harden themselves, say, against pity for their victims—was itself considered as a mark of greatness which was used as part of the self-justification toward the victims, who, through their suffering proved their own inferiority. No conflict of aim, no conflicting claims; unless you take the need for such justifications, their very utterance, as an indication of an albeit dim awareness that not all may be well with one's aim and claims.[3]

Fritz and I can only speculate about what constitutes our cultural identity to declare both the experience of the Third Reich and the rise and fall of socialism/communism in East Germany as peculiarly German, "our own." All we can agree upon readily—if painfully aware of the inadequacy in view of the scope of the question—is the tendency to yield to authority, including the authority of people who claim to have the "grand picture" in terms of a thought system they suggest to "implement."

After Fritz had overcome his professional and also profoundly personal crisis, they decided to take the production to the West. "We did not want you to feel complacent, and the reactions of our audiences showed us how close we really are in terms of the play's questions. We did not want to expose the GDR to the FRG as if we were the problem and you were not. Because in many ways the main work is still to be done. The reactions of our audiences showed that."

Fritz looks back in anger now. His father was a socialist who joined the National Socialist movement relatively early. He believed that the "workers of head and hand" had to join, as the movement claimed. He was short-tempered, good-humored, dangerously authoritarian, and very destructive as a family man. He was domineering and Fritz's mother had little joy in the marriage. They were poor and his father did not provide for them; on the contrary, he would usurp what privilege the patriarchal system offered him. He turned away from fascism slowly and after having served in the war he tried to tell his son that something different was needed.

"I had learnt to stand up for my mother against my father, whom I hated. For a long time my attitude toward life was shaped by my rebellion against him and by my loyalty with my mother. We struggled as socialists to make socialism work. We debated everything, but it took me thirty years to realize that we did not effect what we claimed we were doing. The corruption began to dawn on me only in the eighties. Stalin had been no problem for us; we discussed Stalinism long after Stalin had died and thought we had understood. Only very slowly did I become disgusted with what I began to see around me. But I turned away, focused on my private life, gave up on bringing together theory and everyday reality."

Fritz and I can agree that we have not understood well enough the dimension of what Fritz calls "the psychology of the unpolitical" which we both can see as a legacy in East and West Germany. At the same time his crisis marks an awareness that political ideals—perhaps those in East and West—themselves may be tainted with that very "psychology of the unpolitical."

Upon reflection I realize that Fritz has followed his father's beckoning through his rebellion, through his desire to realize socialism, through his wish to be the better man in a better society, the son who conscientiously tried to bring about what the father in his own time once had hoped for. The son took on the father's role in more than one sense; he, too, had been an actor, who, like Fritz, found a socially accepted expression of some of his passions through (dramatic) art. It occurs to me that my skepticism towards Fritz's summing up of the political crises he has lived through is related to the fact that the realm of the aesthetic for me is as clearly "political" as all the other spaces we occupy. No refuge I am ready to take or grant.

Fritz and Bertha agree that unification as it happens does not allow them to slowly and creatively work out such questions, neither amongst themselves, nor between "us" siblings, East and West. We agree that we are brothers and sisters to Eichmann and Hitler, that socialism has not undone patriarchy; nor had it begun to really transform capitalism. Shame and pain work in Fritz when he admits that they failed in this effort, not simply because capitalism was always there as the lure, as the "evil other," but also because it, too, obviously remained part of their own identities as individuals, as a culture, as an emerging state.

What causes Fritz pain and shame does not disturb another friend of mine. He, let's call him Hans, was born before World War II in Saxony, a region which after the war became part of East Germany. He came to the Federal Republic soon after its creation, and he is profoundly happy that unification made it possible for him to go back, renew old ties, reconnect with the cultural space he left deliberately yet not willingly. As a youngster he was penalized for belonging to a class which, in the new self-understanding of the then ruling class, had been responsible for fascism and its destructive successes. So Hans chose to go west because the education which would have befitted his talents was denied him in East Germany. He is a successful lawyer who has no truck with socialism, but, in spite of that, he is entirely free from the often fiercely aggressive anticommunist attitudes of many of our fellow Germans in the West. He could fit the notion that the safer you feel, the more liberal you can be(come). Applied to some of the disturbing aspects of life in contemporary Germany that would also indicate that both the people who resort to violence against others and those who resort to stricter legal punitive measures, for instance against asylum seekers, do not feel safe. The issues of such feelings of unsafety remain open, but only for the actors and spectators, not for the victims of aggression.

When I ask Hans after unification what his sense of identity comprises, he speaks of different facets which he finds difficult to keep separate and prioritize. When Hans thinks himself as "I" he identifies this "I" with his first name and then with his paternal family name. The first name marks him as unique, while the latter indicates his (paternal) kin, which he immediately associates with the kin of the family whose male head he is himself. Patriarchal norms are truly "normal" for him, they belong to his worldview as "naturally" as his kin.

Without being able to name the characteristics, he insists that part of his kinship identity is his being a Saxon. That label, however, is relevant foremost to define and demarcate his sense of identity toward fellow Germans. Toward non-Germans his being German takes precedence. Toward people generally he claims then that his sense of identity is most clearly shaped by the material possibilities his family of origin provided. He was born into a family of well-to-do entrepreneurs, and he is convinced that that has shaped not only his own aspirations and attitudes but also those of his blood relatives, some of whom still live in the former East Germany. Social class and your profession override in his mind any systemic, institutional shaping of identity. That conviction makes him declare that his relatives will make the best out of the challenges of unification as his family has been wont to under former systems and their challenges.

For Hans the fact that the nation has been united into a state again is satisfying. It brings together "what belongs together." But when I probe the notion of what this "belonging" consists of, he has a hard time defining it. It cannot simply be the German language because too many groups speaking German have lived in different cultural and political contexts. It cannot really be blood relations. And it is his wife who knows about the fragility of the notion of "belonging." She lives just beyond the German-Polish border before the war and had to flee west after the war. She became an outsider in a German community simply by being from "elsewhere," speaking a dialect that was unfamiliar to her new neighbors, and being relatively poorer than they. Hans is too clearheaded, too liberal and sensitive to the issues—especially regarding asylum seekers and xenophobia within Germany and elsewhere—not to see the dangers of the claims of nationalism when combined with economic and military power. Yet he does not share Fritz's or my sense that there is anything questionable either about the process of unification itself or about its immediate or possible long-term consequences. True to his optimistic entrepreneurial attitude he believes that "somehow things will work out." Weighing Fritz's and Hans's words, reviewing the encounters I have had in the recent past trying to understand processes of identity creation, I keep wondering what work that will (turn out to) be.

Notes

1. Borneman 1992 is a differently focused and much more extended discussion of related issues.

I see relevant differences of underlying premises between Borneman's account and mine, especially in his effort to give yet another "grand narrative" to explain through reconstruction what the shaping of East and West Berlin identities were. I can see the function of such metanarratives to be to order the narrative of the anthropologist organizing his material, not showing conclusively that such narrative strategies were indeed inscribed into the activities, let alone intended by the agents under discussion. Another is his relatively contradictory account of East and West German agendas, while he can identify certain effects of measures taken, for example, by Hilde Benjamin, the minister in charge of family laws in East Germany, he finds it hard to acknowledge similar effects for measures taken in the West, claiming that the "capitalist agenda" "avoid[ed] public regulation, instead relying on "market forces or agents acting in the private sphere" (82) as if legislation regulating family life—for instance defining the family after World War II as a traditional unit of father as wage-earning head of the household, mother as housewife, and at least one dependent child—excluding and hence discriminating against all the units which did not fit this category—notably single mothers—had no effects. I interpret this contradiction as bespeaking the narrator's value system; looking from a capitalist system makes it hard to let go of the notion of private space as opposed to the public space, the values connected with the former are opposed to those of the latter, thus replicating in small detail the larger effects of these ideological creations and encounters for which Borneman tries to account. The fact that he casts the West in the satiric genre and the East in the romantic I read as a further sign of his unrecognized privileging of the Western mode if one follows Western attitudes of privileging rationality over emotion. Both the grand narrative gesture and its cultural and more specifically male bias lessen for me the value of Borneman's otherwise fascinating and illuminating study.

2. Arnim 1991 and Sichrovsky 1986, 1988.

3. See Nussbaum 1986, especially part 1.

Bibliography

Arnim, Bettina von. 1991. *Das grosse Schweigen. Von der Schwierigkeit, mit den Schatten der Vergangenheit zu leben*. Munich: Knaur.

Borneman, John. 1992. *Belonging in the Two Berlins. Kin, State, Nation*. Cambridge: Cambridge University Press.

Nussbaum, Martha. 1986. *The Fragility of Goodness. Luck and Ethics in Greek Tragedy and Philosophy*. Cambridge: Cambridge University Press.

Sichrovsky, Peter. 1986. *Strangers in Their Own Land. Young Jews in Germany and Austria Today*. London: Tauris Basic Books.

———. 1988. *Born Guilty. The Children of the Nazis*. London: Tauris Basic Books.

A Look Forward: A Preview of Volume 3

Christopher Pound

18

Imagining In-formation: The Complex Disconnections of Computer Networks

The rise of the computer, its infiltration of all facets of life, and its apparent convenience and efficiency are some of the more well-known stories of late twentieth-century modernity, but a less publicized and perhaps more interesting story is that of the history and present futures of the communication networks that link computers and users in complex, obscure, and nonmachinic ways. This is the Net, known to a few of its users as the world's "Numinous Electronic Tangle," and it is an always plural object with neither inside nor out, as much an artifact of discourse and imagination as a thing of glass, metal, and plastic.

The "Net" and Ethnographic Authority—An Aside

Borrowing from a famous on-line dictionary of popular computer terminologies, the Jargon File v.2.9.6, we have:

> network, the: n. 1. The union of all the major noncommercial, academic, and hacker-oriented networks, such as Internet, the old ARPANET, NSFnet, {BITNET}, and the virtual UUCP and {USENET} 'networks,' plus the corporate in-house networks and commercial time-sharing services (such as CompuServe) that gateway to them. A site is generally considered 'on the network' if it can be reached through some combination of Internet-style (@-sign) and UUCP (bang-path) addresses. See {bang path}, { {Internet address} }, {network address}. In sense 1, 'network' is often abbreviated to 'net.' (Raymond 1991)

When I call this thing, these people, and their words, the "Net," I mean to include myself among them, and in writing the Net, I am inscribing my name into its immense figurations, always becoming only an assemblage of already-saids and inconsequentials. For me, the Net includes a great deal of unmappable territory: networks I have never heard of, networks I have no access to, networks that to me are only names, and the interlinkages of these networks that I can only postulate based on my experiences as a user, a system manager, a network programmer, and a native. Imagining the Net this way (as something decentered, disguised, dislocated, and disconnected) will have been most appropriate for the way it evokes the users' forgetful consciousness of the occurent spatiality that they have been led to believe separates them.

In the past two or three years, the future of computer networking has become a high-profile issue in government, corporate, and academic circles. The problem lies in deciding how to effect improvements in the United States' "information infrastructure," a label that tells all by itself much of the way this debate has unfolded, considering the fact that, in government-speak, "infrastructure" is typically a code word for those physical products of civil engineering that civilians are supposed to use but not have to think about. Unfortunately, this metaphoric association of computer networks with such things as highways, dams, railroads, and power lines completely glosses over the Net's self-referential character as a medium of communication; it obscures the fact that the Net supports a system of signification defined, at least in part, only in terms of itself. Plus, thinking of the Net as "infrastructure" often involves an almost millenarian rhetoric of decadence followed by progress, as in "we're now drowning in information," but "we can see the vistas opening up on the other side" (Gore 1991:21, 23).

According to many, the solution to America's supposed electronic languidness is to build the National Information Superhighway (to use Al Gore's original name for it; "National Information Infrastructure," or NII, has become the standard term), a fiber-optic hyperstructure thousands of times faster and more capacious than the networks currently in place. Several questions about civil rights on the proposed glass-fiber network remain unclear, but for this paper, the imagined history and imagined community implied by those questions, the narrative structures that organize them, and their stakes in the near and far futures of the present are really the matter at hand. As a strategy for addressing this matter while recognizing the decentered and disseminating quality of these discursive formations, I want, first, to reconsider the Net.

In popular culture (that is, in popular media such as magazines, science fiction, or the Usenet), the Net has sometimes been constructed as an alluring space of neon iconics, rogue technopunks, and insidious machine intelligences, accessed by means of a prosthetic sensorium. It is this psychotropically enhanced vision of the Net (one which can only be grounded in the everyday experiences of the Net-workers through an extremely tenuous chain of metonymic substitutions) that continues to be chased down by today's retro-Futurist critics of virtual reality. However, "cyberspace," as envisioned by William Gibson and his devotees, is not at all an appropriate figure for evoking the world-image world of the Net, and even though it may accurately represent the scientific imaginary governing much of the research on virtual worlds, there are already several competing theories of how the Net might be transformed in coming years. I will return to these alternative futures once the grounds on which they are predicated have become clearer.

Other popular constructions of the Net include stories about hackers, paranoid surveillance fantasies, and recent concerns over "computer porn." All of these popular notions about the Net, including cyberspace, are, of course, embedded in the Net itself, in discussions on the Usenet, in Multiuser Domain settings, in conversations in electronic mail and on Internet Relay Chat; they figure intricately in Net-worked discourse on the Net, sometimes forming up as counterdiscourses to the rigidly realist authority of the system managers or government agencies, but more often (re)appearing with all the political guile of a newspaper's Sunday feature. In any case, there is always something else to them.

Invisibility

> Millions of people now inhabit the social spaces that have grown up on the world's computer networks, and this previously invisible global subculture has been growing at a monstrous rate recently (e.g., the Internet growing by 25% per month).
>
> —Howard Rheingold

> The construction of order in every day (modern or traditional) life depends on the postulation of at least dual, parallel worlds, one of which is unseen.
>
> —George Marcus

> Years ago, I created a new word called "ex-formation," information that exists outside the conscious awareness of any living being but that exists in such enormous quantities that it sloshes around and changes the context and the weight of any problem one addresses.
>
> —Albert Gore, Jr.

The Net, because of its largely invisible, acephalic, and even amorphous organization, has a difficult and shadowy history. A particular poetic encounter with the electromechanical technoculture in which the Net has its roots begins in the 1850s: "This is the beginning of a period in which the strategy of using shock to deal with fragmentation is transformed into seeing the multiplicity of codifications of municipal (or urban) reality" (Theall 1992:n.9). If the Net has any traditional or commonsensical narrative that shapes the way stories are told about it, it is the myth of the metropolis, the free and democratic city necessarily burdened by (yet in some way distanced from) some amount of bureaucracy. But, to begin again, the Net is also constituted through the myth of the synapse, the "I join together" of neurology, cognitive science, and Larry Roberts:

> Convinced that computer networking was important, the first task was to set up a test environment to determine where the problems were. Thus, in 1966, I set up two computer networks between Lincoln Laboratory's TX-2 computer and System Development Corporation's Q-32 computer using a 1200 bps dial channel (high speed in those days). The experiment showed that there was no problem getting the computers to talk to each other and use resources on the other computer." (Roberts 1988: 145)

This first coupling/conversation of computer with computer gave birth in time to ARPANET (owned and operated by the Advanced Research Projects Agency, a government office attached to the Department of Defense). Although the origin story of the ARPANET is not well known, its form, focus, and mythical undertones are unsurprising: "It look an individual and it took the experiment to get it to happen. This is one of the cases where the thing wouldn't have happened by itself" (Bell 1988: 170). At first, each new connection appears as a new origin for a new branch of the network, but this illusion (founded on the distinction of some boundary separating cables, multiplexors, servers and routers from "hosts," the computers themselves) eventually dissolves when networks overrun one another and begin to fill themselves in. ARPANET was planned in 1967 and grew from four nodes in 1969 to over four hundred nodes in 1983, but its ramification was negotiated in multiple ways throughout those early years.

During the same period of time, several other networks were established in the United States, Canada, Europe, and Japan, each with a different degree of bureaucratic, corporate, and state controls. Network connections accumulated in a rapid and extremely complicated manner. The networks, themselves, were interconnected and occasionally overlapped; for example, the French network, CYCLADES, was connected to other international networks such as Telenet and also contained within itself a communications subnet named CIGALE, "a pure datagram network, moving only disconnected packets and delivering them in whatever order they arrive without any knowledge or concept of messages, connections, or flow control" (Roberts 1988: 158).

Similarly, the networkers—the people responsible for designing the networks—moved from network to network in a familiar pattern of intellectual and capital migration, founding new companies and new networks yet often remaining attached socially and electronically to their former networks. Larry Roberts, one of the architects of ARPANET, describes how

> the success of the ARPANET led Bolt, Beranek, and Newman (the primary contractor for the ARPANET) to establish a commercial network company, Telenet, in late 1972. In October 1973 I joined Telenet as president, . . . and Telenet started the world's first public service in August 1975 with 7 nodes. By April 1978 Telenet's net-

> work had grown to 187 nodes . . . supporting direct terminal access service in 156 cities and interconnections to 14 other countries. (Roberts 1988: 159)

Demonstrating, in one sense, how networks beget networks. Another example (one that prefigures current ideas regarding how the National Information Superhighway will develop) is that of the French public network service TRANSPAC. This network, implemented by the French PTT administration in 1978, began with an experimental testbed network, RCP, that became operational in 1974. Similarly, the NII is to be built on top of a very large testbed, the NREN (National Research and Education Network), which is itself to be built parallel to the NSFNET (which was built on top of ARPANET in the mid-eighties). The simple, historical relationships of these networks to one another are fairly widely known on the Net, but the ineradicable omens and future traces of Net-works are often masked by a rhetoric of "standards."

In fact, the whole project of computer networking is pretty much an exercise in palimpsestics—a long story of upgrades, standardizations, expansions, and (attempted) erasures, punctuated occasionally by synaptic origin stories like the following:

> Usenet came into being in late 1979. . . . Two Duke University grad students in North Carolina, Tom Truscott and Jim Ellis, thought of hooking computers together to exchange information with the Unix community. Steve Bellovin, a grad student at the University of North Carolina, put together the first version of the news software using shell scripts and installed it on the first two sites: 'unc' and 'duke.'" (Spafford 1993)

The Usenet, also known as "news," is a collaborative, popular medium; it is basically a distributed electronic bulletin board, an enormous collection of articles written by different authors, "published" or "broadcast" throughout much of the Net by those authors, organized by news group and subject heading, and normally (but not always, thanks to spatiotemporal irregularities in the texture of the Net) stored in the order in which the articles were written. Two things make it unusual, compared to other bulletin boards. First, there is no single depository for the articles on Usenet; when an article is sent out, it is relayed by the author's host machine both upstream and downstream (that is, to all machines connected to it that feed other machines and to all machines connected to it that get their feed there). The Usenet is, thus, always dispersed and disseminating, upsetting both the point-to-point rationality of electronic mail and the bounded cumulativity of the database. But, the second important feature of the Usenet is its readership, which is estimated at over two million, worldwide (Kapor 1992).

The Usenet and the "Net"—Another Aside

Among Usenet readers, the "Net" is often used, inappropriately according to some other readers, as a synonym for just those people and machines that get a news feed or, worse yet, for just the Internet. If my Net nomenclature is in some way misleading, this is not the direction in which it is going. The Net, as presented here, is not identical to any particular computer network; rather, it is a figure designed to evoke, contingently, a sense of a particular kind of rationality that hinges on the invisible world of information networks. The Usenet is without a doubt an important index of what might be called Net-work, but it is by no means the thing itself, which is not a thing but a work of translation.

Early on, the Usenet became layered over the ARPANET. Initially, the connectivity seemed attentuated; there was a special hierarchy for ARPANET messages, fa.* (meaning, "from ARPANET"), that existed in parallel to the comp.*, sci.*, soc.*, and other Usenet news group hierarchies. Nowadays, "transparency" rules the Usenet's organization; it has learned to assimilate new hosts and new networks effectively. Special, local hierarchies (for example, rice.* and houston.*) and foreign groups (de.*, for Deutschland) are usually not carried by hosts outside the region they name. There are, however, no special groups for users of the different networks (Internet, Bitnet, FidoNet, and so on) that contribute to the Usenet; although it is possible to identify the "path" of an article (the list of machines through which it traveled in order to reach the reader), articles are not segregated by their origination points. These qualities of the Usenet (among others) contribute to the impression that it is an open forum, that it is the open forum for the Net; a significant number of Net-workers share that impression, making the Usenet an important topic in itself, because it illustrates how popular ideas about public space may unfold technologically on the NREN and, eventually perhaps, the NII.

Network Architecture

> Building the NREN has frequently been described as akin to building a house, with various layers of the network architecture compared to parts of the house. In an expanded view of this analogy, planning the NII is like designing a large, urban city.
>
> The NREN is a big new subdivision on the edge of the metropolis, reserved for researchers and educators. It is going to be built first and is going to look lonely out there in the middle of the pasture for a while. But the city will grow up around it in time, and as construction proceeds, the misadventures encountered in the NREN subdivision will not have to be repeated in others.
>
> —Michael M. Roberts

The newly evolving global metropolis arising in the age of cyberspace is a site where people are intellectual nomads: differentiation, difference, and decentering characterize its structure.

—Donald Theall

Urban redevelopment has converted once vital pedestrian streets into traffic sewers and transformed public parks into temporary receptacles for the homeless and wretched. The American city is, as many critics have recognized, being systematically turned inside out—or, rather, outside in. The valorized spaces of the new megastructures and super-malls are concentrated in the center, . . . public activity is sorted into strictly functional compartments, and circulation is internalized in corridors under the gaze of private police.

The privatization of the architectural public realm, moreover, is shadowed by parallel restructurings of electronic space.

—Mike Davis

Privatization. In some respects, the Internet is already substantially privatized. The physical circuits are owned by the private sector, and the logical networks are usually managed and operated by the private sector. The nonprofit regional networks of the NSFNET increasingly contract out routine operations, including network information centers, while retaining control of policy and planning functions. This helps develop expertise, resources, and competition in the private sector and so facilitates the development of similar commercial services.

—RFC 1192

Senator Gore calls it the "National Information Superhighway" . . . I call it the National Public Network (NPN), in recognition of the vital role information technology has come to play in public life and all that it has to offer, if designed with the public good in mind.

—Mitchell Kapor

The Heavenly City stands for our state of wisdom, and knowledge . . . for our transcendance of both materiality and nature . . . for the world of enlightened human interaction, form and information. . . . The Heavenly City, though it may contain gardens, breathes the crystalline gleam of its own lights, sparkling, insubstantial, laid out like a beautiful equation. The image of The Heavenly City, in fact, is . . . a religious vision of cyberspace.

—Michael Benedikt

A nationwide network of information superhighways has been an insurmountable opportunity, but we are now nearing the crest of the hill. We can see the vistas opening up on the other side. It is once again time to steer by the stars.

—Albert Gore, Jr.

If usage implies necessity, the need for a new high-capacity network is hard to question. ARPANET's 56 kilobits per second lines were retired in 1987 and replaced with NSFNET's original 1544 kbps lines (Karraker 1991). In 1991, those lines were supplanted by new 45,000 kbps lines, and they will not hold up for long, if network traffic continues to increase at anywhere near the same rate it has in recent years. NREN, when completed in 1995, will transmit at the rate of three million kilobits per second. These are the numbers that fire the imagination of the Net's past, present, and near future. Transmitting information at three billion bits per second generates an almost inconceivably large amount of bandwidth, but Net-workers have been quick to envision the possible expansion of voice and video applications, which could overload the Net in a very short amount of time.

However, from a user's perspective, this is only an argument in favor of a new, hypermodern, electronic megastructure—the main line of the metropolis. More and more, it looks as though it's only a matter of time:

> The Congress finds that—
>
> (2) Federal programs, such as the High-Performance Computing Program and National Research and Education Network established by Congress in 1991, are vital to the maintenance of United States leadership in high-performance computing and high-speed network development, particularly in the defense and research sectors;
>
> (3) high-performance computing and high-speed networking have the potential to expand dramatically access to information in many fields, including education, libraries, government information dissemination, and health care, if adequate resources are devoted to the research and development activities needed to do so;
>
> (4) high-performance computing and high-speed networking have the potential to expand opportunities for participation for Americans who have disabilities and to improve equality of opportunity, full participation, independent living, and economic self-sufficiency for Americans with disabilities;
>
> (7) a coordinated, interagency undertaking is needed to identify and promote applications of computing and networking advances developed by the High-Performance Computing Program which will provide large economic and social benefits to the Nation, including new tools for teaching, the creation of digital libraries of electronic information, the development of standards and protocols to make the stores of government information readily accessible by electronic means, and computer systems to improve the delivery of health care. (Boucher 1993).

Scientific advancements, national security, improvements in education and healthcare, and economic prosperity are all visible, there, in the Heavenly City just on the other side of the hill. Of course, the utopian halo surrounding most of the official discourse on the NII is understandable. Propaganda seems to have replaced communication as the popular model for the way in which we receive ideas from other people; and if we cynically understand what we want to say as propaganda, then we must above all behave as though nothing has happened, as though we are always only speaking of the true and the good. Thinking in terms of this insidious sort of cynicism, we think we know that they know they're lying, or at least exaggerating, which means we would also know that a critique of their dissimulations would always fail to convince either them or their also-cynical audience.

Thus the rhetoric of utopia is not really the matter in hand. By reading across and against these threads of the Net, I want to build a textual gathering point with and for something of its users' experience. Some of the most troublesome aspects of the proposed NII are actually the issues to which the technocratic populace of the Net eternally return. Net-workers write endlessly about access, anarchy, censorship, privacy, and property, and I suspect that this returning is one of the primary reticulating features of the Net. That is, as reproductive processes, these never-ending simulated dialogues, on and about the Net, network the Net.

And Amnesic Subjectivity

> The backbone of these communications networks [the high-speed data highways of the future] will be built of fiber optics, hair-thick strands of glass, transmitting digital pulses thousands of times faster than ever before.
>
> —Roger Karraker

> Design the NPN for Transparency. . . . "Transparency," in computer circles, is a subjective state of awareness—and a desirable one. When a program is perfectly transparent, people forget about the fact that they are using a computer. The most successful computer programs are nearly always transparent: a spreadsheet, for instance, is as self-evident as a ledger page.
>
> —Mitchell Kapor

> The same drive of realism that led in cinema to the "invisible style" of Hollywood narrative films, and to the occulation of the production process in favor of a consumption of the product as if it were "natural," is at work again in computing. . . . It [is] clear that the efforts of critique to expose the oppressive effects of "the suture" in cinema (the effect binding the spectator to the illusion of a complete reality) had made no impression on the computer industry. . . . The "twin peaks" of American ideology—realism and individualism—are built into the computing machine (the computer as institution).
>
> —Greg Ulmer

Clearly, there is no such thing as cyberspace. Or perhaps, cyberspace has always existed, concurrent with every moment that thinking has been thought spatially, occurrent, and on a grid or from a point. In any case, the goal of computer networking is, first of all, to establish a grid to cross some space—in fact, to create that space in a telecommunicative sense. Then, at the same time, computer networks operate to erase or repress every inch of their own gridness, their spatiality. On the computer screen, there is to be no distance or difference between the user and the data; the relation into which they enter is to be transparent. Thought of this way, space-time compression (one of the hallmarks of modernity) isn't compression at all, but amnesia. Cyberspace, if it ever had any cartographic potential, becomes imperceptible.

On the other hand, there is a secondary but perhaps more dangerous aspect to this erasure of space, because at some point the user's body was put in or onto that grid, and when it is erased, the body follows; the cyberspatial body is ultimately realized as a representation in space at the end of all spatiality. We were warned of this effect, in part, by Allucquére Rosanne Stone (1991), but the structure of forgetfulness on which this cyberspace effect plays remains to be opened or questioned in detail. However, it may be possible to have some experience with this forgetfulness ourselves by examining precisely those points in Net discourse where the "twin peaks" of American ideology break down. Controversies and controversial communiques appear frequently on the Net, and whether they come in the form of calm, private exchanges in e-mail or explode across the Usenet with vitriolic fury, they are inevitably accompanied by a sort of immediate repetitiveness that simultaneously reinforces the sensus communis and destroys any hope of consensus.

Take for example the recent debate over anonymity on the Internet. In the fall of 1992 something new appeared on the Internet—an anonymous server, an always-running program that upon receiving electronic mail messages from anyone anywhere in the world would post those messages to the Usenet (or mail them to another user on the Internet) under an anonymous account name. Anonymous messages, trailed by a banner instructing readers on how they might send anonymous messages, too, cropped up only occasionally in most news groups, but became regular features of the more confessional and/or salacious news groups. The anonymous server was written by a user in an academic environment and set up on a machine there with the permission of a single local system manager. It operated smoothly for about two months, right up until complaints from all over the Net started to pour in regarding the abuse of the server by three of its eight thousand registered users. This first anonymous server was shut down with a minimum of excitement or commotion. System administrators at the university in question ordered the shutdown, and that seemed to be just about the end of open anonymity on the

Usenet from the perspective of many of its users. The author of the anonymous posting software posted an announcement to the news groups in which the anonymous service was used most frequently reporting its demise, resulting in many frustrated grumblings (group discussions in which that software author took part), but there were no large-scale protests.

However, knowing that many of the machines connected to the net are small, privately owned (that is, identifiable with reference to a particular set of individuals and forming up as a "site," a position of clear technical but debatable legal status on the Net), and apparently unhampered by any bureaucratic controls, the software's author made the source code for the anonymous server (the program as it was written, capable of being modified by other programmers) available via anonymous FTP (a very common service that allows computer users anywhere on the Internet to connect to a public archive without having to request any files in particular and usually without leaving any record of what files they downloaded). Shortly thereafter, an anonymous service was established in Finland on someone's personal (privately owned) computer. Over the course of a few short months, this new service automatically created more than eighteen thousand anonymous accounts (usually at the request of the account owners, though many of them were never actually used).

Several news groups came to rely on the anonymous service, especially those groups, like alt.sexual.abuse.recovery or alt.transgendered, in which the discussions tended to stick to topics of a highly sensitive and personal nature (that is, revealing information that could potentially upset the posters' political realities if it were to emerge publicly there). Contrastingly, some other news groups occasionally suffered wildly inappropriate anonymous messages (for example, articles complaining about cats posted to sci.physics or, more completely, articles that showered insults on relatively mild-mannered users whom the poster had probably never met); in this case, anonymity protects the poster only from the opprobrium of the Usenet's readership. The administrator of this Finnish anonymous service, J., who remained relatively quiet throughout this affair, said little about these (so-called) abuses of the anonymous server. His laconism makes him difficult to figure, in multiple senses of the word, but he apparently did try to instruct his users on the "proper" use of the anonymous server; however, he was clearly reluctant to revoke access to the server and never made it a (public) issue of policy on the Usenet.

He could have just locked them out. Anonymous posting was a privilege, a side effect of his generosity, and something which should suffer no abuse whatsoever. Nor should it intrude on "serious" critical-rational debate of the sort found on the sci.* groups. Those were the reasons for ARMM (Automated Retroactive Minimal Moderation), the monster brainchild of an ex-hippie programmer (D.) from Ohio:

"AUTOMATED CENSORSHIP ON THE NET!"

D., the Net's would-be savior, had complained to J., the Finnish system administrator, through e-mail (in other words, in private) about the abuse of the anonymous service, and when J. refused to block all anonymous posts to the sci.* news groups, D. decided to personally go public with the issue in a politically volatile way that brings into full view the haziness of how the Net has been constructed in legal discourse.

Before explaining exactly what D. did, I should make clear the technological conditions that enabled his actions. It just so happens that anonymous messages made their first appearance on the Usenet long before the appearance of the anonymous server. Authors' real names actually appear on Usenet articles as a social convention (albeit one reinforced by most newsreader software), not as an absolute requirement of the Usenet's technical protocols. Users can, if they understand how news articles are propagated, "forge" messages to the Usenet. Such messages are not entirely untraceable; in fact, the site from which they originated is usually obvious and always discoverable, again, if the reader understands a few things about news's technical protocols. However, D. did not use his ability to forge messages as a way to post anonymous articles; instead, he forged messages that instructed the automated news servers at other sites to cancel anonymous messages, to remove them from the message base and not to propagate them further.

But D. did not forge these cancel messages in secret. In some ways of thinking (ones which distinguish between the actions of a person or author and the actions of "software" or between openly declaring oneself not to be the person named by one's signature and "forging"), he did not actually forge any messages at all. He gave the Usenet "public," or at least the readers of the sci.* and news.admin.policy news groups, several weeks of advance notice that he had written ARMM, a program that would automatically forge cancel messages for all anonymous messages posted to the sci.* news groups, and that he would begin running the program continuously if J. failed to recognize his demands that anonymous messages to sci.* news groups be prohibited before a certain deadline.

D.'s announcements and proclamations touched off a furious and epic-scale flame war (a long and drawn-out exchange of usually vituperative articles that passes for debate on the Usenet) over anonymity, the legal rights of system managers to the Usenet message base, the basis for rigid border enforcement on the Usenet (why sci.* news groups deserved to be excepted from anonymous posts), the status of free speech on the Usenet, and the supposedly democratic conventions that govern the establishment and use of individual news groups. This debate spanned dozens of news groups and took place in many, many different subject chains. Clearly, most of the issues at stake had been debated on the Usenet before, and some, especially the question of whether anonymous messages were appropriate in any context, had probably been simmering continuously since the appearance of the first anonymous server (if not since the beginning of the Usenet, when the possibility of forging anonymous messages first emerged). These arguments are by no means finished and, indeed, may be unfinishable, ensconced as they are in the Networkings of the Usenet.

The Net eludes closure. Notable examples include several "events" (informations that recurred more frequently than others) that followed shortly after D.'s announcements regarding ARMM. First of all, ARMM was activated, and its power to silence not only anonymous users but all users was demonstrated clearly over its first weekend of operation, during which time ARMM cancelled two messages in the sci.* news groups: one anonymous article from Finland and one perfectly innocent article that ARMM canceled by "accident." This event would seem to demonstrate that the whole debate over ARMM was a tempest in a teacup, but the arguments, if anything, grew more heated. Similar claims could be made about J.'s abrupt announcement that he was shutting down his anonymous server due to "external political pressure" from a "well-known Net personality"; a landslide of articles followed, consoling J., lamenting the loss of the anonymous server, and demanding to know the name of the person responsible for shutting it down. Old arguments about Finland's uncertain political position on the Internet resurfaced (some speculated that Finland's link had been threatened again, because of the server; Finland had been shut off in the past, because it was mistakenly believed that a certain group of West German hackers were really Finns), and these questions were never really resolved (in part because J. never tried to answer any of them). J. brought his anonymous server back online with no explanation a few weeks later.

These sorts of digressions and discursive intrusions go on forever on the Net. In this case, the story gets quite a bit stranger. In response to ARMM, system managers at many sites began to automatically refuse cancel messages from all users, mostly out of spite for D.'s "censorship" of the Usenet. Other system managers threatened to forge cancel messages for D.'s cancel messages, predicting robotic "cancel wars" that might clog the Internet. Back on the news groups, several other old stories were dug up and reanimated in terms of the argument over anonymity, and seemingly out of nowhere a new version of D.'s ARMM ran amok, blasting out a few hundred giant test messages and thereby crippling the news software on many machines around the globe and instantly making D. a minor Usenet legend.

But what is important about this story is that it is just this sort of narrative that informs Net-workers' arguments and theories about public and private spaces of computer networks; it strains the cracks in local ideologies of realism and individualism in an unceasingly generative manner. The scale of this particular argument was very large, but more significantly, this is exactly the kind of discussion that is being forever replayed on local bulletin boards, mailing lists, and even multiplayer games throughout the Net. Privacy, property, and free speech are eternal concerns on the Net. The effects of this one debate were very real to many people, such as the readers of alt.sexual.abuse.recovery, who needed the anonymity of the Usenet to talk out their problems, but perhaps even more importantly, this sort of debate and the way it has played out may have long-lasting repercussions in the world of computer networking, because it and the precedents it has set are sure to resurface in debate after debate on the Net—that is, everywhere Net-workers work, not just on the Usenet.

"The imminent death of the net is predicted."

Whenever, as happens often, some event exposes the fragility of the Usenet, by threatening Net Death for a particular site or for all of Finland, for example, the demise of the Usenet is often cynically predicated. Most users understand that the Usenet exists because system managers at the various sites set up software to store the articles as they come in and to propagate them further. However, the users also know that people have come to depend heavily on net news for technical reports, announcements, and so on, encrusting the tenuous and provisional links around which the Usenet is organized in a thick layer of status quo preserving, bureaucratic sediment. Thus, merely suggesting the death of the Usenet negates such a vision as much as it maintains it as a feature of popular discourse.

Other sectors of the Net present alternative visions of what is both imminent and immanent in computer networking. For example, the national computer network Prodigy (an IBM Sears joint venture), hints at the future of online communications systems as computer home-shopping networks:

> Where others see a new way for people to communicate and even create "virtual communities," Prodigy sees vast potential profits from people shopping through their keyboards. "We are an information service," Prodigy spokesman Steve Hein says. "We are not an e-mail service." Although other national services have "malls" and advertising, only Prodigy puts ads on almost every screen a user sees. (Gaffin 1991)

Prodigy is also notorious on the Net for its automated censoring of its message base. "One poster to a musical discussion had a posting returned with a note that he was using language 'inappropriate to the Prodigy service.' He was discussing Bach's B Minor mass—the movement titled 'Cum Sancto Spiritu''' (Tavares 1993). For similar but more politically volatile reasons, the B'nai B'rith's Anti-Defamation League has formally protested against Prodigy's practice of disallowing messages directly to other users (Godwin 1992); this policy prevents Usenet-style flame wars but also happens to shield hateful and grossly anti-Semitic messages from any critical comments.

Clearly, collaborative media (including Prodigy, the Usenet, other bulletin boards, message services, multiuser domains, and mailing lists) vary widely not just in terms of their forms and sociolinguistic instrumentalities but also in the ways they are constructed as bureaucratic apparatus. Returning to the question of the National Information Infrastructure, the circumstances under which it takes place and its overall stakes should now be somewhat more apparent, but their further exegesis (in the epicritical mode of the Usenet and other collaborative media that sustain an illusion of delayed dialogue) may yet be instructive.

In "Highways of the Mind or Toll Roads between Information Castles?" (a magazine article made available for FTP that is designed to introduce readers to all the issues in the superhighway debate as they are being fought out "in Congress, in major corporate boardrooms, and in universities"), Roger Karraker writes:

> At issue are vastly different visions of the roles of government,
> education and corporations.

True, but it is important to realize here that those visions are already being enacted and practiced in the everyday life of the Net. They are constructed in discourse, but that's not the end of it; they are inevitably constituted through their community of operations. Also, what we see of these "visions" actually conceals the unseen and unseeable second nature of the Net—its complex (dis)connections or Net-works that conjoin and conflate these various "roles": governments, educational institutions, corporations, sites, hosts, users, and accounts.

> Who will build the network? (Will the federal government create
the
> infrastructure or will it be left to private enterprise?)

This question is founded on three misleading presuppositions: (1) that the government has the ability to build a fiber-optic network without the cooperation of private contractors, expert advisors working in industry or academia, experienced network designers, managers and programmers that would have to be hired from outside the government, and the operators of existing networks with which the new network would be linked; (2) that the network must always only be one monolithic and autonomous network, built according to some rigid, hypermodernist plan; in fact, several major privately owned fiber-optic networks are already in place in this country, which should suggest the possibility that an information superhighway could have multiple origins and emerge as an already imbricated structure rather than as a wholly new and fundamentally synaptic project; and (3) that the history of the network could be charted as a linear progression from ARPANET to NSFNET to NREN to NII; as earlier portions of this paper have shown, this history is actually interrupted at several points with the emergence of foreign, private, and informal networks; and the erasure and encapsulation of earlier networks is always incomplete.

Thus this question instantiates a whole series of modernist myths about the state and society, about politics and economics, and about the history of technological design and progress. Addressing the issue in this way reinforces a particular hegemonic construction of the Net and obscures its local and practical dimensions.

> Who will have access to network services? The debate here is between
> those who would restrict the network's services to the nation's
> research leaders and those who believe in access to everyone with
> a modem.

D., that programmer from Ohio, thought that Usenet news groups devoted to scientific matter should only be accessible to users who were willing to reveal their identity, and many people agreed with him. Prodigy permits access to its databases for a fee and charges an additional fee for heavy use of electronic mail. Every network incorporates an ideology of access, and the national information superhighway will be no different. That this ideology is contested should come as no surprise, because it exposes a fundamental paradox between democratic and capitalist ideals. However, this question falls into many of the same traps as the preceding question. Research networks, as they exist today, solve the problem of access in multiple ways and permit a wide range of approaches to the problem. The information highway will unquestionably be linked to these networks and to other fiber optic networks, and this immediately opens up many possibilities for gaining access to the highway's resources. To fail to make these connections (which would be impossible if the highway is administered in a manner similar to the way the ARPANET and NSFNET have been) would thwart the highway's entire reason for being. Nevertheless, it does seem unlikely that "access" could ever be unproblematic in the idiom of the Net-work, for reasons I have already described.

> Who will pay for all this? Everyone concedes that the federal
> government will pay the lion's share of getting the network

> underway. But should it do so by directly funding the infra-
structure
> or by paying the user fees of just the big research organizations
> working on federal projects?

This question, too, has three misleading presuppositions: (1) that funding originates from one source and can be directed toward a single destination; (2) that government grants would only be available to "big research organizations"; and (3) that the network cannot be woven together out of the thousands and thousands of miles and dark fiber (unused fiber-optic cable) currently in place. These assumptions, if widely held, would determine much of the outcome of this debate, blinding participants to countless economic and practical strategies for improvising the network.

> What kind of information will be allowed on the network? If the
> federal government owns the network, the First Amendment is in
place
> and unpopular speech and art will be protected. If private en-
terprise
> owns and runs the network, freedom of electronic speech is less
> clear. Conceivably, a corporation owning the network could refuse
to
> allow discussion of controversial topics.

The question of what information will be allowed on computer networks, at least in the United States, must always be resolved in a way such that someone remains unsatisfied. On the one hand, users are at least always constrained by the physical and sociolinguistic instrumentalities of the electronic apparatus; their impossible demands have given rise to innumerable jokes in programming circles. There is simply no limit to what users will want to be able to do. On the other hand, few users enjoy being flamed to kingdom come or digging through the reams of messages that many users judge to be hateful, pornographic, or insipid that circulate on (mostly) unmoderated networks like the Usenet. According to some users, "I can be shown that for any nutty theory, beyond-the-fringe political view or strange religion there exists a proponent on the Net. The proof is left as an exercise for your kill-file" (unattributed quote from rec.games.frp, in Faber 1993). Someone like D. will always ask for some kind of administrative control of the network (despite the availability of something like the kill-file mentioned above, a facility included in most newsreader software that automatically filters messages containing certain character strings, for example, some string indicating the message was posted anonymously, that the reader does not want to see), and in that user's idiom, justice can only be effected through such controls, even though another user can only understand those controls as a limitation of their rights as a citizen or as a paying customer.

Returning to the larger of the Net, I believe that this debate over the information superhighway is an important feature of the present future of computer networking in general. The idea of an invisible information megastructure and the welfare and prosperity that might attend it has captured the imagination of the world's technocrats and has become a momentous issue in Europe, North America, and Asia, as a recent news item, dateline New Delhi, 17 March 1993, illustrates: "The government of India has approved a Rs 10 billion (around $33.3 million) high speed, high performance 'information highway' network connecting all academic and research institutes and laboratories" (Mahabharat 1993).

The ramifications of such monumental expenditures are not at all clear. The possibilities form a contested present future, populated by many different figments of the technosocial imagination: cyberspace, the electropolis, the telecosm/fibersphere, and even the Net itself. Of course, each of these constructions of the Net is historically conditioned, and each necessarily has only a limited usefulness for evoking the multitudinous, dispersed, and dislocated experiences of the Net; but each one also (by being grounded in an experience of the Net and being reincorporated there) sustains and preserves the very fragmentedness of the Net that precludes its ultimate figuration.

Futures Present

> In the center of Fedora, that gray stone metropolis, stands a metal building with a crystal globe in every room. Looking into each globe, you see a blue city, the model of a different Fedora. These are the forms the city could have taken if, for one reason or another, it had not become what we see today. In every age someone, looking at Fedora as it was, imagined a way of making it the ideal city, but while he constructed his miniature model, Fedora was already no longer the same as before, and what had been until yesterday a possible future became only a toy in a glass globe.
>
> —Italo Calvino

I first encountered cyberspace in the mid-1980s in William Gibson's novel, *Neuro-mancer,* which was just about the first place anyone other than William Gibson or Tom Maddox could have encountered the word, if not the idea. Not long after I finished reading the sequel, *Count Zero,* as it was serialized in a magazine, I discovered the Usenet and other Internet services that ignored and rudely violated the atemporal, mostly nonlinguistic, and bodiless rationality that governs Gibson's electronic paraworld. Although this first encounter with the Net led directly to its deexoticization for me, I know this is not at all a universal experience (though I don't know that it's rare, either). In the years that followed, I frequently found myself in electronic debates with friends over things like the purpose or feasibility of adding a visible third dimension to most or all electronic representations. Cyberspace, it seemed, had infiltrated the popular self-consciousness of the Net, despite the obvious contradictions between cyberspace and the Net.

Today, cyberspace and virtual reality are discussed in several news groups on the Usenet (Alt.cyberpunk, alt.cyberpunk.chatsubo, alt.cyberpunk.tech, alt.cyberpunk.movement, alt.cyberspace, bit.listserv.cyber-1, alt.uu.virtual-worlds.misc, sci.virtual-worlds, sci.virtual-worlds.apps, and rec.games.frp.cyber, for example). Most of these groups are quite active, and they reveal in their multiplicity the many possible meanings and (con)figurations that are attached to the label *cyberspace*. The recent appropriation of this popular notion about the Net by researchers in the social sciences, like Sandy Stone of David Tomas, might have reflected the multiple articulations of cyberspace on the Net, but instead these sociological or anthropological constructions of the Net cum cyberspace either pursue Gibson's technological fantasy to its "postorganic" limits (Tomas 1991:46) or collapse the complex problem of a dispersed and disseminating Net into a sociologically comprehensible model of the Net based on "virtual communities" and "imaginal bodies" (Stone 1991:110–13)

Of course, cyberspace both competes and overlaps with several other popular notions about the future of the Net, most of which are inspired by Internet services like the Usenet or some Multiuser Domains, interactive multimedia, the telemedia (television, telegraph, and telephone), and/or accounts of the potential uses of fiber optics. A recent discussion on comp.org.eff.talk, for example, suggested that voice, text, and video would be integrated into nearly all facets of electronic communication with the advent of fiber networks. An even more recent AT&T commercial demonstrated what such a system might look like to users in a hospital (it showed a sort of robust video telephone with several image and text displays, a networked multimedium), but the commercial images revealed little or no concern with adding a third dimension to the user's electronic reality and positively refused to dispense with the textual (as opposed to wholly iconic or audiovisual) nature of electronic communication.

Finally, in just the past few months, a number of articles by George Gilder have appeared in *Forbes* magazine describing recent developments in communications technology that might be imagined to have as great an impact on the world as the microchip. One constellation in this new "telecosm" is called the "fibersphere." In one article devoted to the fibersphere, Gilder (1992) argues that it is possible, using fiber-optic cables, to inexpensively bring up to twenty two-way HDTV channels into every home. Two-way channels would allow anyone with a camcorder or even just a VCR to become owner and operator of their own television network. The fibersphere, then, is supposed to substitute for the atmosphere on an electronic planet.

Now, in conclusion what I would like to bring back into focus are the present circumstances under which these and other alternative visions of the future (including Gore/Boucher's information superhighways) compete on the Net. I have written this collage of histories and present futures from the Net to show how it is vented, scribed, and wired and to illustrate its conflicts and confluences. I have chosen for my collage to be (dis)connected. I have tried to address consistently no particular thing—no object, no identifiable community, no localizable place, and no discursive surface that maps directly to any of these things. This is my way of working out a Net, contingently, prosopopoetically, and sedimentarily addressing the puppet discourses and invisible practices of the Net. The claim that goes along with this is that this is one of the most important ways in which the Net works.

Obviously, to write about computer networks ethnographically demands the refunctioning of conventional terms of social analysis such as community and communication, or form and information. Writing about "virtual communities" only allows me to say strange things like, "Virtual communities are an illusion, but they are nevertheless communities," which is a statement founded on an essentialist and realist notion of community, a suspension of disbelief from which there is no easy escape. If, on the other hand, I write against the etymology of the word *information* and parse it as "the thing resulting from making something that has no form" (as a happeningness or a kind of moment), I might elude for a time the hegemony of word-things and yet begin to write the Net. These stories I have told are just such a beginning to inform (about) the Net. My hope has been to dis-scribe and evoke the workings of this figure, the Net, which is so densely and intricately linked to the global processes of modernity.

References

Benedikt, Michael, ed. 1991. *Cyberspace: First Steps*. Cambridge, Mass.: MIT Press.

Boucher, Representative. 1993. H.R. 1757. National Information Superstructure Act.

Calvino, Italo. 1972. *Invisible Cities*. New York: Harcourt Brace Jovanovich.

Davis, Mike. 1990. *City of Quartz*. London: Verso.

Faber, Ted. 1993. "Re: Usenet Is . . . " Posted 19 Feb. to alt.culture.usenet, alt.quotations, news.future.

Gaffin, Adam. 1991. "Prodigy: Where Is It Going?" FTP: ftp.eff.org.

Gilder, George. 1993. "The Coming of the Fibersphere." Posted 12 Apr. to comp.org.eff.talk. Appeared in shorter form in *Forbes,* 7 Dec. 1992.

Godwin, Mike. 1992. "Prodigy Stumbles as a Forum . . . Again. FTP: ftp.eff.org.

Gore, Albert, Jr. 1991. "Information Superhighways: The Next Information Revolution." *The Futurist* 25.

Kahin, Brian. 1990. Request for Comments 1192: Commercialization of the Internet.

Kapor, Mitchell. 1991a. Testimony summary for "Networks of the Future," FCC hearing. FTP: ftp.eff.org.

———. 1991b. "Building the Open Road: The NREN as Test-Bed for the National Public Network." FTP: ftp.eff.org.

Karraker, Roger. 1991. "Highways of the Mind or Toll Roads between Information Castles?" FTP: ftp.eff.org.

Mahabharat, C. T. 1993. "India—Information Highway Plan Approved 03/17/93." Posted 17 Mar. to clari.nb.telecom.

Marcus, George. 1992. *Lives in Trust.* Boulder, Colo.: Westview.

Raymond, Eric S. 1991. *The Jargon File,* version 2.9.6. 16 Aug.

Rheingold, Harold. 1992. "A Slice of Life in My Virtual Community." FTP: ftp.eff.org.

Roberts, Larry. 1988. "The ARPANET and Computer Networks." In Adele Goldberg, ed. *A History of Personal Workstations.* New York: ACM Press.

Spafford, Gene. 1993. "USENET Software: History and Sources." Posted 2 Apr. to news.announce.newusers.

Stone, Allucquére Rosanne. 1991. "Will the Real Body Please Stand Up? Boundary Stories about Virtual Cultures." In Michael Benedikt, ed. *Cyberspace: First Steps.* Cambridge, Mass.: MIT Press.

Tavares, C. D. 1993. "And Now a Note from Intercourse, Pa." Posted 3 Apr. to rec.humor.funny.

Theall, Donald. 1992. "Beyond the Orality/Literacy Dichotomy: James Joyce and the Prehistory of Cyberspace." *Postmodern Culture* 2.

Tomas, David. 1991. "Old Rituals for New Space: Rites de Passage and William Gibson's Cultural Model of Cyberspace." In Michael Benedikt, ed. *Cyberspace: First Steps.* Cambridge, Mass.: MIT Press.

Ulmer, Greg. 1991. "Grammatology Hypermedia." *Postmodern Culture* 1.

Contributors

Mario Biagioli, author of *Galileo, Courtier* (1993), is participating in the formation of a graduate program in cultural and gender studies of science within the Department of History, University of California, Los Angeles.

Maria E. Carson is a research assistant in the Department of Social Medicine, Harvard Medical School.

Gary Lee Downey, formerly an engineer, is an anthropologist teaching in the Graduate Program in Science and Technology Studies, Virginia Polytechnic Institute and State University. He conducted the bulk of these interviews as part of a larger project called "CAD/CAM Culture: An Excavation of Cyborg Practices," which is a cultural critique of theorizing technology outside of society.

Joseph Dumit is a Ph.D. candidate in the History of Consciousness Program at the University of California, Santa Cruz. Currently a fellow at the Smithsonian National Museum of American History, Medical Sciences Division, he is writing a history and ethnography of PET scanning and objective self-fashioning.

Michael M. J. Fischer, formerly of the Rice University anthropology department, now teaches in the Science, Technology, and Society Program at the Massachusetts Institute of Technology. He is charged with developing a concentration in the cultural studies of science. He was a contributor to Late Editions 1.

Mary-Jo DelVecchio Good teaches in the Department of Social Medicine at Harvard Medical School. She is coeditor-in-chief of the journal *Culture, Medicine, and Psychiatry,* and has published widely on cultural studies of biomedicine, including most recently *The Quest for Competence in American Medicine* (1995) and *Pain as Human Experience* (1992), coedited with P. Brodwin, A. Kleinman, and B. Good.

Hugh Gusterson is assistant professor of anthropology and science studies at the Massachusetts Institute of Technology. His research on nuclear weapons scientists was developed when he was a doctoral student in the Stanford anthropology department. University of California Press will publish his book, *Testing Times: A Nuclear Weapons Laboratory at the End of the Cold War,* in 1995.

Diana L. L. Hill is a doctoral candidate in the Department of Anthropology at Rice University. Her interviews were undertaken especially for this volume.

James Holston, trained as both an architect and an anthropologist, teaches in the anthropology department of the University of California, San Diego. He writes on modern and postmodern urbanism and is the author of *The Modernist City* (1989). He also studies citizenship, law, and democratic change in the Americas, especially Brazil. The dialogue he prepared for this volume began during the 1992 Los Angeles uprising as a way to consider both the collapse and the reinvention of the city.

Herbert C. Hoover, Jr. is associate professor of surgery at Harvard Medical School and associate visiting surgeon at Massachusetts General Hospital. He has written on surgery and breast cancer.

Gudrun Klein teaches in the humanities program at Rice University. Her contribution is part of a research project that is an extended inquiry into processes of identity formation in Germany after World War II. Her contribution is based on conversations during the summer months of 1992, held in some settings of her interlocutors.

Leszek Koczanowicz, a philosopher at the University of Wroclaw, was Michael M. J. Fischer's interlocutor in a contribution for the first volume of this series. His main interest is the historical philosophy of the social sciences, and he is now completing a work on the concept of the self in American pragmatism. Andrzej Staruszkiewicz is Koczanowicz's wife's cousin. The interview was conducted in South Carolina, where Staruszkiewicz was a visiting professor. Koczanowicz and his wife have spent many summers with Staruszkiewicz at their country house in the south of Poland.

Irene Kuter teaches at Harvard Medical School and Massachusetts General Hospital. She coauthored with S. A. Weitzman and H. F. Pizer *Confronting Breast Cancer: New Options in Detection and Treatment* (1987).

Kim Laughlin teaches in the Department of Science and Technology Studies at Rensselaer Polytechnic Institute. Her work on Indian scientists who became political and environmental activists in connection with the Bhopal disaster was developed when she was a doctoral student in the Rice anthropology department.

Rita Linggood teaches radiation therapy at Harvard Medical School in the Joint Center for Radiation Therapy.

George E. Marcus is chair of the Department of Anthropology at Rice University and the founding editor of this annual.

Kathryn Milun is a scholar of comparative literature in the Rice University anthropology department. She was a contributor to Late Editions 1, reporting on conversations she had in Hungary.

Livia Polányi teaches in the Department of Linguistics and Semiotics at Rice University. Her memoir was prepared specifically for this volume.

Christopher Pound is a doctoral candidate in the Department of Anthroplogy at Rice University and a student network programmer at the Rice computer center.

Simon Powell teaches radiation oncology at Harvard Medical School and is a visiting fellow in the Department of Radiation Oncology at Massachusetts General Hospital.

Paul Rabinow teaches in the Department of Anthropology at the University of California, Berkeley.

Kathleen Stewart teaches in the Department of Anthropology at the University of Texas, Austin.

Allucquére Rosanne Stone, a graduate of the History of Consciousness Program at the

University of California, Santa Cruz, teaches in the Department of Radio, TV, and Film at the University of Texas, Austin.

Sharon Traweek, formerly of the Rice University anthropology department, is participating in the formation of a cultural and gender studies of science graduate program within the Department of History at the University of California, Los Angeles.

INDEX